I0885022

Biodiversity of Alpine Vegetation:

Conservation, Challenges, Prospects, and Solutions

Muhammad Nauman Khan
https://orcid.org/0000-0003-3790-0902
Department of Botany, University of Chakwal, Pakistan

IGI Global
Scientific Publishing
Publishing Tomorrow's Research Today

Vice President of Editorial	Melissa Wagner
Director of Acquisitions	Mikaela Felty
Director of Book Development	Jocelynn Hessler
Production Manager	Mike Brehm
Cover Design	Jose Rosado

Published in the United States of America by
IGI Global Scientific Publishing
701 East Chocolate Avenue
Hershey, PA, 17033, USA
Tel: 717-533-8845 | Fax: 717-533-7115
Website: https://www.igi-global.com E-mail: cust@igi-global.com

Copyright © 2026 by IGI Global Scientific Publishing. All rights reserved. No part of this publication may be reproduced, stored or distributed in any form or by any means, electronic or mechanical, including photocopying, without written permission from the publisher. Use of this publication to train generative artificial intelligence (AI) technologies is expressly prohibited. The publisher reserves all rights to license its use for generative AI training and machine learning model development.

Product or company names used in this set are for identification purposes only. Inclusion of the names of the products or companies does not indicate a claim of ownership by IGI Global Scientific Publishing of the trademark or registered trademark.

Library of Congress Cataloging-in-Publication Data

Names: Khan, Muhammad Nauman, 1997- editor
Title: Biodiversity of alpine vegetation : conservation, challenges, prospects, and solutions / edited by Muhammad Nauman Khan.
Description: Hershey, PA : IGI Global Scientific Publishing, [2026] | Includes bibliographical references and index. | Summary: "This book aims to describe the composition, structure, and ecological functions of alpine plant communities, including endemic and threatened species, drawing insights from studies such as those on high mountain ecosystems"-- Provided by publisher.
Identifiers: LCCN 2025052981 (print) | LCCN 2025052982 (ebook) | ISBN 9798337367569 hardcover | ISBN 9798337367576 paperback | ISBN 9798337367583 ebook
Subjects: LCSH: Mountain plants--Ecology | Mountain biodiversity
Classification: LCC QK937 .B56 2026 (print) | LCC QK937 (ebook)
LC record available at https://lccn.loc.gov/2025052981
LC ebook record available at https://lccn.loc.gov/2025052982

British Cataloguing in Publication Data
A Cataloguing in Publication record for this book is available from the British Library.

All work contributed to this book is new, previously-unpublished material.
The views expressed in this book are those of the authors, but not necessarily of the publisher.
This book contains information sourced from authentic and highly regarded references, with reasonable efforts made to ensure the reliability of the data and information presented. The authors, editors, and publisher believe the information in this book to be accurate and true as of the date of publication. Every effort has been made to trace and credit the copyright holders of all materials included. However, the authors, editors, and publisher cannot assume responsibility for the validity of all materials or the consequences of their use. Should any copyright material be found unacknowledged, please inform the publisher so that corrections may be made in future reprints.

I humbly dedicate this book to my beloved father, Master Abdul Qayyum Khan (Late), my mentor, my teacher, and my guiding light. His prayers, values, and lifelong dedication to knowledge remain the foundation of my achievements.

Editorial Advisory Board

Nabeel Ahmad, *Dev Bhoomi Uttarakhand University Dehradun, India*
Baber Ali, *Western Sydney University, Australia*
Sezai Ercişli, *Ataturk University, Erzurum, Turkey*
Shah Fahad, *Abdul Wali Khan University, Pakistan*
Steve Harakeh, *King Abdulaziz University, Jeddah, Saudi Arabia*
Shabab Hussain, *University of Milan, Italy*
Majid Iqbal, *University of Chinese Academy of Sciences, Beijing, China*
Alevcan Kaplan, *Dicle University, Diyarbakır, Turkey*
Arafat Abdel Hamed Abdel Latef, *South Valley University, Qena, Egypt*
Tabarak Malik, *Jimma University, Ethiopia*
Fethi Ahmet Ozdemir, *Bingol University, Bingol, Turkey*
Sarah Abdul Razak, *Universiti Malaya, Kuala Lumpur, Malaysia*
Abdul Razzaq, *Qarshi University, Lahore, Pakistan*
Abdul Bari Shah, *Korea University, Sejong Campus, South Korea*
Mohd. Tariq, *Independent Researcher, Vadodara, India*
Barkat Ullah, *Islamia College, Peshawar, Pakistan*
Sajid Umair, *University of Missouri Kansas City (UMKC), USA*

LIST OF REVIEWERS

Muhammad Adnan, *Bacha Khan University, Charsadda, Pakistan*
Sidra Ahmad, *The University of Agriculture, Peshawar, Pakistan*
Amjad Ali, *Universita Cattolica del Sacro Cuore, Parmense, Italy*
Barkat Ali, *Sichuan Agricultural University Chengdu, China*
Usman Ali, *University of Peshawar, Pakistan*
Nasir Assad, *University of Sargodha, Pakistan*
Sajid Ali Khan Bangash, *The University of Agriculture, Peshawar, Pakistan*
Maryam Bibi, *University of Malakand, Pakistan*

Esra Gökçe Gündüzer, *Gazi University, Ankara, Turkey*
Shabab Hussain, *University of Milan, Italy*
Ammar Javed, *The University of Western Australia, Perth, Australia*
Haris Ahmed Khan, *University of Mianwali, Pakistan*
Zahida Nasreen, *University of Mianwali, Pakistan*
Sikandar Shah, *Warsaw University of Life Sciences, Warsaw, Poland*
Syed Mukaram Shah, *University of Peshawar, Pakistan*
Tabbasum Yaseen, *Bacha Khan University, Charsadda, Pakistan*

Table of Contents

Detailed Table of Contents

Alpine floristics provides insight into the composition, diversity and ecological functions of plant species inhabiting ecosystems above the tree line. Despite their limited extent, alpine regions are disproportionately rich in biodiversity, endemism and functional specialization, making them essential to both theoretical and applied ecology. This study frames alpine floristics within vegetation science, distinguishing it from subalpine and nival zones. It synthesizes key methodological approaches including floristic inventories, phytosociology, functional trait analysis and geospatial tools to document and interpret biodiversity patterns across major mountain systems such as the Himalayas, Andes, Alps & Tibetan Plateau. It also highlights how species composition influences ecosystem processes, conservation planning and climate change resilience. Overall, the study develops a conceptual framework demonstrating the practical value of alpine floristics in ecological monitoring, restoration strategies and policy interventions, emphasizing its importance in a changing global environment.

Alpine biodiversity is among the most ecologically significant and vulnerable components of mountain flora, providing critical ecosystem services. Pakistan, located in the western Himalayas, harbors diverse alpine systems that are increasingly threatened by climate change and anthropogenic pressures. This study assessed altitudinal biodiversity patterns and drivers of community composition. Multivariate analyses showed strong site-specific heterogeneity rather than simple elevational segregation. Bray-Curtis dissimilarity indicated that 83.55% of community variation was concentrated in five sites, identifying biodiversity hotspots. NMDS and β-diversity analyses revealed high species turnover (βsim = 0.79) and low nestedness. Poisson GLM showed richness increased with elevation, peaking at mid-altitudes. Indicator species, CCA, and Mantel tests highlighted altitude and grazing as key drivers, emphasizing the need for site-based conservation strategies.

Alpine vegetation in the Alps represents a highly sensitive ecological system influenced by altitude, climatic variability, and complex topography. This chapter examines modern approaches to alpine vegetation mapping using advanced remote sensing and geospatial techniques. High-resolution satellite imagery, hyperspectral sensors, LiDAR, and UAV-based surveys, integrated with GIS and digital elevation models, enable detailed assessment of vegetation patterns with high spatial and thematic accuracy. Object-based image analysis and machine-learning algorithms improve vegetation classification and allow monitoring of temporal changes in response to environmental drivers. Phenological shifts and habitat transformations are identified through time-series data and field validation, providing critical insights into biodiversity distribution and ecosystem resilience. The chapter highlights the importance of advanced mapping technologies for conservation planning, climate change adaptation, and sustainable land-use management in vulnerable alpine environments worldwide.

The Irano-Turanian region, one of the largest phytogeographical zones of the Old World, supports exceptional alpine plant diversity shaped by complex topography, climatic variability, and historical biogeographic processes. Alpine ecosystems within this region act as major biodiversity hotspots, harboring high species richness along with numerous endemic and relict taxa. Identifying hotspots of alpine flora is essential for understanding ecological dynamics, evolutionary patterns, and conservation priorities. This chapter provides an overview of alpine plant diversity hotspots across the Irano-Turanian region, highlighting centers of endemism, ecological significance, and the influence of environmental and historical drivers. It also addresses key conservation challenges, including habitat degradation, overgrazing, and the growing impacts of climate change, emphasizing the need for coordinated conservation efforts to protect these unique alpine ecosystems.

Two major anthropogenic stresses i.e. habitat fragmentation and overgrazing have profound impacts on alpine plant communities, leading to biodiversity loss, ecosystem degradation, and reduced ecological functioning. Alpine ecosystems are particularly vulnerable due to their high elevation, unique biodiversity, and sensitivity to environmental fluctuations. Habitat fragmentation, driven by infrastructure development, tourism, and climate change, isolates plant populations, disrupts ecological connectivity, and reduces genetic diversity. Fragmented landscapes also intensify grazing pressure, as limited refuge habitats and restricted dispersal increase the vulnerability of palatable species. Conversely, overgrazing accelerates fragmentation by increasing spatial heterogeneity and degrading habitat quality. These combined pressures reduce primary productivity, weaken ecosystem resilience, and impair key services such as soil stabilization and water regulation. Climate change further exacerbates these stresses by altering species distributions and phenological patterns.

Majid Iqbal, Institute of Geographic Sciences and Natural Resources Research, Chinese Academy of Sciences, China
Abdul Razzaq, Qarshi Industries (Pvt.) Ltd., Hattar, Pakistan
Syed Nasar Shah, Quaid-i-Azam University, Islamabad, Pakistan
Sana Wahab, Quaid-i-Azam University, Islamabad, Pakistan
Alevcan Kaplan, Dicle University, Diyarbakır, Turkey

Alpine grasses are key components of high-altitude ecosystems, thriving under extreme conditions such as low temperatures, intense radiation, strong winds, short growing seasons, and poor soils. Their ecological success is driven by specialized physiological, morphological, and reproductive adaptations, including dwarf growth forms, dense root systems, photoprotective pigments, and efficient resource-use strategies. These grasses play vital roles in primary productivity, soil stabilization, carbon sequestration, and nutrient cycling. Their interactions with other alpine plants, herbivores, and soil microbes shape community structure and resilience. However, climate change through warming, glacial retreat, and altered precipitation poses serious threats to alpine grasslands. Understanding their ecology is essential for biodiversity conservation, ecosystem stability, and climate change mitigation in fragile alpine landscapes.

The alpine ecosystems are located in one of the harshest places where plants have to survive with severe cold in winter, extreme winds, short growing seasons, and nutrient-lacking soil. Despite these severe conditions, an extensive variety of organisms, like cushion plants, mosses, and sedges, have been able to spread because of their unique structural, functional, and ecological adaptation characteristics. This chapter integrates a comparative analysis of these three most widespread alpine life-forms, highlighting their roles in the ecosystem. Cushion plants are dense, low and rounded in shape, thereby reducing heat loss and wind stress, providing conducive microhabitats to other organisms. Sledge is predominantly found in the alpine fields and wetlands, where it assists in stabilizing the soil and regulating water control. Mosses are, however, exceedingly resistant to desertification and freezing, retain nutrients, and can develop on barren substrates such as rocks, slopes of scree and snow beds.

Alpine zones are characterized by harsh, extremely cold environmental conditions that have traditionally limited human access. However, increasing human activities in recent years have made these fragile landscapes vulnerable to soil erosion, topographic alterations, and vegetation degradation. Climate change poses an additional threat, causing glacier-induced floods, nutrient loss, and accelerated soil erosion. These pressures have led to significant ecological consequences, including biodiversity loss, disruption of soil–vegetation interactions, altered hydrological systems, and reduced carbon sequestration capacity. Damage to soil and vegetation undermines habitat stability and weakens ecosystem resilience, making alpine environments more susceptible to future disturbances. Due to the slow natural recovery rates in alpine regions, such degradation is often long-lasting or irreversible. Therefore, sustainable conservation strategies such as controlled grazing, restoration using native plant species, and effective protected area management are essential to mitigate these impacts.

The alpine summits have long been central to botanical exploration, serving as reference points for documenting flora at the highest elevations. Historical surveys from early expeditions of the 18th and 19th centuries to standardized 21st century monitoring programs provide valuable insights into floristic patterns, species richness and ecological change on mountain summits. This chapter reviews these surveys as the foundation of alpine floristic knowledge and biodiversity assessment. It includes case studies from major mountain systems such as the Alps, Himalayas, Andes, Karakoram and Rockies and highlights the role of programs like GLORIA. Key findings reveal upward species migration, shifts in community structure and loss of cold-adapted taxa with warming climates. By linking historical records with modern ecological monitoring, this chapter emphasizes the relevance of summit floristics for conservation planning, ecological forecasting and policy guidance. Alpine summits act as natural laboratories, illuminating biodiversity dynamics amid rapid environmental change.

Alpine ecosystems are facing unprecedented challenges mainly due to the interaction of anthropogenic and climatic factors. Among the various impacts of climate change, the gradual movement of vegetation zones upwards is one of the most significant, which ultimately results in a shift of the species present, changes in the timing of life cycles, and a decrease in the productivity of the ecosystem. The spread of heat-tolerant species, the extinction of mountain-top endemics, and plant invasions are some of the changes that are occurring. On the contrary, some species can survive the bad conditions through microclimatic buffering, which can be in conflict with the overall predictions. Furthermore, the application of global databases such as GLOBALP and deep learning for the monitoring of shrub lines and carbon fluxes in response to global warming are some of the ways through which new and more efficient technologies are coming about. The combination of rising temperatures, retreating glaciers, and the vertical movement of both heat-loving and invasive species, along with the impact of humans, is the situation that severely threatens the vegetation that is exclusive to these alpine areas. Human activities, including overgrazing, uncontrolled tourism, road expansion, and hydropower development, exacerbate ecosystem degradation by causing soil destabilization, habitat fragmentation, and decreased resilience in plant communities. Additional stressors such as soil nutrient depletion, erosion, and altered hydrological regimes further disrupt plant-soil interactions and hinder ecosystem functionality. These threaten the biodiversity of alpine vegetation and put the ecological services at risk, such as carbon sequestration, watershed management, and cultural significance for indigenous communities.

Alpine plants inhabit some of the most extreme environments on Earth, characterized by low temperatures, high solar radiation, strong winds, and nutrient-poor soil. To survive and reproduce under these conditions, they exhibit a wide range of physiological and ecological adaptations. These include morphological traits such as dwarfism, cushion growth forms, and dense pubescence; physiological mechanisms such as cold hardiness, antifreeze protein production, and efficient photosynthetic pathways; and ecological strategies such as delayed flowering, seed dormancy, and mutualistic interactions with pollinators and mycorrhizal fungi. Together, these adaptations enable alpine plants to maintain metabolic activity, optimize resource use, and ensure reproductive success despite harsh abiotic and biotic pressures in their environment. Understanding these adaptive strategies is essential for predicting the responses of alpine vegetation to climate change and for developing effective conservation measures.

Alpine ecosystems are distinguished by nutrient-poor soils and low temperatures, short growing seasons, and harsh environmental conditions that have strong influence on plant growth and survival. Soil nutrient dynamics, especially the cycling of carbon (C) and the nutrients (N), play a central role in controlling the productivity, diversity and resilience of alpine vegetation. Microbial communities are important mediators of nutrient transformations that influence nutrient availability under the conditions of limited decomposition and mineralization. This chapter discusses the role of soil microbes, nutrient constraints and biogeochemical processes in determining plant distribution in the alpine communities. It highlights how changes in C and N cycling with climate change affect vegetation resilience and ecosystem functioning. This chapter provides evidence-based insights into soil–plant interactions and the role of microbes in enhancing biodiversity and informing conservation strategies to improve nutrient stability in alpine landscapes.

Foreword

Globally, mountain ecosystems are undergoing unprecedented change, and alpine vegetation is one of the most sensitive components reflecting environmental change. Alpine flora is adapted to the extreme limits of physiological tolerance, where low temperatures, intense radiation, short growing seasons, and poor soils create challenging and highly specialized communities. Although ecologically significant, these systems remain among the least studied and most vulnerable elements of global biodiversity. The book "Biodiversity of Alpine Vegetation: Conservation, Challenges, Prospects, and Solutions" arrives at a crucial time, providing timely and up-to-date information on the ecological study of alpine environments and conservation science.

The book compiles the work of leading researchers who examine alpine vegetation from various perspectives, including biodiversity trends, ecology, conservation, and emerging threats. The chapters collectively highlight the importance of alpine ecosystems in sustaining vital ecosystem services, supporting a high degree of endemism, and regulating climatic and hydrologic systems. The book also addresses deeper issues affecting these ecosystems, particularly those related to climate change, land-use change, overgrazing, invasive species, and human activities in high-altitude areas. Its integrative approach is one of the book's greatest strengths, as contributors place alpine biodiversity within broader ecological, socio-environmental, and conservation contexts.

The readers are guided through case studies of various regions and the world to show how alpine vegetation responds to environmental pressures and how conservation efforts can be developed to balance the two. The section on future prospects and solutions is particularly useful, as it highlights adaptive management, community involvement, and evidence-based policy formulation.

In contrast to most edited volumes, which are fragmented or merely descriptive, this book presents a logical flow of information based on up-to-date scientific studies and field experience. The author's experience ensures that the content is both scholarly and applicable to conservation practitioners, policymakers, and graduate students.

Although the main focus is the vegetation of the Alps, the ideas and techniques presented throughout the chapters are generalized for application to other delicate ecosystems that have been subjected to similar pressures from global changes.

The book is both an action guide and a reference. It highlights the importance of protecting alpine biodiversity while offering viable solutions for sustainable management and long-term sustainability. This volume serves as a valuable resource for researchers, educators, and decision-makers, enhancing their knowledge base and motivating action. I believe that Biodiversity of Alpine Vegetation: Conservation, Challenges, Prospects, and Solutions will make a lasting contribution to the field and stimulate further research and discussion on the conservation of mountain ecosystems in a global context.

Alevcan Kaplan
Faculty of Pharmacy, Dicle University, Diyarbakır, Turkey

Preface

Alpine ecosystems represent some of the most fragile, dynamic, and ecologically significant landscapes on Earth. Rising above the tree line and shaped by extreme climatic conditions, alpine zones harbor unique plant communities that demonstrate remarkable adaptations to cold temperatures, intense radiation, nutrient-poor soils, and short growing seasons. Despite these harsh environments, they sustain notable biodiversity, high levels of endemism, and valuable ecosystem services that support both natural systems and human livelihoods.

This volume presents a comprehensive exploration of alpine vegetation, progressing from foundational concepts to emerging scientific frontiers. It begins with an overview of alpine vegetation and floristics, establishing the conceptual and ecological framework necessary to understand alpine biodiversity. Particular emphasis is placed on regional perspectives, including the alpine biodiversity of Pakistan and the Irano-Turanian region, alongside comparative insights such as vegetation mapping in the European Alps. Through discussions of life forms, microhabitats, grasses, vascular flora, and historical plant surveys, the book integrates classical plant ecology with modern analytical approaches.

A central theme of this work is the increasing vulnerability of alpine ecosystems. Habitat fragmentation, overgrazing, soil erosion, anthropogenic pressures, and climate change are reshaping these landscapes at unprecedented rates. Chapters addressing Red-List classification, ecosystem services, land-use drivers, and emerging threats underscore the urgency of conservation action. The book further examines ecological succession, nutrient cycling, plant–microbe interactions, and physiological adaptations that enable alpine species to persist under environmental stress.

Beyond ecological theory, this volume bridges biodiversity with applied and interdisciplinary perspectives. Ethnobotany, wild food plants, honey bee flora, medicinal plant strategies, phytochemistry, and nanotechnology-based research highlight the socio-economic and scientific significance of alpine plant resources. Discussions on national parks, nurseries, in situ and ex situ conservation, and advances in micropropagation provide practical pathways for sustaining threatened species and preserving alpine endemism under changing climatic and human pressures.

By integrating phytogeography, ecology, conservation biology, biotechnology, and traditional knowledge systems, this book offers a holistic perspective on alpine vegetation. It is intended for researchers, students, conservationists, policymakers, and practitioners seeking to understand both the scientific foundations and contemporary challenges of alpine ecosystems. At a time when alpine biodiversity stands at a crossroads, this compilation emphasizes the need for informed research, collaborative management, and nature-based solutions to ensure the resilience and sustainability of these high-altitude landscapes for future generations.

CHAPTER OVERVIEW

By combining theoretical insights with practical applications, this book serves both as an academic reference and a guide for applied conservation strategies. *Biodiversity of Alpine Vegetation: Conservation, Challenges, Prospects, and Solutions* is a carefully crafted scholarly work tailored to the environmental and ecological research community. Spanning thirty-one comprehensive chapters, the book explores foundational principles, emerging scientific advancements, and critical issues that shape the study and conservation of alpine ecosystems. It provides an in-depth examination of contemporary challenges, innovative applications, and forward-looking strategies essential to sustaining biodiversity in fragile mountain landscapes. Chapter 1 *(Defining Alpine Floristics: Concepts and Scope)* explains the meaning of floristics (study of plant species composition), its methods, scope, and importance in understanding alpine plant diversity. Chapter 2 *(Alpine Biodiversity in Pakistan)* discusses alpine plant diversity in Pakistan (Himalaya, Karakoram, Hindu Kush), endemic species, ecological importance, and conservation concerns. Chapter 3 *(Mapping Alpine Vegetation in the Alps)* describes techniques used to map alpine vegetation (GIS, remote sensing), distribution patterns, and vegetation classification in the European Alps. Chapter 4 *(Hotspots of Alpine Plants in the Irano-Turanian Regions)* identifies biodiversity hotspots in the Irano-Turanian region and discusses phytogeographic significance and endemic richness. Chapter 5 *(Habitat Fragmentation and Overgrazing of Alpine Plant Communities)* examines human pressures such as livestock grazing and land fragmentation, and their effects on plant diversity and ecosystem stability. Chapter 6 *(Ecology of Alpine Grasses)* focuses on ecological roles of alpine grasses, their adaptations, distribution, and importance in grazing systems. Chapter 7 *(Adaptation in Alpine Flora Highlighting Cushions, Sedges, and Mosses)* describes structural and physiological adaptations like cushion growth form, compact morphology, antifreeze compounds, and water retention strategies. Chapter

8 *(Soil Erosion and Vegetation Succession in the Alpine Zone)* explains causes of soil erosion and natural processes of plant colonization and succession after disturbance. Chapter 9 *(Historical Plant Surveys and Floristic Trends on Alpine Summits)* reviews early botanical explorations, long-term monitoring data, and changes in plant composition over time. Chapter 10 *(Emerging Threats and Challenges for Alpine Vegetation: Integrated Perspectives on Shifting Dynamics, Biotic Pressures, and Ecosystem Vulnerability)* highlights shifting ecological dynamics, invasive species, biotic pressures, and increasing ecosystem vulnerability. Chapter 11 *(Physiological and Ecological Adaptations of Alpine Plants)* explains physiological mechanisms such as cold tolerance, photosynthetic efficiency, stress resistance, and nutrient use efficiency. Chapter 12 *(Nutrient Cycling and Plant–Microbe Interactions in Alpine Ecosystems)* discusses soil microbes, nitrogen fixation, mycorrhizae, and their role in maintaining soil fertility and plant growth. This book contributes to the growing body of literature on alpine biodiversity by offering a multidisciplinary and integrative perspective. It combines classical ecological approaches with emerging scientific innovations, bridging gaps between conservation biology, phytochemistry, ethnobotany, climate science, and biotechnology. By addressing both theoretical frameworks and applied solutions, this volume provides a roadmap for sustaining alpine ecosystems under mounting environmental pressures.

In an era marked by biodiversity decline and climate uncertainty, understanding and conserving alpine vegetation is not merely an academic pursuit—it is an ecological necessity. This book aspires to inspire researchers, guide policymakers, and empower conservationists to protect alpine landscapes for future generations.

Muhammad Nauman Khan
University of Chakwal, Punjab, Pakistan

Acknowledgments

The completion of this book, *Biodiversity of Alpine Vegetation: Conservation, Challenges, Prospects, and Solutions*, would not have been possible without the support, guidance, and contributions of numerous individuals who assisted in various stages of this project.

I would like to express my sincere appreciation to Ms. Hansa Gul (University of Mianwali, Punjab, Pakistan) for valuable assistance and cooperation throughout the development of this book. I am deeply grateful to Dr. Abdul Razzaq (Qarshi University, Lahore, Punjab, Pakistan), Dr. Barkat Ullah (Islamia College Peshawar, Pakistan), Dr. Sarah Abdul Razak (Universiti Malaya, Kuala Lumpur, Malaysia), Dr. Alevcan Kaplan (Dicle University, Diyarbakır, Turkey), Dr. Baber Ali (Western Sydney University, New South Wales, Australia), and Dr. Nasir Assad (University of Sargodha, Pakistan) for their academic insights, encouragement, and scholarly support.

Special thanks are extended to Mrs. Sana Wahab and Syed Nasir Shah (Quaid-i-Azam University, Islamabad, Pakistan), Dr. Muhammad Adnan (Bacha Khan University Charsadda, Pakistan), Dr. Majid Iqbal (University of Chinese Academy of Sciences, Beijing, China), Dr. Haris Ahmed Khan (University of Mianwali, Punjab, Pakistan), Dr. Sajid Ali Khan Bangash (Institute of Biotechnology and Genetic Engineering, The University of Agriculture, Peshawar, Pakistan), Dr. Zahida Nasreen (University of Mianwali, Punjab, Pakistan), Dr. Maryam Bibi (University of Malakand, Pakistan), and Dr. Shah Fahad (Abdul Wali Khan University, Pakistan) for their constructive discussions, expert opinions, and continued encouragement.

I would also like to sincerely acknowledge Dr. Sajid Umair (University of Missouri Kansas City (UMKC), Missouri, USA) for introducing and discussing the initial idea of this project in earlier years, which ultimately contributed to its development. His intellectual input is highly appreciated. Furthermore, I am thankful to Mr. Wajid Ullah for his careful editing and reviewing of the English language in the final version of the manuscript.

Finally, I extend my appreciation to all colleagues, contributors, and reviewers whose efforts, directly or indirectly, supported the successful completion of this book.

Muhammad Nauman Khan
University of Chakwal, Punjab, Pakistan

Introduction

This book presents an authoritative synthesis of alpine plant biodiversity drawing from global ecosystems above the treeline, spanning, the Himalayas, Tibetan Plateau, Andes, and beyond. It examines the ecological patterns, adaptive traits, and functional significance of alpine vegetation in sustaining mountain ecosystems and human communities. Alpine vegetation represents one of Earth most captivating and fragile ecosystems shaped by extreme cold, intense ultraviolet radiation, high winds, and short growing seasons. Though they occupy only about 3–4% of the world's land surface, alpine zones support remarkable species richness and endemism: up to 10,000–15,000 vascular plant species globally within a single square meter of alpine terrain, species counts can reach 50–60, offering micro scale diversity unmatched elsewhere. This book offers a comprehensive exploration of alpine plant biodiversity, detailing the physiology and ecological adaptations of cushion plants, grasses, sedges, rosettes, lichens, bryophytes and more organisms uniquely evolved to thrive in high-mountain environments. It examines the dynamic interactions between geology, microclimate, and life, and emphasizes alpine ecosystems' significance in water regulation, carbon storage, and climate mitigation including their role as critical freshwater sources for millions worldwide.

An overview of ecological services provided by alpine regions, the drivers of habitat formation, and the legal and institutional frameworks for protecting mountain biodiversity such as the Alpine Convention's protocols and ecological corridor initiatives. A thorough examination of climate-driven impacts (glacial retreat, shifting vegetation zones, heightened warming at elevation), land-use fragmentation, invasive species, tourism pressures, agriculture and overgrazing. Reflections on ecological connectivity, regional and transboundary protected-area planning, and future pathways for species migration and ecosystem adaptation under changing climatic regimes. Practical strategies for restoration and conservation: establishing protected zones, promoting sustainable tourism, implementing community-based land-practices, ecological monitoring, and restoration trials such as pasture rehabilitation and ecological networks. This is essential reading for ecologists, conservationists, policymakers, and anyone interested in mountain ecosystems or global biodiversity. Readers will come away with a holistic understanding of what makes alpine vegetation so unique, why it's under threat, and how we can act today to protect and restore these vital habitats for tomorrow.

Chapter 1
Defining Alpine Floristics: Concepts and Scope

Gul Nawaz
http://orcid.org/0009-0000-5328-0454
University of the Punjab, Pakistan

Hafeeza Amna Saleem
http://orcid.org/0009-0005-8109-5893
University of the Punjab, Pakistan

ABSTRACT

Alpine floristics provides insight into the composition, diversity and ecological functions of plant species inhabiting ecosystems above the tree line. Despite their limited extent, alpine regions are disproportionately rich in biodiversity, endemism and functional specialization, making them essential to both theoretical and applied ecology. This study frames alpine floristics within vegetation science, distinguishing it from subalpine and nival zones. It synthesizes key methodological approaches including floristic inventories, phytosociology, functional trait analysis and geospatial tools to document and interpret biodiversity patterns across major mountain systems such as the Himalayas, Andes, Alps & Tibetan Plateau. It also highlights how species composition influences ecosystem processes, conservation planning and climate change resilience. Overall, the study develops a conceptual framework demonstrating the practical value of alpine floristics in ecological monitoring, restoration strategies and policy interventions, emphasizing its importance in a changing global environment.

DOI: 10.4018/979-8-3373-6756-9.ch001

ght © 2026, IGI Global Scientific Publishing. Copying or distributing in print or electronic forms without written permission of IGI Global Scientific Publishing is prohibited. Use of this chapter to generative artificial intelligence (AI) technologies is expressly prohibited. The publisher reserves all rights to license its use for generative AI training and machine learning model development.

1. INTRODUCTION

Alpine floristics is a foundational subdiscipline of vegetation ecology that examines the composition, structure and diversity of plant communities above the climatic tree line (Körner, 1995; Körner, 2023). Alpine regions host a unique set of species that exhibit specialized adaptations, such as dwarf growth forms, protective leaf structures and extended freezing tolerance, to cope with low temperatures, short growing seasons, high ultraviolet radiation and steep microclimatic gradients (Polechová & Barton, 2015; Waterman, 2001). Despite occupying a comparatively small proportion of Earth's surface, these ecosystems are important reservoirs of biodiversity and host many endemics, relict in a world where the global warming is a rapidly increasing phenomenon, the evaluation of floristic patterns and dynamics concerning alpine vegetation is crucial and stress-tolerant taxa. Alpine floristics provides the scientific basis for studying species–environment relationships, biogeographic processes and ecological mechanisms that maintain high-altitude vegetation (Gentili et al., 2013; Rawat, 2007; Walker et al., 2001).

In a wider ecological context alpine floristics has been important in connecting a variety of sub-disciplines of the systematics of botany, biogeography and functional ecology by recording the taxa of plants and their distribution patterns across elevation gradients (Decocq, 2016; Mi et al., 2021). Its integrative quality enables researchers to determine the effects of the combination of geological, climatic and anthropogenic factors on shaping the composition of species and structure of communities. Moreover, alpine floristics has immediate consequences to the environmental policy and biodiversity, because alpine ecosystems are one of the most climate-sensitive on the planet. In a world experiencing rapid climate warming, evaluating floristic patterns and dynamics of alpine vegetation is crucial in relation to forecasting upcoming changes of species distribution and ecosystem stability (Körner & Hiltbrunner, 2021).

1.1 Background of Alpine Floristics

Alpine floristics has developed out of the early botanical discoveries, which attempted to provide a catalogue of the mountain flora on Earth, up to the scientifically organized field, which investigates the ecological and evolutionary mechanisms of the vegetation in high altitude (Muellner-Riehl, 2019; Séverine Fauquette et al., 2018; Stöckli et al., 2011). Traditionally, alpine regions were viewed as marginal lands, which were characterized by low productivity but recent studies show them to be the location of adaptive radiation and speciation (Abeli et al., 2018). Isolation, the heterogeneity of microhabitats and environmental stress have been used as natural laboratories where the evolutionary processes have been accelerated in mountain ranges like Himalayas, Andes, Alps and Tibetan Plateau (Cortés & Wheeler, 2018; Muellner-Riehl, 2019; Tito et al., 2020).

Early botanists, including Alexander von Humboldt, Joseph Dalton Hooker and Asa Gray laid the groundwork for floristic zonation and plant geography (Reveal, 1991; Turrill, 2013). Their groundbreaking surveys laid down the theoretical basis that the gradients of elevation are parallel to the latitudinal gradients in determining the type of vegetation. The development of ecological modeling, geospatial analysis, and molecular systematics in the 20th and 21st centuries transformed the study of alpine flora, allowing scientists to map the biodiversity and study the relationship between the alpine taxa, their phylogeny.

1.2 Importance of Studying Alpine Floristics

Knowledge of alpine floristics has critical importance on several scientific as well as practical grounds (Grabherr, 2009; Körner, 1995; Stöckli et al., 2011). First, it facilitates the identification of spatial distributions of biodiversity and endemism in vulnerable mountain ecosystems that are usually hotspots of unique genetic resources (Noroozi et al., 2018). Second, floristic studies provide a point of reference to track environmental change, record of how plant communities react to glacial recession, permafrost thawing and changing snowlines (Kullman, 2010). Third, alpine floristics helps in making conservation plans because it determines areas of importance when they are to be safeguarded, restored or to be used in a sustainable manner (Grêt-Regamey et al., 2008).

On social-ecological level, the value of alpine plants is the contribution to the local livelihoods in terms of ethnobotanical value, medicinal value and ecosystem services such as soil stabilization and water control (Ianni et al., 2015). Additionally, records of alpine flora are used in the international biodiversity databases and are useful in global efforts including the Convention on Biological Diversity (CBD) and the Global Strategy for Plant Conservation (GSPC). Therefore, not only academic knowledge is enriched through alpine floristics, but also policy and practical conservation interventions are oriented at the preservation of high-elevation biodiversity that is subject to growing climate pressure (Körner et al., 2006).

2. CONCEPT OF FLORISTICS

Floristics is a science of plant ecology and taxonomy, which studies the naming, grouping, and geographical distribution of plant species in a specific territory (McLaughlin, 1994). It is the key to ecological, evolutionary and biogeographical studies by giving us a descriptive base on which to identify trends of species richness, composition and turnover. Floristics has been called the beginning of the study of the structure of the ecosystem, as proper recording of the species of plants is the foundation of the subsequent ecological synthesis (Dansereau, 1951). In the alpine case, floristics gains between-wholesomeness, as it can be used to identify delicate biotic and abiotic drivers underpinning the dynamics of vegetation under steep environmental gradient conditions.

2.1 Definition of Floristics

Floristics may be described as the scientific discipline, which records and analyzes the plant species structure of a given region, their taxonomic identity, abundance and distribution associations (Heywood, 2001; McLaughlin, 1994). It is a qualitative and quantitative evaluation of vegetation with a specific focus on the systematic data on plant diversity (Ashby & Stevens, 1935; Silva et al., 2014). Floristics, in ecological contexts, is the intermediary between taxonomy and vegetation science in the sense that it connects the lists of species to their habitat types, environmental conditions and biogeographical patterns (Turrill, 1964). In alpine habitats, floristics is not merely a matter of species listings anymore, it entails the realization of how elevation, microclimate, geology and disturbance history regulate floristic assemblies (Ibrahim et al., 2019; Nagy & Grabherr, 2009; Zaman et al., 2019).

2.2 Historical Development of Floristic Studies

The history of floristic science can be traced to the classical era of natural history when an initial collection of descriptive accounts of local flora had been drawn up by early explorers and botanists (Reveal, 1991). The term floristics came into a form in the 18th and 19th centuries as herbarium collections were put in place and binomial names were introduced by Carl Linnaeus. Later, Humboldt ecological observations formed the basis of the correlation of the distribution of plants with environmental gradients, which further developed to become contemporary phytogeography.

In the 20th century, floristics was extended to dynamic analysis based on community composition, life forms and species turnover broadening its scope beyond static inventories. With the introduction of multivariate statistical methods and geospatial technology in late 20th century, floristic research became a more analytical science. In the modern world, floristic research has become able, by integrating with molecular phylogenetics, remote sensing and GIS-based modeling, to determine not only the spatial and temporal occurrence of species, but also the cause of that occurrence, i.e., the relationship between patterns and evolutionary history and ecological role.

2.3 Key Principles and Approaches in Floristics

The research into flora is directed by a number of principles (Frodin, 2001). First is comprehensiveness, floristic surveys strive to represent the complete species richness of an area being studied; this is to ensure that coverage is made in terms of habitat and microclimate (Gillespie et al., 2004). Second, taxonomic precision/ specific use and conservation of vouchers this is important to ensure scientific integrity. Third, ecological interpretation analyzing floristic data should be carried out in the background of environmental gradients, soil types and soil disturbance regimes (Gentry, 1988). Fourth, temporal relevance- species composition varies with time therefore repeated floristic surveys can give useful information on ecological succession and climate induced changes (MacMahon, 1980).

Floristics has a methodological mix of field surveys, herbarium taxonomy, ecological sampling and more current geospatial tools (GIS) to map species distributions (Raes, 2009; Reddy, 2017). DNA barcoding and molecular systematics now form part of the solution to taxonomic ambiguity, especially of complex alpine genera (Lavergne et al., 2025; Montagna et al., 2016). Through integrating old-fashioned taxonomy with technological advances, floristics keeps changing and developing as both a descriptive and analytical science, which is important in the assessment of biodiversity and conservation planning.

3. ALPINE FLORISTICS: A SPECIALIZED BRANCH

Alpine floristics is a very specific and specialized field of floristic science, which focuses on the flora that grows above the climatic tree line, with environmental extremes characterizing the specific ecological and evolutionary processes (Körner & Change, 2018; Körner & Hiltbrunner, 2021). The discipline is a combination of the botanical taxonomy, ecology and the biogeography to gain insights into the diversity, adaptations and geographical arrangement of the alpine plants. The high-mountain areas are geographically isolated, but serve as biodiversity storehouses and haven of many cold-adapted, relict and endemic taxa (Pauli & Halloy, 2019). Alpine floristics is not, then, just a taxonomic game but rather a holistic method of comprehending the role of environmental gradients, historical processes and evolutionary pressures in the make-up and structure of plant communities in high-elevation environments.

3.1 Definition and Characteristics of Alpine Floristics

It is possible to define alpine floristics as the systematic description and explanation of the composition of plant species, their diversity and distribution in alpine ecosystems, those environments that exist on the one hand above the tree line, and on the other hand below the permanent snowline (Körner, 1995; Körner, 2021). Its main features are; (1) great environmental specialization, such that alpine plants need to live in low temperature, high radiation and short growing periods; (2) scarcity of space with many often being confined to small patches along mountain ridges and slopes; and (3) great endemism and functional diversity which is a condition of evolutionary isolation and niche differentiation.

In contrast to general floristics, alpine floristics concerns the evaluation of topography, microclimate and soil heterogeneity effects on plant assemblages on small spatial scales (Ford et al., 2013; Opedal et al., 2015). It is not about taxonomic diversity but also about forms of life, phenological characteristics and ecological plans, which provide survival benefits. As an example, cushion plants and rosette-forming species are small-scaled plants which reduce heat loss and exposed areas to wind, whereas perennial grasses and sedges prevail in more stable substrates. Alpine floristics is a combination of floristic inventory and ecological interpretation, which focuses on compositional and functional aspects of the vegetation on the high mountains (Evangelista et al., 2016; Körner, 1995).

3.2 Distinctive Features of Alpine Flora Compared to Lowland Flora

The alpine flora is not similar to the lowland flora, but its structure, functionality and adaptive method are different (Körner, 1995). Although the conditions in lowland plant communities are rather stable, the alpine species can withstand a wide range of temperatures, unproductive soils and mechanical damage caused by frost and snow (Billings & Mooney, 1968). As a result, the alpine plants have a number of characteristic features:

- Morphological changes including dwarfism cushion development, hairy or waxy leaves and short stature that restrict desertification and defend meristems.
- Physiological adaptations such as increased photosynthetic efficiency in low temperatures, frost tolerance and the ability to restart metabolism fast after freezing.
- Reproductive mechanisms that is maximally vegetative, clonal and perennial in nature and offsetting brief flowering periods.
- Phenological synchrony, in which development and reproduction are closely synchronized with short averages of snowmelt and favorable thermal regimes.

Also, the floristic structure of the alpine areas displays high quality biogeographical filtering only those species with wide ecological range and cold-tolerance survive (Nagy & Grabherr, 2009). Conversely, lowland vegetation has a higher diversity based on competition and high turnover of succession (Zhang et al., 2018). These contrasts render alpine ecosystems to be very important natural laboratories of studying the ecological processes under the extreme climatic conditions in adaptation, resilience and evolutionary ecology. Comparative summary of key morphological, physiological and ecological traits distinguishing alpine from lowland plant species, reflecting adaptations to high-altitude environments are given in Table 1.

Table 1. Characteristic features differentiating alpine and lowland flora.

Feature category	Alpine flora characteristics	Lowland flora characteristics
Growth form	Dwarf, cushion or rosette habit; compact architecture	Tall, erect or sprawling growth
Leaf morphology	Small, thick and pubescent leaves with reduced surface area	Broad, thin and glabrous leaves
Root system	Extensive fibrous or taproot system for anchorage in shallow soils	Moderate root development adapted to deep fertile soils
Reproduction	Slow growth; delayed flowering; high investment in seed protection	Rapid flowering and reproductive cycles
Physiological adaptation	Enhanced cold and UV tolerance; antifreeze compounds	High photosynthetic rate at moderate temperatures
Habitat	Above tree line; steep slopes; rocky substrates	Below tree line; fertile plains and valleys

3.3 Phytogeographical Significance of Alpine Floristics

Phytogeographical importance of alpine floristics is based on the possibility to explain spatial distributions patterns of certain species, endemism and biotic exchange among mountain systems (Muellner-Riehl, 2019; Testolin et al., 2020). Alpine areas tend to serve as both locations of speciation and refugia of cold adapted taxa that withdrew with the onset of warmer temperatures in the post-glacial periods (Gentili et al., 2015). Comparative study of floristics among ranges, including Himalayas, Alps, Rockies and Andes, has shown convergent evolution, that is independent lineages that evolved to have similar adaptive morphologies in response to similar environmental constraints.

Biogeographically, alpine floristics has been informative with regard to floristic connection, and disjunction, with some of the processes involved in the formation of the existing distributions being investigated through orogenic processes, glacial cycles and climatic variations (Muellner-Riehl, 2019). It also contributes to the definition of phytogeographical areas, detection of endemic areas and conservation priorities according to floristic diversity (Médail & Baumel, 2018). Combination of molecular phylogeny and floristic mapping has enhanced our knowledge on the species diversification, colonization pattern and gene exchange among the isolated alpine populations and thus enhanced the use of floristics as an intermediary between taxonomy and evolutionary biogeography.

4. SCOPE OF ALPINE FLORISTICS

Alpine floristics has a wide and multifaceted scope which comprises geographical, taxonomic, ecological and temporal aspects (Körner, 2023). It goes further to encompass species recording to ecological processes, adaptation, and long-term vegetation dynamics characterizing the alpine biodiversity (Hotaling et al., 2017). The holistic nature of the field enables it to treat both the basic scientific issues and the practical conservation problems in the high mountainous regions of the world.

4.1 Geographic Scope (Global Alpine Regions)

The alpine floristics is found in the mountain systems of all continents, starting by the tropical zones and to the polar regions (Körner, 1995). The Himalayas and Tibetan Plateau in Asia, European Alps and Carpathians in Western Europe, the Andes in South America, the Rocky Mountains of North America, Southern Alps of New Zealand and the East Africa highlands of Mount Kenya and the Rwenzori are major alpine regions in the world. All these regions have a different geological history, climatic regimes and floristic makeup, but are ecologically similar due to their adaptation to cold, wind, and short growing seasons (Ziegler, 1990).

The comparative floristic research within these ranges, helps to determine the patterns of alpine diversity and endemism in the world (Körner, 1995). As an example, Himalayas have been observed to support one of the most prosperous alpine flora and the species turnover has been observed to be high with elevation gradient. On the same note, Andes show excellent levels of north and south floristic continuity and great local specialization. Through its involvement in the world alpine systems, floristics can help in macroecological evaluations and lead transboundary conservation efforts meant to preserve the mountain biodiversity corridors. Overview of the principal global alpine regions and their representative mountain systems, showing altitudinal range, climatic conditions and dominant vegetation types are given in Table 2.

Table 2. Global distribution of major alpine regions.

Alpine region	Major mountain systems	Typical elevation range (m.a.s.l)	Climate characteristics	Dominant vegetation
Himalayas (Asia)	Hindu Kush, Karakoram, Eastern Himalaya	3000-5500	Cold, arid to humid; strong monsoon influence	Cushion plants, sedges, dwarf rhododendrons
Andes (South America)	Central and Southern Andes	3500-4800	Dry cold, high solar radiation	Bunch grasses, azorellas, alpine rosettes
Alps (Europe)	Swiss, French and Italian Alps	2200-3600	Humid temperate; seasonal snow cover	Gentians, saxifrages, alpine meadows
Rockies (North America)	Colorado and Canadian Rockies	2500-4000	Continental cold; strong altitudinal gradient	Alpine tundra, sedges, cushion plants
East African Highlands	Mount Kenya, Rwenzori, Kilimanjaro	3000-4500	Equatorial diurnal climate	Giant lobelias, senecios, tussock grasses

4.2 Taxonomic Scope (Vascular and Non-Vascular Plants)

The alpine floristics comprise taxonomic values of both plants (vascular plants, i.e., angiosperms, gymnosperms, pteridophytes) and non-vascular plants, e.g., bryophytes, lichens, and algae (Mucina et al., 2016). Whereas vascular plants are the structural foundation of alpine vegetation, non-vascular taxa are important in soil stabilization, moisture retention and nutrient cycling. Nival and scree alpine habitats are commonly dominated by cryptogamic crusts, in which the major biomass is the mosses and lichens.

Floristic surveys should thus be more inclusive in that they record all the diversity of plants in terms of the spectrum of diversity which should be inclusive of the macroscopic and microscopic taxa (Sosef et al., 2017). With recent developments in molecular taxonomy and environmental DNA (eDNA) methods, the floristics field has gained more capacity to detect cryptic and morphologically indistinguishable species. This level of taxonomic coverage increases the quality of biodiversity evaluations and guides management of alpine habitats ecologically and communally (Dainese et al., 2015).

4.3 Ecological Scope (Habitats, Elevation Gradients, Microclimates)

Ecologically, the floristics of the alps covers the entire variety of habitats present in the mountain ecology, such as grasslands, scree slopes, snow beds, bogs, rocky outcrops and fellfields (Nagy & Grabherr, 2009). The habitats are dominated by high microclimatic variability where temperature, radiation and soil moisture differences are found in a short scale. The ecological filtering through these habitats occurs in floristic composition in terms of height, steepness, exposure and the type of the substrate (Poff, 1997).

The knowledge of the variation of species assemblage in an elevation gradient is a primary subject of alpine floristics. The result of such analyses can show how the patterns of zonation are formed, as well as how the process of adaptation and competition occur. Microclimatic research further explains the fact that low levels of heterogeneity in the environment support high levels of species co-occurrence even in closed macroclimatic environments. As a result, ecological domain of alpine floristics is broader to cover macroecological mapping up to fine-scale habitat differentiation, where the diversity of flora has a relation to environmental processes, which maintain the alpine ecosystems. Figure 1 shows how habitats, elevation gradients and microclimates vary along a low to high-elevation range.

Figure 1. Ecological scope across elevation gradients.

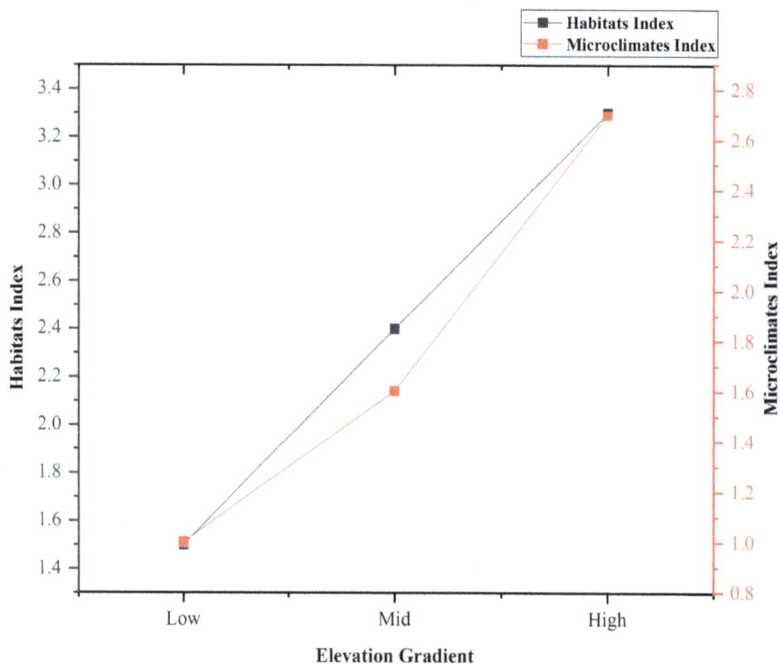

4.4 Temporal Scope (Seasonal and Long-Term Variations)

The time frame of alpine floristics refers to the short-term and long-term attitude towards vegetative dynamics. The alpine vegetation has a very close growth and reproductive cycle that is controlled by the temperature and time of snow melting. Transient community structures and phenological changes that are usually overlooked by single-time surveys are observed in floristic studies that can be done over multiple seasons.

Monitoring over a long period of time, however, gives important information with regards to ecological succession, vegetation changes and the migration of species with regard to climate change. Alpine floristics therefore plays a role in the measurement of temporal variation in the loss of biodiversity, community turnover and range shrinkage of cold-adapted species (Bektaş, 2024). The plots can be measured to determine the rate of biotic response to changing environmental conditions by keeping permanent plots and re-assessing the floristic inventory over decades, making alpine floristics an invaluable instrument in monitoring the future of high-elevation ecosystems in a warming world.

5. METHODOLOGICAL APPROACHES IN ALPINE FLORISTICS

Alpine floristic methodological methods are characterized by a combination of classical botanical methods and modern analytical and technological instruments, which allow to comprehensively evaluate the diversity of flora, its distribution and ecological interactions (Stöckli et al., 2011). Since the alpine ecosystems are complex and remote, methodological rigor is essential to produce valid, repeatable and geographically explicit data. Floristic studies are generally headed on to field reconnaissance and specimen gathering to taxonomic identification, space mapping and synthesis of data. The development of molecular biology, remote sensing and geospatial analysis in the recent decades has extended the extent and accuracy of alpine floristic research and this has enabled researchers to understand vegetation patterns within a multidimensional ecological context.

5.1 Field Surveys and Specimen Collection

The basis of any floristic investigation is based on field surveys. In the alpine, fieldwork should consider such logistical issues as steep slopes, inaccessibility and brief growing periods (Körner, 1995). Systematic surveys are usually performed as altitudinal transects or stratified sampling plots across habitats (i.e., grasslands, scree slopes, snow beds). At every sampling unit, the entire vascular and non-vascular vegetation is noted, photographed and harvested to make vouchers.

To make the specimens useful in the long term, they are pressed, dried and labeled with the necessary metadata, e.g., elevation, coordinates, slope, aspect, substrate type and other associated species (Bowden et al., 2020). Different replicates are usually taken to reflect intra-specific variation. The collection is strengthened by the inclusion of phenological notes (flowering or fruiting stage) which adds interpretive value to the collection. More and more floristic scientists not only use field data but also in situ measures of the environment (e.g., temperature, soil pH and moisture) and allow correlations between the presence of species with an environmental gradient to be made. Field surveys then constitute the empirical basis on the explanation of species composition, indices of diversity and spatial turnover between alpine regions.

5.2 Herbarium and Taxonomic Identification

Field identifications are only validated by herbarium study and taxonomic verification necessary to ensure the scientific credibility of the floristic data (Bowden et al., 2020). The collected specimens are curated, mounted and compared to the authenticated reference material within regional and international herbaria. The morphological identification is based on the characteristics of the diagnostic features (arrangement of leaves, structure of inflorescence, reproductive organs, morphology of trichomes) observed under the microscope with the help of stereomicroscopes, in some cases.

The taxonomic nomenclature is based on internationally accepted databases like Plants of the World Online (POWO), The Plant List, or the International Plant Names Index (IPNI). Work based on herbarium also helps in the identification of synonyms, misidentifications and new records of distribution. Digitized herbarium data is commonly used in complementing field collections in a variety of alpine research studies to enable historical comparison, which can provide insights on how species distributions have changed through time. Documenting herbarium of the appropriate kind will mean that voucher specimen may be made a permanent reference to be used in future taxonomic revisions and phylogenetic studies. Therefore, not only do herbarium studies confirm the identifications made in the field, but also enhance the archival history of the alpine biodiversity over the long term.

5.3 Use of GIS and Remote Sensing in Floristic Mapping

Geographic Information System (GIS) and remote sensing technology integration has transformed research of the alpine flora. Using GIS provides the opportunity to make distributions of species visually and analytically in the space as it allows researchers to map the distributions of diversity, the type of habitat and the possible hotspots in species richness (Salem, 2003). Satellite imagery and drone-based survey is a remote sensing technology that offers high-resolution environmental information regarding land cover, topography, vegetation indices (e.g., Normalized Difference Vegetation Index) and snow cover dynamics.

When the maps of the distribution patterns of the species are modeled by overlapping the floristic records with the environmental overlay, which includes temperature, precipitation and soil data, the researchers will be able to predict possible changes in the habitat of the species under the conditions of varying climate. These spatial analyses come in handy especially in the alpine regions where ecological heterogeneity is highly ecological due to finer micro-habitats and elevation gradients. Moreover, geospatial analyses help in conservation planning because they can determine ecological corridors, isolated refugia and endemism areas. A blend of field-collected information and GIS-modeling offers a combination of the approaches to thinking about the underlying processes of both spatial and temporal diversity of the alpine flora. Some uses of GIS and RS in floristic mapping are shown in Figure 2.

Figure 2. GIS and RS integration the field of alpine floristics.

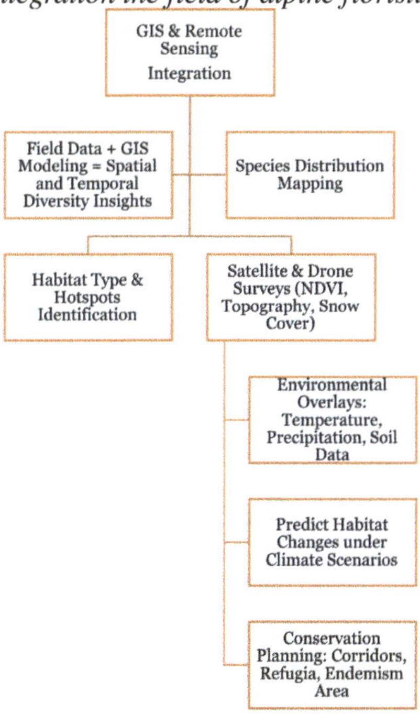

5.4 Integrating Molecular Techniques in Floristic Studies

In the area of modern floristics, there is the growing use of molecular methods to clarify taxonomic uncertainties, determine phylogenetic relationships and estimate genetic variance between alpine taxa (Bell et al., 2018). Essentially, DNA barcoding (based on markers rbcL, matK and ITS) can be used to accurately identify morphologically similar or cryptic species. The evolutionary lineages and speciation processes can also be found in the molecular data touching on geographic isolation and climatic oscillations.

Phylogeographic studies can be used to trace past migration patterns of alpine flora and explain how past glaciations and orogenic events contributed to the formation of modern distributions. Moreover, microsatellites or single nucleotide polymorphisms (SNPs) based on population-level molecular analyses are useful in determining genetic connectivity and resilience of the alpine populations. Integrating both the traditional and the molecular systematics thus gives a total picture of the alpine biodiversity-both in its taxonomic form of species and also in its evolutionary or ecological interactions, which will increase the scientific strength of the floristic studies. Summary of major methodological tools and their applications in alpine floristic research, integrating traditional and modern analytical techniques are presented in Table 3.

Table 3. Methodological approaches in alpine floristics.

Methodological category	Description	Key applications	Limitations
Field surveys & specimen collection	Direct vegetation sampling along elevation transects	Species inventory, habitat classification	Accessibility constraints; seasonal variation
Herbarium & taxonomic identification	Morphological examination using reference collections	Accurate species verification	Requires expert taxonomists
GIS & remote sensing	Geospatial mapping using satellite imagery	Habitat modeling, distribution mapping	Resolution limits in rugged terrain
Molecular techniques	DNA barcoding, phylogenetics	Species delimitation, evolutionary studies	Cost-intensive; DNA degradation at altitude

6. APPLICATIONS OF ALPINE FLORISTICS

The broad scope of application of alpine floristics is not exclusive to the academic research, but also other more practical aspects of biodiversity conservation, environmental management and climate change adaptation. Floristic knowledge supports evidence-based policy, conservation management and sustainability land management by explaining the patterns, drivers and functional roles of alpine flora diversity. Its uses also cross the boundary between ecological science and human welfare by incorporating the ethnobotanical and ecosystem service point of view.

6.1 Understanding Biodiversity and Endemism

The evaluation and measurement of biodiversity and endemism in the high-mountain ecosystems is one of the main uses of alpine floristics (Khan et al., 2013). Floristics determines the total count of species, assemblage and turnover of species through systematic inventories and taxonomic examination among elevation gradients. The analysis of endemism, especially, shows areas of evolutionary distinctiveness that are used as conservation priorities.

The recording of species that live in very specific ecological niches or in isolated mountain ranges allows the identification of evolutionary hotspots which are also important to support world biodiversity through floristics. Such data also have an input into global databases like the Global Biodiversity Information Facility (GBIF) which helps to improve international collaboration and comparative studies. Knowledge in patterns of diversity and endemism using floristics is the ecological underpinning needed in making long term monitoring biodiversity and ecosystem management.

6.2 Conservation Planning and Priority Setting

The design of conservation strategies and the development of the protected areas in mountain regions rely on floristic bit of information (Khan et al., 2013). Identification of key biodiversity areas (KBAs) is informed by mapping spatial location of rare, threatened and endemic species and can be used to establish ecological networks and corridors, which maintain species connectivity. Using floristic data in conjunction with landscape-level studies, conservation planners are able to measure the vulnerability of habitats, set priorities in areas requiring restoration and adopt adaptive management systems (Rudnick et al., 2012). Floristic datasets are also used in transboundary mountain ranges, such as the Himalayas or Andes, to support regional collaboration and harmonization of policies, such as in the alpine convention or Himalayan Biodiversity Corridor. Therefore, alpine floristics offers the scientific knowledge necessary to distribute the conservation resources effectively and preserve the ecological integrity of the mountain systems.

6.3 Assessing Climate Change Impacts on Alpine Flora

The alpine ecosystems are the least resistant to climate change indicators and the study of flora is essential in identifying and explaining the vegetation reaction (Pauli et al., 2003). Numerous decadent floristic surveys have shown positive changes in species range, changed phenological timing, and community structure due to warming temperatures and a decrease in snow cover. Long-term floristic data enable the researcher to model the change in species distribution and determine how vulnerable to extinction cold-adapted taxa are (Lagerholm et al., 2017). This kind of insight is crucial in anticipating the possible alpine biodiversity loss and coming up with some adaptation strategies. Also, floristic information combined with climate models can be used to locate possible refugia, i.e., those areas, which are likely to maintain favorable microclimates in future conditions. As a result, alpine floristics is actively involved in the climate-change studies and offers conservation plans to increase the ecosystem resilience as represented in Figure 3.

Figure 3. Assessing climate change impacts on alpine flora using floristic indicators.

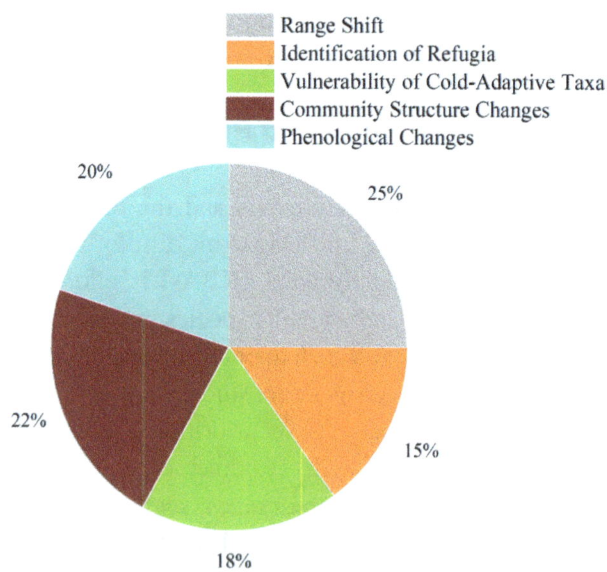

6.4 Linking Floristics with Ethnobotany and Ecosystem Services

Alpine floristics is also related to ethnobotany and ecosystem service valuation, in addition to its ecological and taxonomic aspects (Poncet et al., 2015). Numerous alpine flora has considerable cultural, medicinal and economic importance to the local people who rely on them as a source of healthcare, grazing as well as traditional livelihoods. Floristic documentation gives a basis of identifying, conserving and the management of these biological resources in a sustainable manner (Khan et al., 2013). Besides, floristic diversity also provides ecosystem services including soil stabilization, water regulation, carbon sequestration and support of pollination. By combining the information on floristic data with the socio-ecological models, the researchers will be able to evaluate the role biodiversity plays in supporting human well-being in the mountainous areas. It is in light of these interdisciplinary applications that alpine floristics is not only important as a scientific endeavor but also as an instrument in ensuring sustainable development and conservation of mountain ecosystems.

7. CHALLENGES IN ALPINE FLORISTIC STUDIES

The ecological and conservation value of alpine floristic research even though it is a critical field of study is sometimes limited by various logistical, methodological and scientific impediments. These challenges are mainly caused by the harsh conditions of the environment, inaccessibility and complicacy of the alpine biodiversity. In addition, lack of long-term datasets and taxonomic skills do not allow forming a global picture of the composition of alpine vegetation and its ecological interactions. It is crucial to identify and implement solutions to these limitations in order to make the field more progressive and come up with more efficient conservation models.

7.1 Accessibility and Harsh Environmental Conditions

The steep slopes, rugged landscape and weather variations and the scanty growing seasons of the alpine ecosystems are made challenging by field work (Nagy & Grabherr, 2009). The researchers are inclined to face certain difficulties like the inaccessibility of transportation, the insufficiency of oxygen and the swiftly shifting weather conditions which limit the time of sampling. In addition, the alpine geography makes it difficult to repeat observations, do long-term monitoring, and have permanent plots due to the remoteness of most areas. These limitations do not only influence the data collection but also restrict the depiction of the temporal variability in the floristic inventories.

7.2 Taxonomic Complexities and Identification Issues

The morphological plasticity of alpine species of plant life is frequently high because of the dramatic environmental gradients to which they are subjected (Körner, 2023). This plasticity makes it very hard to define the species correctly because the individuals belonging to the same species may vary greatly in shape and size according to the conditions in the microhabitat. Numerous alpine species are of complex genera whose morphological differences are slight and which thus demand knowledge on an expert level together with access to detailed herbarium data. Moreover, cryptic species and hybridization are also frequent in alpine environments and so, the combination of molecular methods to delimit species is required.

7.3 Data Gaps and Underexplored Regions

Although the mountain biodiversity has become an issue of interest, much of the alpine areas on the earth still remain underexplored especially in the Himalayas, Hindu Kush, Andes, and highlands of East Africa (Payne et al., 2020; Sharma et al., 2010). The outcome of these data gaps is incomplete floristic records which bias the global evaluations of diversity and endemism. Much of the historical data is not complete and the methodology and the metadata of the information are irregular. The unification of standardized protocols and digitized herbarium collections further hinder comparing the data across regions. Making improvements to increase global databases and encouraging open-access repositories are pivotal moves towards reducing these knowledge gaps.

7.4 Need for Collaborative and Interdisciplinary Research

Alpine ecosystems are complex and therefore need to work collectively across disciplinary boundaries (Hotaling et al., 2017). Ecologists, geneticists, climatologists, geographers, and botanists have to collaborate to combine floristic, functional and genomic data. Multinational research initiatives and collaborative networks are required to attain a complete coverage of the space and standardization of the methods (Freshwater et al., 2006). In addition, the local communities and indigenous knowledge systems may provide useful information on the use of alpine plants, their distribution and adaptive mechanisms. Developing interdisciplinary relations improves the depth, fidelity and practicality of alpine floristic research.

8. FUTURE PERSPECTIVES

A combination of traditional botanical survey and modern technology and analysis devices is the future of alpine floristics (Stöckli et al., 2011). Some of the current developments in molecular phylogenetics, remote sensing, geospatial modeling and big data analytics are transforming how scholars document and interpret alpine biodiversity. The use of long-term monitoring schemes and climate modeling may enable forecasting the changes in the composition of species and their range limits in the climate conditions of the future (Karl et al., 1995). Both citizen science projects and community-based surveillance can provide a potential solution to collecting data in the most inaccessible locations and digital herbaria services can make specimen data global and democratic. Enhancement of international partnerships, especially in mountain ranges like the Himalayas, Andes, Alps and Rockies will bring out a better perception of global alpine biodiversity trends. It is also important that the floristic knowledge should be translated into conservation planning and policy making so that fragile alpine ecosystems get the attention they merit in the global biodiversity agendas. With the increased rate of climate change, alpine floristics will be utilized more and more extensively to monitor ecological changes, design recovery programs, and conserve the genetic and functional diversity of high-mountain plants. The further development of this field will define the scientific knowledge of alpine ecosystems and their stability in the world of rapid changes.

9. CONCLUSION

The alpine floristics is one of the staples of comprehending the complexity, dynamics and conservation requirements of high-elevation eco systems. These delicate landscapes, which stand above the tree line, are the home to a special collection of plant life that has adapted to harsh environmental conditions. By systematic exploration, not only the diversity of species is included by the floristic studies but also ecological relations, biogeographical dynamics and functional features underlying the resilience of the alpine vegetation. As has been highlighted in this chapter, alpine floristics is a science that goes beyond being descriptive; it is a crucial scheme to connect taxonomy, ecology and conservation. Since the early botanical research, the discipline has developed into a multidisciplinary methodology taking classical herbarium studies to a new level with modern tools of analysis. It now has an area covering global alpine regions, vascular and non-vascular flora and its spatial, temporal and ecological dimensions. The alpine floristics is one of the staples of comprehending the complexity, dynamics and conservation requirements of high-elevation eco systems. These delicate landscapes, which stand above the tree line, are the home to a special collection of plant life that has adapted to harsh environmental conditions. By systematic exploration, not only the diversity of species is included by the floristic studies but also ecological relations, biogeographical dynamics and functional features underlying the resilience of the alpine vegetation. As has been highlighted in this chapter, alpine floristics is a science that goes beyond being descriptive; it is a crucial scheme to connect taxonomy, ecology and conservation. Since the early botanical research, the discipline has developed into a multidisciplinary methodology taking classical herbarium studies to a new level with modern tools of analysis. It now has an area covering global alpine regions, vascular and non-vascular flora and its spatial, temporal and ecological dimensions.

REFERENCES

Abeli, T., Vamosi, J. C., & Orsenigo, S. (2018). The importance of marginal population hotspots of cold-adapted species for research on climate change and conservation. *Journal of Biogeography*, *45*(5), 977–985. DOI: 10.1111/jbi.13196

Ashby, E., & Stevens, W. (1935). The quantitative analysis of vegetation. *Annals of Botany*, *49*(196), 779–802. DOI: 10.1093/oxfordjournals.aob.a090535

Bektaş, B. (2024). *Transient dynamics of Alpine grasslands linking biodiversity and ecosystem functions in a warming world.* Universite Grenoble Alpes.

Bell, N., Griffin, P. C., Hoffmann, A. A., & Miller, A. D. (2018). Spatial patterns of genetic diversity among Australian alpine flora communities revealed by comparative phylogenomics. *Journal of Biogeography*, *45*(1), 177–189. DOI: 10.1111/jbi.13120

Billings, W. D., & Mooney, H. A. (1968). The ecology of arctic and alpine plants. *Biological Reviews of the Cambridge Philosophical Society*, *43*(4), 481–529. DOI: 10.1111/j.1469-185X.1968.tb00968.x

Bowden, D. A., Rowden, A. A., Chin, C. C., Hempel, S., Wood, B., Hart, A., & Clark, M. R. (2020). *Best practice in seabed image analysis for determining taxa, habitat, or substrata distributions.* Ministry for Primary Industries.

Cortés, A. J., & Wheeler, J. (2018). The environmental heterogeneity of mountains at a fine scale in a changing world. *Mountains, Climate, and Biodiversity, 584.*

Dainese, M., Lepš, J., & de Bello, F. (2015). Different effects of elevation, habitat fragmentation and grazing management on the functional, phylogenetic and taxonomic structure of mountain grasslands. *Perspectives in Plant Ecology, Evolution and Systematics*, *17*(1), 44–53. DOI: 10.1016/j.ppees.2014.09.002

Dansereau, P. (1951). Description and recording of vegetation upon a structural basis. *Ecology*, *32*(2), 172–229. DOI: 10.2307/1930415

Decocq, G. (2016). Moving from Patterns to Processes: A Challenge for the Phytosociology of the Twenty-First Century? In *Vegetation Structure and Function at Multiple Spatial, Temporal and Conceptual Scales* (pp. 407-424). Springer. DOI: 10.1007/978-3-319-21452-8_18

Evangelista, A., Frate, L., Carranza, M. L., Attorre, F., Pelino, G., & Stanisci, A. (2016). Changes in composition, ecology and structure of high-mountain vegetation: A re-visitation study over 42 years. *AoB Plants*, *8*, plw004. DOI: 10.1093/aobpla/plw004 PMID: 26819258

Ford, K. R., Ettinger, A. K., Lundquist, J. D., Raleigh, M. S., & Hille Ris Lambers, J. (2013). Spatial heterogeneity in ecologically important climate variables at coarse and fine scales in a high-snow mountain landscape. *PLoS One*, *8*(6), e65008. DOI: 10.1371/journal.pone.0065008 PMID: 23762277

Freshwater, D., Sherwood, G., & Drury, V. (2006). International research collaboration: Issues, benefits and challenges of the global network. *Journal of Research in Nursing : JRN*, *11*(4), 295–303. DOI: 10.1177/1744987106066304

Frodin, D. G. (2001). *Guide to standard floras of the world: an annotated, geographically arranged systematic bibliography of the principal floras, enumerations, checklists and chorological atlases of different areas*. Cambridge University Press. DOI: 10.1017/CBO9780511541803

Gentili, R., Armiraglio, S., Sgorbati, S., & Baroni, C. (2013). Geomorphological disturbance affects ecological driving forces and plant turnover along an altitudinal stress gradient on alpine slopes. *Plant Ecology*, *214*(4), 571–586. DOI: 10.1007/s11258-013-0190-1

Gentili, R., Bacchetta, G., Fenu, G., Cogoni, D., Abeli, T., Rossi, G., Salvatore, M. C., Baroni, C., & Citterio, S. (2015). From cold to warm-stage refugia for boreo-alpine plants in southern European and Mediterranean mountains: The last chance to survive or an opportunity for speciation? *Biodiversity (Nepean)*, *16*(4), 247–261. DOI: 10.1080/14888386.2015.1116407

Gentry, A. H. (1988). Changes in plant community diversity and floristic composition on environmental and geographical gradients. *Annals of the Missouri Botanical Garden*, *75*(1), 1–34. DOI: 10.2307/2399464

Gillespie, T., Brock, J., & Wright, C. (2004). Prospects for quantifying structure, floristic composition and species richness of tropical forests. *International Journal of Remote Sensing*, *25*(4), 707–715. DOI: 10.1080/01431160310001598917

Grabherr, G. (2009). Biodiversity in the high ranges of the Alps: Ethnobotanical and climate change perspectives. *Global Environmental Change*, *19*(2), 167–172. DOI: 10.1016/j.gloenvcha.2009.01.007

Grêt-Regamey, A., Walz, A., & Bebi, P. (2008). Valuing ecosystem services for sustainable landscape planning in Alpine regions. *Mountain Research and Development*, *28*(2), 156–165. DOI: 10.1659/mrd.0951

Heywood, V. (2001). Floristics and monography—An uncertain future? *Taxon*, *50*(2), 361–380. DOI: 10.2307/1223886

Hotaling, S., Finn, D. S., Joseph Giersch, J., Weisrock, D. W., & Jacobsen, D. (2017). Climate change and alpine stream biology: Progress, challenges, and opportunities for the future. *Biological Reviews of the Cambridge Philosophical Society*, *92*(4), 2024–2045. DOI: 10.1111/brv.12319 PMID: 28105701

Ianni, E., Geneletti, D., & Ciolli, M. (2015). Revitalizing traditional ecological knowledge: A study in an alpine rural community. *Environmental Management*, *56*(1), 144–156. DOI: 10.1007/s00267-015-0479-z PMID: 25847107

Ibrahim, M., Khan, M., Ali, S., Razzaq, A., Zaman, A., Iqbal, M., & Jan, F. (2019). Floristic composition and species diversity of plant resources of rural area "Takht Bhai" district Mardan, Khyber Pakhtunkhwa, Pakistan. *Medicinal and aromatic Plants (Los Angeles)*, *8*(338), 2167. DOI: 10.35248/2167-0412.19.8.338

Karl, T. R., Derr, V. E., Easterling, D. R., Folland, C. K., Hofmann, D. J., Levitus, S., Nicholls, N., Parker, D. E., & Withee, G. W. (1995). Critical issues for long-term climate monitoring. *Climatic Change*, *31*(2), 185–221. DOI: 10.1007/BF01095146

Khan, S. M., Page, S. E., Ahmad, H., & Harper, D. M. (2013). Sustainable utilization and conservation of plant biodiversity in montane ecosystems: The western Himalayas as a case study. *Annals of Botany*, *112*(3), 479–501. DOI: 10.1093/aob/mct125 PMID: 23825353

Körner, C. (1995). Alpine plant diversity: a global survey and functional interpretations. In *Arctic and alpine biodiversity: patterns, causes and ecosystem consequences* (pp. 45–62). Springer. DOI: 10.1007/978-3-642-78966-3_4

Körner, C. (2021). Alpine treelines. In *Alpine plant life: functional plant ecology of high mountain ecosystems* (pp. 141–173). Springer.

Körner, C. (2023). Concepts in alpine plant ecology. *Plants*, *12*(14), 2666. DOI: 10.3390/plants12142666 PMID: 37514280

Körner, C., & Change, G. (2018). Alpine ecosystems and the high-elevation treeline. *Encyclopedia of ecology.*, 407-413.

Körner, C., & Hiltbrunner, E. (2021). Why is the alpine flora comparatively robust against climatic warming? *Diversity*, *13*(8), 383. DOI: 10.3390/d13080383

Körner, C., Nakhutsrishvili, G., & Spehn, E. (2006). High-elevation land use, biodiversity, and ecosystem functioning. In *Land use change and mountain biodiversity* (pp. 3–36). CRC Press. DOI: 10.1201/9781420002874-1

Kullman, L. (2010). Alpine flora dynamics–a critical review of responses to climate change in the Swedish Scandes since the early 1950s. *Nordic Journal of Botany, 28*(4), 398–408. DOI: 10.1111/j.1756-1051.2010.00812.x

Lagerholm, V. K., Sandoval-Castellanos, E., Vaniscotte, A., Potapova, O. R., Tomek, T., Bochenski, Z. M., Shepherd, P., Barton, N., Van Dyck, M. C., Miller, R., Höglund, J., Yoccoz, N. G., Dalén, L., & Stewart, J. R. (2017). Range shifts or extinction? Ancient DNA and distribution modelling reveal past and future responses to climate warming in cold-adapted birds. *Global Change Biology, 23*(4), 1425–1435. DOI: 10.1111/gcb.13522 PMID: 27762483

Lavergne, S., Pouchon, C., Roquet, C., Alberti, A., Boleda, M., Perrier, C., Renaud, J., Boyer, F., Bzeznik, B., Rome, M., Douzet, R., Lammers, Y., Orvain, C., Gavory, F., Labadie, K., Aubert, S., Consortium, P. N., Consortium, P. A., Wincker, P., & Coissac, E. (2025). Towards a comprehensive barcoding and phylogenomic reference for the European Arctic–alpine flora. *Botany Letters, 172*(4), 1–13. DOI: 10.1080/23818107.2025.2549565

MacMahon, J. A. (1980). Ecosystems over time: succession and other types of change. *Forests: fresh perspectives from ecosystem analysis*, 27-58.

McLaughlin, S. P. (1994). Floristic plant geography: The classification of floristic areas and floristic elements. *Progress in Physical Geography, 18*(2), 185–208. DOI: 10.1177/030913339401800202

Médail, F., & Baumel, A. (2018). Using phylogeography to define conservation priorities: The case of narrow endemic plants in the Mediterranean Basin hotspot. *Biological Conservation, 224*, 258–266. DOI: 10.1016/j.biocon.2018.05.028

Mi, X., Feng, G., Hu, Y., Zhang, J., Chen, L., Corlett, R. T., Hughes, A. C., Pimm, S., Schmid, B., Shi, S., Svenning, J.-C., & Ma, K. (2021). The global significance of biodiversity science in China: An overview. *National Science Review, 8*(7), nwab032. DOI: 10.1093/nsr/nwab032 PMID: 34694304

Montagna, M., Mereghetti, V., Lencioni, V., & Rossaro, B. (2016). Integrated taxonomy and DNA barcoding of alpine midges (Diptera: Chironomidae). *PLoS One, 11*(3), e0149673. DOI: 10.1371/journal.pone.0149673 PMID: 26938660

Mucina, L., Bültmann, H., Dierßen, K., Theurillat, J. P., Raus, T., Čarni, A., Šumberová, K., Willner, W., Dengler, J., García, R. G., Chytrý, M., Hájek, M., Di Pietro, R., Iakushenko, D., Pallas, J., Daniëls, F. J. A., Bergmeier, E., Santos Guerra, A., Ermakov, N., & Tichý, L. (2016). Vegetation of Europe: Hierarchical floristic classification system of vascular plant, bryophyte, lichen, and algal communities. *Applied Vegetation Science, 19*(S1), 3–264. DOI: 10.1111/avsc.12257

Muellner-Riehl, A. N. (2019). Mountains as evolutionary arenas: Patterns, emerging approaches, paradigm shifts, and their implications for plant phylogeographic research in the Tibeto-Himalayan region. *Frontiers in Plant Science*, *10*, 195. DOI: 10.3389/fpls.2019.00195 PMID: 30936883

Nagy, L., & Grabherr, G. (2009). *The biology of alpine habitats*. Oxford University Press. DOI: 10.1093/oso/9780198567035.001.0001

Noroozi, J., Talebi, A., Doostmohammadi, M., Rumpf, S. B., Linder, H. P., & Schneeweiss, G. M. (2018). Hotspots within a global biodiversity hotspot-areas of endemism are associated with high mountain ranges. *Scientific Reports*, *8*(1), 10345. DOI: 10.1038/s41598-018-28504-9 PMID: 29985437

Opedal, Ø. H., Armbruster, W. S., & Graae, B. J. (2015). Linking small-scale topography with microclimate, plant species diversity and intra-specific trait variation in an alpine landscape. *Plant Ecology & Diversity*, *8*(3), 305–315. DOI: 10.1080/17550874.2014.987330

Pauli, H., Gottfried, M., & Grabherr, G. (2003). Effects of climate change on the alpine and nival vegetation of the Alps. *Journal of mountain ecology*, *7*(Suppl), 9-12.

Pauli, H., & Halloy, S. R. (2019). High mountain ecosystems under climate change. In *Oxford research encyclopedia of climate science*. DOI: 10.1093/acrefore/9780190228620.013.764

Payne, D., Snethlage, M., Geschke, J., Spehn, E. M., & Fischer, M. (2020). Nature and people in the Andes, East African Mountains, European Alps, and Hindu Kush Himalaya: Current research and future directions. *Mountain Research and Development*, *40*(2), A1. DOI: 10.1659/MRD-JOURNAL-D-19-00075.1

Poff, N. L. (1997). Landscape filters and species traits: Towards mechanistic understanding and prediction in stream ecology. *Journal of the North American Benthological Society*, *16*(2), 391–409. DOI: 10.2307/1468026

Polechová, J., & Barton, N. H. (2015). Limits to adaptation along environmental gradients. *Proceedings of the National Academy of Sciences of the United States of America*, *112*(20), 6401–6406. DOI: 10.1073/pnas.1421515112 PMID: 25941385

Poncet, A., Vogl, C. R., & Weckerle, C. S. (2015). Folkbotanical classification: Morphological, ecological and utilitarian characterization of plants in the Napf region, Switzerland. *Journal of Ethnobiology and Ethnomedicine*, *11*(1), 13. DOI: 10.1186/1746-4269-11-13 PMID: 25971246

Raes, N. (2009). *A quantitative analysis of botanical richness, endemicity and floristic regions based on herbarium records*. Leiden University.

Rawat, G. (2007). *Alpine vegetation of the western Himalaya: species diversity, community structure, dynamics and aspects of conservation* D. Sc Thesis, Kumaun University, Nainital, 239pp].

Reddy, C. S. (2017). Applications of GIS in plant taxonomy, species distribution and ecology. *Journal of Economic and Taxonomic Botany*, *41*(3-4).

Reveal, J. L. (1991). Botanical explorations in the American West-1889-1989: An essay on the last century of a floristic frontier. *Annals of the Missouri Botanical Garden*, *78*(1), 65–80. DOI: 10.2307/2399591

Rudnick, D., Ryan, S. J., Beier, P., Cushman, S. A., Dieffenbach, F., Epps, C., Gerber, L. R., Hartter, J. N., Jenness, J. S., & Kintsch, J. (2012). *The role of landscape connectivity in planning and implementing conservation and restoration priorities.* Issues in Ecology.

Salem, B. (2003). Application of GIS to biodiversity monitoring. *Journal of Arid Environments*, *54*(1), 91–114. DOI: 10.1006/jare.2001.0887

Séverine Fauquette, J.-P. S., Médail, F., Muller[1], G. J.-M., & Bertini, E. (2018). The Alps: a geological, climatic and human perspective on vegetation history and modern plant diversity. *Mountains, climate and biodiversity*, 413.

Sharma, E., Chettri, N., & Oli, K. P. (2010). Mountain biodiversity conservation and management: A paradigm shift in policies and practices in the Hindu Kush-Himalayas. *Ecological Research*, *25*(5), 909–923. DOI: 10.1007/s11284-010-0747-6 PMID: 32214652

Silva, H. C. H., Caraciolo, R. L. F., Marangon, L. C., Ramos, M. A., Santos, L. L., & Albuquerque, U. P. (2014). Evaluating different methods used in ethnobotanical and ecological studies to record plant biodiversity. *Journal of Ethnobiology and Ethnomedicine*, *10*(1), 48. DOI: 10.1186/1746-4269-10-48 PMID: 24916833

Sosef, M. S., Dauby, G., Blach-Overgaard, A., van Der Burgt, X., Catarino, L., Damen, T., Deblauwe, V., Dessein, S., Dransfield, J., Droissart, V., Duarte, M. C., Engledow, H., Fadeur, G., Figueira, R., Gereau, R. E., Hardy, O. J., Harris, D. J., de Heij, J., Janssens, S., & Couvreur, T. L. P. (2017). Exploring the floristic diversity of tropical Africa. *BMC Biology*, *15*(1), 15. DOI: 10.1186/s12915-017-0356-8 PMID: 28264718

Stöckli, V., Wipf, S., Nilsson, C., & Rixen, C. (2011). Using historical plant surveys to track biodiversity on mountain summits. *Plant Ecology & Diversity*, *4*(4), 415–425. DOI: 10.1080/17550874.2011.651504

Testolin, R., Attorre, F., & Jiménez-Alfaro, B. (2020). Global distribution and bioclimatic characterization of alpine biomes. *Ecography*, *43*(6), 779–788. DOI: 10.1111/ecog.05012

Tito, R., Vasconcelos, H. L., & Feeley, K. J. (2020). Mountain ecosystems as natural laboratories for climate change experiments. *Frontiers in Forests and Global Change*, *3*, 38. DOI: 10.3389/ffgc.2020.00038

Turrill, W. (1964). *Plant taxonomy, phytogeography, and plant ecology* (pp. 187-224).

Turrill, W. B. (2013). *Pioneer plant geography: the phytogeographical researches of Sir Joseph Dalton Hooker* (Vol. 4). Springer.

Walker, M. D., Walker, D., Theodose, T., Weber, P., Bowman, W., & Seastedt, T. (2001). *The vegetation hierarchical species-environment. Structure and function of an alpine ecosystem: Niwot Ridge, Colorado*. Oxford University Press.

Waterman, T. H. (2001). Evolutionary challenges of extreme environments (part 2). *The Journal of Experimental Zoology*, *291*(2), 130–168. DOI: 10.1002/jez.1065 PMID: 11479914

Zaman, A., Asadullah, L. B., Muhammad, Z., Razzaq, A., Jelani, G., Ali, U., Raees, N., & Khan, M. N. (2019). Floristics of weeds in Triticum aestivum L. fields of Tehsil Shabqadar, District Charsadda, KP, Pakistan. *International Journal of Botany Studies*, *4*(5), 37–44.

Zhang, W., Ren, C., Deng, J., Zhao, F., Yang, G., Han, X., Tong, X., & Feng, Y. (2018). Plant functional composition and species diversity affect soil C, N, and P during secondary succession of abandoned farmland on the Loess Plateau. *Ecological Engineering*, *122*, 91–99. DOI: 10.1016/j.ecoleng.2018.07.031

Ziegler, A. M. (1990). Phytogeographic patterns and continental configurations during the Permian Period. *Geological Society, London, Memoirs, 12*(1), 363-379.

Chapter 2
Alpine Biodiversity in Pakistan

Muhammad Manzoor
http://orcid.org/0000-0002-5118-3957
Quaid-i-Azam University, Islamabad, Pakistan

Syed Waseem Gillani
http://orcid.org/0009-0009-2926-0723
Quaid-i-Azam University, Islamabad, Pakistan

Mushtaq Ahmad
Quaid-i-Azam University, Islamabad, Pakistan

Maqsood Alam
The University of Azad Jammu and Kashmir, Muzaffarabad, Pakistan

Amir Sohail
Quaid-i-Azam University, Islamabad, Pakistan

Najeeb Ullah
Quaid-i-Azam University, Islamabad, Pakistan

Ahmed Ali
http://orcid.org/0009-0008-8786-2952
The University of Azad Jammu and Kashmir, Muzaffarabad, Pakistan

Abdul Basit Mehmood
Quaid-i-Azam University, Islamabad, Pakistan

ABSTRACT

Alpine biodiversity is among the most ecologically significant and vulnerable components of mountain flora, providing critical ecosystem services. Pakistan, located in the western Himalayas, harbors diverse alpine systems that are increasingly threatened by climate change and anthropogenic pressures. This study assessed altitudinal biodiversity patterns and drivers of community composition. Multivariate analyses showed strong site-specific heterogeneity rather than simple elevational segregation. Bray-Curtis dissimilarity indicated that 83.55% of community variation was concentrated in five sites, identifying biodiversity hotspots. NMDS and β-diversity analyses revealed high species turnover (βsim = 0.79) and low nestedness.

DOI: 10.4018/979-8-3373-6756-9.ch002

ght © 2026, IGI Global Scientific Publishing. Copying or distributing in print or electronic forms without written permission of IGI Global Scientific Publishing is prohibited. Use of this chapter to generative artificial intelligence (AI) technologies is expressly prohibited. The publisher reserves all rights to license its use for generative AI training and machine learning model development.

Poisson GLM showed richness increased with elevation, peaking at mid-altitudes. Indicator species, CCA, and Mantel tests highlighted altitude and grazing as key drivers, emphasizing the need for site-based conservation strategies.

1. INTRODUCTION

The alpine biomes are found in mountainous areas around the world between the perpetual snow region and the sub-alpine forest, characterized by low temperatures, short growing seasons, intense solar radiation, and strong winds. These ecosystems support specialized plant communities adapted to harsh climatic conditions and fragile soils (Quinn, 2008; Shaheen et al., 2019), and are represented by the Himalayas, which is one of the biodiversity ecoregions in the world (Salick et al., 2014). The Himalayan alpine ecosystem is characterized by diverse vegetation, high species richness, and remarkable endemism, making it a critical region for ecosystem services and conservation priorities. These ecosystems regulate water resources, stabilize soils, support unique biodiversity, and are highly sensitive to climate change and anthropogenic disturbances, necessitating focused conservation and sustainable management strategies (Ahmed et al., 2020; Khan et al., 2025; Manzoor et al., 2025; Verrall & Pickering, 2020). It has a higher plant diversity than the world average, that of the alpine biomes, with the Western Himalayan zone containing 1800-1900 species of vascular plants (Rawat, 2007). Temperature, rainfall, and elevation are key environmental factors influencing the distribution and diversity of alpine vegetation by shaping microclimatic conditions, soil development, and moisture availability. Variations in these factors across altitudinal gradients determine species composition, productivity, and adaptive strategies of alpine plant communities (Gillani et al., 2026; Grabherr et al., 2010; Manzoor et al., 2026). The alpine vegetation is adapted to cold conditions with perennial grasses, sedges, cushion plants, mosses, forbs, and lichens, as well as useful aromatic and medicinal plants (Khan et al., 2021; Shaheen et al., 2023). Most of these plants are stunted or spiny in their growth forms, beginning to grow early with a short.lifecycle (Hamid et al., 2020; Manzoor et al., 2025). The seasonality in communities demonstrates the effect of the changes in climatic variation (Körner, 2007, 2016). The alpine areas also have coarser soils that are combined with stones and gravel with weak secondary mineralization and a high proportion of organic carbon because of low production rates (Shaheen et al., 2019). Water exists on a variable basis, which depends on the factors of the altitude, aspect, and steepness of the slope (Leng et al., 2023; Nepal et al., 2014). The plants in these regions exhibit early growth and a limited growing season and are mainly of the Hemicryptophytes species, which is characteristic of the tough climatic conditions (Hamid et al., 2020; Khan et al., 2021; Manzoor et al., 2026;

Shaheen et al., 2023). The ongoing climate change poses a serious threat to alpine biodiversity, particularly within mountainous alpine ecosystems. Rising temperatures, altered precipitation patterns, and the increased frequency of extreme weather events disrupt the delicate balance of these high-altitude habitats. Such changes can lead to habitat loss, shifts in species distributions, reduced endemism, and increased vulnerability of specialized plant and animal communities, highlighting the urgent need for targeted conservation and adaptive management strategies.

It is estimated that global warming will have enormous effects on the alpine ecosystems where the growing season is very inadequate (Kudo & Hirao, 2006). Already, climate change is impacting biodiversity and is anticipated to deteriorate in the coming decades unless significant mitigation is taken (Steinbauer et al., 2018). The Alpine ecosystems are also highly susceptible to climate change, and extensive impacts have already been noticed throughout the world (Hock et al., 2019). In past research, climate change has been discovered to pose a serious threat to the plant diversity beyond the tree line. The most recent model calculations using climate change scenarios forecast that there will be a significant reduction in the habitat areas suitable for the vegetation of alpine systems (Engler et al., 2011). Ensuring high-altitude species are sufficiently endangered by extinction due to climate change (Dirnböck et al., 2011). The mountainous ecosystems are very sensitive to climatic change, and thus, as a result, they are degraded in terms of habitat, landscape changes, and biodiversity is lost. Such environments are highly vulnerable and are at a great risk because of global warming and the decline in habitat (Manzoor et al., 2026). Up to 44% of their habitat could be lost by many species or be subjected to the alteration of appropriate temperatures (Warren et al., 2018). Species also experience the impact of climate change, which is a global threat to biodiversity, by changing the composition and functions of ecosystems that ultimately cause instability in the alpine ecosystem and the extinction of the alpine species (Verrall & Pickering, 2020). The Himalayan alpine zone is characterized by its floral diversity and endemism because of the complicated geology that creates special microclimates and environmental conditions (Manzoor et al., 2026). The alpine plants, which are in most cases threatened by their small geographical ranges, are important in the determination of biodiversity hotspots (Gillani et al., 2026). Consequently, they can be easily eliminated through the extinction processes in the susceptible ecosystems, and that is why these species need efficient conservation strategies (Manzoor et al., 2025).

The preservation of the alpine ecosystems in Pakistan is critical because of the occurrence of greater biodiversity that has certain habitats and narrow geographic locations (Khan et al., 2021; Manzoor et al., 2026; Shaheen et al., 2023; Shaheen et al., 2019). These communities are ecologically viable, and they aid in maintaining the stability of soils, ecosystems, and have social and aesthetic importance (Stirling & Wilsey, 2001). Himalayan alpine is thought to be hosting more biological diversity than the average alpine habitat (Hamid et al., 2020; Salick et al., 2019; Shaheen et al., 2019; Verrall & Pickering, 2020). Alpine grasslands provide various environmental benefits, including species conservation and terrain preservation (Gibbon et al., 2010; Pornaro et al., 2017). It is worth noting that no detailed studies have been conducted with regard to the alpine biodiversity in Pakistan based on its significance in the alpine ecosystem. The Pakistan alpine zone is a biodiversity hotspot that also has a variety of flora and numerous endemic species, which are important in both ecological and economic ecosystems (Manzoor et al., 2026). Nevertheless, they also face the threat of global change, such as climate change, socioeconomic shifts, land use changes, and human disturbances (Gillani et al., 2025). This study focuses on the biodiversity patterns of the alpine zone at high altitudes in the Himalayas, assessing the impact of climate change and anthropogenic disturbance on their potential effects on abundance, distribution, and long-term survival.

2. METHODOLOGY

2.1. Study Area

In Northern Pakistan, the State of Azad Jammu and Kashmir (AJK) is located in the Pir Panjal Subrange of the western Himalayan Mountain and has an area of 13,297 km^2, and a latitude of 73° to 73° East longitude and 33° to 36° North latitude (Gillani et al., 2026). AJK is also characterized by a mountainous and hilly topography with forested mountain slopes and deep valleys that are crossed by many streams and rivers (GOAJK, 2019; Shaheen et al., 2016). The area is a biodiversity hotspot, with numerous agro-climatic regions and habitats because of a high degree of altitudinal gradient (Shaheen et al., 2024), with the lowest point of 488 m in the southern parts of the area bordering the Punjab plains and the highest point of 6,212 m in the north (Manzoor et al., 2025). The climate in the alpine region is very cold, and there is massive snowfall between the months of October and March, which covers the landscape with snow. Summer temperatures vary between 0-20 °C and 10 °C, whereas winter temperatures drop down to -20 °C. Snowfall may reach 4m or more, and the growing season extends between June and September, with average daytime temperatures of 7 °C to 9 °C. The area receives 1100-1300 mm of rainfall per year and has porous and weathered soils, the pH of which is prone to erosion and loss of nutrients due to steep slopes and high run-offs (Pak-Met, 2018; Shaheen et al., 2023).

2.2. Sampling Methodology

To study the trends of alpine biodiversity in Azad Jammu and Kashmir (AJK), the study area was sampled across a range of altitudes between 3000 and 4400 m, encompassing three major districts: Neelum Valley, Muzaffarabad, and Hattian. We divided the entire site into three altitudinal belts: lower, middle, and upper. This was selected to have a wide geographical coverage of the alpine zone in the region. Field work was performed between May and September 2023-2024, comprising pre-assessment surveys, with a detailed sampling procedure to record the structural and diversity features of the alpine vegetation.

In the above districts, a total of 15 alpine sites were selected. The sampling procedure was conducted based on a transect design, and a systematic quadrat sampling was applied throughout the altitudinal gradient. Particularly, a sample of shrubs was done using quadrats with a size of 25 m^2 (5 m × 5 m), and a sample of grasses and herbs was obtained using quadrats 1 m^2 (1 m × 1 m). An extensive survey of vegetation data was made, and the variables included cover, frequency, density, and importance values of the various species, following the standard phytosociological rules provided by (Cox, 1996) and (Mueller-Dombois & Ellenberg, 1974).

Geographic factors that are important in studying the alpine locations, such as altitude, aspect, latitude, and longitude, were taken using a GPS handset. Also, slope gradients were measured using a clinometer. Anthropogenic effects on biodiversity were also to be studied. The degree of grazing and erosion in these alpine regions was determined by measuring visual cues in the form of bare soil, the presence of browsed vegetation, and other signs of trampling, i.e., the presence of hoof tracks and animal feces. In addition to that, the information on non-timber forest products (NTFP) was obtained among the locals, and the soil compaction attributed to the presence of hiking trails was also captured to see a more accurate image of the human impact on the alpine biodiversity in the area (Manzoor et al., 2026).

2.3. Data Analysis

The data were gathered at 15 locations of 3 different altitudinal belts within the western Himalayas: low, middle, and high. The presence and abundance of the species were recorded, and the environmental variables of the species were measured, including grazing, erosion, slope, NTFP extraction, trampling, and elevation. The contributions of sites to community dissimilarity at the site level were assessed by using Bray-Curtis distances, and the sites that make a significant contribution to the overall heterogeneity were located. The significance of the contributions was determined by PERMANOVA (9999 permutations) (Gillani et al., 2026). NMDS was used to visualize the patterns of species composition across belts, and 80%

confidence ellipses and group differences were compared using PERMANOVA and ANOSIM to break down 80% of the relative species richness was then computed, and the correlation between the species richness and elevation was analysed using the Spearman rank correlation as well as Poisson generalized linear models. Richness, diversity, evenness, and maturity index were analysed using multiple regression of polynomials in order to identify the altitude trends and non-linear relationships. Canonical correspondence analysis (CCA) was carried out to test relationships between species and altitudes, using permutation tests of significance and the Mantel tests to test the relationship between the diversity indices and the environmental factors pointing at the key factors, including altitude and grazing (Manzoor et al., 2026).

3. RESULTS

3.1. Floristic Diversity

We identified a total of 219 species from the study area belonging to 136 genera and 56 families. Asteraceae was identified as dominant family with 18 genus and 25 species, followed by Poaceae (10 genus, 16 species), Rosaceae (7 genus, 15 species), Polygonaceae (8 genus, 10 species), Ranunculaceae (6 genus, 10 species), Apiaceae (5 genus, 8 species), Fabaceae (3 genus, 8 species) and Lammiaceae (7 species, 8 species). Seventeen families had only one species, while 12 families had only two species. Additionally, species are unevenly distributed across the families. Life form results revealed that Hemicryptophytes were the leading category, accounting for 44.75%, followed by Geophyte 24.66%, Therophyte 20.55%, Nanophenarophyte 6.39%, Chaemophyte 3.20%, and lianas 0.46%. Leaf spectra results showed that Microphyll was the leading category, accounting for 34.25%, Nanophyll 25.11%, Mesophyll 18.72%, leptophyll 15.07%, and Megaphyll 6.85%.

3.2. Site-wise Contribution to Community Dissimilarity

The contribution analysis of sites to community dissimilarity employed Bray-Curtis distances to evaluate the contribution of each site to the overall dissimilarity. The contribution analysis identified five sites (Sites 6, 13, 1, 12, and 16) that contributed most of the total observed dissimilarity (83.55%). Site 6 itself had a contribution of 23.08%, implying that highly distinct species assemblages exist in the site as opposed to other sites, which are closely followed by Site 13 (22.81%) and Site 1 (18.02%), having a contribution of 23.08% and 63.91%, respectively. This trend implies that these locations are major contributors to the general heterogeneity of the community in the study area. Sites 12 (10.63%) and 16 (9.01%) are the additional contributors to the total dissimilarity with other sites, making moderate contributions. The dissimilarity was 2.80 on average, with an average within-group dissimilarity of 2.13, with a significant dissimilarity in the major contributing sites. Results of the PERMANOVA test proved that there were significant differences between sites ($F = 2.224$, $p = 0.0353$, 9999 permutations), which confirms that the pattern of dissimilarity that is observed is statistically significant. The enormous contribution of Sites 6 and 13, along with the important PERMANOVA values, suggests that there are places that are hotspots of compositional uniqueness, which could be attributed to environmental heterogeneity, differences in microhabitat, or historical biogeographical influences.

3.3. NMDS of Species Assemblages across Altitudinal Belts

NMDS ordination by Bray-Curtis dissimilarity was used in the analysis of the patterns of species composition across the lower, middle, and high-altitude belts (Fig. 1). The two-dimensional solution has provided a value of 0.188 of stress, implying a reasonable representation of multivariate structure in reduced space. Although there was environmental heterogeneity across the elevation, the ordination process showed a high overlap of sites across the three belts, which indicated that there was not much compositional segregation. The belt of lower altitude produced a fairly wide ellipse, which suggested more within-group heterogeneity and increased dispersion of sites along ordination space. The mid-altitude belt had a fairly stable ellipse, indicating a relatively uniform series of species composition within the locations. The belt of high elevation was also dispersed, and the ellipse of the belt was more or less consistent with both lower and middle belts, meaning that there were common species assemblages in all elevations. The confidence elliptical areas of all three groups were crossing over significantly, which indicated the lack of compositional demarcations among belts. These visual patterns were supported by the statistical test. According to PERMANOVA, the altitude belts did not affect the composition of the species ($R^2 = 0.144$, $F = 1.01$, $p = 0.429$). ANOSIM also confirmed extremely low group separation ($R = 0.023$, $p = 0.40$). Mean goodness-of-fit ($R^2 = 0.048$) implied moderate predictive ability of the ordination axes.

Figure 1.

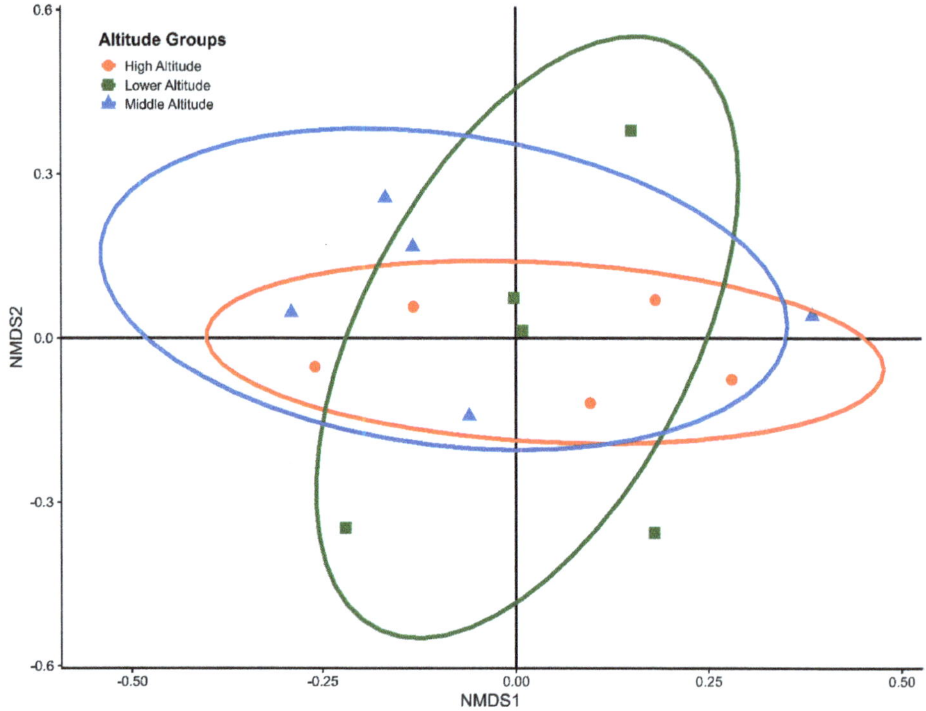

Figure 1 shows NMDS ordination of the composition of species at lower, middle, and high-altitude belts by Bray-Curtis dissimilarity. The points denote sampling points, and the ellipses denote 80% intervals of confidence intervals for every altitude group. High overlaps between the ellipses imply a lack of compositional distinction between altitudinal belts.

3.4. β-Diversity Partitioning Across Sites

Results of β-Diversity showed that the overall species turnover is high with a comparatively low level of nestedness, which means that the distribution of the compositional differences is driven mainly by the replacement of the species (Fig. 2). The overall Sørensen turnover (βsim 0.79) indicates that there was a lot of species changing across sites. Conversely, the average nestedness (βsne) was 0.03, which indicated little species loss or gain pattern to create compositional variation. The mean 0.82 diversity (βsim) indicates that most dissimilarity can be explained by turnover and not by nestedness. Paired β-diversity matrices illuminated interesting site-specific patterns. The values of turnover (βsim) were between 0.62 and 1.00, the maximum dissimilarity between Site 1 vs Site 2 (1.00) and Site 1 vs Site 4 (1.00), which signified different species composition. The values of nestedness (βsne) were minimal in all pairs, 0.00-0.11, with slightly larger values among sites with a difference in richness of species (Site 03 vs Site 14, 0.083). The total diversity (βsim) had a significant correlation with turnover, and values of 0.53-1.00 again showed that community differentiation is dominated by species replacement. On the whole, the evidence suggests that spatial heterogeneity in species composition between sites is caused in large part by turnover as opposed to nestedness, and this could be due to variations in environmental gradients, elevation, or even microhabitat heterogeneity. These results highlight the significance of site-specific conservation in preserving the general biodiversity because each site contains an assemblage that is unique.

Figure 2.

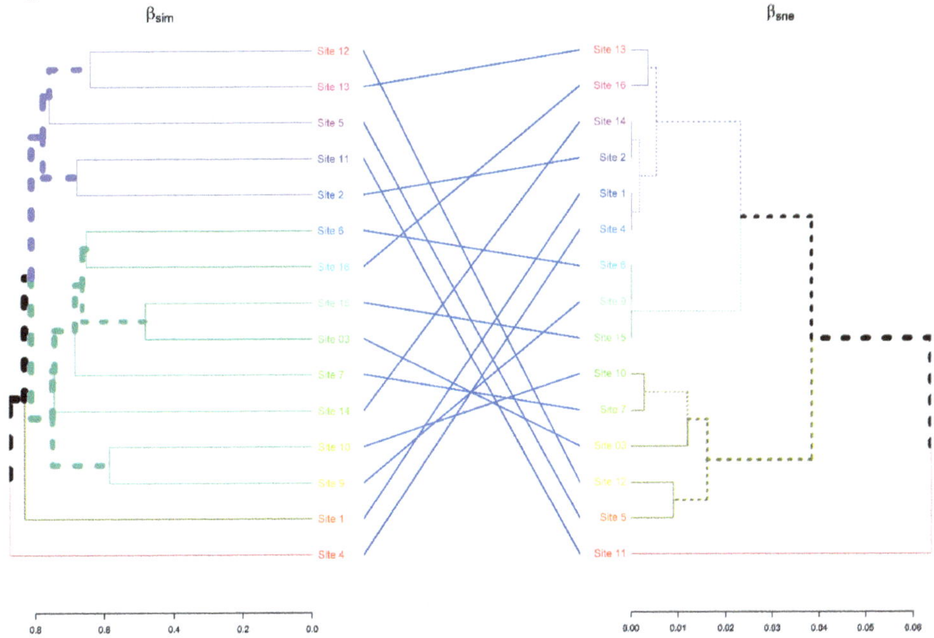

Figure 2 shows the comparison of 15 sites belonging to the study area. The upper triangle displays βsim (turnover), the lower triangle displays βsne (nestedness), and the bolded diagonal displays site identity. The high values of turnover imply different species compositions between sites, whereas the low specificity of the nestedness implies the lack of similarity between compositions.

3.5. Species Richness across Altitudes

The richness of the study area was determined in three different altitudinal belts, which were lower, middle, and high. The overall species richness per belt indicated a definite rising pattern with an increase in altitude, and the low belt provided the lowest richness, the middle belt denoted the medium richness, and the high belt indicated the richest species richness. The rank correlation analysis of the mean elevation against the total species richness showed that there was a moderate positive relationship (rho = 0.50), but the correlation was not significant ($p = 1$), indicating that with elevation, the richness was more likely to increase; however, the further number of elevation points was too few to reach a statistically significant relationship. In order to further test the correlation between species richness and elevation, a Poisson generalized linear model (GLM) was used. The model also indicated a strong positive relationship between elevation and species richness (Estimate = 0.00066, SE = 0.00029, $z = 2.25$, $p = 0.024$), which implies that there is a significant positive relationship between elevation and the number of species among the study sites. This was a reasonable model fit with a residual deviance of 6.35 with 1 degree of freedom and an AIC of 26.35, indicating that the elevation is a notable predictor of species richness in this area. Furthermore, these findings indicate a clear altitudinal gradient in species richness, whereby elevations with greater heights have more diversity. This trend can be an indicator of the impact of environmental heterogeneity, climatic conditions, and habitat conditions through the elevation gradient.

3.6. Polynomial Regression Analysis across Altitudinal Belts

Analysis of six types of ecological measurements from the 15 study sites, using non-parametric regression analysis of the polygons, revealed an evident altitudinal pattern and a pattern by belt. The trend in the species richness and number of species showed a positive correlation in the mid altitude belts, with the highest correlation of these factors in the mid altitude (mean richness = 1.59, mean number of species = 30.6), and the low altitude sites were slightly lower in richness and species number, and showed an increasing trend of the number in high altitude. Diversity indices of Simpson and Shannon were quite high in all belts, and the overall means were 0.92 and 2.95, respectively, which showed a high degree of species diversity. Interestingly, the diversity was highest when Simpson was 0.96, and Shannon diversity was greatest at mid-altitude areas of 3.47, indicating high evenness of species. There was moderate variation in species evenness across belts (mean = 0.87), with minimum and maximum evenness of 0.77 and 0.93, respectively, with more even distribution in mid-altitude communities. The health of the alpine ecosystem measured by maturity index precented the highest in low- and mid-altitude belts (mean = 24.2; max = 30.99) and the lowest in high-altitude sites (min = 11.49), implying the slow rate of succession or the harsh environment at high altitudes. In general, the polynomial regression indicated the presence of altitudinal gradients, with the mid-altitude sites demonstrating the highest level of species richness, diversity, and maturity, low-altitude sites demonstrating the middle level, and the high-altitude sites demonstrating the lower level, which proved the significant impact of altitude on community structure and forest ecosystem dynamics (Fig. 3).

Figure 3.

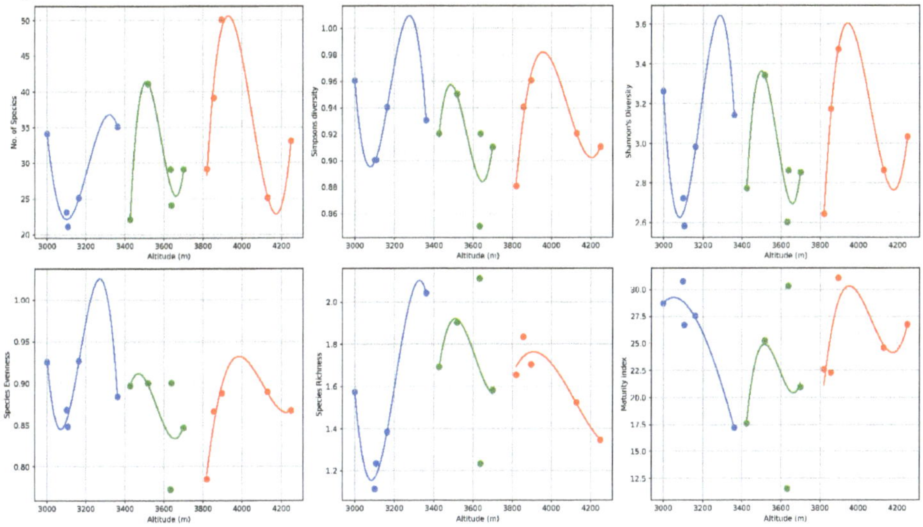

Figure 3 shows six ecological metric plots of 15 sites through the use of a regression model that depicts the ecological metrics as a function of site. The sites are classified according to the altitude: low (blue), mid (green), and high (red). The values of individual sites are represented by dots. The non-linear change over elevation is depicted through curvature, with peaks in species richness and diversity at medium elevations.

3.7. Indicator Species Analysis

The analysis on indicator species across the three altitudinal belts was performed to find out the top 20 species with the highest percentage IndVal% and indicated the species that were most representative of which belt. Out of these, *Iris kashmiriana* (IndVal = 49.58%, Middle Belt) and *Thymus linearis* (IndVal = 45.83%, Middle

Belt) showed the most significant association with their mid-altitude region, which indicates their narrow and specialized distributions. The distinct features of high-altitude belts in the current study include the presence of species like *Iris hookeriana, Trifolium repens, Trifolium pretense*, and *Veronica himalensis*, which have percentages of IndVal of between 41.67 and 44.17, showing their tendency to occur in high altitudes. Several strongly indicative species were present in the lower belt, such as *Blysmus compressus, Allium humile, Saxifraga jacquemontiana, Potentilla atrisanguinea, and Aconitum violaceum*, which had an IndVal range of 40-41.67 per cent, indicating that these species are probably limited to the communities of the lower altitudes. The high IndVal percentages of other species, including *Aconitum chasmanthum, Saxifraga stenophylla, Polygonaum alpinum, Eurphobia wallichii, and Poa alpina*, also supported the idea that the assemblage of plants depended on the elevation. Overall, indicator species are indicators of high-altitude specialization and highlight the compositional signals of low altitude belts, the middle, and the high, and the importance of elevation in organizing Himalayan alpine plant communities.

3.8. CCA Analysis of Species Composition

Canonical Correspondence Analysis (CCA) indicated that altitudinal belts had a very significant effect on the arrangement of plant communities among the study locations (permutation test, $p < 0.001$) and that altitude is one of the most significant ecological factors in the distribution of species (Fig. 4). The two initial canonical axes jointly accounted for 14.12% of the constrained variation (CCA1 = 7.65%, eigenvalue = 0.4516; CCA2 = 6.47%, eigenvalue = 0.3818), which effectively captures the large ecological gradient that is related to elevation. There was a strong site separation on CCA1, where negative scores denoted communities at lower altitudes and positive scores denoted communities at high altitudes, with an intermediate range between sites showing a gradual compositional shift. The ordination scores showed that there was a high degree of directional clustering with a large number of taxa having high absolute values of CCA1 (>1.0), indicating high altitudinal specialization. At the lower altitudes, *Aconitum violaceum* (CCA1 = -1.523, CCA 2 = -0.664), *Dryopteris stewartii* (-1.523, -0.664), *Saxifraga jacquemontiana* (-1.523, -0.664), *Verbascum thapsus* (-1.523, -0.664), and *Arabis amplexicaulis* (-1.523, -0.66). Belt High-altitude species were closely related to *Bergenia stracheyi* (1.067, -0.944), *Allium carolinianum* (1.067, -0.944), *Rheum australe* (1.067, -0.944), *Pedicularis pectinata* (0.599, 0.132), and *Potentilla atrisanguinea* (1.046, -0.892). The low interaction between site clusters, discrete ellipses, and sharp segregation of species vectors on the triplot together points to the fact that the vertical position, or altitude, serves as a powerful environmental filter, favouring pronounced species turnover and ecological zonation over nested community cycles between environments along the altitude gradient.

Figure 4.

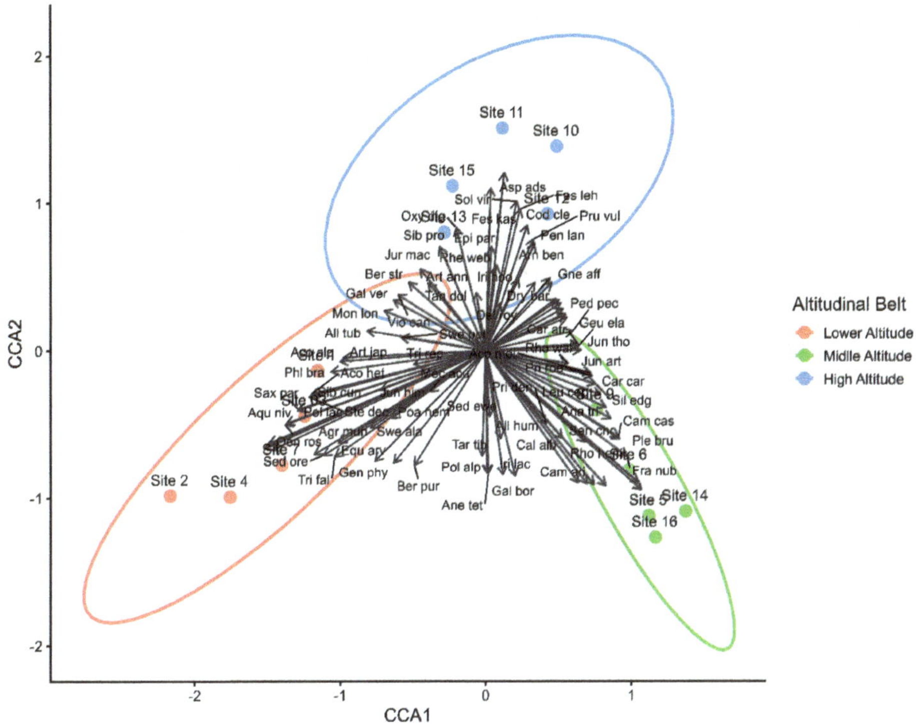

Figure 4 shows Canonical Correspondence Analysis (CCA) triplot of the correlation between the plant species composition and altitudinal belts. Sites are classified under lower, middle, and high-altitude belts whose 95% confidence ellipses are taken. Species vectors represent the direction and strength of correlation with the altitude. The two initial canonical axes (CCA1 = 7.65% and CCA2 = 6.47) indicate a clear separation of the low-altitude, mid altitudes, and the high-altitude alpine communities, showing that vegetation structure is heavily altitude-controlled (permutation test, $p < 0.001$).

3.9. Relationships between Plant Diversity and Environmental Factors

Mantel test was used to test the correlations between plant diversity indices (Simpson, Shannon, Evenness, and the Richness) and the environmental variables (Grazing, Erosion, Slope, NTFP extraction, Trampling, and Altitude) in the 15 study sites (Fig. 5). The results have shown that environmental gradients had moderately moderated diversity indices and several statistically significant relationships. Altitude had the highest positive correlation with Simpson's diversity (Mantel r=0.68, p=0.003), and this showed that diversity was higher at the most elevated sites. Grazing intensity was also significantly correlated with Shannon diversity (r = 0.54, p = 0.012), which indicates that there is a correlation between grazing pressure and the patterns of species diversity. The Evenness of the Species showed a medium relationship with the Slope (r = 0.41, p = 0.037), and Species Richness was positively correlated with NTFP extraction (r =0.32, p= 0.048). In general, 7 out of 24 of the significant relationships were significant at p ≤ 0.05 in the interplay between environmental factors, especially the Altitude and Grazing, which shows that the environmental factor is a significant determinant of the plant diversity in these alpine Himalayan locations. All these trends support the sensitivity of plant communities to abiotic and anthropogenic stress.

Figure 5.

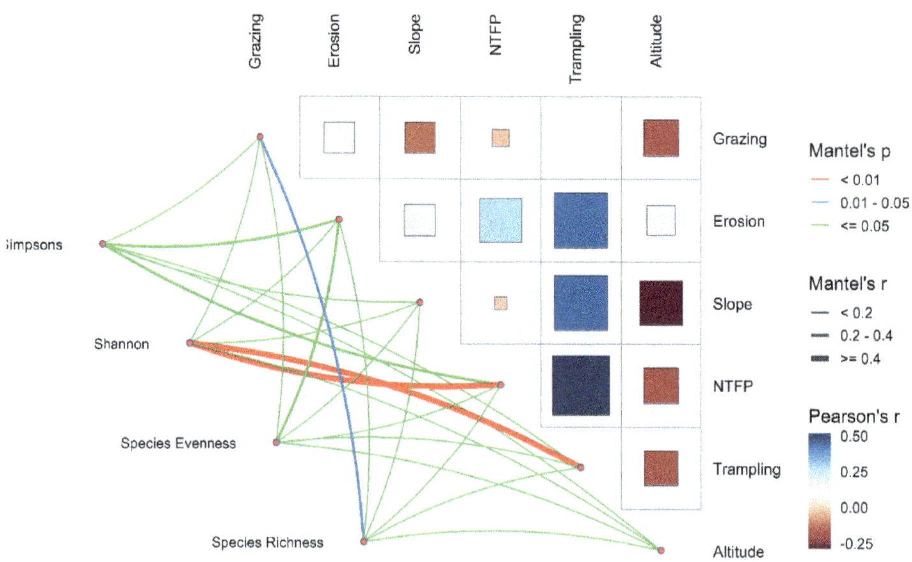

Figure 5 shows the Mantel correlation network: Indicating that there is a relationship between the indices of plant diversity (Simpson, Shannon, Evenness, and

Richness) and the environmental attributes (Grazing, Erosion, Slope, NTFP, Trampling, Altitude) at 15 sites. The width of the link is a value of Mantel (r value), and color is a level of significance ($p < 0.01$, 0.01-0.05, and above 0.05). The highest correlation was found in Simpson diversity/ Altitude and Shannon diversity/ Grazing.

3.10. Impact of Climate Change on Alpine Biodiversity

The significant responses of the species to climate change are geographic range changes, and broad-range species are less susceptible to such changes because they can seek an appropriate habitat (Manes et al., 2021). Conversely, species with limited distributions (i.e., restricted species) are susceptible to local effects that are intensified by climate change, such as habitat alteration and invasion by other species (Shukla et al., 2019). The alpine species tend to occupy certain environmental niches and exhibit low levels of dispersal, reduced population size, and other adaptive capacities, which make regions of high endemism highly vulnerable to the climate-related issues (Enquist et al., 2019). According to projects, future climate conditions are going to aggressively impact these areas, which will negatively affect the biodiversity even in less difficult situations. In the montane, the continuous changes in climate are expected to cause an upward movement of species and thus habitat to alpine biodiversity will be decreased (Verrall & Pickering, 2020). The rearrangements of the land use have increased forest zones, which have further endangered the alpine ecosystems and may have increased the level of extinction in alpine vegetation (Manzoor et al., 2026). Variations in mountain species dynamics can be seen to represent differences that are dependent on ecological domains and the rate of climate change. The higher-elevation species are vulnerable to greater competition and extinction pressure due to rising temperature, especially those species limited by altitude (Freeman et al., 2018). Plant species adapted morphologically and functionally to the harsh environment have a higher chance of survival, and adaptation and phenotypic plasticity play a vital role in their survival (Shaheen et al., 2023). Effective conservation planning necessitates understanding these dynamics and developing tools for climate adaptation and mitigation.

4. FUTURE CONSERVATION PLANS OF ALPINE FLORA OF PAKISTAN UNDER CLIMATE AND ANTHROPOGENIC PRESSURES

The alpine ecosystem of Pakistan is very unique, with rich floral diversity and endemism, and contains major glaciers that control the overall hydrology of the region (Senese et al., 2018; Shaheen et al., 2011), but it is vulnerable to the impact of climate change and anthropogenic pressures. Located in the Hindu Kush- Karakoram-Himalaya ranges and impacted by some of the biggest non-polar glaciers on the earth, these high landscape terrains carry unique floral communities to climate conditions featuring brief growing seasons, intense radiation, sharp environmental gradients, and rough soils (Manzoor et al., 2026). Yet, the same conditions that create endemism and ecological specialization also make alpine flora very vulnerable to climate change and human disturbance. The tendency towards warming, the retreat of glaciers, the transformation of snow regimes, and the growth of grazing pressure, uncontrolled tourism, trampling, erosion, and non-timber forest product (NTFP) extraction are gradually transforming alpine vegetation in Pakistan (Manzoor et al., 2023). Conservation of alpine floral biodiversity in the future is thus necessitated by the combined factor that addresses climatic vulnerability as well as anthropogenic pressures within the alpine mountain ecosystems.

Climate change and Human activities are the most significant threats to the viability of alpine biodiversity, and the increased temperature in the Himalayan region is taking place at a higher rate than the world average, causing glaciers to retreat, destabilization of permafrost, a change in hydrology, and a subsequent rise in vegetation zones. Those alterations can disturb alpine habitats into narrower and narrower elevational zones, and cold-adapted and endemic species would not have many areas to spread upwards. Climate changes also influence the snow melting process and soil moisture regimes and phenology cycles, which interfere with the reproduction and survival of most alpine plants. This means that conservation planning should be climate sensitive and should encompass climate-sensitive measures such as long-term tracking of the change in species elevation, and preservation of climatic refugia, as well as the model of future habitat suitability using GIS and species distribution models (SDMs) to predict threatened species.

Climate change is further worsened by anthropogenic disturbances. The effects of seasonal grazing of livestock, treading on walking tracks, soil erosion on steep slopes, and harvesting of medicinal and aromatic plants have direct impacts on species composition and regeneration (Gillani et al., 2026). These stresses tend to be ecologically sensitive mid and high-altitude belts where there is a high plant and ecological specialization. To ensure that alpine grassland is not degraded, it is vital to have controlled grazing systems, rotation of pastures, and management of the zones by the community. Equally, there should be the establishment of sustainable harvesting regimes of NTFPs and medicinal plants with the local communities in order to balance livelihoods with conservation of biodiversity.

One of the main conservation suggestions can be made in the future based on the identification and prioritization of biodiversity hotspots and places of high compositional uniqueness. Special conservation efforts should be given to sites that contain unique species assemblages and that play a disproportionately great role in the total beta diversity. This is because conserving such areas guarantees as much heterogeneity of flora as possible among landscapes. Policies on conservation should thus go beyond protection that is uniform, and proceed to site-based management approaches based on the ecological data.

Prolonged ecological monitoring programmes are very important in the alpine regions of Pakistan. Interpretations of the time-change in the richness, diversity, and community structure of a vegetation patch can be performed by permanent vegetation plots across the elevational gradients. These datasets can be incredibly useful when it comes to comprehending the principles of alpine flora's reaction to gradual changes in climate, as well as sudden disruptions. A combination of field observations, remote sensing, and geospatial analysis can be used to monitor the vegetation change, glacier retreat, and habitat transformation at a landscape scale.

There is also a need to strengthen the management of a protected area. Numerous alpine zones are located in national parks and conservation zones, but few rules are applied in the areas regarding grazing, tourism, and plant harvesting. Increasing management capacity, training local personnel in monitoring the ecological situation, and including indigenous people in the program of participative conservation can make a big contribution. At the local level, stewardship can be addressed with community awareness campaigns addressing the ecological and economic importance of alpine plants.

Additionally, conservation activities in the future should be science-based and policy combined. Conservation planning should be based on ecological research, such as research on the functional characteristics of plants, the relationships between species and the environment, as well as research on patterns of diversity across altitudinal gradients. The involvement of researchers, conservation agencies, and policymakers is essential in order to convert scientific discoveries into management actions. Finally, the proactive and multi-scale strategy is the key to conservation of alpine floral biodiversity in Pakistan, as it is important to focus on climate change as well as anthropogenic disturbances. These vulnerable mountain ecosystems, through long-term ecological monitoring, site-specific protection, sustainable land-use practices, community involvement, and predictive ecological models, can be preserved and prevent the extinction of the unique plant diversity in these ecosystems for the next generations.

5. CONCLUSION

The alpine zone in Pakistan containing a complex pattern of biodiversity depending on the elevation, the environmental heterogeneity, and Human pressures. This study reveals that an intense degree of species turnover and compositional distinctiveness on a local level are the main causes of community differentiation, and not just altitudinal isolation. Some of them may be regarded as biodiversity hotspots and play a significant role in total heterogeneity, which underlines the significance of special conservation efforts. Even though the species assemblage between the altitudinal belts is significantly overlapping, the ordination and statistical analysis demonstrate that the altitude is an important factor in ecological selection, which determines the species distribution, richness, and functional composition. The emergence of mid-altitude areas as zones of greatest richness and ecosystem maturity, and higher altitudes on a gradient-enhanced richness increase is revealed. The analysis of indicator species and Canonical Correspondence Analysis (CCA) revealed that there is high altitudinal specialization, which represents ecological zonation in the alpine zones. Notably, the existence of strong correlations among the diversity indices, altitude, and grazing intensity indicates the dual effects of the climate-related factors and human activities on the alpine vegetation dynamics. These results indicate that the Himalaya's biodiversity is vulnerable to the current climate change and land-use disturbances. To effectively conserve Himalayan alpine biodiversity, it is essential to implement site-specific management strategies, conduct long-term ecological monitoring, and adopt measures that address environmental gradients and anthropogenic influences. This approach will help protect these unique and sensitive mountain ecosystems.

REFERENCES

Ahmed, M. J., Murtaza, G., Shaheen, H., & Habib, T. (2020). Distribution pattern and associated flora of Jurinea dolomiaea in the western Himalayan highlands of Kashmir: An indicator endemic plant of alpine phytodiversity. *Ecological Indicators*, *116*, 106461. DOI: 10.1016/j.ecolind.2020.106461

Cox, G. W. (1996). Laboratory manual of general ecology. Wm. C. *Brown, Dubuque*, *1*, 232.

Dirnböck, T., Essl, F., & Rabitsch, W. (2011). Disproportional risk for habitat loss of high-altitude endemic species under climate change. *Global Change Biology*, *17*(2), 990–996. DOI: 10.1111/j.1365-2486.2010.02266.x

Engler, R., Randin, C. F., Thuiller, W., Dullinger, S., Zimmermann, N. E., Araujo, M. B., & Albert, C. H. (2011). 21st century climate change threatens mountain flora unequally across Europe. *Global Change Biology*, *17*(7), 2330–2341. DOI: 10.1111/j.1365-2486.2010.02393.x

Enquist, B. J., Feng, X., Boyle, B., Maitner, B., Newman, E. A., Jørgensen, P. M., & Corlett, R. T. (2019). The commonness of rarity: Global and future distribution of rarity across land plants. *Science Advances*, *5*(11), eaaz0414. DOI: 10.1126/sciadv.aaz0414 PMID: 31807712

Freeman, B. G., Scholer, M. N., Ruiz-Gutierrez, V., & Fitzpatrick, J. W. (2018). Climate change causes upslope shifts and mountaintop extirpations in a tropical bird community. *Proceedings of the National Academy of Sciences of the United States of America*, *115*(47), 11982–11987. DOI: 10.1073/pnas.1804224115 PMID: 30373825

Gibbon, A., Silman, M. R., Malhi, Y., Fisher, J. B., Meir, P., Zimmermann, M., Dargie, G. C., Farfan, W. R., & Garcia, K. C. (2010). Ecosystem carbon storage across the grassland–forest transition in the high Andes of Manu National Park, Peru. *Ecosystems (New York, N.Y.)*, *13*(7), 1097–1111. DOI: 10.1007/s10021-010-9376-8

Gillani, S. W., Ahmad, M., Manzoor, M., Al-Andal, A., Khan, R. W. A., Ayoola, J. O., & Waheed, M. (2025). Integrating environmental variables and biotic interactions to predict the future distribution of the endangered Aconitum heterophyllum Wall. ex Royle. *BMC Plant Biology*, *26*(1), 45. DOI: 10.1186/s12870-025-07859-y PMID: 41339780

Gillani, S. W., Ahmad, M., Manzoor, M., Waheed, M., Khan, R. W. A., & Sohail, A. (2026). Geographic distribution patterns, population dynamics, and habitat vulnerability of endangered and endemic Trillium govanianum. *Environmental Monitoring and Assessment*, *198*(2), 162. DOI: 10.1007/s10661-026-15013-w PMID: 41569443

GOAJK. (2019). *2019. AJK at Glance; A report by Planning & Development Department Muzaffarabad.* Azad Jammu & Kashmir.

Grabherr, G., Gottfried, M., & Pauli, H. (2010). Climate change impacts in alpine environments. *Geography Compass, 4*(8), 1133–1153. DOI: 10.1111/j.1749-8198.2010.00356.x

Hamid, M., Khuroo, A. A., Malik, A. H., Ahmad, R., & Singh, C. P. (2020). Assessment of alpine summit flora in Kashmir Himalaya and its implications for long-term monitoring of climate change impacts. *Journal of Mountain Science, 17*(8), 1974–1988. DOI: 10.1007/s11629-019-5924-7

Hock, R., Rasul, G., Adler, C., Cáceres, B., Gruber, S., Hirabayashi, Y., . . . Kutuzov, S. (2019). High mountain areas. In *IPCC special report on the ocean and cryosphere in a changing climate* (pp. 131-202). H.-O. Pörtner, DC Roberts, V. Masson-Delmotte, P. Zhai, M. Tignor, E....

Khan, R. W. A., Shaheen, H., Qayyum, S., Awan, S. N., Shah, T. A., Alsahli, A. A., Younous, Y. A., & El-Sheikh, M. A. (2025). Predicting habitat refugia of the medicinal plant Dolomiaea costus under two climate change scenarios. *Discover Life, 55*(1), 2. DOI: 10.1007/s11084-024-09670-5

Khan, S., Shaheen, H., Aziz, S., & Nasar, S. (2021). Diversity and distribution of Genus Primula in Kashmir region: An indicator genus of the western Himalayan mountain wetlands and glacial forelands. *Biodiversity and Conservation, 30*(6), 1673–1688. DOI: 10.1007/s10531-021-02163-1

Körner, C. (2007). The use of 'altitude'in ecological research. *Trends in Ecology & Evolution, 22*(11), 569–574. DOI: 10.1016/j.tree.2007.09.006 PMID: 17988759

Körner, C. (2016). Plant adaptation to cold climates. *F1000 Research*, 5. PMID: 27990251

Kudo, G., & Hirao, A. S. (2006). Habitat-specific responses in the flowering phenology and seed set of alpine plants to climate variation: Implications for global-change impacts. *Population Ecology, 48*(1), 49–58. DOI: 10.1007/s10144-005-0242-z

Leng, R., Harrison, S., & Anderson, K. (2023). Himalayan alpine ecohydrology: An urgent scientific concern in a changing climate. *Ambio, 52*(2), 390–410. DOI: 10.1007/s13280-022-01792-2 PMID: 36324019

Manes, S., Costello, M. J., Beckett, H., Debnath, A., Devenish-Nelson, E., Grey, K.-A., & Krause, C. (2021). Endemism increases species' climate change risk in areas of global biodiversity importance. *Biological Conservation, 257*, 109070. DOI: 10.1016/j.biocon.2021.109070

Manzoor, M., Ahmad, M., Gillani, S. W., Waheed, M., Grießinger, J., Shaheen, H., & Meier, W. J.-H. (2025). Biodiversity patterns and environmental drivers of fragile alpine glacial lake ecosystems in the Himalaya: Implications for conservation and management. *Ecological Indicators, 178*, 114028. DOI: 10.1016/j.ecolind.2025.114028

Manzoor, M., Ahmad, M., Gillani, S. W., Waheed, M., Shaheen, H., Haq, S. M., & Almutairi, K. F. (2026). Alpine Endemism under Threat: Ecological Patterns and Conservation Strategies in the Changing Himalayas. *Global Ecology and Conservation*, 04051. DOI: 10.1016/j.gecco.2025.e04051

Manzoor, M., Ahmad, M., Gillani, S. W., Waheed, M., Shaheen, H., Mehmood, A. B., Fonge, B. A., & Al-Andal, A. (2025). Population dynamics, threat assessment, and conservation strategies for critically endangered Meconopsis aculeata in alpine zone. *BMC Plant Biology, 25*(1), 358. DOI: 10.1186/s12870-025-06361-9 PMID: 40102752

Manzoor, M., Ahmad, M., Zafar, M., Gillani, S. W., Shaheen, H., Pieroni, A., & Makhkamov, T. (2023). The local medicinal plant knowledge in Kashmir Western Himalaya: A way to foster ecological transition via community-centred health seeking strategies. *Journal of Ethnobiology and Ethnomedicine, 19*(1), 56. DOI: 10.1186/s13002-023-00631-2 PMID: 38037066

Mueller-Dombois, D., & Ellenberg, H. (1974). *Aims and methods of vegetation ecology*. Wiley.

Nepal, S., Krause, P., Flügel, W. A., Fink, M., & Fischer, C. (2014). Understanding the hydrological system dynamics of a glaciated alpine catchment in the Himalayan region using the J2000 hydrological model. *Hydrological Processes, 28*(3), 1329–1344. DOI: 10.1002/hyp.9627

Pak-Met. (2018). The normals of climatic data of Azad Jammu & Kashmir. In: Pakistan Meteorological Department, Regional Centre Muzaffarabad, Pakistan.

Pornaro, C., Schneider, M., Leinauer, B., & Macolino, S. (2017). Above-and belowground patterns in a subalpine grassland-shrub mosaic. *Plant Biosystems, 151*(3), 493–503. DOI: 10.1080/11263504.2016.1187679

Quinn, J. A. (2008). Arctic and alpine biomes.

Rawat, G. (2007). Pastoral practices, wild mammals and conservation status of alpine meadows in western Himalaya. *Journal of the Bombay Natural History Society, 104*(1), 5.

Salick, J., Fang, Z., & Hart, R. (2019). Rapid changes in eastern Himalayan alpine flora with climate change. *American Journal of Botany*, *106*(4), 520–530. DOI: 10.1002/ajb2.1263 PMID: 30934119

Salick, J., Ghimire, S. K., Fang, Z., Dema, S., & Konchar, K. M. (2014). Himalayan alpine vegetation, climate change and mitigation. *Journal of Ethnobiology*, *34*(3), 276–293. DOI: 10.2993/0278-0771-34.3.276

Senese, A., Maragno, D., Fugazza, D., Soncini, A., D'Agata, C., Azzoni, R. S., & Asif Khan, M. (2018). Inventory of glaciers and glacial lakes of the Central Karakoram National Park (CKNP–Pakistan). *Journal of Maps*, *14*(2), 189–198. DOI: 10.1080/17445647.2018.1445561

Shaheen, H., Attique, A., Riaz, M. T., Manzoor, M., Khan, R. W. A., & Riaz, M. T. (2024). From biodiversity hotspot to conservation hotspot: Assessing distribution, population structure, associated flora and habitat geography of threatened Himalayan Yew in temperate forest ecosystems of Kashmir. *Biodiversity and Conservation*, *33*(2), 1–25. DOI: 10.1007/s10531-023-02758-w

Shaheen, H., Azad, B., Mushtaq, A., & Khan, R. W. A. (2016). Fuelwood consumption pattern and its impact on forest structure in Kashmir Himalayas. *Bosque (Valdivia)*, *37*(2), 419–424. DOI: 10.4067/S0717-92002016000200020

Shaheen, H., Aziz, S., Nasar, S., Waheed, M., Manzoor, M., Siddiqui, M. H., Alamri, S., Haq, S. M., & Bussmann, R. W. (2023). Distribution patterns of alpine flora for long-term monitoring of global change along a wide elevational gradient in the Western Himalayas. *Global Ecology and Conservation*, *48*, e02702. DOI: 10.1016/j.gecco.2023.e02702

Shaheen, H., Ibrahim, M., & Ullah, Z. (2019). Spatial patterns and diversity of the alpine flora of Deosai plateau, western Himalayas. *Pakistan Journal of Botany*, *51*(1), 205–212. DOI: 10.30848/PJB2019-1(39)

Shaheen, H., Khan, S. M., Harper, D. M., Ullah, Z., & Qureshi, R. A. (2011). Species diversity, community structure, and distribution patterns in western Himalayan alpine pastures of Kashmir, Pakistan. *Mountain Research and Development*, *31*(2), 153–159. DOI: 10.1659/MRD-JOURNAL-D-10-00091.1

Shukla, P. R., Skeg, J., Buendia, E. C., Masson-Delmotte, V., Pörtner, H.-O., Roberts, D.,… Van Diemen, S. (2019). Climate Change and Land: an IPCC special report on climate change, desertification, land degradation, sustainable land management, food security, and greenhouse gas fluxes in terrestrial ecosystems.

Steinbauer, M. J., Grytnes, J.-A., Jurasinski, G., Kulonen, A., Lenoir, J., Pauli, H., & Barni, E. (2018). Accelerated increase in plant species richness on mountain summits is linked to warming. *Nature*, *556*(7700), 231–234. DOI: 10.1038/s41586-018-0005-6 PMID: 29618821

Stirling, G., & Wilsey, B. (2001). Empirical relationships between species richness, evenness, and proportional diversity. *American Naturalist*, *158*(3), 286–299. DOI: 10.1086/321317 PMID: 18707325

Verrall, B., & Pickering, C. M. (2020). Alpine vegetation in the context of climate change: A global review of past research and future directions. *The Science of the Total Environment*, *748*, 141344. DOI: 10.1016/j.scitotenv.2020.141344 PMID: 32814293

Warren, R., Price, J., Graham, E., Forstenhaeusler, N., & VanDerWal, J. (2018). The projected effect on insects, vertebrates, and plants of limiting global warming to 1.5 C rather than 2 C. *Science*, *360*(6390), 791–795. DOI: 10.1126/science.aar3646 PMID: 29773751

Chapter 3
Mapping Alpine Vegetation in the Alps

Sarah Abdul Razak
http://orcid.org/0000-0003-3364-936X
Universiti Malaya, Malaysia

Chang Xue
Universiti Malaya, Malaysia

Hansa Gul
http://orcid.org/0009-0001-5554-617X
University of Mianwali, Pakistan

Maryam Bibi
University of Malakand, Pakistan

Nathan Wise
Western Sydney University, Penrith, Australia

Khizar Hayat
Islamia College, Peshawar, Pakistan

Syed Nasar Shah
http://orcid.org/0000-0002-9002-6110
Quaid-i-Azam University, Islamabad, Pakistan

ABSTRACT

Alpine vegetation in the Alps represents a highly sensitive ecological system influenced by altitude, climatic variability, and complex topography. This chapter examines modern approaches to alpine vegetation mapping using advanced remote sensing and geospatial techniques. High-resolution satellite imagery, hyperspectral sensors, LiDAR, and UAV-based surveys, integrated with GIS and digital elevation models, enable detailed assessment of vegetation patterns with high spatial and thematic accuracy. Object-based image analysis and machine-learning algorithms improve vegetation classification and allow monitoring of temporal changes in response to environmental drivers. Phenological shifts and habitat transformations are identified through time-series data and field validation, providing critical insights into biodiversity distribution and ecosystem resilience. The chapter highlights the

DOI: 10.4018/979-8-3373-6756-9.ch003

ght © 2026, IGI Global Scientific Publishing. Copying or distributing in print or electronic forms without written permission of IGI Global Scientific Publishing is prohibited. Use of this chapter to generative artificial intelligence (AI) technologies is expressly prohibited. The publisher reserves all rights to license its use for generative AI training and machine learning model development.

importance of advanced mapping technologies for conservation planning, climate change adaptation, and sustainable land-use management in vulnerable alpine environments worldwide.

1. INTRODUCTION

Change in vegetation cover is also regarded as an important method to understand the interface between the human and natural world, and therefore as a measure of ecological well-being and land-use dynamics. Information about past and current alterations in vegetation cover is highly significant for land-use planning, sustainable resource management, and environmental evaluation. Projecting future trends provides a long-term view that can be used to make decisions regarding resource allocation, ecosystem restoration and adaptation measures to the effects of climate change. Therefore, there is an urgent need to monitor, plan, and manage changes in vegetation cover by maintaining reliable updated information at both local and global levels (Almalki et al., 2022).

The Alps are alpine regions which are ecologically special landscapes, with vegetation patterns closely related to topography, climate, and elevation gradients. High-altitude ecosystems are among first to be affected by climate change due to rising temperatures, loss of snow cover from glacial retreat, and changes in precipitation patterns. Understanding vegetation dynamics in these areas is essential for assessing ecological resilience, biodiversity distribution, and carbon sequestration potential. Both remote sensing and GIS have become indispensable tools because they enable large-scale analyses efficiently, even in rugged terrain with limited access (Matyukira & Mhangara, 2024).

Climatic stress and vegetation degradation are highly sensitive to climatic conditions and the level of adaptive capacity in arid, semi-arid and high-altitude alpine areas where natural resources to support vegetation are minimal. GIS and remote sensing techniques are very useful for detecting changes in density, distribution, and composition because the spatial and temporal patterns of vegetative cover can be accurately determined using these techniques. This makes GIS an excellent choice, as it is versatile and includes tools for storing, analyzing, and visualizing environmental data. The combination of satellite imagery, elevation models, and ecological parameters with GIS will help identify areas that have experienced or are likely to experience major changes in vegetative cover (Liping et al., 2018; Reis, 2008; Wu et al., 2006).

Remote sensing images reflect the electromagnetic radiations released by the Earth surface in various spectral bands and they offer data in measuring the environmental changes because of natural or man-made causes (Kennedy et al., 2007;

Yang et al., 2021). Sensors used in the visible, near-infrared ranges and shortwave infrared are most suitable for mapping vegetation, analyzing canopy vigor, and assessing productivity (Rundquist, 2001). These sensors are used in alpine ecosystems to identify finer-scale vegetation processes based on topography and changes in microclimatic conditions.

In the past, various remote sensing methods have been used to detect changes in land cover and vegetation. For instance, Cui et al., (2013) analyzed a series of Landsat images for semi-arid systems to quantify changes in vegetation, demonstrating the effectiveness of this proved how effective temporal image analysis for environmental monitoring. Other studies have combined remotely sensed datasets with GIS- derived parameters such as slope, elevation, topography, and land use to improve the accuracy of vegetation mapping vegetation (Liping et al., 2018; Reis, 2008). The integration of multispectral and hyperspectral data has further monitoring capabilities by enabling detailed spectral analysis and classification of plant communities based on wavelength-dependent reflectance.

Multispectral and hyperspectral imagery typically cover the 0.52 to 0.90 µm range (green to near-infrared), which is important for studying vegetation characteristics (Allbed & Kumar, 2013). The near-infrared (NIR) region is highly indicative of the type health status, and density of vegetation. Healthy plants exhibit high NIR reflectance and low visible reflectance because they are actively photosynthesizing while stressed or unhealthy vegetation shows the opposite pattern. Since photosynthesis mainly depends on chlorophyll absorption in the red and blue wavelengths, more visible reflectance usually indicates less photosynthetic activity and greater plant stress (Karabulut, 2018). Consequently, using NIR (0.7–1.1 µm) together with red (0.6–0.7 µm) spectral data has become common practice in studies monitoring vegetation (Foley et al., 1998; Xue & Su, 2017). This chapter primarily reviews and assesses contemporary GIS and remote sensing practices applied to mapping alpine vegetation in the Alps.

2. IMPORTANCE OF MAPPING ALPINE VEGETATION

Mapping alpine vegetation is important for ecology, sustainable land use, and conservation efforts in sensitive mountain ecosystems such as the Alps. The alpine flora helps maintain biodiversity, regulate environmental hydrology, and provide ecosystem services that are crucial to human health and environmental stability.

2.1 Ecological and Environmental Significance

The plants of the Alps are characteristic of specific ecosystems that survive extreme weather conditions, including low temperatures, short growing seasons, and intense solar radiation. These ecosystems are highly sensitive to environmental conditions, especially climate warming and land-use stressors. The steep environmental gradients and fragmented habitats of alpine landscapes mean that even minor alterations in temperature or precipitation can significantly affect species composition, distribution, and ecosystem functioning (Nagendra et al., 2013a).

These changes can be identified through vegetation visualization, as it may provide spatially explicit data on the distribution and condition of plants. This information is essential for examining ecosystem resilience, vulnerable species, and feedbacks, such as the effects of vegetation dynamics on the soil stability and carbon transfer (Pepin et al., 2015).

2.2 In Favor of Conservation and Land-Use Planning

Conservation planning in alpine areas is impossible without accurate vegetation maps. These maps help identify significant habitats, ecological corridors, and other biodiversity hot spots that should be conserved. The data they provide support the design and management of protected areas, ensuring that conservation efforts are focused on the most precious and endangered ecological sites.

Moreover, alpine regions often face competing land uses, such as tourism, agriculture, and infrastructure development. Vegetation mapping informs sustainable land-use decisions by identifying areas at risk of degradation, guiding restoration projects, and preventing habitat fragmentation (Melis et al., 2020).

2.3 Climate Change Monitoring and Adaptation

The alpine ecosystems are recognized as early warning signs of climate change because of the rapid ecological impacts of warming. Climate-related changes detectable through vegetation mapping, particularly with multi-temporal satellite data and phenological indices, include the upward movement of treelines, shrub encroachment into meadows, and changes in the duration of growing seasons.

These understandings can enable adaptive management by offering baseline information and tracking data over time. As an example, mapping can be used to locate refugia where sensitive species are likely to survive under changing climatic conditions, as well as areas that become more vulnerable (Shaheen et al., 2023).

2.4 Integration of Remote Sensing and GIS Technologies

Remote sensing and Geographic Information Systems (GIS) are effective and complementary tools for mapping alpine vegetation. GIS enables the integration of various types of data, including satellite imageries, topographic variables, and climatic data, and allows for detailed spatial analysis. Digital Elevation Models (DEMs) included in GIS account for altitude, slope, and aspect, which are key factors determining vegetation patterns in mountainous areas (Tompolidi et al., 2025). Remote sensing offers repeatable and timely observations that overcome the logistical constraints of fieldwork in the Alps.

Multispectral and hyperspectral satellite sensors can record vegetation reflectance at different spectral bands enabling the identification of vegetation types and the evaluation of plant health. UAVs provide additional value by offering ultra-high spatial resolution information, which could be unavailable from ground surveys (Wallace et al., 2016b).

Combined, these technologies make it possible to map and monitor vegetation cover in more detail, track ecological changes over time, and implement active management.

2.5 Importance for Regional and Global Biodiversity

The alpine plants contribute significantly to global biodiversity, and endemic and specialist species are often restricted to very small ecological niches. Mapping plant communities at small spatial scales is important for capturing biodiversity patterns, assessing conservation status, and meeting international biodiversity obligations, including those of the Convention on Biological Diversity (Nagendra et al., 2013a).

Furthermore, alpine ecosystems serve as important carbon sinks, and understanding vegetation distribution and productivity helps estimate carbon stocks and fluxes relevant to global climate regulation.

3. REMOTE SENSING DATA AND CLASSIFICATION METHODS FOR VEGETATION

Mapping remote sensing in conjunction with Geographic Information Systems has significantly revolutionized vegetation mapping, especially in challenging alpine environments. These technologies enable rapid, large-scale, and cost-effective acquisition and analysis of spatial data, overcoming the limitations of traditional field surveys in difficult terrain. The heart of vegetation mapping is the accurate classification of land cover types based on spectral, spatial, and contextual information derived from remotely sensed imagery (Figure 1).

Figure 1.

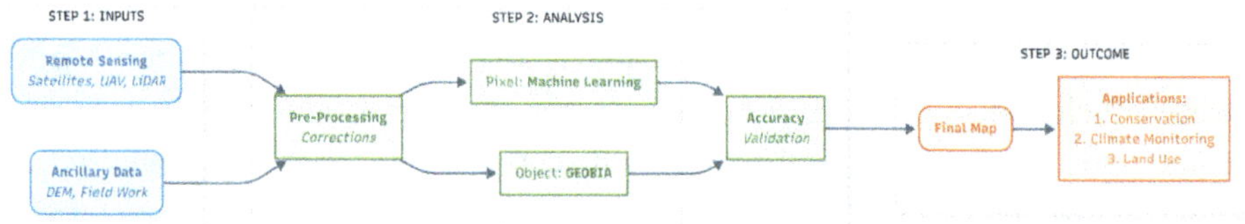

Figure 1 shows the methodological framework for mapping alpine vegetation in the Alps. The workflow illustrates the integration of multi-source remote sensing data (satellite, UAV, LiDAR) and ancillary datasets with advanced classification strategies (Machine Learning and GEOBIA) to derive ecological insights for conservation and land-use planning.

3.1 Sources of Remote Sensing Data

Various remote sensing platforms and sensor types offer distinct benefits for mapping alpine vegetation Multispectral sensors such as: WorldView, Sentinel-2, and Landsat satellites collect data in a limited number of broad spectral bands, including visible light, near-infrared (NIR), and shortwave infrared (SWIR). These data are useful for differentiating vegetation types and for calculating indices like NDVI, which indicate plant health and coverage. (Jędrych et al., 2017b; Mohseni, 2024).

3.2 Hyperspectral Sensors:

EO-1 Hyperion and EnMap instruments acquire data across hundreds of narrow spectral bands, allowing detailed spectral characterization of species and conditions within vegetation. This spectral richness enables fine discrimination between plant communities and detection of subtle signals indicating stress (Fassnacht et al., 2016b).

3.3 LiDAR (Light Detection And Ranging):

LiDAR gives three-dimensional structural information about vegetation, like canopy height and density which is very important in mountainous alpine areas where vertical structure affects habitat quality and ecosystem processes (Wallace et al., 2016b)

3.4 Unmanned Aerial Vehicles:

UAVs equipped with multispectral or hyperspectral cameras provide ultra-high spatial resolution data, enabling detailed mapping of small-scale vegetation patches in hard-to-access areas. The most influential predictor variables for shrub and sapling cover were structural LiDAR-derived metrics, followed by spectral vegetation indices and topographic variables.

4. IMAGE CLASSIFICATION TECHNIQUES

The use of vegetation mapping relies on the ability to categorize remote sensing data into meaningful categories. The primary methods of classification are:

4.1 Supervised Classification

This classification requires an analyst to choose representative training samples from known vegetation classes. Each pixel is then categorized by algorithms based on its similarity to the training data's spectral signature. The most common methods are: Maximum Likelihood Classification (MLC), Support Vector Machines (SVM), Random Forest (RF), and Artificial Neural Networks (ANN). SVM and RF machine learning models are effective for handling non-linear and mixed pixels in heterogeneous alpine landscapes (Maxwell et al., 2018).

4.2 Unsupervised Classification

Unsupervised methods occur without prior information about land cover types, and cluster pixels based on their spectral similarity. K-Means and Iterative Self-Organizing Data Analysis (ISODATA) are among the algorithms used to form classes, which are then interpreted by analysts. This method is faster but may not distinguish different vegetation types when their spectral signatures are similar (Jensen, 2009).

4.3 Geographic Object-Based Image Analysis (GEOBIA

Building on pixel-based analysis, GEOBIA segments images into objects groups of nearby pixels -based on radiometric spatial, and textural properties. This method minimizes noise and increases classification accuracy by incorporating shape, size, and neighborhood context. GEOBIA is effective for high-resolution imagery and complex alpine vegetation patterns (Blaschke, 2010).

To illustrate how these advanced methods integrate different data sources in a real-world scenario, the findings from Bulluck et al., (2022) offer a compelling case study. In their successful effort to map the notoriously difficult shrub/sapling class in the Appalachian Mountains, they used a Random Forest model and quantified the importance of each data type. The results show that a LiDAR-derived canopy height model was the most important predictor, demonstrating the power of combining structural and spectral data (Table 1).

Table 1. The Power of Data Fusion: Key Predictor Variables for Classifying Difficult Shrub/Sapling Cover

Data Category	Most Important Predictor in Category	Overall Rank	Contribution to Classification
Structural	LiDAR-derived canopy height model	1st	Directly measures vegetation height, decisively separating low-stature shrubs (< 4 m) from tall, spectrally similar trees.
Spectral	NDVI (from NAIP imagery)	2nd	Measures vegetation "greenness" and photosynthetic activity, helping to distinguish vegetation from bare ground and developed areas.
Topographic	Elevation (from DEM)	4th	Contextualizes habitat, as specific shrub communities often occur at particular elevation ranges (e.g., high-elevation pastures).
Spatial / Textural	Texture metric (from VBMP imagery)	6th	Captures the patchy, heterogeneous spatial pattern characteristic of shrub cover, which differs from smooth pastures or uniform forests.

5. ALGEBRA-BASED TECHNIQUES

Algebra-based change detection relies on simple mathematical operations applied to two or more images taken at different times. These operations include image differencing, image ratioing, and Change Vector Analysis (CVA). These methods produce difference images that indicate the magnitude and direction of spectral change, enabling rapid detection of altered areas. Even though algebra-based approaches are computationally efficient and easy to apply, they may not capture finer vegetation changes or changes in regions with low signal-to-noise ratios. However, they remain a useful initial method for large-scale vegetation surveillance and serve as a foundation for more sophisticated hybrid methods (Almalki et al., 2022) (Table 2).

Table 2. Comparative analysis of remote sensing classification techniques and geospatial integration strategies for alpine vegetation mapping

Approach	Principle/Mode of Action	Advantages	Limitations
Supervised Classification	Uses labeled training data for pixel categorization	High accuracy; error correction possible	Time-consuming; depends on training quality
Unsupervised Classification	Clusters pixels by spectral similarity	Fast; no prior data needed	Requires post-interpretation; may merge distinct classes
GEOBIA	Groups pixels into meaningful objects	Incorporates spatial context; suitable for high-resolution data	Complex setup; computationally demanding
Transform-Based Techniques	Applies mathematical transforms (PCA, Tasseled Cap) to enhance change detection	Reduces redundancy; highlights key changes	Requires careful interpretation
GIS Integration	Combines spatial datasets for temporal analysis	Enables large-scale monitoring and modeling	Data heterogeneity may affect accuracy
Algebra-Based Techniques	Uses arithmetic operations (difference, ratio) to detect change	Simple and efficient	May overlook subtle or mixed changes

6. APPLICATIONS OF VEGETATION MAPPING

Remote sensing and GIS technology in alpine vegetation mapping offer valuable data that support various ecological, conservation, and land management objectives in alpine terrains, such as the Alps. With improved spatial data and analysis tools, researchers and poverty alleviation professionals can address significant and urgent environmental problems more specifically and efficiently.

6.1 Conservation Planning and Protected Area Management

The alpine areas have high-resolution vegetation maps that are crucial for designing, managing, and monitoring protected areas. Detailed spatial definitions of vegetation types and habitat areas can be used to identify biodiversity hotspots, habitats of rare or endemic species, and ecological corridors that enable species movement of species (Nagendra et al., 2013b; Ul Hassan Chaudhary & Sivertun, 2015). For example, multispectral satellite data from platforms such as Sentinel-2 and WorldView can differentiate vegetation by elevation ranges and help prioritize the conservation of fragile habitats (Jędrych et al., 2017a; Matyukira & Mhangara, 2024).

Machine learning classifiers (SVM and Random Forests) and object-based image analysis (OBIA) are used to detect and classify the complex alpine vegetation structure (meadows, shrublands, and coniferous forests) more effectively than previous methods (Blaschke, 2010; Maxwell et al., 2018). Hyperspectral sensors on UAVs offer ultra-fine spatial resolution, enabling the detection of small and patchy vegetation communities that cannot be detected by satellite-based imaging (Melis et al., 2020).

Three-dimensional vegetation structure data obtained with LiDAR are also useful in enhancing habitat assessments by providing information on canopy height and density, which help assess the quality and structural diversity of habitat in alpine forests and shrublands (Wallace et al., 2016a).

6.2 Climate Change Monitoring and Phenological Analysis

The alpine ecosystems are highly sensitive to climatic changes and the distribution of species, phenology, and vegetation composition can change rapidly (Pepin et al., 2015; Steinbauer et al., 2018). Remote sensing enables vegetation mapping that can be monitored using time-series data to identify changes in vegetation cover and productivity both seasonally and inter-annually. For example, satellites collect vegetation indexes such as NDVI and EVI, which can be used to monitor snowline regression, increases in treeline elevation, and the encroachment of alpine meadows by shrubs. Hyperspectral sensors record spectral data in greater detail to detect signs of stress, such as drought or nutrient deficiencies, providing early warning of climate-induced vegetation stress. Multi-source data (a combination of optical, radar, LiDAR, and environmental variables) can also be used to support robust assessments of vegetation responses to climate drivers, aided by scenario analysis and adaptive management strategies (Linqing et al., 2025) (Matyukira & Mhangara, 2024).

6.3 Habitat Fragmentation and Connectivity Assessment

The alpine flora is found in most discontinuities, which are restricted by steep landscapes and human settlements. Vegetation mapping can be used to examine species survival and system stability, as it supports research on habitat fragmentation, ecological connectivity, and landscape organization (Nagendra et al., 2013b). High-resolution remote sensing data and GIS-based connectivity models help identify the locations of isolated vegetation patches, corridors, and potential barriers to wildlife movement (Ul Hassan Chaudhary & Sivertun, 2015). This spatial knowledge is applied to support conservation efforts such as restoration, corridor development, and land use zoning. LiDAR and UAV surveys enhance connectivity studies by combining them with structural vegetation surveys, including vertical habitat complexity assessments that are relevant to most alpine species (Wallace et al., 2016a).

6.4 Sustainable Land-Use and Ecosystem Services

Vegetation maps guide sustainable land-use planning by identifying areas of productive land use, grazing, and those prone to erosion or degradation (Melis et al., 2020). Observing vegetation dynamics supports restoring balance between conservation and human activities in alpine areas, such as agriculture, tourism, and infrastructure development.

Mapping is useful in ecosystem service evaluations because it measures vegetation cover and condition, which are important for carbon sequestration, soil stabilization, water control, and the provision of biodiversity (Körner & Kèorner, 1999).

The combination of remote sensing, GIS, and ecological analysis can support evidence-based decisions to protect biodiversity in the Alps, control climate effects, and promote sustainable development.

7. TRADITIONAL APPROACHES TO ALPINE VEGETATION MAPPING

Conventional methods for mapping alpine vegetation have primarily relied on field surveys, interpretation of aerial photographs, and the integration of topographic and climatic information. Although these methods are fundamentally important, they are labor-intensive and often spatially constrained with restricted temporal resolution. The introduction of remote sensing and machine learning, has revolutionized the field; however, understanding the advantages and disadvantages of traditional methods remains critical for ecological studies and conservation management (Dirnböck et al., 2003a; Pedrotti, 2018; Thomas et al., 1986; Verrall & Pickering, 2020; Zimmermann & Kienast, 1999). The initial mapping of alpine vegetation was conducted using manual field surveys and interpretation of aerial photographs, along with topographic indices such as slope, aspect, and elevation, which significantly affect the distribution of alpine plants (Dirnböck et al., 2003a; Thomas et al., 1986; Zimmermann & Kienast, 1999). Researchers would outline the vegetation units on aerial photographs and confirm them through extensive fieldwork, identifying plant communities based on floristic composition and physiognomy. This method was detailed and site-specific, but had the disadvantage of being subjective and time and resource-consuming, particularly in remote or rugged alpine settings. These maps were refined with the help of the Pedrotti, (2018) and Thomas et al., (1986).

However, such maps were often inaccurate due to the resolution of existing elevation models and the inability to capture microhabitat heterogeneity or historical land use impacts (Dirnböck et al., 2003b; Hoersch et al., 2002). Although these were difficult, conventional mapping techniques provided a critical foundation for understanding alpine vegetation patterns and their environmental drivers. They also served as the basis for developing more objective, scalable, and repeatable mapping methods, including remote sensing based and automated classification systems. Today, collective monitoring of alpine ecosystems is more reliable and precise due to the integration of historical field data with contemporary technologies, which support both scientific research and environmental preservation (Dirnböck et al., 2003a; Pedrotti, 2018; Thomas et al., 1986; Verrall & Pickering, 2020; Zimmermann & Kienast, 1999).

A critical component of traditional vegetation mapping is understanding the cultural and practical significance of flora. For indigenous communities in alpine regions, the landscape is mapped not only by its physical features but also by the utility of its plant life. This ethnobotanical knowledge is a deep, living repository of information on how to sustainably use natural resources for health and well-being. The following table provides a case study from the Himalayan region, based on the research of Naz et al. (2022), illustrating how key alpine plant species are traditionally used as therapeutic medicines (Naz et al., 2022) Table 3.

Table 3. Alpine medicinal plants, their families, part used and medicinal uses

S. No	Plant specie	Family	Part used	Medicinal uses or other uses	References
1.	*Berberis lycium* Royle.	Berberidaceae	Bark	Used to treat jaundice	(Naz et al., 2022)
2.	*Betula utilis* D.Don.	Betulaceae	Bark	Used internally and as a paste to treat jaundice and other skin diseases.	
3.	*Saussurea lappa (Decne.) Sch.Bip.*	Asteraceae	Roots	Used for asthma and in powdered form for various skin diseases.	
4.	*Podophyllum emodi* Wall. ex Hook.f. & Thomson	Berberidaceae	Roots	A lotion is made from the roots, which is used as a purgative (to induce bowel movements).	
5.	*Geranium wallichianum* D.Don ex Sweet.	Geraniaceae	Roots	The roots are used in powdered form for wound healing and as a tea to treat rheumatism.	
6.	*Thymus linearis* Benth.	Labiatae	Whole plant	An extract of the whole plant is used as a diuretic and is also applied externally (as *khameera*) for back pain.	
7.	*Berginia virgata* Harv. ex Benth. & Hook.f.	Saxifragaceae	Root	The root is powdered and applied to chronic ulcers and for wound healing. This plant had the highest Use Value in the study.	
8.	*Achillea herba-rotta* All.	Asteraceae	Dried flowering aerial parts	Treating toothache and headaches	(Pieroni & Giusti, 2009)
9.	*Artemisia absinthium* L.	Asteraceae	Dried aerial parts	Tea used as an anti-hypertensive, digestive, and an antihelminthic	
10.	*Artemisa genipi* Weber.	Asteraceae	Dried flowering aerial parts	It is used against coughs and as a digestive	
11.	*Chelidonium majus* L	Papaveraceae	Fresh latex	Topically applied for treating warts	

continued on following page

Table 3. Continued

S. No	Plant specie	Family	Part used	Medicinal uses or other uses	References
12.	*Allium stracheyi* Baker.	Amaryllidaceae	Shoot	Stomached-ache	(Thapliyal et al., 2024)
13.	*Anaphalis contorta* (D.Don) Hook.f.	Asteraceae	Leaves	Paste of fresh leaves is applied on cuts, wounds and boils	
14.	*Arnebia benthamii* (Wall. ex G.Don) I.M.Johnst.	Boraginaceae	Roots	Remedy to treat hairfall	
15.	*Rhodiola imbricate* Edgew.	Celastraceae	Roots	Utilized to treat cough, fever, headache and anaemia	
16.	*Ephedra intermedia* Schrenk & C.A.Mey.	Ephedraceae	Dried aerial parts	Used to treat asthma	
17.	*Actea spicata* L.	Ranunculaceae	Roots and berries	Snakebites, Nervous disorders, Goiter, Asthma, Joints pain, Fever	(Kayani et al., 2015)
18.	*Androsace rotundifolia* Hardwicke.	Primulaceae	Leaves	Stomach disorders, Skin diseases, **Menstrual problems**	
19.	*Berberis pseudumbellata* Parker.	Berberidaceae	Leaves, Fruits, Roots	Indigestion, Typhoid, Muscular pains, Jaundice, Urinary problems	
20.	*Vaccinium myrtillus* L.	Ericaceae	Leaves	Spices	(Abbet et al., 2014)
21.	*Taraxacum officinale* F. H. Wiggers.	Asteraceae	Flowers	Jams	
22.	*Rosa canina* L.	Rosaceae	Fruits	Jams	

8. GROUND TRUTHING AND ACCURACY ASSESSMENT

The accuracy assessment of all vegetation maps produced by remote sensing depends on the validity of the product's credibility assessment, which in turn relies on the quality of ground-truthing data. Nevertheless, gathering such data in alpine areas is infamously difficult due to challenging geography and poor accessibility, which has remained a persistent obstacle to large-scale ecological analysis (Matyukira & Mhangara, 2024) . Field surveys offer the best data, but logistical limitations often result in only sparse validation data being collected, which may not represent the entire heterogeneity of the landscape. This data sparsity is a bottleneck that affects the accuracy of even the most sophisticated classification models.

To address this issue, new methodologies are emerging that leverage citizen science and other publicly available data sources to supplement traditional fieldwork (Pettorelli et al., 2014). One example is the novel ground truthing technique called the Instagram ground-truthing approach, developed by Werner et al. (2024) to create a large-scale distribution map of the treeline species *Nothofagus pumilio* (Poepp. & Endl.) Krasser in the remote Southern Andes. This method uses social media as passive or opportunistic citizen science by utilizing the huge number of publicly shared geotagged photos available online.

The procedure involved systematically searching Instagram for posts featuring recognizable species traits (e.g. autumn coloring and leaf shape), and recognizable landscape elements (e.g. mountain peaks, lakes, and glaciers)). By overlaying these visual signals, scientists could precisely geolocate species occurrences, creating a validation dataset of more than 1,200 points. The key finding of the study was that this method produced a much larger dataset than the established Global Biodiversity Information Facility (GBIF) for the same region and exhibited a lower level of spatial bias. Whereas the traditional data collection methods tend to focus on easily accessible tourist hubs, the Instagram approach effectively captured events in the backgrounds of photos, which served as validation markers for the locations of more remote and inaccessible treelines (Werner et al., 2024).

There is a direct challenge between social media data mining and controlled classification of Sentinel-2 images, and the fundamental issues of spectral confusion and landscape heterogeneity typical of alpine regions (Fassnacht et al., 2016a; Nagendra et al., 2013b). These approaches will increase the accuracy of large scale vegetation maps and provide an effective, affordable, way to model future monitoring in the most demanding mountain topography in the world due to the large number of validation points they contain.

9. CHALLENGES AND LIMITATIONS IN ALPINE VEGETATION MAPPING

It presents several challenges that create critical problems for mapping alpine vegetation using remote sensing and GIS, and these issues affect the accuracy, reliability, and applicability of the results.

9.1 Spectral Confusion and Landscape Heterogeneity

The alpine environment is also characterized by sparse, patchy, mixed vegetation with bare soil, rocks, snow, and water bodies, all of which can contribute to similar spectral signatures and cause spectral confusion (Fassnacht et al., 2016a). The sparse vegetation structure leads to mixed pixels, especially at the middle spatial resolution of satellites, which hinders vegetation-type discrimination and reduces classification accuracy (Nagendra et al., 2013b). Additionally, species level discrimination is difficult due to the overlapping reflectance properties of alpine plant communities.

9.2 Climatic variability and Temporal

Phenological controls of vegetation in alpine ecosystems change significantly with season and interannual climatic changes (Reed et al., 1994). Spectral responses and vegetation indices are influenced by the duration of snow cover, the freezing and thawing of frozen periods, and precipitation variability. As a result, single-date images might fail to reveal the entire phenological cycle, resulting to misclassification or underestimation of vegetation extent. Multi-temporal data are necessary but are often limited by cloud cover, data availability, and costs (Pettorelli et al., 2014).

9.3 Topographical and Atmospherical Effects

Topography causes complex lighting conditions including shadows and changing sun angles, which negatively affect reflectance measurements (Long & Shen, 2013). Haze and aerosol scattering in the atmosphere also reduce image quality. Although topographic normalization methods exist (Richter, 1997), residual artifacts may remain, affecting vegetation indices and classification.

9.4 Data and Methodological Constraints

Hyperspectral and LiDAR data are better for vegetation discrimination; however, they are costly and not always available in large alpine regions (Fassnacht et al., 2016a; Wallace et al., 2016a). UAVs are flexible for data collection, but they are weather-dependent and subject to flight rules and range limitations. Alpine regions are remote, so training data might be hard to obtain, making machine learning models less transferable (Maxwell et al., 2018).

10. FUTURE DIRECTIONS AND INNOVATIONS

Further advancements in technology and methodology are sure to eliminate most modern constraints of alpine vegetation mapping.

10.1 Advanced Sensor Technologies

Future satellite missions such as NASA's Surface Biology and Geology (SBG) and ESA's EnMAP hyperspectral satellites will provide high-resolution hyperspectral images globally and significantly enhance species-level vegetation discrimination and stress signal detection in alpine regions (Fassnacht et al., 2016a; Ustin et al., 2009).

With the development of hyperspectral, LiDAR, and thermal sensors as part of UAV payloads, fine-scale monitoring of vegetation in challenging environments will become possible (Li et al., 2018; Wallace et al., 2016a).

10.2 Data Fusion and Machine Learning

Deep learning algorithms are increasingly able to detect more complex spectral-spatial patterns, which can improve the classification of vegetation and change. Optical, radar, and LiDAR sensors are used for data fusion, enhancing resistance to atmospheric and topographic noise and improving classification performance (Li et al., 2018; Maxwell et al., 2018).

10.3 Phenological and Functional Trait Mapping

The projection of vegetation response to climate change can be predicted only by mapping remote sensing data to plant functional traits and phenology. There will be an increase in time-series analysis including high-frequency satellite data to monitor growing season dynamics, stress responses and ecosystem resilience (Li et al., 2018).

10.4 Integrating Participatory and Citizen Science

Integrating remote sensing and ground-based observation through citizen scientists and other local stakeholders can augment training data, enhance model validation, and support adaptive management. Participatory methods can help address data gaps in remote alpine areas and increase the relevance of vegetation monitoring to society (Pettorelli et al., 2014).

11. CONCLUSION

Mapping alpine vegetation within the Alps has played a significant role in assessing ecological health in one of the most sensitive and topographically complex regions on Earth. This chapter has demonstrated how combined geospatial technologies can improve our ability to observe high-altitude ecosystems by moving beyond traditional, labor-intensive field surveys. The combination of mixed multispectral and hyperspectral remote sensing techniques, LiDAR-based structural measurements, and Geographic Information Systems (GIS) provides a powerful toolset for recording complex patterns of alpine vegetation across steep environmental slopes. The introduction of new computational methods is a key focus of this review. The shift to Geographic Object-Based Image Analysis (GEOBIA), which utilizes machine learning algorithms such as Random Forest and Support Vector Machines, has made it much easier to address issues of spectral confusion and landscape heterogeneity. With these instruments, it is now possible to correctly map fragmented habitats, meadows, and shrublands even in the rugged and shaded regions of the Alpine range. Also, Digital Elevation Models (DEMs) are important for understanding local microclimatic variations that influence plant distribution. Ecologically, multi-temporal satellite images and phenological indices have provided direct evidence that the Alps are vulnerable to climate change. The upward migration of the treeline, the expansion of shrubs into high altitude meadows, and changes in growth periods have all been observed, highlighting the role of alpine vegetation as an early warning system for global environmental changes. These mapping products are not merely academic exercises; rather, they are fundamental to conservation planning, helping to identify biodiversity hotspots, assess ecological connectivity, and support international commitments like the Convention on Biological Diversity (CBD). Despite these advancements, challenges remain, including the high cost of high-resolution data, atmospheric interference, and the logistical limitations of UAVs during harsh weather. Finally, large-scale monitoring of alpine vegetation through innovative remote sensing is essential for promoting regional resilience, informing sustainable land-use management tactics, and ensuring the long-term preservation of the unique biological legacy of the Alps amid significant climatic uncertainty.

REFERENCES

Abbet, C., Mayor, R., Roguet, D., Spichiger, R., Hamburger, M., & Potterat, O. (2014). Ethnobotanical survey on wild alpine food plants in Lower and Central Valais (Switzerland). *Journal of Ethnopharmacology, 151*(1), 624–634. DOI: 10.1016/j.jep.2013.11.022 PMID: 24269246

Allbed, A., & Kumar, L. (2013). Soil salinity mapping and monitoring in arid and semi-arid regions using remote sensing technology: A review. *Advances in Remote Sensing, 2*(4), 373–385. DOI: 10.4236/ars.2013.24040

Almalki, R., Khaki, M., Saco, P. M., & Rodriguez, J. F. J. R. S. (2022). Monitoring and mapping vegetation cover changes in arid and semi-arid areas using remote sensing technology. *RE:view, 14*(20), 5143.

Blaschke, T. (2010). Object based image analysis for remote sensing. *ISPRS Journal of Photogrammetry and Remote Sensing, 65*(1), 2–16. DOI: 10.1016/j.isprsjprs.2009.06.004

Bulluck, L., Lin, B., & Schold, E. J. R. S. (2022). Fine resolution imagery and LIDAR-derived canopy heights accurately classify land cover with a focus on shrub/sapling cover in a mountainous landscape. *14*(6), 1364.

Cui, X., Gibbes, C., Southworth, J., & Waylen, P. (2013). Using remote sensing to quantify vegetation change and ecological resilience in a semi-arid system. *Land (Basel), 2*(2), 108–130. DOI: 10.3390/land2020108

Dirnböck, T., Dullinger, S., Gottfried, M., Ginzler, C., & Grabherr, G. (2003a). Mapping alpine vegetation. 209-219. DOI: 10.1007/978-3-642-18967-8_10

Dirnböck, T., Dullinger, S., Gottfried, M., Ginzler, C., & Grabherr, G. (2003b). Mapping alpine vegetation based on image analysis, topographic variables and Canonical Correspondence Analysis. *Applied Vegetation Science, 6*(1), 85–96. DOI: 10.1111/j.1654-109X.2003.tb00567.x

Fassnacht, F. E., Latifi, H., Stereńczak, K., Modzelewska, A., Lefsky, M., Waser, L. T.,...Ghosh, A. J. R. s. o. e. (2016b). Review of studies on tree species classification from remotely sensed data. *186*, 64-87.

Fassnacht, F. E., Latifi, H., Stereńczak, K., Modzelewska, A., Lefsky, M., Waser, L. T., Straub, C., & Ghosh, A. (2016a). Review of studies on tree species classification from remotely sensed data. *Remote Sensing of Environment, 186*, 64–87. DOI: 10.1016/j.rse.2016.08.013

Foley, W. J., McIlwee, A., Lawler, I., Aragones, L., Woolnough, A. P., & Berding, N. (1998). Ecological applications of near infrared reflectance spectroscopy–a tool for rapid, cost-effective prediction of the composition of plant and animal tissues and aspects of animal performance. *Oecologia, 116*(3), 293–305. DOI: 10.1007/s004420050591 PMID: 28308060

Hoersch, B., Braun, G., & Schmidt, U. (2002). Relation between landform and vegetation in alpine regions of Wallis, Switzerland. A multiscale remote sensing and GIS approach. *Computers, Environment and Urban Systems, 26*(2-3), 113–139. DOI: 10.1016/S0198-9715(01)00039-4

Jędrych, M., Zagajewski, B., & Marcinkowska-Ochtyra, A. (2017a). Application of Sentinel-2 and EnMAP new satellite data to the mapping of alpine vegetation of the Karkonosze Mountains. *Polish Cartographical Review, 49.*

Jędrych, M., Zagajewski, B., & Marcinkowska-Ochtyra, A. J. P. C. R. (2017b). Application of Sentinel-2 and EnMAP new satellite data to the mapping of alpine vegetation of the Karkonosze Mountains. *49.*

Jensen, J. R. (2009). *Remote sensing of the environment: An earth resource perspective 2/e*. Pearson Education India. 1-239.

Karabulut, M. (2018). An examination of spectral reflectance properties of some wetland plants in Göksu Delta, Turkey. *Journal of International Environmental Application and Science, 13*(4), 194–203.

Kayani, S., Ahmad, M., Sultana, S., Shinwari, Z. K., Zafar, M., Yaseen, G., & Bibi, T. (2015). Ethnobotany of medicinal plants among the communities of Alpine and Sub-alpine regions of Pakistan. *Journal of Ethnopharmacology, 164*, 186–202. DOI: 10.1016/j.jep.2015.02.004 PMID: 25680839

Kennedy, R. E., Cohen, W. B., & Schroeder, T. A. (2007). Trajectory-based change detection for automated characterization of forest disturbance dynamics. *Remote Sensing of Environment, 110*(3), 370–386. DOI: 10.1016/j.rse.2007.03.010

Körner, C., & Kèorner, C. (1999). *Alpine plant life: functional plant ecology of high mountain ecosystems* (Vol. 4). Springer. DOI: 10.1007/978-3-642-98018-3

Li, Y., Zhang, H., Xue, X., Jiang, Y., & Shen, Q. (2018). Deep learning for remote sensing image classification: A survey. *Wiley Interdisciplinary Reviews. Data Mining and Knowledge Discovery, 8*(6), e1264. DOI: 10.1002/widm.1264

Linqing, Z., Tian, F., Liang, T., Fensholt, R., He, T., & Schaepman-Strub, G. (2025). Topographic effects on vegetation greening and area expansion in global alpine zones under climate change. *International Journal of Applied Earth Observation and Geoinformation, 142*, 104727. DOI: 10.1016/j.jag.2025.104727

Liping, C., Yujun, S., & Saeed, S. (2018). Monitoring and predicting land use and land cover changes using remote sensing and GIS techniques—A case study of a hilly area, Jiangle, China. *PLoS One*, *13*(7), e0200493. DOI: 10.1371/journal.pone.0200493 PMID: 30005084

Long, Y., & Shen, Z. (2013). Disaggregating heterogeneous agent attributes and location. *Computers, Environment and Urban Systems*, *42*, 14–25. DOI: 10.1016/j.compenvurbsys.2013.09.002

Matyukira, C., & Mhangara, P. (2024). Advances in vegetation mapping through remote sensing and machine learning techniques: A scientometric review. *European Journal of Remote Sensing*, *57*(1), 2422330. DOI: 10.1080/22797254.2024.2422330

Maxwell, A. E., Warner, T. A., & Fang, F. (2018). Implementation of machine-learning classification in remote sensing: An applied review. *International Journal of Remote Sensing*, *39*(9), 2784–2817. DOI: 10.1080/01431161.2018.1433343

Melis, M., Dessì, F., Loddo, P., Maccioni, A., Gallo, M., Ul Hassan, R., & Aurang Zaib, M. (2020). ESA Sentinel 2 Imagery and GBGEOApp: Integrated Tools for the Deosai National Park Management Plan. *The International Archives of the Photogrammetry, Remote Sensing and Spatial Information Sciences*, *43*, 145–151. DOI: 10.5194/isprs-archives-XLIII-B3-2020-145-2020

Melis, M., Dessì, F., Loddo, P., Maccioni, A., Gallo, M., Ul Hassan, R., & Sciences, S. I. (2020). ESA Sentinel 2 Imagery and GBGEOApp. *Integrated Tools for the Deosai National Park Management Plan.*, *43*, 145–151.

Mohseni, P. (2024). *Mapping snow and vegetation coverage using multitemporal open satellite Imagery: the case study of the Maritime Alps*. Politecnico di Torino.

Nagendra, H., Lucas, R., Honrado, J. P., Jongman, R. H., Tarantino, C., Adamo, M., & Mairota, P. J. E. I. (2013a). Remote sensing for conservation monitoring: Assessing protected areas, habitat extent, habitat condition, species diversity, and threats. *33*, 45-59.

Nagendra, H., Lucas, R., Honrado, J. P., Jongman, R. H. G., Tarantino, C., Adamo, M., & Mairota, P. (2013b). Remote sensing for conservation monitoring: Assessing protected areas, habitat extent, habitat condition, species diversity, and threats. *Ecological Indicators*, *33*, 45–59. 10.1016/j.ecolind.2012.09.014. DOI: 10.1016/j.ecolind.2012.09.014

Naz, A., Amjad, M. S., Umair, M., Altaf, M., & Ni, J. (2022). Medicinal plants used as therapeutic medicine in the Himalayan region of Azad Jammu and Kashmir. *South African Journal of Botany*, *150*, 194–216. 10.1016/j.sajb.2022.07.007. DOI: 10.1016/j.sajb.2022.07.007

Pedrotti, F. (2018). Three-dimensional maps of vegetation series and geoseries with the use of stereograms. *52*, 55-67. DOI: 10.24193/Contrib.Bot.52.4

Pepin, N., Bradley, R., Diaz, H., Baraer, M., Caceres, E., Forsythe, N., & Liu, X. (2015). Elevation-dependent warming in mountain regions of the world. [In.]. *Nature Climate Change*, *5*(5), 424–430. DOI: 10.1038/nclimate2563

Pettorelli, N., Laurance, W. F., O'Brien, T. G., Wegmann, M., Nagendra, H., & Turner, W. (2014). Satellite remote sensing for applied ecologists: Opportunities and challenges. *Journal of Applied Ecology*, *51*(4), 839–848. DOI: 10.1111/1365-2664.12261

Pieroni, A., & Giusti, M. E. (2009). Alpine ethnobotany in Italy: Traditional knowledge of gastronomic and medicinal plants among the Occitans of the upper Varaita valley, Piedmont. *Journal of Ethnobiology and Ethnomedicine*, *5*(1), 32. DOI: 10.1186/1746-4269-5-32 PMID: 19895681

Reed, B. C., Brown, J. F., VanderZee, D., Loveland, T. R., Merchant, J. W., & Ohlen, D. O. (1994). Measuring phenological variability from satellite imagery. *Journal of Vegetation Science*, *5*(5), 703–714. DOI: 10.2307/3235884

Reis, S. (2008). Analyzing land use/land cover changes using remote sensing and GIS in Rize, North-East Turkey. *Sensors (Basel)*, *8*(10), 6188–6202. DOI: 10.3390/s8106188 PMID: 27873865

Richter, R. (1997). Correction of atmospheric and topographic effects for high spatial resolution satellite imagery. *International Journal of Remote Sensing*, *18*(5), 1099–1111. DOI: 10.1080/014311697218593

Rundquist, D. C. (2001). Field techniques in remote sensing: Learning by doing. *Geocarto International*, *16*(1), 85–90. DOI: 10.1080/10106040108542186

Shaheen, H., Aziz, S., Nasar, S., Waheed, M., Manzoor, M., Siddiqui, M. H.,...Conservation. (2023). Distribution patterns of alpine flora for long-term monitoring of global change along a wide elevational gradient in the Western Himalayas. *48*, e02702.

Steinbauer, M. J., Grytnes, J.-A., Jurasinski, G., Kulonen, A., Lenoir, J., Pauli, H., & Barni, E. (2018). Accelerated increase in plant species richness on mountain summits is linked to warming. *Nature*, *556*(7700), 231–234. DOI: 10.1038/s41586-018-0005-6 PMID: 29618821

Thapliyal, N., Bhojak, P., Arya, D., & Sekar, K. C. (2024). Ethnobotanical assessment of medicinal plants used in alpine regions of Uttarakhand, west Himalaya. *Ethnobotany Research and Applications*, *27*, 1–24. DOI: 10.32859/era.27.44.1-24

Thomas, Frank, & Isard, S. (1986). Alpine vegetation classification using high resolution aerial imagery and topoclimatic index values. *Photogrammetric Engineering and Remote Sensing*, *52*, 381–388.

Tompolidi, A.-M., Mantovani, L., Frigeri, A., & Nazzareni, S. J. G. (2025). Remote Sensing Applications for Geological Mapping in the Mediterranean Region. A Review. *Geosciences*, *15*(11), 425.

Ul Hassan Chaudhary, W., & Sivertun, Å. (2015). Object-Based Analysis of Multispectral RS Data and GIS for Detection of Climate Change Impact on the Karakoram Range Northern Pakistan. *Journal of Environmental Science and Engineering A*, *4*(4), 303–310. DOI: 10.17265/2162-5298/2015.06.005

Ustin, S., Gitelson, A., Jacquemoud, S., Schaepman, M. E., Asner, G., Gamon, J., & Zarco-Tejada, P. (2009). Retrieval of foliar information about plant pigment systems from high resolution spectroscopy. *Remote Sensing of Environment 113 (*2009*) Suppl.1, 113*. DOI: 10.5167/uzh-23317

Verrall, B., & Pickering, C. (2020). Alpine vegetation in the context of climate change: A global review of past research and future directions. *The Science of the Total Environment*, *748*, 141344. DOI: 10.1016/j.scitotenv.2020.141344 PMID: 32814293

Wallace, L., Lucieer, A., Malenovský, Z., Turner, D., & Vopěnka, P. (2016a). Assessment of forest structure using two UAV techniques: A comparison of airborne laser scanning and structure from motion (SfM) point clouds. *Forests*, *7*(3), 62. DOI: 10.3390/f7030062

Wallace, L., Lucieer, A., Malenovský, Z., Turner, D., & Vopěnka, P. J. F. (2016b). Assessment of forest structure using two UAV techniques: A comparison of airborne laser scanning and structure from motion (SfM) point clouds. *7*(3), 62.

Werner, M., Weidinger, J., Böhner, J., Schickhoff, U., & Bobrowski, M. (2024). Instagram data for validating Nothofagus pumilio distribution mapping in the Southern Andes: A novel ground truthing approach. *Frontiers of Biogeography*, *17*, e140606. DOI: 10.21425/fob.17.140606

Wu, Q., Li, H., Wang, R., Paulussen, J., He, Y., Wang, M., Wang, B., & Wang, Z. (2006). Monitoring and predicting land use change in Beijing using remote sensing and GIS. *Landscape and Urban Planning*, *78*(4), 322–333. DOI: 10.1016/j.landurbplan.2005.10.002

Xue, J., & Su, B. (2017). Significant remote sensing vegetation indices: A review of developments and applications. *Journal of Sensors*, *2017*(1), 1353691. DOI: 10.1155/2017/1353691

Yang, Z., Shen, Y., Jiang, H., Feng, F., & Dong, Q. (2021). Assessment of the environmental changes in arid and semiarid mining areas using long time-series Landsat images. *Environmental Science and Pollution Research International*, *28*(37), 52147–52156. DOI: 10.1007/s11356-021-14417-0 PMID: 33997931

Zimmermann, N., & Kienast, F. (1999). Predictive mapping of alpine grasslands in Switzerland: Species versus community approach. *Journal of Vegetation Science*, *10*(4), 469–482. DOI: 10.2307/3237182

Chapter 4
Hotspots of the Alpine Plants In The Irano-Turanian Regions

Khizar Hayat
http://orcid.org/0009-0005-9273-4873
Islamia College, Peshawar, Pakistan

Abid Ullah
Islamia College, Peshawar, Pakistan

Muhammad Nauman Khan
http://orcid.org/0000-0003-3790-0902
Department of Botany, University of Chakwal, Pakistan

Sana Wahab
Quaid-i-Azam University, Islamabad, Pakistan

Nouman Ahmad
http://orcid.org/0009-0009-3370-2862
University of Mianwali, Punjab, Pakistan

Iqra Malik
University of Sargodha, Pakistan

Ghulam Jelani
University of Peshawar, Pakistan

ABSTRACT

The Irano-Turanian region, one of the largest phytogeographical zones of the Old World, supports exceptional alpine plant diversity shaped by complex topography, climatic variability, and historical biogeographic processes. Alpine ecosystems within this region act as major biodiversity hotspots, harboring high species richness along with numerous endemic and relict taxa. Identifying hotspots of alpine flora is essential for understanding ecological dynamics, evolutionary patterns, and conservation priorities. This chapter provides an overview of alpine plant diversity hotspots across the Irano-Turanian region, highlighting centers of endemism, ecological significance, and the influence of environmental and historical drivers. It also

DOI: 10.4018/979-8-3373-6756-9.ch004

ght © 2026, IGI Global Scientific Publishing. Copying or distributing in print or electronic forms without written permission of IGI Global Scientific Publishing is prohibited. Use of this chapter to generative artificial intelligence (AI) technologies is expressly prohibited. The publisher reserves all rights to license its use for generative AI training and machine learning model development.

addresses key conservation challenges, including habitat degradation, overgrazing, and the growing impacts of climate change, emphasizing the need for coordinated conservation efforts to protect these unique alpine ecosystems.

1. INTRODUCTION

1.1 Background of the Irano-Turanian Regions

The Irano-Turanian floristic region (IT region) was first clearly defined by Edmond Boissier in Flora Orientalis (1867) and has since been named and delimited in different ways by various authors (Takhtajan, 1986; Zohary, 1973). The Irano-Turanian phytogeographic region encompasses Anatolia, parts of Jordan, Syria, the Israeli-Palestinian region, the Sinai Peninsula, upper Mesopotamia, a large part of the Armenian Highlands, southern and eastern Transcaucasia, the Caspian shore of Iran, the Iranian Plateau (without the tropical deserts), the westernmost Himalayas, and the arid territory of south-eastern European Russia extending to the Gobi Desert (Takhtajan, 1986). The IT region in Southwest Asia is one of the richest floristic areas of the Holarctic Kingdom, with most species diversity concentrated in the Iranian plateau, Anatolian plateau, and Central Asia (Takhtajan, 1986; Zohary, 1973). The Irano-Turanian region is notable for high rates of plant speciation and endemism, especially in large genera like *Astragalus, Cousinia, Acantholimon, Acanthophyllum*, and *Allium* (Takhtajan, 1986; Zohary, 1973). Its climate differs significantly from surrounding areas, particularly in continentality and seasonal differences, whereas the region's flora is characterized by a large number of endemic genera and an extremely high degree of species endemism (Djamal et al., 2012). The Irano-Anatolian region and the mountains of Central Asia are well- known among the 34 global biodiversity "Hotspots" (Mittermeier et al., 2005). On a large-scale, the IT region has traditionally been seen as a refuge for many taxa found in neighboring regions, most notably the Mediterranean (Akhani et al., 2010; Blondel, 2010). Nonetheless, this idea has been based mainly on floristic observations and not on explicit phylogenetic or biogeographical data.

1.2 Importance of Alpine Plant Hotspots

Alpine plant hotspots are small, high-elevation areas characterized by exceptionally high species richness, endemism, and ecological uniqueness, and they play a crucial role in global biodiversity and ecosystem functioning. These areas serve as centers of speciation, refuges during climate change, sources of ecosystem services, and natural indicators of environmental transformations (Testolin et al., 2021; Yang et al., 2025; Zhang et al., 2021). Protecting alpine hotspots is essential for sustaining both biodiversity and human well-being downstream (Grabherr et al., 2010; Körner, 2023). Alpine hotspots contain an exceptional concentration of endemic and range-restricted species. The isolation of mountain "Sky Islands" and steep altitudinal gradients promote speciation and the persistence of relict taxa (Testolin et al., 2021).

The alpine plants play a significant role in providing critical ecosystem services. These ecosystems regulate hydrology by influencing snowfall and melting, filtering mountain water, stabilizing soil, and sequestering carbon in alpine meadows and peatlands (Grêt-Regamey et al., 2012; Körner, 2023). Dense vegetation cover also reduces erosion and helps prevent landslides, directly benefiting local communities. The topographic heterogeneity of alpine areas creates microclimatic refugia, small habitat areas that protect species from regional warming (Nadeau et al., 2017). These microrefugia enable cold-adapted species to persist despite climate change, acting as temporary "arks" for biodiversity (Scherrer & Körner, 2011). Because of their sensitivity to climatic variables, alpine ecosystems serve as sentinels for detecting environmental change. Shifts in species composition, phenology, and elevation range are already observable in alpine hotspots across Europe and Asia (Benito Alonso & Villar Pérez, 2012). Long-term alpine monitoring networks, such as GLORIA (Global Observation Research Initiative in Alpine Environments), demonstrate that alpine flora respond rapidly to temperature increases, making these hotspots early indicators of broader ecosystem change (Pauli et al., 2012).

2. OVERVIEW OF THE IRANO–TURANIAN REGION

2.1. Geographic Extent and Boundaries

The Irano-Turanian Region is a vast Southwest–Central Asian floristic realm of roughly 16,000,000 km^2, characterized by a strongly continental, xeric climate and a mix of mountain woodlands, steppes, semi-deserts and halophytic communities. Its exact limits vary depending on the author and whether climatic or floristic criteria are used.

2.1.1. Geographic Extent

The section summarizes the location of the Irano-Turanian Region and available spatial data describing its extent. It also notes the precise geographic coordinates or area estimates reported in the supplied literature.

- *Area Estimate:* The region is estimated to be about 16,000,000 km^2, roughly 30% of Eurasia (Djamal et al., 2012).
- *Broad Span:* The region extends across much of southwestern and central Asia, from Anatolia eastward into continental Central Asia according to floristic reviews (Djamal et al., 2012).
- *Precise Coordinates:* Precise latitude and longitude bounding coordinates are not provided in the supplied sources, so exact corner coordinates are insufficient evidence.
- *Countries Explicitly Reported:*

  I. **Turkey:** Central, eastern and southeastern Anatolia are included in the Irano-Turanian phytogeographical region (Manafzadeh et al., 2017)
  II. **Iran:** is a core area of the region and has been the subject of bioclimatic mapping which associates the Irano-Turanian domain with a distinct Mediterranean-type macrobioclimate in Iran (Djamal et al., 2012).
  III. **Southern Turkmenistan:** (as part of the Khorassan-Kopet Dagh province) is reported to be within the Irano-Turanian floristic province (Golvan & Rioux, 1961).

- *Broader Statement on Coverage:* Additional parts of Central Asia and adjacent continental arid lands are implicated by floristic syntheses, but explicit country lists beyond those mentioned above are not fully enumerated in the supplied excerpts (Djamal et al., 2012).

2.1.2. Major Boundaries

This section describes the major directional limits described in the literature and explains how adjacent phytogeographical regions or bioclimates are used to delimit the Irano-Turanian Region. It highlights the northern, southern, eastern and western contacts supported by the sources.

- *North Boundary:* The Irano-Turanian domain in northern Iran borders a temperate/Euro-Siberian bioclimate (e.g., Hyrcanian/Euro-Siberian elements), so the northern boundary is typically defined where temperate/Euro-Siberian vegetation and climate replace Irano-Turanian elements (Djamal et al., 2012).
- *South Boundary:* In Iran the southern margin transitions toward a Tropical/ Saharo-Sindian macrobioclimate (tropical desert zones) and the Saharo-Sindian phytogeographic influence, so the southern limit is often placed where tropical or desert elements dominate (Djamal et al., 2012).

- *West Boundary:* Western limits are commonly placed in Anatolia or long the eastern Mediterranean margin, where Mediterranean or Anatolian floristic influences increase; Anatolian central and eastern provinces are explicitly treated as Irano-Turanian in regional treatments (Djamal et al., 2012; Manafzadeh et al., 2017).
- *East Boundary:* Eastern limits are considered the transition into the continental Central Asian steppes and deserts. Some provincial studies, such as those on the Khorassan-Kopet Dagh, emphasize transitional corridors that link Irano-Turanian cores to neighboring Euro-Siberian and Central Asian elements, illustrating gradual eastern transitions rather than sharp boundaries (Golvan & Rioux, 1961).
- *Nature of Boundaries:* Authors note that many boundaries are transitional, ecotonal or defined differently depending on whether floristic, climatic, or physiographic criteria are used, so sharp cartographic delimitations differ by approach (Djamal et al., 2012).

2.2. Vegetation and Ecology

This section summarizes the characteristic plant communities, major habitat types and the dominant climatic regime that define the Irano-Turanian Region in the cited literature, emphasizing the multi-habitat mosaic reported by specialists. The Irano-Turanian Region is characterized by a continental, highly seasonal climate with cold winters and hot summers and by a mosaic of open woodlands, mountain steppes, semi-deserts and saline or halophytic habitats. These ecological features are repeatedly emphasized in regional floristic and bioclimatic studies (Djamal et al., 2012).

2.2.1. Major Vegetation Types (listed from regional floristic inventories)

- ***Juniperus* Woodlands**: Montane woodlands occur on rocky, well-drained slopes (Golvan & Rioux, 1961).
- ***Pistacia* and Shrub Woodlands**: Low woodlands and scrub often on xeric soils (Golvan & Rioux, 1961).
- **Mountain Steppe and Grasslands:** Montane and submontane steppes that host many Irano-Turanian elements (Golvan & Rioux, 1961).
- **Semi-desert Steppes and Desert Fringe Communities:** Xeric steppe grading to semi-desert.

- **Halophytic and Saline Vegetation:** Salt-tolerant communities are widespread on degraded and saline soils, with numerous Irano-Turanian halophytes in Turkey and adjacent areas (Manafzadeh et al., 2017).
- **Isolated Forest Enclaves and Hygrophilous Communities**: Small Hyrcanian or Euro-Siberian refugia occur as local enclaves within the broader region (Golvan & Rioux, 1961).
- **Climatic Characterization:** The region is often treated as a distinct continental Mediterranean-type or "Xero-estival-continental" bioclimate (characterized by seasonal precipitation and strong continentality), distinguishing it from maritime Mediterranean and tropical desert climates (Djamal et al., 2012).
- **Biodiversity Features:** High levels of xerophytic diversity and numerous regional endemics are reported; the region is considered a major Old-World source of xerophyte taxa and a hotspot of evolutionary diversification in several plant groups.

Definitions and Debates

This section explains how different scholars have defined or delimited the Irano-Turanian Region, summarizes methodological differences (floristic versus climatic delimitation), and notes key areas of disagreement or uncertainty.

The literature records substantial variation in delimitations because authors have used different criteria, such as historical floristic ties, climatic classification, physiography and provincial floras. Reviewers emphasize that borders have been outlined differently over the past two centuries and remain a topic of active study.

Primary basis for delimitation	Key implication for boundaries
Historical and floristic synthesis and review	Emphasizes large Irano-Turanian core across SW–Central Asia and documents that many boundary proposals exist in the literature (Manafzadeh et al., 2017)
Climatic determinism tests in phytogeography	Evaluates whether climate alone can delimit phytogeographic regions and shows climatic signals matter but do not fully resolve all floristic boundaries (Djamal et al., 2012)
Global Bioclimatic Classification applied to Iran	Proposes an "Irano-Turanian" or "xero-estival-continental" bioclimate distinct from Mediterranean and tropical deserts, thus providing a climate-based delimitation in Iran (Djamali et al., 2011)
Provincial floristic inventory (Khorassan-Kopet Dagh)	Treats transitional provinces as corridors with mixed Irano-Turanian, Euro-Siberian and local endemics, underlining ecotonal limits (Golvan & Rioux, 1961)

3. CLIMATE, TOPOGRAPHY, SOIL, AND HYDROLOGY

3.1 Climate and Topography

Climatically, the IT domain is characterized by continentality and aridity. Lowland basins commonly receive very limited precipitation (often <100 mm yr^{-1} in hyper-arid depressions), whereas windward mountain slopes receive substantially more (several hundred mm yr^{-1}), producing distinct rainfall and vegetation belts (Beaumont et al., 1976; Djamali et al., 2011). The seasonal precipitation cycle is strongly skewed toward late autumn–spring (with winter and spring maxima), accompanied by long, hot, dry summers and cold winters (Djamali et al., 2011). High potential evapotranspiration and intermittent precipitation lead to soil salinization, episodic river flow and extensive semi-desert and desert systems across plateau interiors (Roozitalab et al., 2018).

Topography is the primary modulator of microclimate and hydrology across the IT region: orographic uplift on windward mountain slopes enhances precipitation and creates refugial montane habitats, while rain shadows on leeward sides drive desertification in intermontane basins (Memariani et al., 2016). Elevation-driven climatic stratification supports a mosaic of xeric shrub-steppe and semi-desert at low–mid elevations, montane steppes and woodlands at intermediate elevations, and alpine communities at the highest altitudes. This steep environmental heterogeneity underpins the region's high plant endemism and phytogeographic complexity (Memariani, Zarrinpour, et al., 2016).

Recent remote-sensing and climate modeling have refined our understanding of these patterns and their trajectories under climate change. Country-wide MODIS NDVI analyses reveal spatially variable vegetation trends in Iran (2000–2020), with greening hotspots along certain mountain belts (e.g. western Zagros) but declines elsewhere, highlighting topography-mediated vegetation responses and the utility of time-series remote sensing for monitoring functional vegetation change (Ghorbanian et al., 2022). Snow-cover and land-cover change studies in basins draining the Alborz and Zagros confirm strong seasonality of snow resources and trends that affect runoff seasonality and water availability for montane ecosystems (Entezami et al., 2024). Regional climate model and CMIP6 assessments show robust warming and increasing extremes across Central and West Asia, with projections of increased heat-wave duration, greater consecutive dry days, and shifts in precipitation extremes all of which interact with IT topography to amplify regional drought risk and alter ecohydrological regimes (Afsari et al., 2024; Fallah et al., 2024; Qiu et al., 2023; Rai et al., 2024).

Taken together, classical phytogeographic syntheses and recent remote-sensing and climate-model literature portray the Irano–Turanian Region as a climatically stressed but topographically complex area where mountains function both as biodiversity refugia and as critical controllers of regional hydrology. This makes the IT region a key natural laboratory for studying how plant functional traits and community composition respond to interacting gradients of temperature, moisture, and elevation and emphasizes the importance of integrating high-resolution topographic data, long-term satellite time-series, and downscaled climate projections in future biodiversity and conservation research.

3.2 Soil Characteristics and Hydrology

3.2.1. Soil Characteristics:

Soils in the IT region are largely aridic and xeric, with low organic carbon content, high carbonate and gypsum accumulations, and widespread salinization. Salt-affected soils are often found in intermontane basins and low-lying plains, where electrical conductivity (EC) and exchangeable sodium are the most common indicators of soil degradation (Fathizad et al., 2020; Matinzadeh et al., 2022). Alluvial deposition, aeolian processes, and microtopography make the soil highly heterogeneous and create steep gradients in texture, salinity and fertility over short distances (Mahmoudabadi et al., 2017; Wang et al., 2024). Carbonate and gypsum layers found in many basins indicate a lack of rainfall and large evapotranspiration, which limit leaching and increase the accretion of soluble salts (Basak et al., 2022). The poor soil organic carbon (SOC) is also a typical limitation particularly in desert plains where there is a low supply of biomass and high soil erosion (Roozitalab et al., 2018). SOC values are somewhat greater in montane regions due to the presence of vegetation cover and minimal evaporation (Memariani, Zarrinpour, et al., 2016). Remote sensing combined with machine-learning algorithms has increasingly been used to map soil salinity, EC, and texture across the region. Studies using Sentinel-2 and Landsat imagery have accurately predicted salinity patterns in central Iran and surrounding deserts, facilitating the identification of areas vulnerable to degradation (Wang et al., 2024; Wang et al., 2021).

3.2.2. Hydrology:

The IT region has low annual precipitation (less than 100mm in desert basins; 400-600mm on mountain slopes), very variable year-to-year rainfall, and strong seasonality with high precipitation during late autumn and spring (Beaumont et al., 1976). According to the latest research, groundwater recharge in Iran is declining at an average rate of around 3.8 mm yr^{-1} on average over 200217, and this is mainly related to the extraction

intensified by irrigation, aggravated by climate variation (Noori et al., 2023). Regional variability has been highlighted using remote sensing (GRACE) and distributed hydrological models (WetSpass-M, MODFLOW): mountain headwaters and northern basins retain relatively high recharge, while central and southern deserts experience extreme losses (Ghorbanian et al., 2022; Safdari et al., 2022).

In most basins, capillary rise is caused by shallow groundwater, which increases surface salinity and forms a feedback loop with soil erosion and vegetative stress (Feizizadeh et al., 2022). The drying of salty lakes (e.g. Lake Urmia) drying and the exposure of salty lakebeds contribute to increased salt redistribution and dust emission, underscoring the interconnection between hydrology, soil salinity, and ecosystem health (Feizizadeh et al., 2022).

3.3. Climate Change Impacts on Soil and Hydrology:

Climate change effects on soil and hydrology, as predicted by CMIP6 and regional modeling, suggest a warming climate, more frequent heat waves, and altered precipitation patterns throughout the IT area (Fallah et al., 2024; Qiu et al., 2023; Rai et al., 2024). Such alterations will increase soil salinization, decrease groundwater recharge, and heighten the threat of desertification. Remote sensing research has shown increased green cover in certain mountainous areas and decreased vegetation in lowland basins, indicating the vulnerability of the plant-soil-water nexus to climate variability extremes (Behroozeh et al., 2025).

Sooner or later, integrated soil-water-vegetation monitoring, a combination of field measurements, remote sensing, and hydrological modeling, tends to be applied to detect groundwater degradation hotspots, quantify the lack of recharge, and inform sustainable land management interventions. Nonetheless, significant data gaps remain regarding deeper soil profiles, historical groundwater abstraction, and the association between the human-water system and soil salinity trends (Norouzi Nazar et al., 2023).

4. ALPINE ZONES WITHIN THE IRANO-TURANIAN REGION

4.1 Definition and Characteristics of Alpine Zones

In the IT context, the alpine zone is defined as the area above the climatic or tree line where woody vegetation is absent and herbaceous or cushion life-forms dominate. In the IT region, the alpine zone refers to the area above the tree line where no woody vegetation grows and herbaceous or cushion plants are the predominant life forms. The high altitudes of major mountain ranges in the region, such as the Zagros, Alborz, Kopet-Dag, Hindu Kush, and Pamir, are covered by alpine zones in the Irano- Turanian (IT) region. These areas are generally referred to as regions above the upper tree line, where the climate is so severe that forest cover cannot grow and instead supports herbaceous, dwarf-shrub, and cushion vegetation (Akhani et al., 2010; Noroozi et al., 2019). The key characteristics of Alpine Zones in the Irano-Turanian region include:

1. Harsh Climate

Harsh climatic conditions that define alpine zones include low mean annual temperatures, prolonged snow cover, extreme freeze-thaw periods, and very brief growing periods, typically lasting only a few weeks (Körner, 2021). Regional precipitation is unevenly distributed; the Alborz Mountains are relatively humid due to the influence of the Caspian, while the Zagros and Central Iranian ranges are drier and more continental (Nazari et al., 2022). The soils in these areas are generally shallow, rocky, poorly developed, and mostly calcareous, having been eroded and formed through cryogenic activities. Alpine soils are normally shallow, coarse, calcareous, and poorly developed as a result of erosion and cryogenic processes (Noroozi et al., 2018).

2. Specialized Vegetation Structure

The alpine belt is highly specialized with vegetation dominated by hemicryptophytes, cushion plants and cold or drought tolerant dwarf shrubs, including the genera, *Acantholimon, Astragalus, Dionysia* and *Aethionema.* These plants are frequently small and excessively pubescent, with deep roots to withstand harsh conditions (Körner, 2021). The IT region exhibits a high level of endemism in alpine flora, with many species found only in isolated high-mountain areas (Forster et al., 2023; Razbani et al., 2023). More than 50% of the vascular plant species in these areas are narrow endemics, indicating long-term isolation and the role of mountains as refugia (Ramzi et al., 2024).

3. Ecological and Hydrological Importance

Alpine areas also provide essential ecological and hydrological services. They are considered water towers because they regulate snowmelt and supply water to rivers and groundwater systems in arid and semi-arid lowlands (Akhani & Ejtehadi, 2022). However, these areas face growing threats from climate change, overgrazing, unsustainable wild plant harvesting, and insufficient protection within current conservation networks. Climate warming is pushing alpine species to higher elevations, shrinking available habitat, particularly for narrow endemics, and threatening their long-term survival (Behroozian et al., 2025; Najafi et al., 2023).

4. 2 Major Alpine Ecoregions

4.2.1. Alborz Mountains Alpine Ecoregion

The Alborz range, which runs northward along the southern bank of the Caspian Sea in Iran, features extensive alpine and subalpine terrain up to 4800 m. Because of the influence of the Caspian region, the area is very wet with humidity that favors cushion plants, perennial herbs and hemicryptophytes that grow in cold, wet, and short seasons. The Alborz AE is considered one of the hotspots of alpine endemism in the Irano Turanian region (Noroozi & Körner, 2018).

4.2.2. Zagros Mountains Alpine Ecoregion

In western Iran, the highlands are found in Alpine habitats located mainly on high ridges that rise over 3,500 m above sea level, in the Zagros Mountains. The vegetation includes alpine meadows, dwarf shrubs, and rosette herbs, and has adapted to a continental climate with dry summers and winter frost. The endemism is also important, making the Zagros AE a major biodiversity hotspot in the region (Yaselyani et al., 2024).

4.2.3. Khorassan–Kopet Dag Alpine Ecoregion

This ecoregion covers northeastern Iran and southern Turkmenistan, including the Kopet Dag and Hezar Masjed massifs. Alpine areas here are remote and continental with extreme seasonal variations and scarce rainfall. The vegetation is characterized by endemic and range-limited species, indicating centuries of isolation and evolutionary divergence (Noroozi et al., 2019).

4.2.4. Yazd–Kerman Massifs Alpine Ecoregion

Central and southern Iran (highlands), including Yazd and Kerman, is drier and more isolated alpine terrain. These massifs have a distinct flora, with some species being local endemics, and were designated as a distinct Alpine Endemic Area, despite the severe arid conditions (Basak et al., 2022).

4.3. Endemic and Rare Flora

The Irano-Turanian (IT) region is a significant center of floral diversity and endemism. Its complex landscape of mountains, plateaus, and dry plains has created a diverse range of ecological habitats, resulting in the development of many endemic and rare plant species. Endemism is especially pronounced in high mountains where remote, high elevation habitats are home species highly adapted to severe climatic conditions (Noroozi et al., 2019).

4.4. Patterns of Endemism and Rare Flora

The main centers of endemism in the IT region are the Alborz, Zagros, and Kopet-Dagh mountain ranges. Geographic isolation and reduced dispersal are examples of geographic barriers, because endemic species are often confined to a single mountain massif or valley. Hemicryptophytes, cushion plants, and dwarf shrubs dominate the alpine and subalpine areas, indicating adaptation to short growing periods, frost, and poor soils (Noroozi & Körner, 2018). Recent research has shown that taxa once believed to be widespread are usually cryptic, narrow-endemic species. For example several new species within the genus *Aethionema* (Brassicaceae) have been described in the Zagros Mountains, each confined to a particular ridge. Similarly, the long-lost alpine plant *Arenaria bulica* Stapf ex F.N.Williams was rediscovered after more than 100 years, underscoring the necessity of conducting floristic surveys in remote mountainous regions (Pahlevani & Falatoury, 2024).

4.5. Drivers of Endemism

A combination of topographic, climatic, historical, and ecological factors is the primary cause of endemism and rarity in the Irano-Turanian region. Isolated habitats in the area's mountain ranges promote allopatric speciation, while variations in precipitation, temperature, and snow cover along elevational gradients create a species-rich mosaic of ecological niches (Noroozi et al., 2019). Moreover, the harsh environmental factors of alpine and arid regions, including extreme temperatures, short growing seasons, and low soil nutrient content–exert strong ecological pressure on the specialization of life-forms, often resulting in ecological narrowness and the presence of endemic and rare taxa (Noroozi & Körner, 2018).

5. HOTSPOT OF ALPINE PLANT DIVERSITY

5.1 Hindu Kush–Karakoram–Pamir Regions

The Hindu Kush Karakoram Pamir (HKKP) region is one of the most significant high-mountain systems in the world, spanning Afghanistan, Pakistan, China, and Tajikistan. It harbors much of the tallest and most widely frozen landscape beyond the polar regions and provides some of the most vital eco system, services particularly water resources, to downstream areas (Chettri et al., 2019). The HKKP is a region of biophysical complexity subject to constant tectonic uplift, and is a key area for studying climate change, water security, and mountain hazards (Schild & Sharma, 2011).

The Hindu Kush Karakoram Pamir (HKKP) is the most ecologically diverse high-mountain system on Earth, formed by steep altitude gradients, glacial geomorphology, and complex climate interactions. It features sharp environmental gradients, i.e., dry valleys alongside permanently glaciated mountains, which support a variety of vegetation belts, each with distinct ecological processes and species pools (Schild & Sharma, 2011; Spehn et al., 2012). This combination of climatic heterogeneity and geographic isolation has resulted in high beta-diversity and the formation of numerous glacial refugia that facilitated speciation during Quaternary climate fluctuations. (Profile, 2017).

The elevation, precipitation gradient and ancient biogeographical barriers strongly structure the HKKP flora, creating remarkable pockets of endemism. Cold-tolerant species like *Saussurea gnaphalodes* (Royle ex DC.) Sch.Bip., *Acantholimon diapensioides* Boiss. and *Rheum tibeticum* Maxim. ex Hook.f. occur in the alpine and subnival belts, many of which evolved under long-term unstable climatic conditions (Liu et al., 2021). The alpine meadows of the region are reported to be among the most species-rich high-altitude ecosystems in the world (Spehn et al., 2012; Sundriyal & Sharma, 2016).

Juniper woodlands, including *Juniperus seravschanica* (K.Koch) Takht., *Juniperus excels* Willd., and *Juniperus macropoda* Boiss., are relics of epiphytic forests with high ecological and cultural value (Rahmonov et al., 2017). Palaeoecological examinations indicate that these juniper stands have survived glacial and interglacial periods, indicating great resilience but also extreme susceptibility to modern warming (Dodonov & Baiguzina, 1995).

The most common cause of ecological change in the HKKP is climate change. Numerous high-mountain regions worldwide have warmed more rapidly than the global average, leading to upslope shifts in plant communities, changes in phenological cycles, and loss of habitat for cold-specialized fauna (Maharjan et al., 2023; Pepin et al., 2015). Model projections predict significant losses of suitable habitats for alpine species such as *Saussurea* subsp. and snow leopard prey (Terzi et al., 2019).

Intensification of land-use, such as overgrazing, fuelwood harvesting, increasing infrastructure, and uncontrolled tourism, is disintegrating habitats and altering local ecological processes (Ali et al., 2025). Also, invasive species like *Cirsium arvense* (L.) Scop. and *Caragana* subsp. are colonizing increasingly disturbed habitats; they compete with native plants and make ecosystems more susceptible (Zargar et al., 2025).

5.2 Zagros Mountains

One of the most important orogenic belts in the Middle East is the Zagros Mountains which extend from the Strait of Hormuz in western Iran along for about 1600km, as a continuation of the southeastern part of Turkey. This range is an active continental collision zone, where deformation is mainly driven by the ongoing northward movement of the Arabian plate into the Eurasian plate (Agard et al., 2011; Berberian & King, 1981). These long-term tectonic processes have created a typical fold-and-thrust belt, large scale anticlines, and sedimentary basins rich in petroleum, making the Zagros region one of the most vital hydrocarbon provinces in the world (Allen & Talebian, 2011).

The Zagros Mountains feature a wide range of biomes, including oak-rich woodlands (*Quercus brantii* Lindl. and *Quercus infectoria* Oliv.) in the central and northern ranges, and the steppe and semi-arid shrublands in the south (Zehzad et al., 2002). The region supports diverse flora and fauna, such as the Persian fallow deer (*Dama mesopotamica*), Persian leopard (*Panthera pardus saxicolor*), and various endemic plant species shaped by the area's climatic gradients and topographic complexity (Akhani, 2006; Kiabi et al., 2002). Nevertheless, the impact of human activities on biodiversity and ecosystem stability is increasing, with overgrazing, deforestation, and climate change as the main threats (Talebi et al., 2014).

The mountains remain an important socio-ecological landscape that sustains pastoralist communities, rain-fed agriculture, and substantial water resources in western Iran today (Fattahi et al., 2019).

5.3 Alborz Mountains

The Alborz Mountains, which extend over 900 km along northern Iran from Azerbaijan in the northwest to the Kopet Dag in the northeast, are one of the most significant orogenic systems in Southwest Asia. This mountain chain represents the natural boundary between the Iranian central plateau and the Caspian Sea basin, creating steep climatic patterns and clear ecological belts (Kehl, 2009; Massah &

Saadi, 2021). Its highest point is Mount Damavand (5,671 m), a potentially active stratovolcano and the tallest peak in the Middle East (Davidson et al., 2004).

Geologically, the Alborz Mountains from an intricate convergence zone resulting from the interaction of the Arabian, Eurasian, and Central Iranian microplates. They are characterized by folded Mesozoic and Cenozoic sedimentary sequences, intrusive igneous rocks, and Neogene volcanic units (Naeimi et al., 2022). The region's tectonic history has produced high seismicity, with frequent earthquakes associated with active thrusting and strike-slip faulting along the North Alborz and South Alborz fault systems (Tourani et al., 2024).

Ecologically, the Alborz Mountains host one of Iran's richest biodiversity zones. The northern slopes, influenced by Caspian Sea humidity, support Hyrcanian broadleaf forests, dominated by *Fagus orientalis* Lipsky, *Quercus castaneifolia* Coss. ex J.Gay, *Carpinus betulus* L., and *Parrotia persica* C.A.Mey., the latter a Tertiary relict species with high evolutionary significance (Akhani et al., 2010). In contrast, the southern slopes transition into semi-arid steppe vegetation, illustrating how orographic effects shape sharp ecological boundaries. The range sustains diverse wildlife, including the Persian leopard (*Panthera pardus saxicolor*), bezoar goat (*Capra aegagrus*), brown bear (*Ursus arctos*), and endemic amphibians such as *Rana pseudodalmatina* (Kiabi et al., 1999; Rastegar-Pouyani et al., 2015). However, increasing anthropogenic pressures, deforestation, road construction, unregulated tourism, and climate-driven shifts in vegetation pose substantial risks to ecosystem integrity and long-term conservation.

5.4 Kopet Dag and Khorasan Highlands

One of the most important mountain systems located in the southwestern part of the Eurasian continent is the Kopet Dag Mountains (also spelled Koppeh Dagh or Kospetdag), which form a natural border between northeastern Iran and southwestern Turkmenistan. The Kopet Dag, approximately 650 km long in a northwest-southeast direction, resulted from the collision between the Arabian and Eurasian plates, with deformation occurring along major thrust and strike-slip faults (Izyumov & Kuzmin, 2014). Its extent consists of folded Mesozoic-Cenozoic sedimentary series and active fault zones, which contribute to the high seismicity of both Iranian Khorasan and Turkmenistan (Tchalenko, 1975).

Ecologically, Kopet Dag is located at the convergence of the Irano-Turanian, Central Asian, and Mediterranean floristic regions, resulting in high landscape heterogeneity and high endemism. The vegetation varies between Artemisia steppe communities at low elevations to juniper woodlands (*Juniperus polycarpos* K.Koch) and sparse montane scrub at higher altitudes (Memariani, Zarrinpour, et al., 2016). The area is home to several endemic and relict species associated with Pleistocene climate fluctuations. It also supports wildlife such as the Persian leopard (*Panthera pardus saxicolor*) and wild goat (*Capra aegagrus*), which depend on mountainous conditions and discontinuous montane ecosystems (Noori et al., 2024; Yusefi et al., 2019).

West of the Kopet Dag lies a larger formation comprising the Khorasan Highlands, Binalud, Hezar-Masjed, and Quchan mountain ranges in northeastern Iran. These highlands create an ecotone between the Iranian Plateau and the Central Asian steppe belt, resulting in creating ecological distances with different climate gradients and biogeographical shifts (Noroozi et al., 2008). The Binalud Mountains, with elevations exceeding 3,200 m, are especially characterized by cold alpine meadows, juniper stands, and a steppe forest mosaic due to the orographic influence of the Caspian and Central Asian weather systems (Memariani, Zarrinpour, et al., 2016).

The Khorasan Highlands are an essential component of hydrology, supplying endorheic basins (Kashafrud and Atrak catchments) that support agriculture and the urban communities of Mashhad, Nishapur, and Bojnurd (Khalili et al., 2016). Nonetheless, soil erosion, habitat fragmentation, and long-term ecological degradation have resulted from overgrazing, deforestation, and frequent drought (Talebi et al., 2014). Changes in weather conditions, rising temperatures, and altered rainfall patterns also threaten the resilience of the region's steppe, montane, and woodland ecosystems (Tabari & Talaee, 2014).

Figure 1. Alpine Plants Hotspots of the Irano-Turanian Region

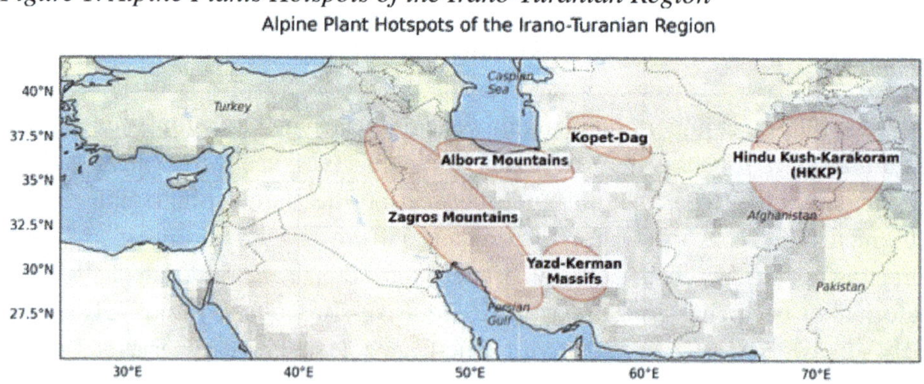

5.5 Other Notable Micro-Hotspots

In addition to the major mountain systems such as the Zagros, Alborz, Kopet Dag, and Khorasan Highlands, several smaller mountain ranges, isolated plateaus, and valley systems act as micro-hotspots of biodiversity and endemism. These areas often harbor species-rich and relict flora and fauna, reflecting historical refugia, unique climatic conditions, and topographic isolation (Nobis et al., 2015).

6. THREATS TO ALPINE PLANT HOTSPOTS

6.1 Climate Change and Glacial Retreat

The Alpine biodiversity hotspots are increasingly threatened as climate warming pushes treelines higher and accelerates glacial retreat, causing habitats to be squeezed by advancing boundaries. Although newly uncovered terrain can temporarily support a surge of pioneer species, these benefits are short-lived. In the long term, woody encroachment, hydrological changes, biological invasions, and collapses in plant-pollinator networks lead to sustained reductions in species richness and ecological processes (Bektaş, 2024). Effective conservation responses therefore require coordinated actions focused on habitat protection, invasive species management, hydrological stabilization, and long-term ecosystems management (BOSTANABAD).

6.2 Habitat Fragmentation and Overgrazing

There is extensive ecological degradation in alpine biodiversity hotspots, which are being fragmented and overgrazed. Species richness is declining, functional composition is changing, and natural regeneration processes are weakening. The main drivers of these pressures are increased road networks, increased livestock activities, and the conversion of alpine habitats to agricultural or residential use. New conservation strategies, especially participatory decision-making, managed grazing regimes, and remote-sensing based monitoring, show promise in reducing these effects and supporting the long-term recovery of ecosystems (Chemini & Rizzoli, 2003).

6.3 Invasive Species and Land Use Change

Combined invasive species pressure, intensified land use, and accelerating climate change are endangering Alpine plant biodiversity hotspots. Increased temperatures and more frequent disturbance regimes associated with rapid land-use expansion have been found to heighten the vulnerability of high-elevation plant communities to biological invasions. For example the upward movement and strong competitive ability of exotic species, like *Senecio inaequidens* DC., in the European Alps illustrate this trend (Haider et al., 2022). The climate change coupled with human activities in the Tibetan-Himalayas region, also increases the threat to endemic alpine plants. Scenarios with high carbon emissions predict disastrous future losses, and the open spaces of current reserves contain less than a quarter of the potential victims, highlighting the need to develop more extensive preservation strategies (Wei & Wang, 2025). Simultaneously, there is an increase in species at high-altitude areas driven by climate-induced phenological advances and elevational range shifts, following accelerating habitat fragmentation and compromising the stability of local alpine communities (Lepcha, 2024). In South Africa, mountain systems exhibit similar patterns, with invasive woody species like *Acacia, Pinus,* and *Prosopis* playing major roles in woody encroachment and ecosystem transformation (Canavan et al., 2021). Together, these observations underscore the increasing susceptibility of alpine ecosystems to combined anthropogenic and climatic stressors, justifying the need to implement conservation actions, enhance the network of protected areas, and pursue overall climate change mitigation to preserve alpine biodiversity.

6.4 Unsustainable Collection of Medicinal Plants

The Alpine biodiversity hotspots face multiple threat drivers that reduce habitat areas, decrease population sizes, and weaken ecosystem resilience. In most mountainous areas, unsustainable harvesting of medicinal plants, combined with land-use pressures and climate-related stressors, is a key factor contributing to biodiversity loss. Large- scale extraction of wild medicinal species occurs due to high commercial demand, and in numerous cases, destructive harvesting methods – including the removal of entire plants or roots and widespread illegal harvesting have been recorded. There is also clear evidence that most medicinal raw materials are still collected from the wild, with a significant portion extracted using destructive techniques, sharply increasing the extinction risk for high-value species (Shinwari et al., 2024). Trade occurs across national borders, and many species are extracted in remote, rugged parts of the Alps, further intensifying pressure on these resources (Mishra et al., 2024).

7. RESEARCH GAPS AND FUTURE DIRECTIONS

Although botanical and ecological studies have been conducted for decades on the Iranian Plateau and neighboring mountain complexes, much remains unknown about the overall distribution, spread, and intensity of biospecies hotspots. To enable effective conservation planning amid accelerating climate and land-use change, these gaps must addressed.

7.1 Need for Comprehensive Floristic Surveys

Among the most significant research gaps across the Zagros, Alborz, Kopet Dag and Khorasan Highlands, and related micro-hotspots, is the lack of accessible, standardized floristic inventories. The current floristic records are often founded on scattered collections or historic expeditions from the middle of the 20 th century, leaving vast areas under-surveyed or completely unrecorded (Akhani, 2006; D. Zohary, 1973).This is especially problematic in isolated alpine areas, border areas, and politically volatile areas, where entrance has long been restricted.

The future research should focus on systematic, geographically abbreviated floristic surveys using standardized sampling protocols, supported by high-resolution georeferencing and open-access data technologies. Integrating herbarium digitization with remote sensing and species distribution models will play a vital role in identifying centers of endemism and conservation priority regions (Makki et al., 2023).

7.2 Integrating Traditional Knowledge with Science

Another significant research agenda concerns the combination of indigenous and local ecological knowledge (LEK) with formal scientific research. Individuals in pastoralist, farming, and local healer systems in the Zagros, Alborz, and Khorasan areas have centuries of experience with plant phenology, medicinal uses, grazing practices and climatic changes, gained through their interaction with the mountain environments (Shinwari & Gilani, 2003).

Nevertheless, this information is not often conventionally recorded or included in biodiversity measurements and conservation plans. Ethnobotanical literature shows that traditional land-use activities e.g. rotational grazing, seasonal migration, and selective harvesting may increase ecosystem resilience and support plant diversity when managed sustainably (Berkes et al., 2000).

To shape future research, participatory and transdisciplinary methodologies that incorporate ethnobotany, landscape ecology and conservation biology should be implemented. The relevance, legitimacy, and sustainability of conservation efforts can be enhanced through co-production of knowledge with local communities, especially in micro-hotspots where formal protection is weak (Lee et al., 2023).

7.3 Monitoring Climate Change Impacts on Hotspots

The largest long-term risk to mountain biodiversity hotspots is climate change, but long-term ecological surveillance coverage is limited in most of Iran and Central Asia. There are signs of increased warming in mountainous areas, resulting in a rise in vegetation zones, changes in phenology, and retreating habitats for cold-adapted species ("Elevation-dependent warming in mountain regions of the world", 2015; Urbach et al., 2024).

Although these trends are present, there are little permanent monitoring plots or climate-biodiversity observatories in the Zagros and Alborz or Kopet Dag. This reduces the ability to detect early warning signs of ecosystem degradation, including decreases in alpine species richness or shifts in community structure (Shrestha et al., 2012).

The future directions will focus on integrated monitoring frameworks that combine *in situ* ecological observation, automated climate stations, remote sensing, and species distribution modeling. Establishing elevational transects and long-term plots in both macro and micro-hotspots will enable robust evaluation of climate effects and adaptive potential (Farinotti et al., 2020). These data are necessary to create climate-sensitive conservation interventions, such as assisted migration to ecological corridors and dynamic networks of protected areas.

8. CONCLUSION

The Irano- Turanian region is one of the largest botanical sources of alpine flora. The region is highly diverse in terms of environment and topography, and it has a long evolutionary background. This chapter has shown that the distribution of alpine plant hotspots in this region is not entirely random but is closely related to mountain systems, altitudinal gradients, climatic transitions, and edaphic specialization. The combination of these factors has facilitated high species richness, spectacular endemism, and the survival of relict taxa in extreme alpine habitats. The discovery of alpine plant hotspots in the Irano–Turanian region suggests the significance of mountain refugia in preserving genetic material and sustaining unique plant assemblages. The ecological amplitude and local distributions of many alpine species are limited, making them highly sensitive to environmental changes. These delicate ecosystems are increasingly threatened by climate warming, changes in precipitation regimes, overgrazing, land-use changes, and unregulated harvesting, which may lead to the contraction and extinction of endemic species. This chapter highlights the value of combining floristic inventories, biogeographical studies and contemporary spatial applications to provide a clear definition of biodiversity hotspots and identify where to focus conservation activities. The conservation plans should consider current trends in species richness as well as address future changes in species ranges due to climate change. Transboundary cooperation, long-term ecological observation, and community-based management are especially important in the Irano-Turanian region, where alpine landscapes cross political boundaries and support ancient ways of life. To sum up, the alpine plant hotspots of the Irano-Turanian region represent irreplaceable reservoirs of biodiversity and evolutionary heritage. Their preservation is critical not only for maintaining role for sustaining of regional and global plant diversity but also in preserving ecosystem services and ecological resilience under rapidly changing climatic conditions. The identification, preservation, and long-term management of these hotspots must therefore become primary concerns for biodiversity conservation and alpine research in the area.

REFERENCES

Afsari, R., Nazari-Sharabian, M., Hosseini, A., & Karakouzian, M. (2024). A CMIP6 multi-model analysis of the impact of climate change on severe meteorological droughts through multiple drought indices—Case study of Iran's metropolises. *Water (Basel), 16*(5), 711. DOI: 10.3390/w16050711

Agard, P., Omrani, J., Jolivet, L., Whitechurch, H., Vrielynck, B., Spakman, W., Monié, P., Meyer, B., & Wortel, R. (2011). Zagros orogeny: A subduction-dominated process. *Geological Magazine, 148*(5-6), 692–725. DOI: 10.1017/S001675681100046X

Akhani, H. (2006). Biodiversity of halophytic and sabkha ecosystems in Iran. In *Sabkha Ecosystems: Volume II: West and Central Asia* (pp. 71-88). Springer. DOI: 10.1007/978-1-4020-5072-5_6

Akhani, H., Djamali, M., Ghorbanalizadeh, A., & Ramezani, E. (2010). Plant biodiversity of Hyrcanian relict forests, N Iran: An overview of the flora, vegetation, palaeoecology and conservation. *Pakistan Journal of Botany, 42*(1), 231–258.

Ali, M. F., Ahmad, M., Mukhtar, L., Ali, N., Raza, M. Y., Qadir, G., . . . Ahmad, S. (2025). Climate Change and Biodiversity. In *Climate Resilient and Sustainable Agriculture: Volume 1: Adaptation and Mitigation Strategies* (pp. 23-47). Springer. DOI: 10.1007/978-3-032-00190-0_2

Allen, M. B., & Talebian, M. (2011). Structural variation along the Zagros and the nature of the Dezful Embayment. *Geological Magazine, 148*(5-6), 911–924. DOI: 10.1017/S0016756811000318

Basak, N., Rai, A. K., Barman, A., Mandal, S., Sundha, P., Bedwal, S., & Sharma, P. C. (2022). Salt affected soils: Global perspectives. In *Soil health and environmental sustainability: Application of Geospatial Technology* (pp. 107–129). Springer. DOI: 10.1007/978-3-031-09270-1_6

Beaumont, P., Blake, G. H., & Wagstaff, J. M. (1976). *The Middle East: a Geographical Study.* Wiley.

Behroozeh, S., Khoorani, A., Damaneh, H. E., & Jokar Arsanjani, J. (2025). Monitoring vegetation dynamics across land use types in Iran: Spatiotemporal relationships with soil temperature and water volume. *Earth Science Informatics, 18*(2), 261. DOI: 10.1007/s12145-024-01645-4

Behroozian, M., Amini, T., Zare, H., & Ejtehadi, H. (2025). Assessing climate change impacts on the geographical distribution of Cupressus sempervirens in the Mediterranean region. *Scientific Reports*, *15*(1), 40127. DOI: 10.1038/s41598-025-23930-y PMID: 41249316

Bektaş, B. (2024). *Transient dynamics of Alpine grasslands linking biodiversity and ecosystem functions in a warming world*. Universite Grenoble Alpes.

Benito Alonso, J. L., & Villar Pérez, L. (2012). Continent-wide response of mountain vegetation to climate change.

Berberian, M., & King, G. (1981). Towards a paleogeography and tectonic evolution of Iran. *Canadian Journal of Earth Sciences*, *18*(2), 210–265. DOI: 10.1139/e81-019

Berkes, F., Colding, J., & Folke, C. (2000). Rediscovery of traditional ecological knowledge as adaptive management. *Ecological Applications*, *10*(5), 1251–1262. DOI: 10.1890/1051-0761(2000)010[1251:ROTEKA]2.0.CO;2

Blondel, J. (2010). *The Mediterranean region: biological diversity in space and time*. Oxford University Press.

BOSTANABAD, A. S. Biodiversity Report: Alpine Plants of the Southern Alps in New Zealand–Preserving High-Altitude Ecosystems.

Canavan, K., Canavan, S., Clark, V. R., Gwate, O., Richardson, D. M., Sutton, G. F., & Martin, G. D. (2021). The alien plants that threaten South Africa's mountain ecosystems. *Land (Basel)*, *10*(12), 1393. DOI: 10.3390/land10121393

Chemini, C., & Rizzoli, A. (2003). Land use change and biodiversity conservation in the Alps. *Journal of Mountain Ecology*, *7*(1), 1–7.

Chettri, N., Shrestha, A. B., & Sharma, E. (2019). Climate change trends and ecosystem resilience in the Hindu Kush Himalayas. In *Himalayan Weather and Climate and their Impact on the Environment* (pp. 525–552). Springer.

Davidson, J., Hassanzadeh, J., Berzins, R., Stockli, D. F., Bashukooh, B., Turrin, B., & Pandamouz, A. (2004). The geology of Damavand volcano, Alborz Mountains, northern Iran. *Geological Society of America Bulletin*, *116*(1-2), 16–29. DOI: 10.1130/B25344.1

Djamali, M., Akhani, H., Khoshravesh, R., Andrieu-Ponel, V., Ponel, P., & Brewer, S. (2011). Application of the global bioclimatic classification to Iran: Implications for understanding the modern vegetation and biogeography. *Ecologia Mediterranea*, *37*(1), 91–114. DOI: 10.3406/ecmed.2011.1350

Djamali, M., Baumel, A., Brewer, S., Jackson, S. T., Kadereit, J. W., López-Vinyallonga, S., Mehregan, I., Shabanian, E., & Simakova, A. (2012). Ecological implications of Cousinia Cass.(Asteraceae) persistence through the last two glacial–interglacial cycles in the continental Middle East for the Irano-Turanian flora. *Review of Palaeobotany and Palynology*, *172*, 10–20. DOI: 10.1016/j.revpalbo.2012.01.005

Djamali, M., Brewer, S., Breckle, S. W., & Jackson, S. T. (2012). Climatic determinism in phytogeographic regionalization: A test from the Irano-Turanian region, SW and Central Asia. *Flora-Morphology, Distribution. Flora (Jena)*, *207*(4), 237–249. DOI: 10.1016/j.flora.2012.01.009

Dodonov, A. E., & Baiguzina, L. (1995). Loess stratigraphy of Central Asia: Palaeoclimatic and palaeoenvironmental aspects. *Quaternary Science Reviews*, *14*(7-8), 707–720. DOI: 10.1016/0277-3791(95)00054-2

Elevation-dependent warming in mountain regions of the world. (2015). *Nature Climate Change, 5*(5), 424-430.

Entezami, H., Mojarrad, F., Shahabi, H., & Ghaderpour, E. (2024). Spatiotemporal Variability in Snow and Land Cover in Sefid-Rud Basin, Iran. *Sustainability (Basel)*, *16*(21), 9381. DOI: 10.3390/su16219381

Fallah, B., Didovets, I., Rostami, M., & Hamidi, M. (2024). Climate change impacts on Central Asia: Trends, extremes and future projections. *International Journal of Climatology*, *44*(10), 3191–3213. DOI: 10.1002/joc.8519

Farinotti, D., Immerzeel, W. W., de Kok, R. J., Quincey, D. J., & Dehecq, A. (2020). Manifestations and mechanisms of the Karakoram glacier Anomaly. *Nature Geoscience*, *13*(1), 8–16. DOI: 10.1038/s41561-019-0513-5 PMID: 31915463

Fathizad, H., Ardakani, M. A. H., Sodaiezadeh, H., Kerry, R., & Taghizadeh-Mehrjardi, R. (2020). Investigation of the spatial and temporal variation of soil salinity using random forests in the central desert of Iran. *Geoderma*, *365*, 114233. DOI: 10.1016/j.geoderma.2020.114233

Fattahi, Z., Beheshtian, M., Mohseni, M., Poustchi, H., Sellars, E., Nezhadi, S. H., & Jamali, P. (2019). Iranome: A catalog of genomic variations in the Iranian population. *Human Mutation*, *40*(11), 1968–1984. DOI: 10.1002/humu.23880 PMID: 31343797

Feizizadeh, B., Omarzadeh, D., Mohammadzadeh Alajujeh, K., Blaschke, T., & Makki, M. (2022). Impacts of the Urmia Lake drought on soil salinity and degradation risk: An integrated geoinformatics analysis and monitoring approach. *Remote Sensing (Basel)*, *14*(14), 3407. DOI: 10.3390/rs14143407

Forster, P. M., Smith, C. J., Walsh, T., Lamb, W. F., Lamboll, R., Hauser, M., & Palmer, M. D. (2023). Indicators of Global Climate Change 2022: Annual update of large-scale indicators of the state of the climate system and human influence. *Earth System Science Data, 15*(6), 2295–2327. DOI: 10.5194/essd-15-2295-2023

Ghorbanian, A., Mohammadzadeh, A., & Jamali, S. (2022). Linear and non-linear vegetation trend analysis throughout Iran using two decades of MODIS NDVI imagery. *Remote Sensing (Basel), 14*(15), 3683. DOI: 10.3390/rs14153683

Golvan, Y.-J., & Rioux, J.-A. (1961). Écologie des mérions du Kurdistan Iranien Relations avec l'épidémiologie de la Peste rurale-(Enquête de l'Institut Pasteur de l'Iran. Rapport préliminaire). *Annales de Parasitologie Humaine et Comparee, 36*(4), 449–588. DOI: 10.1051/parasite/1961364449 PMID: 13900030

Grabherr, G., Gottfried, M., & Pauli, H. (2010). Climate change impacts in alpine environments. *Geography Compass, 4*(8), 1133–1153. DOI: 10.1111/j.1749-8198.2010.00356.x

Grêt-Regamey, A., Brunner, S. H., & Kienast, F. (2012). Mountain ecosystem services: Who cares? *Mountain Research and Development, 32*(S1), S23–S34. DOI: 10.1659/MRD-JOURNAL-D-10-00115.S1

Haider, S., Palm, S., Bruelheide, H., de Villemereuil, P., Menzel, A., & Lachmuth, S. (2022). Disturbance and indirect effects of climate warming support a plant invader in mountains. *Oikos, 2022*(4), e08783. DOI: 10.1111/oik.08783

Izyumov, S., & Kuzmin, Y. O. (2014). Study of the recent geodynamic processes in the Kopet-Dag region. *Izvestiya. Izvestiya, Russian Academy of Sciences. Physics of the Solid Earth, 50*(6), 719–731. DOI: 10.1134/S1069351314060019

Kehl, M. (2009). Quaternary climate change in Iran—The state of knowledge. *Erdkunde, 63*(1), 1–17. DOI: 10.3112/erdkunde.2009.01.01

Khalili, K., Tahoudi, M. N., Mirabbasi, R., & Ahmadi, F. (2016). Investigation of spatial and temporal variability of precipitation in Iran over the last half century. *Stochastic Environmental Research and Risk Assessment, 30*(4), 1205–1221. DOI: 10.1007/s00477-015-1095-4

Kiabi, B. H., Abdoli, A., & Naderi, M. (1999). Status of the fish fauna in the South Caspian Basin of Iran. *Zoology in the Middle East, 18*(1), 57–65. DOI: 10.1080/09397140.1999.10637782

Kiabi, B. H., Dareshouri, B. F., Ghaemi, R. A., & Jahanshahi, M. (2002). Population status of the Persian leopard (Panthera pardus saxicolor Pocock, 1927) in Iran. *Zoology in the Middle East, 26*(1), 41–47. DOI: 10.1080/09397140.2002.10637920

Körner, C. (2021). The alpine life zone. In *Alpine Plant Life: Functional Plant Ecology of High Mountain Ecosystems* (pp. 23-51). Springer. DOI: 10.1007/978-3-030-59538-8_2

Körner, C. (2023). Concepts in alpine plant ecology. *Plants*, *12*(14), 2666. DOI: 10.3390/plants12142666 PMID: 37514280

Lee, H., Calvin, K., Dasgupta, D., Krinner, G., Mukherji, A., Thorne, P., ... & Park, Y. (2023). *IPCC, 2023: Climate change 2023: Synthesis report, summary for policymakers. Contribution of working groups i, II and III to the sixth assessment report of the intergovernmental panel on climate change [core writing team, h. Lee and j. Romero (eds.)]. IPCC, geneva, Switzerland* (pp. 1-34). Intergovernmental Panel on Climate Change (IPCC).

Lepcha, P. (2024). Climate change and its impact on mountainous plant species. *RE:view*.

Liu, J.-Q., Li, J.-L., & Lai, Y.-J. (2021). Plant diversity and ecology on the Qinghai–Tibet Plateau. *Journal of Systematics and Evolution*, *59*(6), 1139–1141. DOI: 10.1111/jse.12813

Maharjan, S. K., Sterck, F. J., Raes, N., Zhao, Y., & Poorter, L. (2023). Climate change induced elevational range shifts of Himalayan tree species. *Biotropica*, *55*(1), 53–69. DOI: 10.1111/btp.13159

Mahmoudabadi, E., Karimi, A., Haghnia, G. H., & Sepehr, A. (2017). Digital soil mapping using remote sensing indices, terrain attributes, and vegetation features in the rangelands of northeastern Iran. *Environmental Monitoring and Assessment*, *189*(10), 500. DOI: 10.1007/s10661-017-6197-7 PMID: 28894961

Makki, T., Mostafavi, H., Matkan, A. A., Aghighi, H., Valavi, R., Chee, Y. E., & Teimori, A. (2023). Impacts of climate change on the distribution of riverine endemic fish species in Iran, a biodiversity hotspot region. *Freshwater Biology*, *68*(6), 1007–1019. DOI: 10.1111/fwb.14081

Manafzadeh, S., Staedler, Y. M., & Conti, E. (2017). Visions of the past and dreams of the future in the O rient: The I rano-T uranian region from classical botany to evolutionary studies. *Biological Reviews of the Cambridge Philosophical Society*, *92*(3), 1365–1388. DOI: 10.1111/brv.12287 PMID: 27349491

Massah, B. A., & Saadi, T. (2021). Detection and attribution of changing in seasonal variability cause of climate change (Case study: hillsides of central Southern Alborz Mountains).

Matinzadeh, Z., López-Angulo, J., Escudero, A., Palacio, S., Abedi, M., & Akhani, H. (2022). Functional structure of plant communities along salinity gradients in Iranian salt marshes. *Plant-Environment Interactions*, *3*(1), 16–27. DOI: 10.1002/pei3.10070 PMID: 37283692

Memariani, F., Akhani, H., Joharchi, M. R., Khoshravesh, R., & Malekmohammadi, M. (2016). Plant diversity of the Khorassan-Kopet Dagh Floristic Province (Irano-Turanian Region). *Phytotaxa, 249*(1), 001-002.

Memariani, F., Zarrinpour, V., & Akhani, H. (2016). A review of plant diversity, vegetation, and phytogeography of the Khorassan-Kopet Dagh floristic province in the Irano-Turanian region (northeastern Iran–southern Turkmenistan). *Phytotaxa*, *249*(1), 8–30. DOI: 10.11646/phytotaxa.249.1.4

Mishra, A. P., Chandra, N., Kumar, A., Pande, C. B., & Singh, G. (2024). *Threatened Medicinal Plants in the Indian Himalayan Region*. Springer. DOI: 10.1007/978-3-031-73687-2

Mittermeier, R. A., Da Fonseca, G. A., Rylands, A. B., & Brandon, K. (2005). A brief history of biodiversity conservation in Brazil. *Conservation Biology*, *19*(3), 601–607. DOI: 10.1111/j.1523-1739.2005.00709.x

Nadeau, C. P., Urban, M. C., & Bridle, J. R. (2017). Climates past, present, and yet-to-come shape climate change vulnerabilities. *Trends in Ecology & Evolution*, *32*(10), 786–800. DOI: 10.1016/j.tree.2017.07.012 PMID: 28844791

Naeimi, A., Alavi, S. A., & Madanipour, S. (2022). Tectonic evolution of the foreland basin of the SE Alborz Mountains, northern Iran. *Journal of Asian Earth Sciences*, *223*, 104981. DOI: 10.1016/j.jseaes.2021.104981

Najafi, S., Sharafati, A., & Moghaddam, H. K. (2023). Impact of climate change adaptation strategies on groundwater resources: A case study of Sari-Neka coastal aquifer, Northern Iran. *Environmental Earth Sciences*, *82*(23), 571. DOI: 10.1007/s12665-023-11205-6

Nazari, Z., Ejtehadi, H., Mirshamsi, O., & Memariani, F. (2022). Modelling potential distribution of an endemic plant (Ferula Microcolea) based on bioclimatic variables in Iran. *Acta Botanica Hungarica*, *64*(1-2), 157–172. DOI: 10.1556/034.64.2022.1-2.8

Nobis, M., Ebel, A. L., Nowak, A., Paszko, B., Bobrov, A. A., Kotukhov, Y. A., & Olonova, M. V. (2015). Contribution to the flora of Asian and European countries: New national and regional vascular plant records, 4. *Acta Botanica Gallica*, *162*(4), 301–316. DOI: 10.1080/12538078.2015.1090329

Noori, R., Maghrebi, M., Jessen, S., Bateni, S. M., Heggy, E., Javadi, S., Noury, M., Pistre, S., Abolfathi, S., & AghaKouchak, A. (2023). Decline in Iran's groundwater recharge. *Nature Communications*, *14*(1), 6674. DOI: 10.1038/s41467-023-42411-2 PMID: 37865681

Noori, S., Zahiri, R., Yusefi, G. H., Rajabizadeh, M., Hawlitschek, O., Rakhshani, E., Husemann, M., & Rajaei, H. (2024). Patterns of zoological diversity in Iran—A review. *Diversity*, *16*(10), 621. DOI: 10.3390/d16100621

Noroozi, J., Akhani, H., & Breckle, S.-W. (2008). Biodiversity and phytogeography of the alpine flora of Iran. *Biodiversity and Conservation*, *17*(3), 493–521. DOI: 10.1007/s10531-007-9246-7

Noroozi, J., & Körner, C. (2018). A bioclimatic characterization of high elevation habitats in the Alborz mountains of Iran. *Alpine Botany*, *128*(1), 1–11. DOI: 10.1007/s00035-018-0202-9 PMID: 29576762

Noroozi, J., Talebi, A., Doostmohammadi, M., Manafzadeh, S., Asgarpour, Z., & Schneeweiss, G. M. (2019). Endemic diversity and distribution of the Iranian vascular flora across phytogeographical regions, biodiversity hotspots and areas of endemism. *Scientific Reports*, *9*(1), 12991. DOI: 10.1038/s41598-019-49417-1 PMID: 31506504

Noroozi, J., Talebi, A., Doostmohammadi, M., Rumpf, S. B., Linder, H. P., & Schneeweiss, G. M. (2018). Hotspots within a global biodiversity hotspot-areas of endemism are associated with high mountain ranges. *Scientific Reports*, *8*(1), 10345. DOI: 10.1038/s41598-018-28504-9 PMID: 29985437

Norouzi Nazar, M. S., Asadolahi, Z., Rabbani, F., Abbaspour, K. C., & Sakieh, Y. (2023). Modeling the integrated effects of landuse and climate change on the hydrologic response of Gorganroud watershed in Iran. *Theoretical and Applied Climatology*, *151*(3-4), 151. DOI: 10.1007/s00704-022-04345-5

Pahlevani, A., & Falatoury, A. N. (2024). Rediscover of the Iranian endemic alpine Arenaria bulica after 139 years and note of the related species (Caryophyllaceae). *Phytotaxa*, *646*(1), 47–57. DOI: 10.11646/phytotaxa.646.1.3

Pauli, H., Gottfried, M., Dullinger, S., Abdaladze, O., Akhalkatsi, M., Alonso, J. L. B., & Calzado, R. F. (2012). Recent plant diversity changes on Europe's mountain summits. *Science*, *336*(6079), 353–355. DOI: 10.1126/science.1219033 PMID: 22517860

Pepin, N., Bradley, R., Diaz, H., Baraer, M., Caceres, E., Forsythe, N., & Liu, X. (2015). Elevation-dependent warming in mountain regions of the world. [In.]. *Nature Climate Change*, *5*(5), 424–430. DOI: 10.1038/nclimate2563

Profile, E. (2017). Mountains of Central Asia Biodiversity Hotspot. *Draft for Submission to the CEPF Donor Council, 13.*

Qiu, Y., Yan, Z., Feng, J., Hua, L., Fan, L., Li, Z., Wang, J., & Qian, C. (2023). Robust historical and future drying trends in Central Asia evidenced by the latest observation and modeling datasets. *Atmospheric Research*, *295*, 107033. DOI: 10.1016/j.atmosres.2023.107033

Rahmonov, O., Rahmonov, M., Opała-Owczarek, M., Owczarek, P., Niedźwiedź, T., & Myga-Piątek, U. (2017). Ecological and cultural importance of juniper ecosystem in the area of Zeravshan valley (Tajikistan) on the background of environmental condition and anthropogenic hazards. *Geographia Polonica*, *90*(4), 441–461. DOI: 10.7163/GPol.0110

Rai, P., Bangelesa, F., Abel, D., Ziegler, K., Huang, J., Schaffhauser, T., Pollinger, F., Disse, M., & Paeth, H. (2024). Extreme precipitation and temperature indices under future climate change in central Asia based on CORDEX-CORE. *Theoretical and Applied Climatology*, *155*(7), 6015–6039. DOI: 10.1007/s00704-024-04976-w

Ramzi, S., Noroozi, J., Gholizadeh, H., Hamzeh'ee, B., Asri, Y., Talebi, A., & Zarezadeh, S. (2024). IranVeg–the Vegetation Database of Iran: Current status and the way forward. *Vegetation Classification and Survey*, *5*, 237–256. DOI: 10.3897/VCS.114081

Rastegar-Pouyani, N., Gholamifard, A., Karamiani, R., Bahmani, Z., Mobaraki, A., Abtin, E., & Sayyadi, F. (2015). Sustainable management of the Herpetofauna of the Iranian plateau and coastal Iran. *Amphibian & Reptile Conservation*, *9*(1), 1–15.

Razbani, E., Ghahremaninejad, F., & Joharchi, M. R. (2023). A Floristic Study of the Northern Slope of Bahar Mountain Region, North Khorasan Province, Iran. *Taxonomy and Biosystematics*, *15*(54), 21–40.

Roozitalab, M. H., Siadat, H., & Farshad, A. (2018). *The soils of Iran*. Springer. DOI: 10.1007/978-3-319-69048-3

Safdari, Z., Nahavandchi, H., & Joodaki, G. (2022). Estimation of groundwater depletion in Iran's catchments using well data. *Water (Basel)*, *14*(1), 131. DOI: 10.3390/w14010131

Scherrer, D., & Körner, C. (2011). Topographically controlled thermal-habitat differentiation buffers alpine plant diversity against climate warming. *Journal of Biogeography*, *38*(2), 406–416. DOI: 10.1111/j.1365-2699.2010.02407.x

Schild, A., & Sharma, E. (2011). Sustainable mountain development revisited. *Mountain Research and Development*, *31*(3), 237–241. DOI: 10.1659/MRD-JOURNAL-D-11-00069.1

Shinwari, A. K., Ahmad, M., Alsahli, A. A., Israr, A., Shinwari, Z. K., Shinwari, S., & Yilmaz, M. E. (2024). Himalayan medicinal plants wealth, threats and conservation for securing the future of biodiversity. *Pakistan Journal of Botany*, *56*(6), 2327–2350. DOI: 10.30848/PJB2024-6(42)

Shinwari, Z. K., & Gilani, S. S. (2003). Sustainable harvest of medicinal plants at Bulashbar Nullah, Astore (northern Pakistan). *Journal of Ethnopharmacology*, *84*(2-3), 289–298. DOI: 10.1016/S0378-8741(02)00333-1 PMID: 12648828

Shrestha, U. B., Gautam, S., & Bawa, K. S. (2012). Widespread climate change in the Himalayas and associated changes in local ecosystems. *PLoS One*, *7*(5), e36741. DOI: 10.1371/journal.pone.0036741 PMID: 22615804

Spehn, E. M., Rudmann-Maurer, K., Korner, C., & MaseLLi, D. (2012). Mountain biodiversity and global change.

Sundriyal, M., & Sharma, B. (2016). Status of biodiversity in Central Himalaya. *Applied Ecology and Environmental Sciences*, *4*(2), 37–43.

Tabari, H., & Talaee, P. H. (2014). Sensitivity of evapotranspiration to climatic change in different climates. *Global and Planetary Change*, *115*, 16–23. DOI: 10.1016/j.gloplacha.2014.01.006

Takhtajan, A. (1986). Florestic regions of the world.

Talebi, K. S., Sajedi, T., & Pourhashemi, M. (2014). Forests of Iran. *A treasure from the past, a hope for the future, 10*.

Tchalenko, J. (1975). Seismicity and structure of the Kopet Dagh (Iran, USSR). *Philosophical Transactions of the Royal Society of London. Series A, Mathematical and Physical Sciences*, *278*(1275), 1–28. DOI: 10.1098/rsta.1975.0019

Terzi, S., Torresan, S., Schneiderbauer, S., Critto, A., Zebisch, M., & Marcomini, A. (2019). Multi-risk assessment in mountain regions: A review of modelling approaches for climate change adaptation. *Journal of Environmental Management*, *232*, 759–771. DOI: 10.1016/j.jenvman.2018.11.100 PMID: 30529418

Testolin, R., Attorre, F., Borchardt, P., Brand, R. F., Bruelheide, H., Chytrý, M., & Haider, S. (2021). Global patterns and drivers of alpine plant species richness. *Global Ecology and Biogeography*, *30*(6), 1218–1231. DOI: 10.1111/geb.13297

Tourani, M., Isik, V., Saber, R., Caglayan, A., & Chitea, F. (2024). Evaluation of seismicity and seismotectonics in the Alborz Mountains: Insights from seismic parameters, Northern Iran. *Journal of Seismology*, *28*(3), 675–706. DOI: 10.1007/s10950-024-10218-3

Urbach, D., Körner, C., & Hilpold, A. (2024). Mountain biodiversity under change. In *Safeguarding Mountain Social-Ecological Systems* (pp. 11–15). Elsevier. DOI: 10.1016/B978-0-12-822095-5.00002-4

Wang, F., Han, L., Liu, L., Bai, C., Ao, J., Hu, H., Li, R., Li, X., Guo, X., & Wei, Y. (2024). Advancements and perspective in the quantitative assessment of soil salinity utilizing remote sensing and machine learning algorithms: A review. *Remote Sensing (Basel)*, *16*(24), 4812. DOI: 10.3390/rs16244812

Wang, J., Peng, J., Li, H., Yin, C., Liu, W., Wang, T., & Zhang, H. (2021). Soil salinity mapping using machine learning algorithms with the Sentinel-2 MSI in arid areas, China. *Remote Sensing (Basel)*, *13*(2), 305. DOI: 10.3390/rs13020305

Wei, J.-D., & Wang, W.-T. (2025). Vulnerability Assessment of Six Endemic Tibetan-Himalayan Plants Under Climate Change and Human Activities. *Plants*, *14*(15), 2424. DOI: 10.3390/plants14152424 PMID: 40805772

Yang, C., Xu, H., Li, Q., Wang, X., Tang, B., Chen, J., & Chen, M. (2025). Global loss of mountain vegetated landscapes and its impact on biodiversity conservation. *Nature Communications*, *16*(1), 8971. DOI: 10.1038/s41467-025-64449-0 PMID: 41068100

Yaselyani, M., Bagheri, A., Saeidi, H., & Noroozi, J. (2024). Vascular plant diversity of the high mountains of Fereydunshahr, Central Zagros, Iran. *Vegetation Classification and Survey*, *5*, 329–346. DOI: 10.3897/VCS.110546

Yusefi, G. H., Faizolahi, K., Darvish, J., Safi, K., & Brito, J. C. (2019). The species diversity, distribution, and conservation status of the terrestrial mammals of Iran. *Journal of Mammalogy*, *100*(1), 55–71. DOI: 10.1093/jmammal/gyz002

Zargar, S. A., Malik, R. A., Khuroo, A. A., Ganie, A. H., & Reshi, Z. A. (2025). Alien flora of the Himalayan highlands: Naturalised and invasive plants in the Trans-Himalayan region of Ladakh, India. *Biological Invasions*, *27*(4), 1–19. DOI: 10.1007/s10530-025-03562-z

Zehzad, B., Kiabi, B. H., & Madjnoonian, H. (2002). The natural areas and landscape of Iran: An overview. *Zoology in the Middle East*, *26*(1), 7–10. DOI: 10.1080/09397140.2002.10637915

Zhang, Y., Chen, J., & Sun, H. (2021). Alpine speciation and morphological innovations: Revelations from a species-rich genus in the northern hemisphere. *AoB Plants*, *13*(3), plab018. DOI: 10.1093/aobpla/plab018 PMID: 34025962

Zohary, D. (1973). The origin of cultivated cereals and pulses in the Near East. *Chromosomes Today*, *4*, 307–320.

Zohary, M. (1973). *Geobotanical foundations of the Middle East*.

Chapter 5
Habitat Fragmentation and Overgrazing of the Alpine Plant Communities

Amjad Ali
http://orcid.org/0000-0001-8157-576X
Università Cattolica Del Sacro Cuore, Italy

Shabab Hussain
http://orcid.org/0000-0002-2537-4792
University of Milan, Italy

Nayab Rehman
COMSATS University, Islamabad, Pakistan

Murtaza Hussain
Government Degree College, Havellia, Pakistan

Wasim Abbas
Quaid-i-Azam University, Islamabad, Pakistan

Mohsin Ali
Chinese Academy of Sciences, Beijing, China

ABSTRACT

Two major anthropogenic stresses i.e. habitat fragmentation and overgrazing have profound impacts on alpine plant communities, leading to biodiversity loss, ecosystem degradation, and reduced ecological functioning. Alpine ecosystems are particularly vulnerable due to their high elevation, unique biodiversity, and sensitivity to environmental fluctuations. Habitat fragmentation, driven by infrastructure development, tourism, and climate change, isolates plant populations, disrupts ecological connectivity, and reduces genetic diversity. Fragmented landscapes also intensify grazing pressure, as limited refuge habitats and restricted dispersal increase the vulnerability of palatable species. Conversely, overgrazing accelerates fragmentation by increasing spatial heterogeneity and degrading habitat quality. These combined

DOI: 10.4018/979-8-3373-6756-9.ch005

ght © 2026, IGI Global Scientific Publishing. Copying or distributing in print or electronic forms without written permission of IGI Global Scientific Publishing is prohibited. Use of this chapter to generative artificial intelligence (AI) technologies is expressly prohibited. The publisher reserves all rights to license its use for generative AI training and machine learning model development.

pressures reduce primary productivity, weaken ecosystem resilience, and impair key services such as soil stabilization and water regulation. Climate change further exacerbates these stresses by altering species distributions and phenological patterns.

1. INTRODUCTION

Habitat fragmentation is a process by which environment divided continuously into smaller and isolated sections. Habitat fragmentation has several elements; i.e. habitat area loss, edge expansion, interior ratio, patch isolation and configuration modifications (Muhammed & Elias, 2021). Both natural (landslides, glacial retreat) and man-made (roads, land development, pasture conversion) factors can produce fragmentation (Sarma et al., 2024). Degradation of fragmentation cause from over-grazing, which occurs when grazing occurs more frequently or with more intensity compared to vegetation can recover (Bardgett et al., 2021; Varga et al., 2021). Habitat fragmentation cause effects in reduced cover, trampling, soil compaction, erosion, reduced biomass as well as cause changed in carbon (C) and nitrogen (N) cycles (Ekka et al., 2023)

Plant communities that are adapted to severe abiotic stresses can be found in alpine habitats, these plants are distinguished by freezing temperatures, brief growing seasons and frequently harsh climatic circumstances (Marian et al., 2022). These ecosystems, which are frequently found in high latitudes or altitudes, are especially sensitive to human disturbances like overgrazing and habitat fragmentation (Haq et al., 2021; Körner, 2019) Fragmentation decreases connectivity across habitat patches due to changing edge effects, microclimate and biotic interactions, (Mendes & Prevedello, 2020). Overgrazing by cattle leads to depletes biomass, compacts the soil, alters nutrient cycles, as well as cause the changes in the species composition (Abdelsalam, 2021). This chapter examines the interactions between habitat fragmentation and overgrazing and how they impact alpine plant communities' structure, function, resilience and biodiversity.

Isolation of fragmented alpine patches can hinder seed distribution, decrease gene flow and reduce the ability to recolonize following disturbance. Additionally, small, isolated patches may have fewer individuals of rare species due to demographic stochastic (Sexton et al., 2024). Edges alter the light, wind and moisture regimes which are more benefited to the species that are more adaptable to changing conditions (Burgess et al., 2022; Shinoda & Akasaka, 2020). Herbivore pressure, pathogen/predator dynamics, and plant pollinator interactions can all be changed by fragmentation. For instance, some models suggest that if natural enemies are less able to colonize fragmented patches, habitat fragmentation may result in herbivore overgrazing (Šerá, 2021).

Overgrazing often eliminates appealing species while favouring less appetizing or grazing-tolerant ones, because above-ground parts are preferentially removed, the percentage of biomass frequently changes towards roots (increasing the root: shoot ratio). Under intense grazing pressure, reveal significant drops in aboveground biomass as well as changes in the relative dominance of functional groups (e.g. Gramineae, Cyperaceae) (Wang et al., 2021) Overgrazing cause severe consequences for instance soil compact, decreased infiltration, increased erosion, loss of organic matter, changes in the C: N ratio as well as altered the availability of C and N. Besides from these consequences overgrazing hinders the flow of nutrients required for healing. Changes in soil accessible nutrients, stoichiometry and root-shoot nutrient allocation were noted in the *Kobresia* meadow investigations (Lin et al., 2022). In general, grazed plant communities exhibit reduced stability, lower biomass and potentially poorer species richness (Zhu et al., 2024). Severe grazing slows recovery after disturbances (such as trampling and biomass removal), especially in abiotic hard conditions (such as snow beds and arid ridges). Slow plant growth, restrictions on seedling recruitment and environmental factors are to blame for this (Sharma et al., 2020).

Local intensification might be caused by concentrated grazing on the remaining areas or by fragmented habitat that limits the movement of animals and herbivores (Kauer et al., 2024). Overgrazed patches that are fractured have a reduced capacity to recolonize from nearby patches, seed banks, or undisturbed areas (Gonzalez & Ghermandi, 2021). Overgrazing alters the nutrient cycles as well as soil structure; fragmentation restricts the flow of soil microbes or the amount of organic matter that can enter from nearby patches, which alternately decreasing ecosystem resilience. The combined effects may benefit highly resistant species (for instance, those with deep roots, low palatability and short stature) while decreasing overall species richness, functional diversity, and ecosystem multifunctionality (Park et al., 2022).

The effects of habitat fragmentation and overgrazing have a strong influence on palatable and slow growing vegetation, which results in severe decreases in species richness and evenness of alpine ecosystems (Shaheen et al., 2024). The occurrence of these disturbances is especially threatening to the special alpine species that is more sensitive to change. Important ecosystem processes, including fodder generation, soil carbon trapping, and water management, and erosion management are also impacted negatively. Besides this, the change in plant characteristics (such as growth form, phenology, root/shoot ratio) may cause a disturbance in the feedback mechanisms in an ecosystem, and the community processes, which would make the system less resilient to disturbances such as climatic extremes, grazing pressure and habitat fragmentation (Burton et al., 2020; Pandian et al., 2024). To mitigate these negative effects, various strategies have been proposed and implemented, including grazing management (Teague & Kreuter, 2020), landscape-level planning to reduce fragmentation (Bigard et al., 2020), soil and vegetation restoration efforts (Zhou et al., 2023), monitoring of ecological thresholds and early warning indicators (Esposito et al., 2022), and adaptive management approaches in response to climate change (Abbass et al., 2022).

1.1. Overgrazing of Alpine Plant Communities

Alpine flora communities are adapted in such a way that they are most at ease with low temperatures, shallow edaphic roots, short phenological periods and often steep microtopographic slopes. There are anthropogenic disturbances which may cause disproportional ecological impacts, including habitat fragmentation and increased grazing. These effects are altered species composition, loss of functional diversity, soil degradation, and decreased landscape connectivity. These systems are naturally patchy and species-poor at the local scale (Körner, 2023).

1.2. Habitat fragmentation

There are many factors that affects habitat of alpine plants that are roads development in the mountains, ski infrastructure, powerlines, and tourism amenities. Fragmentation diminishes patch area, amplifies edge effects, and disrupts dispersal pathways. Small patches support fewer species and are more prone to local extinction; isolation hinders decolonization and promotes genetic drift and breeding in small populations. Functionally, fragmentation frequently removes specialists and directs populations toward generalist or ruderal species, lowering trait variety (for example, the loss of slow-growing stress tolerant perennials) and affecting ecosystem processes like carbon storage and nutrient cycling (Rixen & Rolando, 2013).

Recent empirical syntheses confirm these patterns; a 2024 analysis found that habitat fragmentation substantially reduces plant functional diversity in mountain ecosystems, with measurable declines in trait richness and increased dominance by disturbance-tolerant species. Fragmentation also interacts with climate change isolated alpine patches limit uphill range shifts, trapping cold-adapted taxa in shrinking refugia (Helm et al., 2024).

1.3. Overgrazing: Direct and Indirect Pathways

Grazing has long been a common land use in many alpine regions; low to moderate grazing can help to preserve open ecosystems and biodiversity. Overgrazing, defined as high stocking rates, year round grazing, or short recovery intervals, promotes vegetation loss, soil compaction, erosion, and nutrient redistribution (Hu et al., 2025). Direct consequences include loss in palatable forbs and grasses, seedling trampling, and selection for short, prostrate, or spiky species. Overgrazing has an indirect impact on soil physical structure (compaction, reduced porosity), organic matter, and microbial diversity, as well as soil pH and nutrient cycles, making circumstances less suitable for many alpine specialists (Chen et al., 2025). Frequently grazing reduces aboveground biomass significantly and modifies plant functional composition; heavy grazing is associated with lower species richness and degraded soil properties (compaction, lower C and N pools). According to one summary, grazing reduces aboveground biomass by around half and systematically affects soil C:N:P ratios; other recent experimental evidence linked intensive grazing to reductions in microbial network complexity and long-term meadow degradation (Wang et al., 2024).

1.4. Combined effects and Management Implications

Overgrazing and fragmentation seldom work in isolation. Therefore, smaller patches increase edge effects and decrease refugia where plants can avoid grazing, fragmented landscapes with small, isolated patches are less resistant to grazing pressure (Pelliza et al., 2021). High-value alpine meadows may become barren land or invasive dominated swards because to overgrazing on fractured terrain, which also speeds up erosion. On the other hand, removal of grazing (exclusion experiments) can enhance landscape connectivity and allow vegetation structure to recover (Chang et al., 2022). Restoring or preserving habitat connectivity needs to be the priority. This can be accomplished by establishing ecological corridors, restricting the development of new linear infrastructure across montane landscapes, and using ecological design principles to reduce edge effects (Santoianni et al., 2024).

Implementing rotational grazing systems and reducing cow stocking rates can reconcile the use of the agricultural system with the conservation of biodiversity. On the other hand, overgrazing (continuous and excessive) results in significant losses of biomass, soil fertility and functional plant diversity. Careful grazing selection should be made in accordance with conservation goals. Moderate grazing may preserve or even increase plant diversity through prevention of encroachment by shrubs and promote many vegetation patterns (Peringer et al., 2022).

To discover and solve new concerns, for example, changes in soil chemistry or decreased cold tolerance of recovering plant species, soil parameters such as pH, organic carbon, and nutrient balance need to be monitored on a constant basis. Site specific and adaptive management, to increase the ecological functionality and resilience of alpine plant communities against increasing anthropogenic and climatic pressures. Important features of restoration include re-introduction of native alpine species to replace the plant cover, de-compaction of the soil to help improve soil aeration and water infiltration and a gradual return to grazing at sustainable levels once recovery has started (Nethmini et al., 2025).

1.5. Habitat Fragmentation and Overgrazing

Overgrazing affected the dependency of plant species diversity on environmental variables in alpine wetlands on the Qinghai-Tibet Plateau, increasing the dominance of hydrological regimes as drivers and reducing soil-plant interactions. In the same region, severe meadow degradation linked to overgrazing and climate pressures reduced aboveground biomass by more than 70%, drove species composition shifts (from legumes and grasses to forbs and unpalatable grasses), and caused declines in soil organic matter, porosity, and nutrient availability (Ye et al., 2025a). These nutrient imbalances in the soil also have an impact on microbial communities, which interferes with subterranean activities that are essential to plant growth. Furthermore, due to human encroachment and increasing grazing, extensive evaluations of alpine grassland landscapes have shown a sharp decline in connectedness and an increase in fragmentation along lake edges (Zhang et al., 2025). Alpine plant communities' ability to adapt to climate change may potentially be impacted by the combined effects of overgrazing and fragmentation (Yang et al., 2025).

1.6. Ecological Consequences of Fragmentation

Excessive grazing has been associated with significant drops in soil organic carbon and nitrogen content as well as a 60–70% decrease in aboveground biomass in alpine meadows on the Qinghai–Tibet Plateau. Trampling-induced soil compaction reduces the capacity for infiltration, interferes with root respiration, and prevents the establishment of seedlings. Microbial populations necessary for the mineralization of nutrients and the breakdown of organic matter also decrease as soil porosity decreases (Hou et al., 2024).

Ecosystem multifunctionality the ability of an ecosystem to sustain several ecological processes at once is more broadly affected by the combined effects of fragmentation and overgrazing. Alpine ecosystems that are fragmented and grazed exhibit decreases in functional diversity and redundancy, which means that the extinction of a few important species may result in disproportionate reductions in ecosystem functioning (Chu et al., 2022a). There should be rules to tackle the ecological effects of fragmentation and overgrazing in alpine environments need a comprehensive framework that incorporates land-use management, biodiversity conservation, and climate adaption techniques (Prescott et al., 2019).

1.7. Threats to Alpine Plant Biodiversity

In recent decades, however, their biodiversity has come under increasing pressure from habitat fragmentation and overgrazing; human-driven processes that often reinforce each other; considered as major threats leading to rapid loss of biodiversity (Donovan et al., 2020).

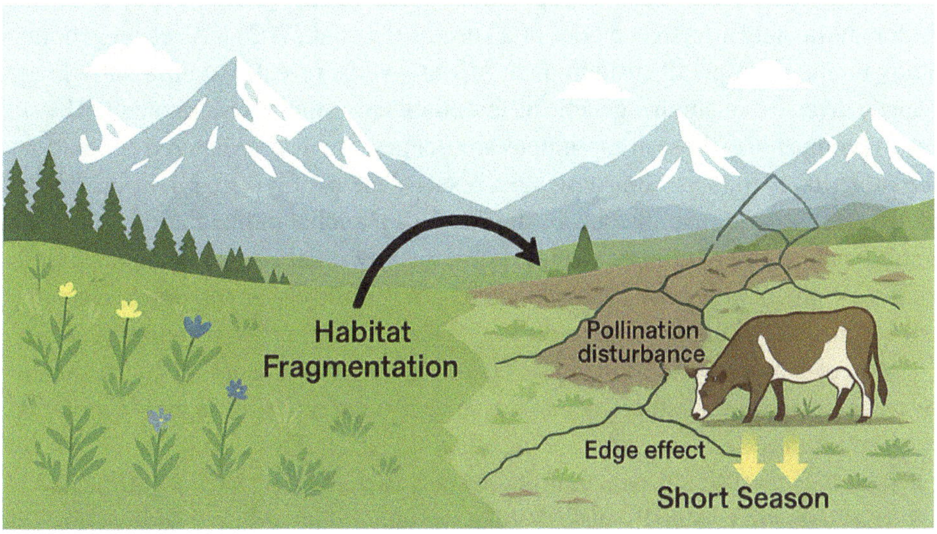

Figure 1 shows alpine plant biodiversity under threat; habitat fragmentation, overgrazing, edge effects, and disrupted pollination collectively undermine the survival of fragile alpine species

2. MECHANISMS AND ECOLOGICAL EFFECTS OF FRAGMENTATION

Habitat fragmentation is the process in which continuous habitats are broken down into smaller, isolated patches. It may be due to natural or human-induced activities such as road construction, deforestation, and land-use changes. However mostly this fragmentation is accelerated by infrastructure development related to tourism, hydropower projects, and agricultural expansion. Fragmentation disrupts the spatial continuity of vegetation, reducing gene flow among populations and increasing the vulnerability of small, isolated plant populations to extinction (Riva et al., 2024).

When habitats divided, then plant populations in remnants are cut-off from each other; hence limited pollination, interrupt seed dispersal, and reduce genetic exchange. As alpine plants, often with limited seed dispersal capabilities rely on landscape connectivity for gene flow. Fragmentation creates physical barriers that isolate populations, leading to genetic drift and inbreeding depression. Eventually isolation restricts gene flow between sub-populations, altering gene pool of a species. These factors ultimately affect gene pool of a species (Li et al., 2021). A recent genomic study on the widespread cushion plant *Silene acaulis* revealed significantly lower genetic diversity and higher genetic differentiation in populations fragmented by ski resorts compared to those in continuous landscapes. Small and scattered populations are more likely to face inbreeding; as a result, gene pool reduces. This erosion of genetic diversity ultimately reduces the adaptive potential of populations, a critical liability in the face of climate change. Subsequently small and isolated populations become vulnerable to random losses and have less ability of adaptation to changing environmental conditions, hence accelerating local extinction (Bonanomi et al., 2023).

Moreover, fragmentation increases edge effect that ultimately alter microclimates by experiencing temperature fluctuations, speedy wind exposure, higher evaporation rates result in soil drying and leaching (Li et al., 2021; Theobald et al., 2024). All the factors synergistically penetrate deeply into the remaining habitat and challenge species adaptation specially make habitat unsuitable for sensitive species, disturb species distribution pattern and community composition. More likely to accelerate local extinction, especially under changing climate conditions (Holzmann et al., 2023).

2.1. Overgrazing and Vegetation Degradation

Grazing by domestic livestock is a common practice in mountainous regions. However, when grazing intensity exceeds the regenerative capacity of vegetation, it is termed as overgrazing. It leads to severe ecological degradation. Overgrazing reduces plant cover, which exposes soil to erosion. Additionally overgrazing acts as a powerful filter, as selectively consuming the most palatable and nutritious plant species. Over time, this shifts the entire community composition by dominating with unpalatable, spiny or toxic plants, especially invasive species, leading to a marked decline in species richness and functional diversity decline Continuous grazing exceeds the land's natural carrying capacity and slow-growing plants decline as well as prevents their regeneration. On other hand unpalatable or invasive species spread aggressively (Tulloch et al., 2023).

2.2. Soil Degradation and Regeneration Failure

Alpine overgrazing is a serious ecological problem that causes vegetation loss, soil deterioration, and the lack of regeneration in delicate high-elevation ecosystems. Alpine soils are extremely susceptible to disturbance since they are usually shallow and deficient in organic matter Constant and heavy grazing decreases the amount of plants on the soil, which exposes it to erosive forces and speeds up the loss of nutrients and physical deterioration of organic carbon and microbiological activity (Ge et al., 2025). Trampling has the worst physical impacts because it compacts the soil, reducing soil porosity, water penetration, soil aeration, and root growth (Yu et al., 2019).

3. PHYSIOLOGICAL STRESS RESPONSES OF ALPINE PLANTS

3.1. Photosynthetic Inhibition and Pigment Changes

Alpine plants are vulnerable to edge effects due to fragmentation where the plant was exposed to more wind, temperature changes, and light intensity. By destroying chloroplast structure and lowering chlorophyll content, these microclimatic changes can interfere with photosynthesis (Park et al., 2025). In a similar way, excessive grazing may change the pigment composition and decrease the amount of leaf area available for photosynthesis, causing plants to shift their energy from growth to defence and tissue repair (Zhou et al., 2024).

Figure 2. Physiological and functional responses to alpine ecosystem

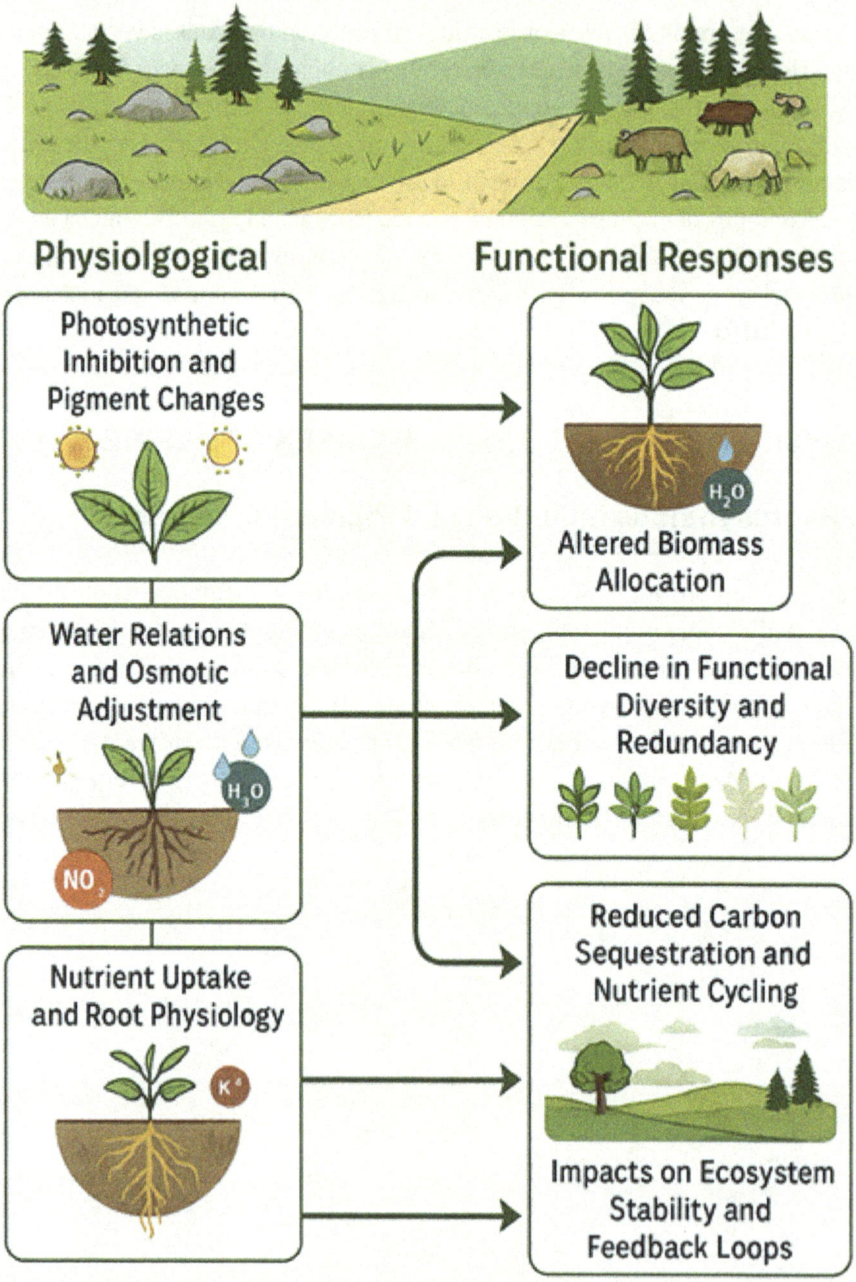

3.2. Water Relations and Osmotic Adjustment

Due to a low plant cover, fragmentation often changes the soil hydrology by raising the rate of evaporation and reducing soil infiltration. Overgrazing accelerates this process further by compressing the soil and making it less porous and limiting root penetration. In response, the Alpine plants undergo an osmotic adjustment that entails accumulation of solutes such as proline, soluble carbohydrates and amino acids to maintain cell turgor in case there is shortage of water. However, this flexibility is lost over time, as a result of continuous stress and results in physiological exhaustion (Zavala et al., 2010).

3.3. Functional Responses at the Community and Ecosystem Level

3.3.1. Altered Biomass Allocation

In plants, stress that lasts over a long period, especially overgrazing reduces the capacity of plants to apply this trick and eventually their energy is depleted. Equally, the plants in solitary ecosystems are found to have reduced biomass build up and altered community composition, where low-stature stress-adaptable organisms prevail (Qi et al., 2019).

3.3.2. Decline in Functional Diversity and Redundancy

The essential features in ecosystem resilience are functional traits of the leaf area, specific leaf area (SLA), depth of roots, and the phenological behaviour. Conversely, fragmentation acts as an obstacle to the pollination, seed dispersal and colonization processes and overgrazing also selectively removes palatable plants. Consequently, when the community is homogenized, functioning redundancy is minimal, and hence, ecosystem systems are easily disturbed (Ning et al., 2004).

3.3.3. Reduced Carbon Sequestration and Nutrient Cycling

The physiological restraints affect the ecosystem. Lower carbon sequestration results from decreased biomass production and photosynthetic efficiency. Meanwhile, soil degradation accelerated by trampling and erosion diminishes organic matter input, affecting microbial activity and nutrient cycling. Grazed and fragmented sites across alpine Asia have shown 20–40% lower soil organic carbon and nitrogen levels compared to undisturbed meadows (Gurau et al., 2025).

4. THE DUAL THREAT: HABITAT FRAGMENTATION AND OVERGRAZING IN ALPINE PLANT COMMUNITIES

Alpine plant communities occupy some of the planet's most climatically extreme environments, high elevation plateaus, cirques, ridges and summits where low temperatures, high irradiance, short growing seasons, and rugged topography define a narrow window for plant growth and reproduction. Because species in these systems are often slow-growing, long-lived, and adapted to a narrow set of environmental conditions, they are inherently vulnerable to disturbance. Two anthropogenic pressures in particular, habitat fragmentation and overgrazing act singly and synergistically to erode biodiversity, reduce ecosystem function, and compromise the resilience and adaptive capacity of alpine plant communities (Zhou et al., 2024)

Alpine plant communities, which exist in some of the World's most sensitive and biodiverse high-altitude ecosystems, are under increasing threat from the combined pressures of habitat fragmentation and overgrazing. Habitat fragmentation, often driven by anthropogenic activities such as infrastructure expansion, agricultural encroachment, tourism development, and climate-induced range shifts, disrupts the continuity of alpine landscapes and isolates plant populations. This loss of connection ultimately reduces the adaptability of alpine flora to changes in the environment and enhances genetic drift as well as the reduction of gene flow (Chen et al., 2017). These stresses have been aggravated by the fact that overgrazing by the domestic animals, especially sheep, goats and the yaks on pastoral lands has been seen to be a major cause of degradation on the alpine ecosystems. The intensive grazing encourages grazing tolerant or unpleasant species in a range of native assemblages, reduces vegetative cover, and changes the composition of plant communities. This leads to an increase in the fragmentation of habitats and the further destruction of the landscape in the long-term, caused by soil compaction, the loss of water permeability, loss of nutrients, and erosion (Gambo et al., 2025; Klein et al., 2004a).

Also, overgrazing is very hard to recover because alpine vegetation has an inherently low rate of regeneration because of short growth periods, low temperatures, and limited nutrient levels. Climate change adds to these limitations by making alpine species migrate to higher altitudes, shrinking their habitat and leading to heightened competition on resources, further weakening ecosystem resilience, biodiversity and posing a threat to ecosystem services such as carbon capturing, water management and soil stabilization. The conservation can involve rotational grazing, controlled grazing, building grazing free areas, regeneration of degraded areas by planting native species and the establishment of biological corridors to enhance connectivity (Körner, 2021; Nagy & Grabherr, 2009a). Addressing these challenges demands an integrated landscape-level approach that balances pastoral livelihoods with the ecological integrity of alpine ecosystems.

5. HABITAT FRAGMENTATION: CAUSES AND CONSEQUENCES

Habitat fragmentation refers to the subdivision of once-continuous habitat into smaller, more isolated patches. In Alpine regions, fragmentation arises from a combination of natural and anthropogenic processes. Natural fragmentation is driven by topography, rock outcrops, and glacial history; however, human activities now intensify fragmentation via road and trail construction, ski-area development, cable cars, mining, tourism-related infrastructure, and agricultural encroachment into formerly marginal areas. Climate change is a further driver of 'climatic fragmentation, thermophilization and upward shifts in species ranges reduce the area of high-elevation habitat, creating island-like patches on mountaintops (Chala et al., 2016; Lin et al., 2023a). Consequences of fragmentation are manifold. Smaller patches support smaller populations, which are more vulnerable to demographic stochasticity, reduced recruitment, and local extinction. Isolation reduces gene flow and seed rain, increasing genetic drift and potentially leading to inbreeding and loss of adaptive variation. Empirical studies in the European Alps and other montane systems demonstrate higher genetic differentiation in marginal, isolated populations compared with central populations patterns observed for long-lived trees (e.g., *Pinus cembra*) and for alpine herbs (Tóth et al., 2019). At the community level, fragmentation increases edge effects (altered microclimate, higher desiccation, wind exposure), encourages invasion by generalists and disturbance-tolerant taxa, and reduces the abundance of specialists and endemics.

6. OVERGRAZING: PATTERNS, MECHANISMS AND PRIMARY IMPACTS

Overgrazing occurs where the intensity or duration of herbivory (by livestock or wild herbivores) exceeds the ability of plant communities to recover. In alpine systems, where growing seasons are short, and plant turnover is slow even episodic overgrazing can have long-lasting effects? Overgrazing patterns vary continuous grazing; high stocking rates near camps and water, and the concentration of grazing along access routes are common in many mountains pastoral systems. Mechanistically, grazing removes photosynthetic tissue, reduces carbohydrate reserves, damages reproductive structures (flowers, seeds), and through trampling, compacts soils and damages seedlings and vegetative propagules (Lin et al., 2023a).

Primary impacts of heavy grazing documented across alpine meadows include reduced aboveground biomass and canopy cover, reduced litter accumulation, shifts in species composition toward grazing-tolerant grasses or sedges, declines of palatable forbs and dwarf shrubs, and decreased soil organic carbon and nitrogen pools. Trampling drives soil compaction, reducing infiltration and increasing surface runoff and erosion processes that degrade seedbeds and exacerbate plant recruitment failure (Zhang et al., 2025).

6.1. Synergistic effects: fragmentation × overgrazing

While fragmentation and overgrazing pose a threat, their interaction often produces outcomes more severe than the sum of their parts. Fragmented landscapes concentrate cattle in accessible sections and edges, increasing grazing pressure locally. Edges provide less protection and buffer, therefore plants in small fragments face increased microclimatic stress as well as stronger herbivory. Reduced patch size and connectedness hinder populations' ability to recolonize following disruption, as well as genetic exchange, which reduces resilience. Furthermore, tiny populations that survive intense grazing may lack adequate seed banks or vegetative propagules to recover, pushing transitions to depleted, homogeneous plant communities dominated by unpleasant species (Tóth et al., 2019).

7. GENETIC AND EVOLUTIONARY CONSEQUENCES

The ecological effects and the evolutionary effects of reduced effective population sizes due to both fragmentation and loss of biomass by overgrazing are both short term and long term. Fragments reduce gene flow and in combination with demographic loss due to overgrazing could lead to disappearance of rare alleles and low adaptive capacity (Zambrano et al., 2019). The disappearance of genetic diversity threatens the capability of adapting to stressors of the future including climatic changes and the emerging pests or diseases, Fragments decrease the size of populations and cause separation of plant groups, leading to gene exchange and a high amount of genetic drift. Inbreeding is more likely to occur in small and isolated populations, resulting in decreased genetic diversity and, potentially, inbreeding depression, decreased fitness, reproductive success, and adaptation to environmental shock. The networks of pollination are broken, and genetic flow is further constrained, and opportunities of dispersion are limited preventing recolonization of appropriate habitats, accelerating local extinction. These stresses are aggravated by overgrazing which selectively removes the palatable species (Nogales et al., 2025).

8. SOIL–PLANT–MICROBE INTERACTIONS AFFECTED BY BOTH PRESSURES

Soil properties and microbial community mediate nutrient cycling and plant growth; they are highly sensitive to both fragmentation-driven edge effects and grazing. Heavy grazing reduces litter inputs and root turnover, lowering soil organic matter; trampling increases bulk density and reduces porosity. These physical changes modify the soil microbiome shifts in bacterial and fungal community composition influence key processes such as nitrogen mineralization, nitrification, phosphorus mobilization, and symbiotic interactions with roots. Recent meta-analyses and regional studies on the Tibetan Plateau and other alpine areas document that grazing intensity alters the relative abundance of functional microbial groups and can reduce microbial diversity under heavy pressure (Zhou et al., 2025).

9. THRESHOLDS, NONLINEARITY AND RESILIENCE LOSS

Alpine systems often exhibit threshold-like responses: low-to-moderate grazing or modest habitat loss may be absorbed with little visible change, but beyond a critical point, feedback (e.g., erosion, seed bank loss, soil compaction) can produce rapid state shifts to low-diversity, low-biomass conditions that are difficult to reverse. These nonlinearities are crucial for management: policy should aim to keep systems well within safe operating spaces (resilience basins) to avoid crossing thresholds that require expensive restoration or that lead to the permanent loss of local endemics (Battaglini et al., 2014).

10. TEMPORAL DYNAMICS: LAG EFFECTS AND CUMULATIVE IMPACTS

Because many alpine plants are perennial with slow growth and infrequent recruitment, impacts from fragmentation and grazing can show substantial lag times. A single severe grazing event may cause immediate biomass loss, but the full demographic consequences reduced recruitment, shifting age structures, and genetic erosion may only become evident years or decades later. Similarly, fragmentation can produce 'extinction debts where populations persist for a time but eventually decline due to small population dynamics and reduced gene flow. Cumulative impacts across multiple years repeated heavy grazing, erosion episodes, or successive years of low snowpack compound stress and lower the probability of recovery (Chala et al., 2016).

The combined effects of habitat fragmentation and overgrazing are posing an increasing danger to alpine plant groups, which are found in some of the most delicate and biodiverse high-altitude ecosystems on Earth. The continuity of alpine landscapes is disrupted, and plant populations are isolated by habitat fragmentation, which is frequently caused by human activities including infrastructure construction, agricultural infringement, tourism growth, and climate-induced range shifts. Alpine flora's ability to adapt to changing environmental conditions is ultimately diminished by this loss of connection, which also increases genetic drift and decreases gene flow (Walther et al., 2002). Ecosystem stability is further compromised by fragmented habitats, which also result in edge effects, modified microclimates, and heightened susceptibility to invasive species. Overgrazing by domestic animals, particularly sheep, goats, and yaks in pastoral areas, has been identified as a major contributor to the degradation of alpine ecosystems, exacerbating these stresses. In addition to reducing vegetation cover and changing the structure of plant communities, intensive grazing favors grazing-tolerant or unpalatable species over a variety of native assemblages (Speed et al., 2012). As a result, habitats become more fragmented and landscape deterioration accelerates over time due to soil compaction, decreased water infiltration, nutrient depletion, and increased erosion (Klein et al., 2004b).

11. ALPINE PLANT COMMUNITIES UNDER PRESSURE: FRAGMENTATION AND OVERGRAZING DYNAMICS

The dynamic temporal and geographical processes that fragment and overgraze use to change alpine plant communities are summarized in alpine plant communities under pressure. We look at functional changes in community composition, spatial patterns of deterioration, the accumulation of impacts over time, and recovery possibilities under various management systems (Ye et al., 2025b).

12. SPATIAL DYNAMICS: PATCH SIZE, ISOLATION AND EDGE EFFECTS

The structure, composition, and resilience of alpine plant communities are significantly shaped by spatial dynamics, especially when fragmentation and overgrazing are present. A major factor in determining the stability of an ecosystem is patch size. While smaller patches are more susceptible to stochastic events, local extinctions, and decreased genetic diversity, larger patches typically support higher species richness, maintain more stable microclimates, and act as a buffer against external disturbances. By restricting gene flow between plant populations, upsetting pollination networks, and decreasing the likelihood of recolonization after disruption, habitat isolation intensifies these consequences. Isolated patches eventually develop into ecological "islands," where small populations experience more inbreeding and are less able to adjust to environmental changes (De Toma et al., 2025).

Bigger patches with refugia and internal variability are better able to withstand grazing pressure and preserve seed supplies. Stress-sensitive species have less access to microhabitat in small patches due to edge conditions (more wind exposure, larger temperature fluctuations). Isolation reduces the potential for recolonization and gene transfer by obstructing seed distribution and pollinator movement (Tóth et al., 2019).

13. FUNCTIONAL RESPONSES AND COMMUNITY REASSEMBLY

Stress-tolerant life histories and characteristics, such as low stature, clonal reproduction, robust leaves, high root allocation, and quick tillering, are usually favored by functional reassembly under grazing and fragmentation. Common results include declines in forb richness and an increase in swords dominated by graminoids (Zhang et al., 2023). These changes affect the suitability of habitat for herbivores, pollinators, and related animals as well as nutrient cycling mechanisms (e.g., increased litter recalcitrance, changed decomposition rates).

14. CASE EXAMPLES AND REGIONAL CONTRASTS

Numerous field research conducted throughout the Qinghai–Tibetan Plateau (QTP) demonstrate that while heavy grazing consistently lowers soil carbon and nitrogen and favors degraded states with increased bare soil and erosion resistance, light-to-moderate grazing can occasionally increase plant diversity and stimulate some nutrient pools (Lin et al., 2023b).

Many alpine specialists' ranges have shrunk, and edge-driven plant community homogenization has risen in European mountain systems due to the fragmentation of montane forests and high-mountain meadows caused by infrastructure development and changes in land. Afroalpine giant rosette plants and other tropical alpine systems are particularly vulnerable to habitat fragmentation and range contraction due to warming (Chala et al., 2016b).

14.1. Management responses: prevention, mitigation and restoration

Three general categories comprise management strategies that lessen the dual threat: prevention (limiting access and reducing new fragmentation), mitigation (modifying grazing commands, rotational grazing, setting seasonal herding rules, and establishing accessible water and salt licks away from sensitive patches), and active restoration (reseeding, soil amendments, erosion control, fencing to enable recovery). Although results vary on the length of the exclusion, the local climate, and baseline degradation, grazing exclusion whether temporary or permanent has been frequently used and typically results in increases in biomass, cover, and soil organic carbon (Chen et al., 2017).

14.2. Monitoring and indicators for adaptive management

Monitoring indicators that are informative, scalable, and sensitive are necessary for effective management. Plant species richness and functional group composition, percentage of bare ground, aboveground biomass, root: shoot ratios, soil bulk density, soil organic carbon and available nutrients, microbial community structure (16S/ITS diversity, functional gene abundance), and genetic diversity measurements (where possible) are all suggested indicators. Plot-scale surveys combined with remote sensing (high-resolution satellite or drone photography) allow for the landscape-scale evaluation of fragmentation and recovery trajectories (Damgaard, 2022).

14.3. Research needs and synthesis

Even while there is mounting evidence of how habitat fragmentation and overgrazing affect alpine plant communities, there are still large research gaps that prevent us from creating all-encompassing conservation and management plans. It is difficult to generalize findings across many alpine ecosystems because a large portion of the current information is generated from restricted studies, frequently concentrated on species or locations. To completely comprehend the dynamics of fragmentation and grazing pressures in these delicate environments, long-term, multi-scale research that incorporate ecological, climatic, and socioeconomic aspects are desperately needed (Notarnicola et al., 2023). To understand how plant communities react and adapt to coupled stressors, future research should focus on spatially explicit landscape modeling, functional trait studies, and genetic connectivity evaluations. Furthermore, more multidisciplinary research is needed to investigate how land-use regulations, ecosystem resilience, and conventional grazing methods interact, especially in the context of climate change (Probo et al., 2025).

Long-term manipulative experiments combining fragmentation and grazing treatments, mechanistic studies relating plant demography to genetic outcomes under combined stressors, enhanced comprehension of microbial-mediated nutrient cycling under compounding pressures, and socio-ecological studies on pastoralist practices and incentives to maintain biodiversity-friendly grazing regimes are among the most important research needs (Probo et al., 2025). Crucially, integrating scientific monitoring with local indigenous knowledge can enhance management design and acceptance.

15. CONCLUSION

Alpine plant communities are threatened by habitat fragmentation and overgrazing, which work together to alter vegetation structure, species richness, ecosystem stability, and function. High grazing intensity decreases aboveground biomass and changes the functional content of plants, according to empirical data (such as those from the Qinghai–Tibet Plateau); fragmentation speeds up deterioration and lowers recovery capability. There should be proper rules and regulations by authorized bodies that work on managing and saving alpine plant communities.

REFERENCES

Abbass, K., Qasim, M. Z., Song, H., Murshed, M., Mahmood, H., & Younis, I. (2022). A review of the global climate change impacts, adaptation, and sustainable mitigation measures. *Environmental Science and Pollution Research International*, *29*(28), 42539–42559. DOI: 10.1007/s11356-022-19718-6 PMID: 35378646

Abdelsalam, M. I. (2021). Effects of Overgrazing on Rangeland Resources in Semi-arid Areas and Rangel and Management: A review Article. *Agrica*, *10*(2), 144–151. DOI: 10.5958/2394-448X.2021.00022.5

Bardgett, R. D., Bullock, J. M., Lavorel, S., Manning, P., Schaffner, U., Ostle, N., Chomel, M., Durigan, G., & Fry, L., E., Johnson, D., Lavallee, J. M., Le Provost, G., Luo, S., Png, K., Sankaran, M., Hou, X., Zhou, H., Ma, L., Ren, W., … Shi, H. (. (2021). Combatting global grassland degradation. *Nature Reviews. Earth & Environment*, *2*(10), 720–735. DOI: 10.1038/s43017-021-00207-2

Battaglini, L., Bovolenta, S., Gusmeroli, F., Salvador, S., & Sturaro, E. (2014). Environmental Sustainability of Alpine Livestock Farms. *Italian Journal of Animal Science*, *13*(2), 3155. DOI: 10.4081/ijas.2014.3155

Bigard, C., Thiriet, P., Pioch, S., & Thompson, J. D. (2020). Strategic landscape-scale planning to improve mitigation hierarchy implementation: An empirical case study in Mediterranean France. *Land Use Policy*, *90*, 104286. DOI: 10.1016/j.landusepol.2019.104286

Bonanomi, G., Idbella, M., Allegrezza, M., & Tesei, G. (2023). Dieback of the cushion plant Silene acaulis at its southern limit of distribution in the Apennines. *Alpine Botany*, *133*(1), 57–62. DOI: 10.1007/s00035-023-00292-7

Burgess, T. I., Oliva, J., Sapsford, S. J., Sakalidis, M. L., Balocchi, F., & Paap, T. (2022). Anthropogenic Disturbances and the Emergence of Native Diseases: A Threat to Forest Health. *Current Forestry Reports*, *8*(2), 111–123. DOI: 10.1007/s40725-022-00163-0

Burton, P. J., Jentsch, A., & Walker, L. R. (2020). The Ecology of Disturbance Interactions. *Bioscience*, *70*(10), 854–870. DOI: 10.1093/biosci/biaa088

Chala, D., Brochmann, C., Psomas, A., Ehrich, D., Gizaw, A., Masao, C. A., Bakkestuen, V., & Zimmermann, N. E. (2016). Good-bye to tropical alpine plant giants under warmer climates? Loss of range and genetic diversity in *Lobelia rhynchopetalum. Ecology and Evolution*, *6*(24), 8931–8941. DOI: 10.1002/ece3.2603 PMID: 28035281

Chang, Y., Ding, Y., Zhao, Q., Qin, J., & Zhang, S. (2022). Optimization of canopy resistance models for alpine meadow in the northeastern Tibetan Plateau. *Journal of Hydrology (Amsterdam), 610*, 128007. DOI: 10.1016/j.jhydrol.2022.128007

Chen, J., Li, Y., Yang, Y., & Sun, H. (2017). How cushion communities are maintained in alpine ecosystems: A review and case study on alpine cushion plant reproduction. *Plant Diversity, 39*(4), 221–228. DOI: 10.1016/j.pld.2017.07.002 PMID: 30159515

Chen, L., Zeng, X., & Gao, Y. (2025). The impact of grazing exclusion on phosphorus dynamics in alpine grassland ecosystems. *Plant and Soil, 513*(1), 1215–1230. DOI: 10.1007/s11104-025-07238-4

Chu, L., Hou, M., & Jiang, Z. (2022a). How does the fragmentation of pasture affect grassland ecology?——Evidence from typical pastoral areas in China. *Ecological Indicators, 136*, 108701. DOI: 10.1016/j.ecolind.2022.108701

Damgaard, C. (2022). Adaptive management plans rooted in quantitative ecological predictions of ecosystem processes: Putting monitoring data to practical use. *Environmental Conservation, 49*(1), 27–32. DOI: 10.1017/S0376892921000357

De Toma, A., Malavasi, M., Marzialetti, F., & Cutini, M. (2025). Unveiling spatial patterns and trajectories of shrub dynamics in Mediterranean alpine ecosystems. *Plant Ecology, 226*(2), 149–160. DOI: 10.1007/s11258-024-01479-6

Donovan, V. M., Roberts, C. P., Wonkka, C. L., Beck, J. L., Popp, J. N., Allen, C. R., & Twidwell, D. (2020). Range-wide monitoring of population trends for Rocky Mountain bighorn sheep. *Biological Conservation, 248*, 108639. DOI: 10.1016/j.biocon.2020.108639

Ekka, P., Patra, S., Upreti, M., Kumar, G., Kumar, A., & Saikia, P. (2023). Land Degradation and Its Impacts on Biodiversity and Ecosystem Services. In Raj, A., Jhariya, M. K., Banerjee, A., Nema, S., & Bargali, K. (Eds.), *Land and Environmental Management through Forestry* (1st ed., pp. 77–101). Wiley., DOI: 10.1002/9781119910527.ch4

Esposito, M., Palma, L., Belli, A., Sabbatini, L., & Pierleoni, P. (2022). Recent Advances in Internet of Things Solutions for Early Warning Systems: A Review. *Sensors (Basel), 22*(6), 2124. DOI: 10.3390/s22062124 PMID: 35336296

Gambo, A., Fullen, M. A., & Baldwin, T. C. (2025). Indigenous farmers' knowledge and perception of desertification and soil erosion in Jigawa State, Nigeria. *Discover Soil, 2*(1), 22. DOI: 10.1007/s44378-025-00028-5

Ge, P., Xue, J., Ru, Y., Li, Y., Li, D., Han, P., Li, A., & Huang, J. (2025). Intensive rotational grazing has positive effects on productivity of rangeland. *Agriculture, Ecosystems & Environment*, *384*, 109558. DOI: 10.1016/j.agee.2025.109558

Gonzalez, S. L., & Ghermandi, L. (2021). Overgrazing causes a reduction in the vegetation cover and seed bank of Patagonian grasslands. *Plant and Soil*, *464*(1–2), 75–87. DOI: 10.1007/s11104-021-04931-y

Gurau, S., Imran, M., & Ray, R. L. (2025). Algae: A cutting-edge solution for enhancing soil health and accelerating carbon sequestration – A review. *Environmental Technology & Innovation*, *37*, 103980. DOI: 10.1016/j.eti.2024.103980

Haq, S. M., Yaqoob, U., Calixto, E. S., Kumar, M., Rahman, I. U., & Hashem, A., Abd_Allah, E. F., Alakeel, M. A., Alqarawi, A. A., Abdalla, M., Lone, F. A., Khan, M. A., Khan, U., & Ijaz, F. (. (2021). Long-Term Impact of Transhumance Pastoralism and Associated Disturbances in High-Altitude Forests of Indian Western Himalaya. *Sustainability*, *13*(22), 12497. DOI: 10.3390/su132212497

Helm, N., Chytrý, K., Hülber, K., Moser, D., Wessely, J., Gattringer, A., Hausharter, J., Pauli, H., Winkler, M., Saccone, P., Lamprecht, A., Rutzinger, M., Mayr, A., Kollert, A., & Dullinger, S. (2024). Fine-scale alpine plant community assembly: Relative roles of environmental sorting, dispersal processes and species interactions. *Journal of Ecology*, *112*(12), 2745–2757. DOI: 10.1111/1365-2745.14401

Holzmann, K. L., Walls, R. L., & Wiens, J. J. (2023). Accelerating local extinction associated with very recent climate change. *Ecology Letters*, *26*(11), 1877–1886. DOI: 10.1111/ele.14303 PMID: 37721806

Hou, Y., Zhao, W., Zhan, T., Hua, T., & Pereira, P. (2024). Alpine meadow and alpine steppe plant-soil network in Qinghai-Tibet Plateau, China. *Catena*, *244*, 108235. DOI: 10.1016/j.catena.2024.108235

Hu, X., Sun, Q., Zhang, S., Li, W., Wang, X., Long, R., Jin, G., & Zhang, B. (2025). Effects of No-Till Seeding and Fertilization on Vegetation Restoration and Soil Physicochemical Properties in Alpine Degraded Grazing Grasslands. *Agronomy (Basel)*, *15*(3), 578. DOI: 10.3390/agronomy15030578

Kauer, L., Imholt, C., Jacob, J., & Kuehn, R. (2024). Assessing the effects of land-use intensity on small mammal community composition and genetic variation in Myodes glareolus and Microtus arvalis across grassland and forest habitats. *Landscape Ecology*, *40*(1), 1. DOI: 10.1007/s10980-024-02021-9

Klein, J. A., Harte, J., & Zhao, X. (2004a). Experimental warming causes large and rapid species loss, dampened by simulated grazing, on the Tibetan Plateau. *Ecology Letters*, *7*(12), 1170–1179. DOI: 10.1111/j.1461-0248.2004.00677.x

Klein, J. A., Harte, J., & Zhao, X. (2004b). Experimental warming causes large and rapid species loss, dampened by simulated grazing, on the Tibetan Plateau. *Ecology Letters*, *7*(12), 1170–1179. DOI: 10.1111/j.1461-0248.2004.00677.x

Körner, C. (2019). Mountain Biodiversity, its Causes and Function: An Overview. In Körner, Ch., & Spehn, E. M. (Eds.), *Mountain Biodiversity* (1st ed., pp. 3–20). Routledge., DOI: 10.4324/9780429342585-1

Körner, C. (2021). *Alpine Plant Life: Functional Plant Ecology of High Mountain Ecosystems*. Springer International Publishing., DOI: 10.1007/978-3-030-59538-8

Körner, C. (2023). Concepts in Alpine Plant Ecology. *Plants*, *12*(14), 2666. DOI: 10.3390/plants12142666 PMID: 37514280

Li, J., Li, X., Gao, J., Kazhaocairang, , Ma, G., & Qi, X. (2021). Micro-scale fragmentation of the alpine meadow landscape on the Qinghai-Tibet Plateau under external disturbances. *Catena*, *201*, 105220. DOI: 10.1016/j.catena.2021.105220

Lin, L., Cao, G., Xu, X., Li, C., Fan, B., Li, B., Lan, Y., Si, M., & Dai, L. (2022). Changes and Relationships between Components in the Plant-Soil System and the Dominant Plant Functional Groups in Alpine Kobresia Meadows Due to Overgrazing. *Diversity*, *14*(3), 183. DOI: 10.3390/d14030183

Lin, L., Xu, X., Cao, G., Zhang, F., Li, Y., Fan, B., Li, Q., & Huang, J. (2023a). Synergies between microsites of plant communities and steady-stage alpine meadows on the Qinghai–Tibetan Plateau. *Grassland Research*, *2*(4), 289–298. DOI: 10.1002/glr2.12057

Lin, L., Xu, X., Cao, G., Zhang, F., Li, Y., Fan, B., Li, Q., & Huang, J. (2023b). Synergies between microsites of plant communities and steady-stage alpine meadows on the Qinghai–Tibetan Plateau. *Grassland Research*, *2*(4), 289–298. DOI: 10.1002/glr2.12057

Marian, M., Licciardello, G., Vicelli, B., Pertot, I., & Perazzolli, M. (2022). Ecology and potential functions of plant-associated microbial communities in cold environments. *FEMS Microbiology Ecology*, *98*(1), fiab161. Advance online publication. DOI: 10.1093/femsec/fiab161 PMID: 34910139

Mendes, C. B., & Prevedello, J. A. (2020). Does habitat fragmentation affect landscape-level temperatures? A global analysis. *Landscape Ecology*, *35*(8), 1743–1756. DOI: 10.1007/s10980-020-01041-5

Muhammed, A., & Elias, E. (2021). Class and landscape level habitat fragmentation analysis in the Bale mountains national park, southeastern Ethiopia. *Heliyon*, *7*(7), e07642. DOI: 10.1016/j.heliyon.2021.e07642 PMID: 34377861

Nagy, L., & Grabherr, G. (2009a). *The Biology of Alpine Habitats*. Oxford University PressOxford., DOI: 10.1093/oso/9780198567035.001.0001

Nethmini, R. T., Zhao, H., Pan, L., Qin, X., Huang, J., He, Q., Shi, X., Jiang, G., Hou, Q., Chen, Q., Li, X., Dong, K., Xie, L., & Li, N. (2025). Thermal sensitivity and niche plasticity of generalist and specialist leaf-endophytic bacteria in Mangrove Kandelia obovata. *Communications Biology*, *8*(1), 5. DOI: 10.1038/s42003-024-07446-1 PMID: 39753754

Ning, W., Jian, L., & Zhaoli, Y. (2004). Grazing intensity on the plant diversity of alpine meadow in the eastern Tibetan plateau. *Rangifer*, •••, 9–15. DOI: 10.7557/2.24.4.1664

Nogales, M., Mendes, S. B., Rumeu, B., Marrero, P., Romero, J., Guerrero-Campos, M., Mallorquín, Á., Trujillo, O., Sánchez, C., & Vargas, P. (2025). The potential ecological collapse of the seed dispersal network in an insular high mountain: Teide National Park. *Global Ecology and Conservation*, *62*, e03793. DOI: 10.1016/j.gecco.2025.e03793

Notarnicola, R. F., Nicotra, A. B., Kruuk, L. E. B., & Arnold, P. A. (2023). Effects of warming temperatures on germination responses and trade-offs between seed traits in an alpine plant. *Journal of Ecology*, *111*(1), 62–76. DOI: 10.1111/1365-2745.14014

Pandian, K., Vijayakumar, S., Mustaffa, M. R. A. F., Subramanian, P., & Chitraputhirapillai, S. (2024). Biochar – a sustainable soil conditioner for improving soil health, crop production and environment under changing climate: A review. *Frontiers in Soil Science*, *4*, 1376159. DOI: 10.3389/fsoil.2024.1376159

Park, M., Guzmán, Q. J. A., Dyalsingh, H., Scott, A., & Cavender-Bares, J. (2025). *Forest composition and diversity buffer microclimates and enhance productivity*. Life Sciences. DOI: 10.32942/X2ZP87

Park, S., Kim, J. H., & Lee, E. J. (2022). Resistance of plant communities to invasion by tall fescue: An experimental study combining species diversity, functional traits and nutrient levels. *Basic and Applied Ecology*, *58*, 39–49. DOI: 10.1016/j.baae.2021.12.002

Pelliza, Y. I., Fernandez, A., Saiz, H., & Tadey, M. (2021). Together we stand, divided we fall: Effects of livestock grazing on vegetation patches in a desert community. *Journal of Vegetation Science*, *32*(2), e13015. DOI: 10.1111/jvs.13015

Peringer, A., Frank, V., & Snell, R. S. (2022). Climate change simulations in Alpine summer pastures suggest a disruption of current vegetation zonation. *Global Ecology and Conservation, 37*, e02140. DOI: 10.1016/j.gecco.2022.e02140

Prescott, C. E., Frouz, J., Grayston, S. J., Quideau, S. A., & Straker, J. (2019). Rehabilitating forest soils after disturbance. In *Developments in Soil Science* (Vol. 36, pp. 309–343). Elsevier., DOI: 10.1016/B978-0-444-63998-1.00013-6

Probo, M., Lonati, M., Ravetto Enri, S., Mariotte, P., & Pauler, C. (2025). Influence de la gestion pastorale sur la conservation de la biodiversité des pâturages de montagne. *INRAE Productions Animales, 8436*, 8436. Advance online publication. DOI: 10.20870/productions-animales.2025.38.2.8436

Qi, Y., Wei, W., Chen, C., & Chen, L. (2019). Plant root-shoot biomass allocation over diverse biomes: A global synthesis. *Global Ecology and Conservation, 18*, e00606. DOI: 10.1016/j.gecco.2019.e00606

Riva, F., Koper, N., & Fahrig, L. (2024). Overcoming confusion and stigma in habitat fragmentation research. *Biological Reviews of the Cambridge Philosophical Society, 99*(4), 1411–1424. DOI: 10.1111/brv.13073 PMID: 38477434

Rixen, C., & Rolando, A. (Eds.). (2013). *The Impacts of Skiing and Related Winter Recreational Activities on Mountain Environments*. BENTHAM SCIENCE PUBLISHERS., DOI: 10.2174/97816080548861130101

Santoianni, L. A., Innangi, M., Varricchione, M., Carboni, M., La Bella, G., Haider, S., & Stanisci, A. (2024). Ecological features facilitating spread of alien plants along Mediterranean mountain roads. *Biological Invasions, 26*(11), 3879–3899. DOI: 10.1007/s10530-024-03418-y PMID: 39324107

Sarma, B., Sarmah, N., & Malakar, A. (2024). Degradation of Water and Land Resources in the Himalayan Mountain Ecosystems. In Borthakur, A., & Singh, P. (Eds.), *Addressing the Climate Crisis in the Indian Himalayas* (pp. 31–52). Springer Nature Switzerland., DOI: 10.1007/978-3-031-50097-8_2

Šerá, B. (2021). Tree Regeneration by Seeds in Natural Forests. *Forests, 12*(10), 1346. DOI: 10.3390/f12101346

Sexton, J. P., Clemens, M., Bell, N., Hall, J., Fyfe, V., & Hoffmann, A. A. (2024). Patterns and effects of gene flow on adaptation across spatial scales: Implications for management. *Journal of Evolutionary Biology, 37*(6), 732–745. DOI: 10.1093/jeb/voae064 PMID: 38888218

Shaheen, H., Attique, A., Riaz, M. T., Manzoor, M., Khan, R. W. A., & Riaz, M. T. (2024). From biodiversity hotspot to conservation hotspot: Assessing distribution, population structure, associated flora and habitat geography of threatened Himalayan Yew in temperate forest ecosystems of Kashmir. *Biodiversity and Conservation, 33*(2), 553–577. DOI: 10.1007/s10531-023-02758-w

Sharma, K. P., Tiwari, A., & Shrestha, B. B. (2020). Changes in regeneration and leaf traits of Rhododendron campanulatum along a treeline ecotone in central Nepal. *Journal of Mountain Science, 17*(3), 602–613. DOI: 10.1007/s11629-019-5386-y

Shinoda, Y., & Akasaka, M. (2020). Interaction exposure effects of multiple disturbances: Plant population resilience to ungulate grazing is reduced by creation of canopy gaps. *Scientific Reports, 10*(1), 1802. DOI: 10.1038/s41598-020-58672-6 PMID: 32020019

Speed, J. D. M., Austrheim, G., Hester, A. J., & Mysterud, A. (2012). Elevational advance of alpine plant communities is buffered by herbivory. *Journal of Vegetation Science, 23*(4), 617–625. DOI: 10.1111/j.1654-1103.2012.01391.x

Teague, R., & Kreuter, U. (2020). Managing Grazing to Restore Soil Health, Ecosystem Function, and Ecosystem Services. *Frontiers in Sustainable Food Systems, 4*, 534187. DOI: 10.3389/fsufs.2020.534187

Theobald, D. M., Jacob, A. L., Elsen, P. R., Beever, E. A., Ehlers, L., & Hilty, J. (2024). Evaluating ecosystem protection and fragmentation of the world's major mountain regions. *Conservation Biology, 38*(3), e14240. DOI: 10.1111/cobi.14240 PMID: 38407527

Tóth, E. G., Tremblay, F., Housset, J. M., Bergeron, Y., & Carcaillet, C. (2019). Geographic isolation and climatic variability contribute to genetic differentiation in fragmented populations of the long-lived subalpine conifer Pinus cembra L. in the western Alps. *BMC Evolutionary Biology, 19*(1), 190. DOI: 10.1186/s12862-019-1510-4 PMID: 31623551

Tulloch, A. I. T., Healy, A., Silcock, J., Wardle, G. M., Dickman, C. R., Frank, A. S. K., Aubault, H., Barton, K., & Greenville, A. C. (2023). Long-term livestock exclusion increases plant richness and reproductive capacity in arid woodlands. *Ecological Applications, 33*(8), e2909. DOI: 10.1002/eap.2909 PMID: 37602895

Varga, K., Csízi, I., Monori, I., & Valkó, O. (2021). Threats and challenges related to grazing paddocks: Recovery of extremely overgrazed grassland after grazing exclusion. *Arid Land Research and Management, 35*(3), 346–357. DOI: 10.1080/15324982.2020.1869120

Walther, G.-R., Post, E., Convey, P., Menzel, A., Parmesan, C., Beebee, T. J. C., Fromentin, J.-M., Hoegh-Guldberg, O., & Bairlein, F. (2002). Ecological responses to recent climate change. *Nature, 416*(6879), 389–395. DOI: 10.1038/416389a PMID: 11919621

Wang, J., Soininen, J., & Heino, J. (2021). Ecological indicators for aquatic biodiversity, ecosystem functions, human activities and climate change. *Ecological Indicators, 132*, 108250. DOI: 10.1016/j.ecolind.2021.108250

Wang, Y., Liu, Y., Ding, J., Chen, P., Zhan, T., Yao, Y., Song, J., & Fu, B. (2024). Impact of functional groups on aboveground biomass in alpine grassland communities. *Progress in Physical Geography, 48*(5–6), 698–715. DOI: 10.1177/03091333241280095

Yang, C., Qi, Y., Guo, J., Peng, L., Xiong, N., Zhang, W., & Zhao, W. (2025). Habitat fragmentation increases the risk of local extinction of small reptiles: A case study on Phrynocephalus przewalskii. *Ecotoxicology and Environmental Safety, 290*, 117717. DOI: 10.1016/j.ecoenv.2025.117717 PMID: 39823673

Ye, K., Li, J., Liang, Z., Zhao, X., Luo, H., Chen, Z., Du, P., Wu, K., Qiao, R., & Hu, J. (2025a). Overgrazing impacts plant species diversity in alpine wetlands indirectly by altering its environmental dependency. *Global Ecology and Conservation, 59*, e03532. DOI: 10.1016/j.gecco.2025.e03532

Ye, K., Li, J., Liang, Z., Zhao, X., Luo, H., Chen, Z., Du, P., Wu, K., Qiao, R., & Hu, J. (2025b). Overgrazing impacts plant species diversity in alpine wetlands indirectly by altering its environmental dependency. *Global Ecology and Conservation, 59*, e03532. DOI: 10.1016/j.gecco.2025.e03532

Yu, L., Chen, Y., Sun, W., & Huang, Y. (2019). Effects of grazing exclusion on soil carbon dynamics in alpine grasslands of the Tibetan Plateau. *Geoderma, 353*, 133–143. DOI: 10.1016/j.geoderma.2019.06.036

Zambrano, J., Garzon-Lopez, C. X., Yeager, L., Fortunel, C., Cordeiro, N. J., & Beckman, N. G. (2019). The effects of habitat loss and fragmentation on plant functional traits and functional diversity: What do we know so far? *Oecologia, 191*(3), 505–518. DOI: 10.1007/s00442-019-04505-x PMID: 31515618

Zavala, L. M., Jordán, A., Bellinfante, N., & Gil, J. (2010). Relationships between rock fragment cover and soil hydrological response in a Mediterranean environment. *Soil Science and Plant Nutrition, 56*(1), 95–104. DOI: 10.1111/j.1747-0765.2009.00429.x

Zhang, Q., Zhou, Q., Liu, F., Ma, W., Chen, Q., Wei, B., Li, L., & Zhi, Z. (2025). Progressive Shrinkage of the Alpine Periglacial Weathering Zone and Its Escalating Disaster Risks in the Gongga Mountains over the Past Four Decades. *Remote Sensing (Basel), 17*(14), 2462. DOI: 10.3390/rs17142462

Zhang, Z., Zhao, Y., Lin, H., Li, Y., Fu, J., Wang, Y., Sun, J., & Zhao, Y. (2023). Comprehensive analysis of grazing intensity impacts alpine grasslands across the Qinghai-Tibetan Plateau: A meta-analysis. *Frontiers in Plant Science, 13*, 1083709. DOI: 10.3389/fpls.2022.1083709 PMID: 36733589

Zhou, T., Yang, H., & Du, Y. (2025). Effects of grazing on alpine grassland soil available nutrients across the Tibetan Plateau. *Frontiers in Soil Science, 4*, 1525837. DOI: 10.3389/fsoil.2024.1525837

Zhou, W., Li, C., Wang, S., Ren, Z., & Stringer, L. C. (2023). Effects of vegetation restoration on soil properties and vegetation attributes in the arid and semi-arid regions of China. *Journal of Environmental Management, 343*, 118186. DOI: 10.1016/j.jenvman.2023.118186 PMID: 37224686

Zhou, Z., Su, P., Yang, J., Shi, R., & Ding, X. (2024). Warming affects leaf light use efficiency and functional traits in alpine plants: Evidence from a 4-year in-situ field experiment. *Frontiers in Plant Science, 15*, 1353762. DOI: 10.3389/fpls.2024.1353762 PMID: 38567127

Zhu, J., Zhang, Y., Wu, J., Zhang, X., Yu, G., Shen, Z., Yang, X., He, Y., Jiang, L., & Hautier, Y. (2024). Herbivore exclusion stabilizes alpine grassland biomass production across spatial scales. *Global Change Biology, 30*(1), e17155. DOI: 10.1111/gcb.17155 PMID: 38273528

Chapter 6
Ecology of the Alpine Grasses

Majid Iqbal
http://orcid.org/0000-0003-0561-8494
Institute of Geographic Sciences and Natural Resources Research, Chinese Academy of Sciences, China

Abdul Razzaq
Qarshi Industries (Pvt.) Ltd., Hattar, Pakistan

Syed Nasar Shah
http://orcid.org/0000-0002-9002-6110
Quaid-i-Azam University, Islamabad, Pakistan

Sana Wahab
Quaid-i-Azam University, Islamabad, Pakistan

Alevcan Kaplan
http://orcid.org/0000-0001-6738-7527
Dicle University, Diyarbakır, Turkey

ABSTRACT

Alpine grasses are key components of high-altitude ecosystems, thriving under extreme conditions such as low temperatures, intense radiation, strong winds, short growing seasons, and poor soils. Their ecological success is driven by specialized physiological, morphological, and reproductive adaptations, including dwarf growth forms, dense root systems, photoprotective pigments, and efficient resource-use strategies. These grasses play vital roles in primary productivity, soil stabilization, carbon sequestration, and nutrient cycling. Their interactions with other alpine plants, herbivores, and soil microbes shape community structure and resilience.

DOI: 10.4018/979-8-3373-6756-9.ch006

ght © 2026, IGI Global Scientific Publishing. Copying or distributing in print or electronic forms without written permission of IGI Global Scientific Publishing is prohibited. Use of this chapter to generative artificial intelligence (AI) technologies is expressly prohibited. The publisher reserves all rights to license its use for generative AI training and machine learning model development.

However, climate change through warming, glacial retreat, and altered precipitation poses serious threats to alpine grasslands. Understanding their ecology is essential for biodiversity conservation, ecosystem stability, and climate change mitigation in fragile alpine landscapes.

1. INTRODUCTION

Alpine grasslands are established in mountainous regions at high altitudes, where extreme environmental conditions such as low temperatures, high winds, and frequent snow and ice cover prevent the growth of trees and other vegetation (Bita-Nicolae et al., 2023). The prolong warming of the climate in the alpine area and overgrazing boosts changes in the composition of grasses in the pasture (Wang et al., 2019). These changes in plant diversity are established to affect the above and below ground ecosystem functions, such as soil organism diversity by plant soil feedbacks. Availability of soil nutrients and activity of microbial communities, such as, may either impact positively or negatively on above-ground vegetation, the best quantifiable consequence of which is often a change in the productivity of grassland. One of the ecological properties is plant productivity, which is closely connected to the nutrient circulation, energy flow, and carbon sequestration. The primary indicator of this productivity is above ground biomass. Moreover, the richness of plant species leads to ecosystem stability and assists in maintaining ecosystem functioning as well as the provision of the essential services in the long run (Yingcheng Wang et al., 2023).

In the Himalayan-Karakoram-Hindukush (HKH) ecosystems, alpine grasses in the high mountain systems of Pakistan have important structural and functional components, supporting biodiversity, rangeland productivity, and hydrological regulation under highly stressful climatic conditions. In Pakistan, alpine grasslands above approximately 3,200-3,500 m occur as meadows, steppes, and high altitude pastures, which together help buffer climatic effects, sustain pastoral livelihoods, and maintain carbon and nutrient cycles in the region (Altaf, 2022). Grasslands cover approximately 20% of the Earth land area. They are crucial not only as a source of material resources but also for the conservation of biodiversity and ecosystem services. Grasslands play an important role in ecological stability and the socio-economic growth of regions. However, in recent decades, extreme weather events and human actions have led to a sharp increase in ecological problems, such as large-scale grassland degradation. Between 2000 and 2019, about 39.06% of the world total grasslands experienced some degree of degradation (Liu et al., 2019). In recent decades, climate change coupled with intense anthropogenic pressures, most notably overgrazing has degraded nearly half of the world's alpine grasslands.

This widespread degradation has triggered significant shifts in vegetation composition, plant diversity, productivity, and soil properties, collectively undermining the ecosystem services these fragile habitats provide (Miehe et al., 2019; Teng et al., 2020; Xie et al., 2018). Grassland degradation not only reduces plant productivity but also depletes stocks of soil organic carbon (SOC) and available nitrogen (N) and phosphorus (P), as documented in multiple studies (Dong et al., 2012; Li et al., 2014; Wang et al., 2014). This nutrient loss subsequently creates a negative feedback loop, further constraining plant productivity and ecosystem function in terrestrial systems (Xu et al., 2019).

The alpine grasslands of Pakistan are concentrated in Gilgit-Baltistan and in the Northern Khyber Pakhtunkhwa, located in the Karakoram, western Himalaya, and Hindu Kush ranges, and cover altitudes of about 3,500-5,000 m. These areas have thin soils, well developed only in limited areas, short growing periods, severe solar radiation, freeze thaw cycles, and grass communities with forbs and dwarf shrubs interspersed according to local moisture and topography gradients. The major rangelands include the Deosai Plateau, Khunjerab and neighboring Pamir-type steppes, and the high pastures of Chitral, Upper Dir, and Swat, (Figure 1), which together form some of the key summer grazing fields for transhumant and semi-nomadic peoples (Altaf, 2022; Q. Hou et al., 2022).

The setting of these alpine systems is tightly linked to cryospheric processes, as many grasslands occupy glaciated valleys, moraines, and high plateaus where snowmelt and glacier-fed streams control soil moisture regimes. Continuous transitions from dry alpine steppe to mesic alpine meadow and then to nival vegetation support substantial beta-diversity, with localized endemism reported around lakes such as Karambar and in isolated cirques of the western Himalaya in Pakistan. These landscapes also overlap with protected areas and national parks, such as Central Karakoram National Park and Khunjerab National Park, where alpine grasslands make up a major proportion of the vegetation cover and provide habitat for flagship fauna, including Himalayan ibex and other mountain ungulates that depend on grass and sedge forage (Haq & Al Fouzan, 2017; Ishaq et al., 2016).

1.1 Ecological Drivers and Human Pressures

Grasslands perform a wide range of essential ecological functions, including conserving biodiversity, regulating physical and chemical fluxes within ecosystems, mitigating environmental pollution, and maintaining landscape stability (Pornaro et al., 2017). Alpine grasslands are generally categorized into two main management types: meadows and pastures. They are recognized for supporting exceptionally high species richness compared with forested and shrub-dominated ecosystems. Biodiversity patterns in alpine regions are strongly controlled by key environmental variables such as soil pH, bulk density, nutrient availability, soil moisture, temperature regime, photoperiod, and soil organic carbon (SOC) content. Acidic soil conditions (pH < 7) are often associated with reduced plant species diversity (Rahim et al., 2024; Venterink et al., 2001). Higher SOC levels enhance soil structure and water-holding capacity, enhancing nutrient retention and sustaining soil fertility. In grassland ecosystems, soil bulk density is critical for regulating microbial activity, soil aeration, porosity, nutrient dynamics, and moisture availability, all of which directly influence floristic composition and diversity. In addition to edaphic controls, alpine grassland vegetation is influenced by anthropogenic pressures, particularly grazing intensity; livestock management practices, stocking density, and the use of supplemental feed (Zhou et al., 2023).

In recent decades, Himalayan alpine meadows have experienced increasing grazing pressure. The expansion of livestock populations, combined with the contraction of available grazing land has intensified overgrazing and poses significant threats to the structural integrity and ecological functioning of these sensitive high-mountain ecosystems (Wani et al., 2022). Himalayan pastures have historically served as vital forage resources, sustaining livestock grazing for centuries (Miller, 1999). Traditionally grazed from March - April through September - October, these high-altitude rangelands have experienced a significant reduction in available grazing area per animal over recent decades for instance, declining from 0.15 ha/animal in 1977 to 0.10 ha/animal by 1982 in parts of Kashmir, with continued decreases reported thereafter (Malik, 1988; Qureshi et al., 2007). In light of this mounting pressure, vegetation scientists recommend that grazing in these fragile alpine communities, especially above 2,000 m be strictly limited, controlled, and timed to late summer after full vegetation growth to allow for plant recovery and sustained productivity (Altesor et al., 2005; Shaheen et al., 2011).

2. DIVERSITY AND COMMUNITY STRUCTURE

Alpine grasslands in northern Pakistan host a diverse assemblage of Poaceae, Cyperaceae, and associated forb species, with many taxa showing restricted distributions across the HKH region. Floristic surveys in the Karambar Lake area and other high-altitude basins show that alpine meadows and steppes contain many Himalayan endemics, including several grass species and sedges, which coexist with cushion vegetation and small herbs. This contributes to high local species richness despite extreme abiotic pressures. Community composition changes to bunchgrass and xerophytic grass-shrub steppes on drier slopes and ridges, while graminoid dominated meadows persist in relatively moist microsites. These fine-scale changes are influenced by variations in snow persistence, wind exposure, and soil depth (Joshi et al., 2013; Shaheen et al., 2015; Ullah et al., 2015).

At the landscape level, remote sensing vegetation mapping in Central Karakoram National Park indicates that above 3, 500 m, alpine meadow is probably the most productive type of rangeland, while dry steppe and subalpine areas, though having lower biomass, still provide significant forage value. Environmental filters and biotic interactions shape these grasses dominated communities, resulting in patchy mosaics of sedge mats, turf forming grasses, and dwarf shrubs spaced meters apart, which increases spatial heterogeneity and resource partitioning. The functional trait variation within and between grass species (e.g. plant height, leaf area, rooting depth) influences how a community captures light, water, and nutrients. Recent studies of other alpine systems demonstrate that community-weighted plant height is also strongly related to carbon sequestration potential, which is likely true in the alpine systems of Pakistan as well (Ning et al., 2014; Quan et al., 2024).

Figure 1. Grassland in the Pakistani Hindu Kush range (Gabin Jabba Valley, Swat, Pakistan)

3. ADAPTATIONS FOR SURVIVAL IN EXTREME ENVIRONMENTS

3.1 Morphological and Physiological Adaptations

Current rates of climate change are unprecedented in at least the last two millennia (Legg, 2021), driving widespread reshuffling of ecological communities (Davis & Shaw, 2001; Parmesan & Yohe, 2003). Accurately projecting future community composition and ecosystem function therefore requires understanding which species are vulnerable or advantaged under shifting climates and different species interactions. Such responses are fundamentally governed by species' eco-physiological traits (Beyschlag & Ryel, 2007), knowledge of which can refine predictions of mortality (Anderegg et al., 2016), distribution shifts (Kearney & Porter, 2009), and ecosystem processes (Verheijen et al., 2013). However, because direct physiological measurements are often time-intensive and underrepresented in trait databases, many studies rely instead on morphological proxies such as leaf mass per area and plant height to infer physiological strategies and resource acquisition (Lavorel & Garnier, 2002; Shipley et al., 2016). This reliance on indirect traits limits the mechanistic precision of predictions regarding plant community responses to climate change (Visakorpi et al., 2023). The Pakistani alpine grasses exhibit a complex set of morphological features indicative of convergent adaptation to cold, windy, and low-nutrient environments, including tussock or dwarf growth forms, narrow leaves, and dense tillering that reduce mechanical damage and minimize transpirational water loss. Most species grow in tight tufts or mats that create boundary layers of relatively warmer, less turbulent air, and their low stature is less exposed to abrasive ice and high-speed winds typical of ridgelines and exposed plateaus. Root systems are typically dense and fine, often confined to shallow soil horizons with patchy organic matter, enabling efficient use of thawed surface layers during the brief growing season and allowing rapid regrowth after grazing or frost events (Altaf, 2022).

Physiologically, alpine grasses maintain photosynthetic activity at low temperatures and high irradiance by adjusting pigment composition, enzymatic capacities, and membrane properties. Elevated levels of photoprotective pigments such as carotenoids and anthocyanin-like compounds dissipate excess radiation and limit photo-oxidative damage during clear, cold days, while biochemical acclimation supports sustained carbon gain at temperatures that constrain lowland congeners. Many alpine grasses also exhibit high tissue nutrient-use efficiency and conservative resource strategies, with slow leaf turnover and high N and P resorption before senescence. This trait is particularly advantageous in Pakistan's nutrient-poor alpine soils, where low temperatures and short thaw periods severely constrain mineralization, limiting the release of nutrients from organic matter (Altaf, 2022; Khan et al., 2020; Wani et al., 2022). Cutting-edge research has expanded our understanding of these adaptations, revealing that some alpine angiosperms can grow at soil temperatures of 0°C and with as few as 180 degree-days of warmth, challenging previously assumed thermal limits for plant life. The relationship between specific traits and elevation is a key focus, with studies often using common gardens or reciprocal transplant experiments to distinguish between fixed genetic adaptations and plastic responses to the harsh mountain environment (Hülber et al., 2018).

3.2 Reproductive and Life-History Strategies

Reproductive strategies of Pakistani alpine grasses are shaped by short growing seasons, unpredictable snowfall, and frequent disturbances from frost heave and grazing, favoring a combination of sexual and clonal reproduction that maximizes persistence and colonization ability. Many alpine species reproduce clonally through rhizomes and tillering, which allow rapid lateral spread and gap filling, stabilize soil patches, and ensure local population continuity even though flower and seed production fail due to late frosts or drought. Sexual reproduction, while more variable from year to year, remains critical for maintaining genetic diversity and enabling long-distance dispersal along elevational and microtopographic gradients, especially for species occupying dynamic habitats such as moraines, avalanche tracks, and streamside meadows (Liu et al., 2024a).

Flowering phenology in high-altitude grasslands is tightly synchronized with the onset of snowmelt, so many grasses complete flowering and seed set within a narrow window during midsummer, often under intense grazing pressure. Wind pollination predominates in many graminoids, but zoochorous dispersal by livestock and wild ungulates, as well as hydrochory along meltwater channels, facilitates seed movement across complex terrain in Pakistan's alpine regions. These life-history strategies, which combine clonality with episodic but sometimes abundant seed production, enhance resilience to interannual climatic variability and disturbance, supporting the long-term maintenance of grass-dominated communities in the western Himalaya and Karakoram (Haq & Al Fouzan, 2017).

4. ECOSYSTEM FUNCTIONS AND PROCESSES

4.1 Roles In Primary Productivity and Soil Stabilization

Alpine grasses form the primary photosynthetic base of many high-altitude food webs in northern Pakistan, contributing a large share of the above-ground net primary productivity (ANPP) of alpine meadows and steppe pastures. NDVI-based assessments in the Central Karakoram and adjoining ranges show that alpine meadows, dominated by grasses and sedges, can have substantially higher seasonal productivity than lower-elevation dry steppes, largely due to enhanced snowmelt-driven moisture and cooler summer temperatures that reduce evaporative stress. These productive grasslands are essential for grazing ungulates, including domestic sheep, goats, yaks, and wild species such as the Himalayan ibex, which depend on alpine forage during summer, thereby linking plant productivity directly to pastoral and wildlife health (Ali Shah et al., 2022; Haq & Al Fouzan, 2017).

Beyond biomass production, alpine grasses play a critical role in soil stabilization by forming dense root mats that bind shallow, often skeletal soils on steep slopes, moraines, and cryoturbated surfaces. Their fibrous roots and turf-forming growth reduce erosion from snowmelt runoff and intense rainfall, limiting sediment delivery to rivers and helping maintain downstream water quality in major basins such as the Indus. As grasses trap wind-blown particles and organic debris, they foster the gradual development of soil horizons and microtopography that promote colonization by other plants, thus acting as ecosystem engineers in Pakistan's alpine landscapes (Ali et al., 2024; Ishaq et al., 2016).

4.2 Carbon Sequestration and Nutrient Cycling

One-third of all carbon dioxide (CO_2) emitted by humans is absorbed by terrestrial ecosystems, which serve as important carbon sinks on Earth and significantly help alleviate global warming (Friedlingstein et al., 2022). The relationship between the CO_2 absorbed by plants through photosynthesis and the CO_2 emitted by the ecosystem through respiration is called the net ecosystem exchange (NEE) of CO_2. NEE is negative when the ecosystem acts as a carbon sink and positive when it acts as a carbon source. Quantifying NEE at regional and global scales is necessary for accurate estimates of the terrestrial carbon budget and for a better understanding of the carbon cycle and carbon-climate feedback processes (Friedlingstein et al., 2022; Y Wang et al., 2023). Nevertheless, due to in situ observation, and a lack of mechanistic understanding, there are large gaps of uncertainty in these estimates.

Alpine grasslands are significant but highly variable carbon sinks. Both grass biomass and soil organic matter stores are sensitive to climatic conditions. Long-term regional studies on high altitude plateaus in Asia have shown that alpine meadows, generally serve as better carbon sinks than the drier steppes. The carbon sequestration rates depend greatly on vegetation height and overall productivity, so structurally complex and taller grass communities may increase carbon sequestration under adequate moisture conditions. In Pakistan, site-specific carbon budget data are limited, but alpine grasslands growing on periglacial and permafrost-affected soils are probably storing substantial carbon in deep root networks and slow-decomposing organic horizons.

Temperature, soil moisture and soil development in alpine grasslands closely regulate nutrient cycling. The quality of litter, root exudates, and contact with soil biota are the main factors influencing the development of these cycles and are played by grass species. Decomposition of grass litter is slow in cold conditions, resulting in a slow release of nutrients. Meanwhile, intensive grazing activities promote nutrient recycling through dung and urine, which locally enrich the availability of nitrogen (N) and phosphorus (P), forming small fertility islands on the pastures of Pakistan (Ishaq et al., 2016; Quan et al., 2024).

Emerging experimental studies in other alpine and subarctic ecosystems show that shifts in root traits and community composition under warming can alter rhizosphere microbial activity and nutrient mineralization. These findings suggest that climate-driven changes in grass species composition and functional traits in Pakistan high-altitude grasslands may substantially influence future nutrient cycling and carbon nutrient interactions (Altaf, 2022).

5. BIOTIC INTERACTIONS AND GRAZING DYNAMICS

Biotic interactions play a major role in determining the structure and resilience of alpine grass communities in Pakistan. Competition, facilitation, grazing, and symbiosis together create mosaics of vegetation. Grasses directly compete with forbs, cushion plants, and dwarf shrubs for light, water, and nutrients. However, they can also help other species by reducing wind speed at the surface, trapping snow, and moderating extremes in soil temperature, especially in exposed microsites. At the community scale, these interactions generate fine-grained patterns of dominance and coexistence that change with grazing intensity, topography, and snow cover. This contributes to high landscape-level diversity (Altaf, 2022).

Grazing by livestock and wild herbivores is a defining process in Pakistan's alpine grasslands. Traditional transhumant systems depend on the seasonal movement of herds to high pastures. The scale of this activity is significant; a detailed survey in Kaghan Valley estimated the total livestock population grazing its alpine pastures at more than 163,000 animals. Herds are composed predominantly of goats (56%) and sheep (39%). Both local semi-nomadic grazers and long-distance nomadic grazers manage these herds, relying entirely on this forage for their livelihood (Ali et al., 2021).

Moderate grazing helps maintain productive, species-rich grasslands by preventing shrub encroachment and encouraging grass regrowth. In contrast, chronic overgrazing near settlements and main routes reduces palatable grass cover, compacts soils, and promotes unpalatable species or bare ground. Recent experimental work in Tibetan alpine meadows has demonstrated that excluding grazers can directly increase pollination reliability by altering the complex web of interactions between different plant species (Iqbal & Grichar, 2024). Furthermore, the role of symbiotic associations with soil microbes, particularly mycorrhizal fungi is increasingly recognized as vital for nutrient uptake along steep elevation gradients in alpine environments (Cáceres-Mago et al., 2022).

Symbiotic associations with soil microbes including mycorrhizal fungi and N-cycling bacteria, influence grass performance and ecosystem processes, although detailed studies in Pakistani alpine soils remain limited compared to other regions (Haq & Al Fouzan, 2017; Iqbal et al., 2025; Liu et al., 2024b). Biotic interactions strongly influence the structure and resilience of alpine grass communities in Pakistan, where competition, facilitation, grazing, and symbioses collectively shape vegetation mosaics. Grasses compete with forbs, cushion plants, and dwarf shrubs for light, water, and nutrients, but may also facilitate other species by reducing wind speed at the surface, trapping snow, and moderating soil temperature extremes, particularly in exposed microsites. At the community scale, these interactions generate fine-grained patterns of dominance and coexistence that vary with grazing intensity, topography, and snow cover, contributing to high landscape-level diversity (Altaf, 2022).

6. CLIMATE CHANGE IMPACTS ON ALPINE GRASSLANDS

Grassland ecosystems are highly sensitive to climatic variations. Temperature and precipitation are among the most important climatic factors for identifying the presence, distribution, and general health of grasslands. Variations in these parameters affect soil moisture, soil microbial communities, photosynthesis, and plant respiration. These processes regulate grassland growth, ecosystem productivity, and ecosystem degradation or restoration (Qingqing Hou et al., 2022). One of the most immediate impacts is the timing of snowmelt, a primary trigger for flowering phenology. Studies have shown that while the start of flowering in alpine grasslands is highly responsive to earlier snowmelt, it is only weakly affected by summer drought, highlighting its sensitivity to winter and spring conditions (Rose-Person et al., 2024). Changes in snowmelt timing can also disrupt the delicate synchrony between plants and their pollinators (Rose-Person et al., 2024).

Climate change was observed as a major factor contributing to grassland degradation in this study. The area degraded due to climate effects was larger than areas affected by human activities, especially in the northwest of Qilian County. This finding is consistent with prior studies (Gang et al., 2014), which emphasized climate change as a primary cause of nearly half of global grassland degradation.

Another threat caused by climate change affects the alpine grasslands in Pakistan, primarily due to rising temperatures, shifting precipitation and snowfall patterns, and glacier melt. Recent evaluations of mountainous areas in Pakistan report increasing average temperatures, reduced seasonal snow cover, and more frequent disasters such as glacial lake outburst floods. These factors influence the length of the growing season, soil moisture regimes, and disturbance patterns in high-altitude ecosystems, directly affecting alpine grass phenology, productivity, and species composition.

The cold-adapted species can be driven northward, eventually reaching their limits in the mountains and becoming locally extinct, as has been observed in other alpine systems (Watts et al., 2022). This upward shift is not uniform, as the success of migrating species is often modified by competition from existing resident vegetation. Moreover, long-term experimental warming studies in alpine heath and meadows have confirmed that sustained temperature increases lead to significant changes in plant composition and diversity over time (Parisod, 2025).

Remote-sensing analyses reveal mixed trends in vegetation vigor across Pakistan's highlands. Greening is observed in some regions, associated with afforestation and climate impacts, whereas in other locations, degradation and browning of pastures at higher altitudes are evident. The research on Asian alpine grasslands indicates that warmer and, in some areas, wetter conditions might initially be beneficial, increasing carbon sequestration and productivity. However, these benefits may be offset by increased CO_2 losses during winter and extreme weather fluctuations over time. In Pakistan, projections of warming and changes in monsoon patterns suggest that alpine grasslands may experience growth opportunities in certain areas, but also face threats of drought, erosion, invasion by other species, and rangeland decline in others (Ali Shah et al., 2022; He et al., 2021).

Direct evidence of this upward shift has been documented in Pakistan from Karakoram range. A fourteen-year study (2001-2014) in the Haramosh and Bagrote valleys observed a clear trend of species migrating to higher altitudes. Grazing and drought-resistant species like *Bistorta affinis* (D. Don) Greene and various *Potentilla* species have shifted up from the subalpine zone. Other species, including *Anemone rupicola* and *Pulsatilla wallichiana* have now completely shifted into the alpine zone from lower elevations. This migration is increasing species richness at higher altitudes but also indicates that specialist high-alpine species, like *Saussurea simpsoniana* are being squeezed out as their habitats shrink from below. This provides concrete, local evidence of the profound impact of climate warming on Pakistan's alpine flora (He et al., 2021; Khan et al., 2018).

7. CONSERVATION, MANAGEMENT, AND POLICY IN PAKISTAN

Effective protection of alpine grasses and their ecosystems in Pakistan merges ecological knowledge with rangeland management, protected area planning, and climate adaptation policies. National parks at high altitudes and community-managed conservancies in Gilgit-Baltistan and Khyber Pakhtunkhwa already include large areas of alpine pastures. Still, most grazing systems operate under open access, leading to localized overuse near roads, settlements, and tourism centers. Recent regional action plans, such as the climate adaptation frameworks for Chitral and other mountain districts emphasize the need to manage rangeland health, reduce erosion, and support pastoral livelihoods under changing climatic conditions (Bai et al., 2024; Khan et al., 2016).

Priority actions include improving governance of grazing through rotational grazing and stocking controls, restoring degraded slopes with native grass species, and monitoring vegetation change using field plots integrated with remote sensing to detect early signs of decline or compositional shifts. Strengthening research on the functional traits of alpine grasses, soil-plant-microbe interactions, and carbon budgets in Pakistan will help produce better estimates of ecosystem services and vulnerability, aligning national conservation and climate strategies with global evidence that alpine grasslands are not only sensitive indicators but also active regulators of climate change. Additionally, the cultural significance and economic value that local communities place on these pastures can inform co-managed conservation models that safeguard grassland biodiversity while supporting traditional livelihoods (Bai et al., 2024; Quan et al., 2024).

The economic case for such conservation is very strong. For example, an extensive valuation study of the alpine pastures in Kaghan Valley estimated their economic contribution at over US$10.87 million per year from forage production alone. This highlights how a single ecosystem service can significantly benefit local and regional economies with real monetary value. Therefore, it is urgent to adopt the sustainable management practices recommended by such studies to avoid degradation and maintain these important natural and economic assets for the future sustainability of the region (Ali et al., 2021; He et al., 2021).

This process must address a dual challenge: mitigating a range of direct anthropogenic and natural threats while simultaneously implementing targeted restoration techniques. The primary factors of degradation, including overgrazing, tourism, and climate change, and the key management strategies to counteract them, are summarized in Table 1.

Table 1. Major threats and restoration techniques for alpine pastures in Pakistan

Category	Specific Factor / Technique	Ecological Impact / Mechanism of Action	Relevance to Pakistan's Alpine Pastures
Threats / Factors Of Degradation			
Human Factors	**Overgrazing**	Reduces nutritious forage species; causes soil compaction that lowers water infiltration and accelerates erosion.	Primary cause of degradation due to increasing livestock numbers from pastoral groups such as the Gujjars and Bakarwals.
	Infrastructure & Tourism Development	Landscape fragmentation; road and hotel construction removes vegetation; tourist vehicles cause trampling and soil compaction.	Growing threat in Kaghan, Swat, Hunza, Fairy Meadows where unregulated tourism severely damages alpine meadows.
	Indiscriminate Digging (Medicinal Plants)	Uprooting for the "Jari Boti" trade removes vegetation cover and exposes topsoil to wind and water erosion.	Frequent in Chitral and Diamer where high-value medicinal herbs are harvested unsustainably.
	Over-Cultivation	Expands agriculture into grasslands, reducing grazing area and increasing pressure on remaining pastures.	Occurs in sub-alpine valley zones where farmland is encroaching on traditional summer grazing areas.
Natural Factors	**Climate Change**	Rising temperatures reduce soil moisture; altered precipitation and rapid glacial melt heighten erosion and landslide risk.	Serious concern in Pakistan's high mountains critical for regulating water flow into the Indus River system.
	Wildlife Disturbance (Rodents)	Burrowing species (e.g., marmots) consume forage and create bare soil patches that intensify erosion.	Significant in areas like Deosai National Park where marmot populations are abundant.
Restoration Techniques			
Management	**Enclosure (Fencing) & Grazing Control**	Reduces grazing pressure; supports natural vegetation recovery; basis for rotational grazing.	Widely used in community-managed rangelands across Gilgit-Baltistan and KPK.
Active Intervention	**Sown Grassland (Reseeding)**	Re-establishes vegetation using native, hardy species; controls erosion and restores forage.	Effective on landslide-affected areas, slopes, and heavily overgrazed sites.
	Fertilization	Boosts nutrient-depleted soils, enhancing biomass and productivity.	Organic manure is practical and sustainable for remote alpine restoration projects.
	Rodent Control	Reduces destructive rodent populations through biological/ecological methods.	Necessary where rodents are a major source of pasture degradation

8. FUTURE PERSPECTIVES

Building on the ecological knowledge acquired in this chapter, the long-term conservation and sustainable management of alpine grasslands in Pakistan needs an interdisciplinary approach. Further development should focus on several priority areas to address significant knowledge gaps and translate scientific knowledge into policy and practice:

- **Integrated Monitoring and Early Warning Systems:** Although basic research has been conducted on composition and threats, there is an urgent need for long-term, systematic monitoring programs. The next generation programs must combine advanced remote sensing methods (e.g. satellite and drone-based NDVI analysis) with permanently established on the ground ecological plots. This approach will enable tracking of vegetation phenology, changes in species composition, primary productivity, and provide early warning of degradation hotspots (i.e. brown spots) to support adaptive management.

- **Concentrate on Below-Ground Effects:** A big part of the ecological power and operation of alpine grasslands is concealed in the soil. The next generation of research must focus on below-ground processes as a black box. This includes quantifying the effects of warming and grazing on soil microbial communities, understanding plant, soil-microbe interactions, and precisely mapping the stocks and stability of various soil organic carbon fractions (e.g. POM vs. MAOM) in Pakistani ecosystems. Such research is essential for forecasting the future carbon sink potential of grasslands.
- **Socio-Ecological Systems Research:** Future of these grasslands cannot be understood without considering the livelihoods of pastoral communities who rely on them. The future studies have to go beyond ecological evaluations and adopt a socio-ecological approach. This requires measuring the ecological carrying capacity of various pastures and combining this data with the socio-economic demands of nomadic and semi-nomadic grazers. Co-management models that balance traditional knowledge and scientific concepts should be developed to develop ecologically sustainable and socially equitable grazing systems.
- **Climate Change Adaptation and Restoration Science:** Given the recorded effects of climate change, passive conservation without intervention is insufficient. Active development and experimentation with climate adaptation strategies are necessary for the future. This may include defining and extending native grass ecotypes that are drought and heat tolerant to restore degraded parts, using traditional grazing calendars to align with altered plant phenology, and planning at the landscape level to allow natural species migration pathways. The economic valuation of ecosystem services, as demonstrated in the Kaghan Valley, should be expanded to other regions to build a powerful economic case for investing in these nature-based solutions.

9. CONCLUSION

The high mountain ecosystems in the alpine grasslands of Pakistan support plant life at the extreme limits of the existence, with an astounding number of morphological, physiological, and reproductive adaptations. As this chapter shows, these grasslands are not a passive vegetative cover, but a dynamic and lively component of the Himalayan, Karakoram and Hindukush region. They are engines of productivity that sustain domestic livestock and charismatic fauna, help hold soil on steep terrain to prevent erosion, and serve as critical, though vulnerable stores of terrestrial carbon. This system involves complex interactions between abiotic controls (such as temperature and soil conditions) and biotic interactions, with particular emphasis on the omnipresent role of grazing in maintaining the ecological integrity of these systems. But this strength is being tested. Such ecosystems exist where high levels of anthropogenic stress mainly a long history of grazing, often bordering on overgrazing interact with the newly emerging impacts of global climate change. All the evidence outlined above points to the reality that these landscapes are already being altered by warming trends, which are changing plant phenology, shifting species to higher elevations, and threatening the balance of the carbon cycle. Whether the alpine grasslands of Pakistan can persist in the future depends on how they balance these two challenges. On one hand, it is difficult to avoid local stresses; on the other, it is challenging to adapt to global environmental change. Their continued role as providers of ecosystem services, such as forage production, biodiversity preservation, and hydrological regulation, will not be assured and will depend on recognizing their value and adopting prudent conservation and management plans.

REFERENCES

Ali, A., Khan, M. A., Ashraf, M. I., & Goheer, M. A. J. P. J. o. B. (2021). Economic valuation of ecosystem services in alpine pastures of western Himalayas: a case study of Kaghan valley, Pakistan. *53*(5), 1801-1806.

Ali, I., Shah, A., Alotaibi, B., & Ali, A. (2024). Assessing the impacts of climate change on high mountain land-based livelihoods: An empirical investigation in District Nagar, Gilgit-Baltistan, Pakistan. *Heliyon*, *10*(21), e39877. DOI: 10.1016/j.heliyon.2024.e39877 PMID: 39553690

Ali Shah, I., Muhammad, Z., Khan, H. J. J. o. W., & Change, C. (2022). Impact of climate change on spatiotemporal variations in the vegetation cover and hydrology of district Nowshera. *13*(11), 3867-3882.

Altaf, M. J. E. (2022)... *Ecology of Mountain Ran.*, *6*, 141–149.

Altesor, A., Oesterheld, M., Leoni, E., Lezama, F., & Rodríguez, C. (2005). Effect of grazing on community structure and productivity of a Uruguayan grassland. *Plant Ecology*, *179*(1), 83–91. DOI: 10.1007/s11258-004-5800-5

Anderegg, W. R., Klein, T., Bartlett, M., Sack, L., Pellegrini, A. F., Choat, B., & Jansen, S. (2016). Meta-analysis reveals that hydraulic traits explain cross-species patterns of drought-induced tree mortality across the globe. *Proceedings of the National Academy of Sciences of the United States of America*, *113*(18), 5024–5029. DOI: 10.1073/pnas.1525678113 PMID: 27091965

Bai, Y., Peng, Y., Zhang, D., Yang, G., Chen, L., Kang, L., Zhou, W., Wei, B., Xie, Y., & Yang, Y. (2024). Heating up the roof of the world: Tracing the impacts of in-situ warming on carbon cycle in alpine grasslands on the Tibetan Plateau. *National Science Review*, *12*(2), nwae371. Advance online publication. DOI: 10.1093/nsr/nwae371 PMID: 39872222

Beyschlag, W., & Ryel, R. J. (2007). Plant physiological ecology: An essential link for integrating across disciplines and scales in plant ecology. *Flora-Morphology, Distribution. Flora (Jena)*, *202*(8), 608–623. DOI: 10.1016/j.flora.2007.05.001

Bita-Nicolae, C., Yildiz, F., & Kaya, O. J. S. (2023). Exploring the biodiversity and conservation value of alpine grasslands in the Bucegi Massif. *Romanian Carpathians.*, *15*(16), 12643.

Cáceres-Mago, K., Cáceres, A., & Llambí, L. D. J. A. B. (2022). Effects of nurse shrubs on symbioses between soil fungi and associated plants along a tropical alpine elevation gradient. *132*(2), 285-300.

Davis, M. B., & Shaw, R. G. (2001). Range shifts and adaptive responses to Quaternary climate change. *Science, 292*(5517), 673–679. DOI: 10.1126/science.292.5517.673 PMID: 11326089

Dong, S., Wen, L., Li, Y., Wang, X., Zhu, L., & Li, X. (2012). Soil-quality effects of grassland degradation and restoration on the Qinghai-Tibetan Plateau. *Soil Science Society of America Journal, 76*(6), 2256–2264. DOI: 10.2136/sssaj2012.0092

Friedlingstein, P., O'sullivan, M., Jones, M. W., Andrew, R. M., Gregor, L., Hauck, J., Le Quéré, C., Luijkx, I. T., Olsen, A., & Peters, G. P. J. E. s. s. d. (2022). Global carbon budget 2022. *14*(11), 4811-4900.

Gang, C., Zhou, W., Chen, Y., Wang, Z., Sun, Z., Li, J., Qi, J., & Odeh, I. J. E. E. S. (2014). Quantitative assessment of the contributions of climate change and human activities on global grassland degradation. *72*(11), 4273-4282.

Haq, I. U., & Al Fouzan, K. J. J. o. R. M. C. (2017). Research productivity at King Saud bin Abdul Aziz University for health sciences, Kingdom of Saudi Arabia: A bibliometric appraisal. *21*(2).

He, J.-S., Dong, S., Shang, Z., Sundqvist, M. K., Wu, G., Yang, Y. J. P., & Soil. (2021). Above-belowground interactions in alpine ecosystems on the roof of the world. *458*(1), 1-6.

Hou, Q., Ji, Z., Yang, H., & Yu, X. (2022). Impacts of climate change and human activities on different degraded grassland based on NDVI. *Scientific Reports, 12*(1), 15918. DOI: 10.1038/s41598-022-19943-6 PMID: 36151254

Hou, Q., Ji, Z., Yang, H., & Yu, X. (2022). Impacts of climate change and human activities on different degraded grassland based on NDVI. *Scientific Reports, 12*(1), 15918. DOI: 10.1038/s41598-022-19943-6 PMID: 36151254

Hülber, K., Sonnleitner, M., Haider, J., Schwentenwein, M., Winkler, M., Schneeweiss, G. M., & Schönswetter, P. (2018). Reciprocal transplantations reveal strong niche differentiation among ploidy-differentiated species of the Senecio carniolicus aggregate (Asteraceae) in the easternmost Alps. *Alpine Botany, 128*(2), 107–119. DOI: 10.1007/s00035-018-0209-2

Iqbal, M., Tian, Y., Zhang, Y., Liu, M., Song, M., Zhou, J., Jin, P., Ma, D., Yu, C., & Xu, X. (2025). Grazing directly drives shifts in species composition with minimal mediation by edaphic properties in a temperate grassland. *Plant and Soil*, 1–22. DOI: 10.1007/s11104-025-08114-x

Iqbal, M. A., & Grichar, W. J. (2024). *Grasslands: Conservation and Development*. BoD–Books on Demand. DOI: 10.5772/intechopen.107797

Ishaq, S., Khan, M., Begum, F., Hussain, K., Amir, R., Hussain, A., & Ali, S. J. J. o. M. A. R. (2016). Climate change impact on mountain biodiversity: a special reference to Gilgit-Baltistan of Pakistan. *1*, 53-63.

Joshi, S., Shrestha, R. M., Ismail, M., Qamer, F. M., Jasra, A. W., Dan, Z., Ning, W. J. H.-a. r., & Himalayas, t. i. i. t. H. K. (2013). High-altitude rangelands and land use practices in the Karakoram-Pamir landscape. 37.

Kearney, M., & Porter, W. (2009). Mechanistic niche modelling: Combining physiological and spatial data to predict species' ranges. *Ecology Letters*, *12*(4), 334–350. DOI: 10.1111/j.1461-0248.2008.01277.x PMID: 19292794

Khan, G., Khan, B., Qamer, F. M., Abbas, S., Khan, A., & Xi, C. J. J. o. K. S. U.-S. (2016). Himalayan ibex (Capra ibex sibirica) habitat suitability and range resource dynamics in the Central Karakorum National Park, Pakistan. *28*(3), 245-254.

Khan, M. N., Ali, S., Yaseen, T., Adnan, M., Ullah, S., Zaman, A., Iqbal, M., Shah, S. N., Ali, A., & Razzaq, A. (2020). Assessment of proximate and nutritional contents in selected weedy grasses for potential use as fodder in District Charsadda, KP: Assessment of proximate and nutritional contents in selected weedy grasses. *Proceedings of the Pakistan Academy of Sciences: B. Life and Environmental Sciences, 57*(2), 83-94.

Khan, S., Khatoon, S., & Hussain, A. (2018). Inventorying and monitoring of alpine specific genera and species of haramosh and bagraote valleys., (karakoram range) gilgit-baltistan, Pakistan. *Pakistan Journal of Agricultural Sciences. Pakistan Journal of Agricultural Sciences*, *6*(3), 595–601. Advance online publication. DOI: 10.21162/PAKJAS/18.4998

Lavorel, S., & Garnier, E. (2002). Predicting changes in community composition and ecosystem functioning from plant traits: Revisiting the Holy Grail. *Functional Ecology*, *16*(5), 545–556. DOI: 10.1046/j.1365-2435.2002.00664.x

Legg, S. (2021). IPCC, 2021: Climate change 2021-the physical science basis. *Interaction*, *49*(4), 44–45.

Li, Y.-Y., Dong, S.-K., Wen, L., Wang, X.-X., & Wu, Y. (2014). Soil carbon and nitrogen pools and their relationship to plant and soil dynamics of degraded and artificially restored grasslands of the Qinghai–Tibetan Plateau. *Geoderma*, *213*, 178–184. DOI: 10.1016/j.geoderma.2013.08.022

Liu, P., Jing, X., Degen, A., & Shang, Z. (2024a). Conservation and Sustainable Utilization of Rangelands in the Hindu Kush Himalayan Region. 190–205. DOI: 10.1079/9781800622579.0014

Liu, P., Jing, X., Degen, A., & Shang, Z. (2024b). Conservation and Sustainable Utilization of Rangelands in the Hindu Kush Himalayan Region. In *Sustainable Ecological Restoration and Conservation in the Hindu Kush Himalayan Region: A Comprehensive Review* (pp. 190–205). CABI GB. DOI: 10.1079/9781800622579.0014

Liu, Y., Zhang, Z., Tong, L., Khalifa, M., Wang, Q., Gang, C., Wang, Z., Li, J., & Sun, Z. J. E. I. (2019). Assessing the effects of climate variation and human activities on grassland degradation and restoration across the globe. *106*, 105504.

Malik, M. (1988). Feed Availability, Requirements for Animals and Current Patterns of Utilization in Pakistan. *Non-conventional Feed Resources and Fibrous Agricultural Residues*, 158.

Miehe, G., Schleuss, P.-M., Seeber, E., Babel, W., Biermann, T., Braendle, M., Chen, F., Coners, H., Foken, T., Gerken, T., Graf, H.-F., Guggenberger, G., Hafner, S., Holzapfel, M., Ingrisch, J., Kuzyakov, Y., Lai, Z., Lehnert, L., Leuschner, C., & Wesche, K. (2019). The Kobresia pygmaea ecosystem of the Tibetan highlands–Origin, functioning and degradation of the world's largest pastoral alpine ecosystem: Kobresia pastures of Tibet. *The Science of the Total Environment*, *648*, 754–771. DOI: 10.1016/j.scitotenv.2018.08.164 PMID: 30134213

Miller, D. J. (1999). Herders of forty centuries: nomads of Tibetan rangelands in western China.

Ning, W., Rawat, G. S., Sharma, E. J. H.-a. r., & Himalayas, t. i. i. t. H. K. (2014). High-altitude ecosystem interfaces in the Hindu Kush Himalayan region. 3.

Parisod, C. (2025). The niche of Alpine Botany. *Alpine Botany*, *135*(1), 1–5. DOI: 10.1007/s00035-025-00329-z

Parmesan, C., & Yohe, G. (2003). A globally coherent fingerprint of climate change impacts across natural systems. *Nature*, *421*(6918), 37–42. DOI: 10.1038/nature01286 PMID: 12511946

Pornaro, C., Schneider, M., Leinauer, B., & Macolino, S. (2017). Above-and below-ground patterns in a subalpine grassland-shrub mosaic. *151*(3), 493-503.

Quan, Q., He, N., Zhang, R., Wang, J., Luo, Y., Ma, F., Pan, J., Wang, R., Liu, C., Zhang, J., Wang, Y., Song, B., Li, Z., Zhou, Q., Yu, G., & Niu, S. (2024). Plant height as an indicator for alpine carbon sequestration and ecosystem response to warming. *Nature Plants*, *10*(6), 890–900. DOI: 10.1038/s41477-024-01705-z PMID: 38755277

Qureshi, R. A., Ghufran, M. A., Gilani, S. A., Sultana, K., & Ashraf, M. (2007). Ethnobotanical studies of selected medicinal plants of Sudhan Gali and Ganga Chotti hills, district Bagh, Azad Kashmir. *Pakistan Journal of Botany*, *39*(7), 2275–2283.

Rahim, N. H., Razak, S. A., Chang, X., Saun, F. C., Khan, M. N., Hamzah, S. N., Rohman, F., Ali, B., Kaplan, A., & Iqbal, M. (2024). Ecosystem services by urban forest (UF) towards climate change adaptation: A review. *Polish Journal of Environmental Studies, 33*(4), 3503–3513. DOI: 10.15244/pjoes/177147

Rose-Person, A., Spasojevic, M. J., Forrester, C., Bowman, W. D., Suding, K. N., Oldfather, M. F., & Rafferty, N. E. (2024). Early snowmelt advances flowering phenology and disrupts the drivers of pollinator visitation in an alpine ecosystem. *Alpine Botany, 134*(2), 141–150. DOI: 10.1007/s00035-024-00315-x

Shaheen, H., Khan, S. M., Harper, D. M., Ullah, Z., & Qureshi, R. A. (2011). Species diversity, community structure, and distribution patterns in western Himalayan alpine pastures of Kashmir, Pakistan. *Mountain Research and Development, 31*(2), 153–159. DOI: 10.1659/MRD-JOURNAL-D-10-00091.1

Shaheen, H., Mashwani, Z. U. R., & Dar, M. E. U. I.(2015). Spatial patterns and diversity of alpine vegetation across Langer–Shandur Valley, Hindukush Himalayas. *Current Science,* 1534-1539.

Shipley, B., De Bello, F., Cornelissen, J. H. C., Laliberté, E., Laughlin, D. C., & Reich, P. B. (2016). Reinforcing loose foundation stones in trait-based plant ecology. *Oecologia, 180*(4), 923–931. DOI: 10.1007/s00442-016-3549-x PMID: 26796410

Teng, Y., Zhan, J., Agyemang, F. B., & Sun, Y. (2020). The effects of degradation on alpine grassland resilience: A study based on meta-analysis data. *Global Ecology and Conservation, 24*, e01336. DOI: 10.1016/j.gecco.2020.e01336

Ullah, Z., Ahmad, M., Sher, H., Shaheen, H., & Khan, S. M. J. P. J. B. (2015). Phytogeographic analysis and diversity of grasses and sedges (Poales) of Northern Pakistan. *47*, 93-104.

Venterink, H. O., Wassen, M. J., Belgers, J., & Verhoeven, J. J. J. o. E. (2001). Control of environmental variables on species density in fens and meadows: importance of direct effects and effects through community biomass. 1033-1040.

Verheijen, L., Brovkin, V., Aerts, R., Boenisch, G., Cornelissen, J., Kattge, J., Reich, P. B., Wright, I. J., & Van Bodegom, P. (2013). Impacts of trait variation through observed trait–climate relationships on performance of an Earth system model: A conceptual analysis. *Biogeosciences, 10*(8), 5497–5515. DOI: 10.5194/bg-10-5497-2013

Visakorpi, K., Block, S., Pellissier, L., Levine, J. M., & Alexander, J. (2023). Ecophysiological and morphological traits explain alpine plant species' response to warming. *Functional Ecology, 37*(2), 287–301. DOI: 10.1111/1365-2435.14228

Wang, H., Li, J., Zhang, Q., Liu, J., Yi, B., Li, Y., Wang, J., & Di, H. (2019). Grazing and enclosure alter the vertical distribution of organic nitrogen pools and bacterial communities in semiarid grassland soils. *Plant and Soil, 439*(1), 525–539. DOI: 10.1007/s11104-019-04045-6

Wang, X., Dong, S., Yang, B., Li, Y., & Su, X. (2014). The effects of grassland degradation on plant diversity, primary productivity, and soil fertility in the alpine region of Asia's headwaters. *Environmental Monitoring and Assessment, 186*(10), 6903–6917. DOI: 10.1007/s10661-014-3898-z PMID: 25023744

Wang, Y., Dang, N., Feng, K., Wang, J., Jin, X., Yao, S., Wang, L., Gu, S., Zheng, H., Lu, G., & Deng, Y. (2023). Grass-microbial inter-domain ecological networks associated with alpine grassland productivity. *Frontiers in Microbiology, 14*, 1109128. DOI: 10.3389/fmicb.2023.1109128 PMID: 36760496

Wang, Y., Xiao, J., Ma, Y., Ding, J., Chen, X., Ding, Z., & Luo, Y. (2023). Persistent and enhanced carbon sequestration capacity of alpine grasslands on Earth's Third Pole [In.]. *Science Advances, 9*(20), eade6875. DOI: 10.1126/sciadv.ade6875 PMID: 37196073

Wani, I. A., Verma, S., Gupta, R., Ganaie, M. M., Nigam, G., Shafik, H. M., & Al-Misned, F. A. J. S. (2022). Nutrient analysis and species diversity of alpine grasslands: a comparative analysis of less studied biodiversity hotspots. *14*(2), 887.

Watts, S. H., Mardon, D. K., Mercer, C., Watson, D., Cole, H., Shaw, R. F., & Jump, A. S. (2022). Riding the elevator to extinction: Disjunct arctic-alpine plants of open habitats decline as their more competitive neighbours expand. *Biological Conservation, 272*, 109620. DOI: 10.1016/j.biocon.2022.109620

Xie, H., Wu, Q., Hu, J., Yu, L., Bie, P., Wang, H., & Deng, D. (2018). Changes in soil physical and chemical properties during the process of alpine meadow degradation along the Eastern Qinghai-Tibet Plateau. *Eurasian Soil Science, 51*(12), 1440–1446. DOI: 10.1134/S1064229318130045

Xu, H. P., Zhang, J., Pang, X. P., Wang, Q., Zhang, W. N., Wang, J., & Guo, Z. G. (2019). Responses of plant productivity and soil nutrient concentrations to different alpine grassland degradation levels. *Environmental Monitoring and Assessment, 191*(11), 678. DOI: 10.1007/s10661-019-7877-2 PMID: 31654145

Zhou, H., Yang, X., Zhou, C., Shao, X., Shi, Z., Li, H., Su, H., Qin, R., Chang, T., Hu, X., Yuan, F., Li, S., Zhang, Z., & Ma, L. (2023). Alpine grassland degradation and its restoration in the Qinghai–Tibet plateau. *Grasses, 2*(1), 31–46. DOI: 10.3390/grasses2010004

Chapter 7
Adaptation in Alpine Flora:
Highlighting Cushions, Sedges, and Mosse

Sidra Ahmad
http://orcid.org/0009-0004-4831-5906
University of Agriculture, Peshawar, Pakistan

Nayab Ahmad
University of Agriculture, Peshawar, Pakistan

Md Shahinoor Islam
http://orcid.org/0000-0001-7663-7399
Department of Chemical Engineering, Bangladesh University of Engineering and Technology, Dhaka, Bangladesh

Hansa Gul
http://orcid.org/0009-0001-5554-617X
Department of Zoology, University of Mianwali, Punjab, Pakistan

ABSTRACT

The alpine ecosystems are located in one of the harshest places where plants have to survive with severe cold in winter, extreme winds, short growing seasons, and nutrient-lacking soil. Despite these severe conditions, an extensive variety of organisms, like cushion plants, mosses, and sedges, have been able to spread because of their unique structural, functional, and ecological adaptation characteristics. This chapter integrates a comparative analysis of these three most widespread alpine life-forms, highlighting their roles in the ecosystem. Cushion plants are dense, low and rounded in shape, thereby reducing heat loss and wind stress, providing conducive

DOI: 10.4018/979-8-3373-6756-9.ch007

ght © 2026, IGI Global Scientific Publishing. Copying or distributing in print or electronic forms without written permission of IGI Global Scientific Publishing is prohibited. Use of this chapter to generative artificial intelligence (AI) technologies is expressly prohibited. The publisher reserves all rights to license its use for generative AI training and machine learning model development.

microhabitats to other organisms. Sledge is predominantly found in the alpine fields and wetlands, where it assists in stabilizing the soil and regulating water control. Mosses are, however, exceedingly resistant to desertification and freezing, retain nutrients, and can develop on barren substrates such as rocks, slopes of scree and snow beds.

1. INTRODUCTION

The Alpine ecosystems should be regarded as one of the least thriving ecosystems on the earth, which requires high biological sustainability necessities (Körner et al., 2022). The irregular acclimatization of alpine plants is the response to the undue stress of this ecological environment. Examples of such plants that can survive extreme cold conditions, high speeds of wind, UV rays and a short lifespan are very good illustrations of evolutionary success (Althiab-Almasaud et al., 2024). Algiers is far above the tree line, and the environment becomes biologically extreme. The low temperature of these regions, which is typical of summer frosts and a lot of sun radiation, is reported to be caused by the low altitude of the thin atmosphere (Barrows et al., 2022). Space heterogeneity also occurs in snow cover, soil moisture, and wind exposures due to steep slopes and other topography. Plants are under extreme pressure in such situations. These plants have to adapt to the cold freeze conditions, dry climate, and inadequate resources and duration to reproduce and grow. Changes in the soil and air temperatures in the short term can exceed 20 °C per day. The radical abiotic constraints act as ecological filters, and only species with extremely special morphological and physiological features have been preserved (Spence & Tingley, 2020). The alpine ecosystems are typified by adaptation. The abiotic stress in the alpine zones is more than in the tropical or temperate zones because competition and succession are not the most critical ecological processes in the alpine areas. The evolutionary history of the alpine plants is not one of competitive advantage, but rather of survival and adaptation (Körner, 2021a). These transformations are not only productive to the survival of the individual but also contribute to the arrangement of communities, ecosystem functioning, along the design of the environment (Llambí et al., 2021). They are not a scholarly drill but are rather what is required in the era of global warming. The alpine ecosystems are very fragile, and alterations in the dominant forms of life can conceivably have extensive effects on the integrity of ecological and biodiversity. A comparative ecological roles and survival strategies of alpine life forms are summarized in Table 1. It is then significant how plants have adapted to the high altitudes to predict the future of the alpine ecosystem in the then-changing world (Hou et al., 2024).

Figure 1. Conceptual model illustrating altitudinal zonation and adaptive strategies of alpine flora (cushions, sedges, and mosses)

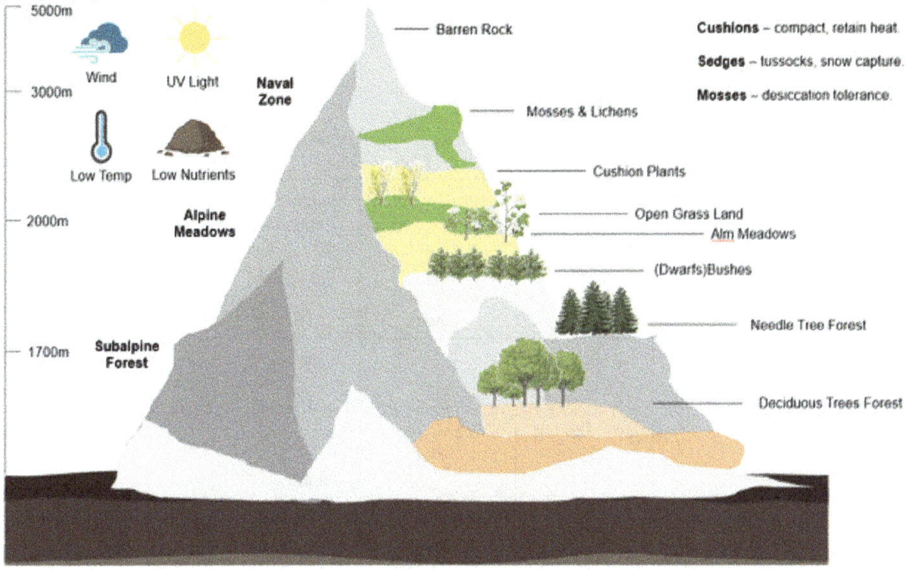

Environmental gradients increase in stress with elevation, while plant life-forms exhibit morphological and physiological adaptations that allow survival under strong UV, low temperature, and nutrient-poor conditions (Wei et al., 2025; Zhao, 2023). While numerous studies describe individual adaptive traits in alpine flora, comparative analyses among different life-forms such as cushion plants, sedges, and mosses remain limited. Most existing research isolates physiological or morphological adaptations within a single species, leaving a gap in understanding how different plant groups coordinate survival strategies under shared alpine stressors. This gap limits our ability to predict ecosystem-level responses to environmental change.

The primary objective of this chapter is to examine how alpine plants have evolved to withstand the extreme environmental pressures characteristic of high-altitude ecosystems. By focusing on three representative life-forms, cushion plants, sedges, and mosses, the chapter aims to explore their distinct structural, functional, and physiological adaptations that enable survival under intense cold, high solar radiation, and limited growing periods. It also seeks to compare their ecological roles and adaptive efficiencies, highlighting how these strategies contribute to the stability and resilience of alpine ecosystems. Furthermore, the chapter intends to evaluate the potential vulnerabilities of these plant groups under projected climate change scenarios and to identify emerging research directions that can enhance understanding of plant persistence and ecosystem sustainability in such fragile habitats.

Table 1. Life-form strategies in alpine plants

Life-form	Growth Form	Microclimate Modification	Cold/Altitude Coping Strategies	Ecological Role on the Mountain	Examples	Reference
Cushion plants	Tight, low domes with very short shoots	Break wind, trap warmth, and retain snow	Can photosynthesize in cold; efficient water use; long-lived leaves	Create safe "nurse" spots for seedling establishment	*Silene acaulis, Azorella spp.*	Chen, J. et al. (2024).
Sedges (Cyperaceae)	Tussocks with rhizomes; spreads via tillers	Elevate growing points above cold/wet ground; trap snow	Thrive in cool climates; air spaces in roots for wet sites; heavy root biomass	Stabilize soil; regulate water flow; build resilient turf	*Carex curvula, Kobresia spp.*	Asghar, A. et al. (2025).
Mosses	Thick mats on rock or soil	Maintain a moist boundary layer; insulate surface	Poikilohydric; tolerate desiccation and freeze–thaw cycles	Pioneer colonizers; initiate soil formation; retain nutrients	*Polytrichum, Grimmia, Racomitrium commune*	Bell-Butler, P. (2024).
Lichens	Crustose, foliose, or fruticose thalli	Moderate temperature extremes; trap fine particles	Fungal-algal symbiosis; extreme drought tolerance	Weather rock; initiate soil; add nitrogen (via cyanolichens)	*Cladonia, Xanthoria, Umbilicaria hyperborean*	Finne, E. A. et al. (2025).
Dwarf shrubs	Very low woody mats; prostrate stems	Warmer near ground; trap snow; reduce wind exposure	Evergreen/ semi-evergreen; mycorrhizal; some fix nitrogen (e.g., Dryas)	Provide structure and litter; support fertility islands; forage	*Dryas octopetala, Empetrum nigrum nigrum*	Buyens, I. P. R. et al. (2023).
Prostrate willows	Creeping stems with ground-level leaves	Enhance boundary layer warmth; provide microhabitats for invertebrates	Cold-hardy photosynthesis; rapid leaf-out post-snowmelt	Offer early-season cover and food; stabilize fine sediments	*Salix herbacea, Salix polaris*	Dumas, K. et al. (2025).
Grasses (Poaceae)	Clumps/tillers; sometimes rhizomatous	Trap snow; reduce erosion; moderate soil temperatures	Primarily C3; cold-adapted; regrows rapidly after frost or grazing	Stabilize slopes; key forage species; contribute to turf development	*Festuca spp., Poa alpine*	Mou, X. M. et al. (2022).
Rosette forbs	Low rosettes with short floral stalks	Leaves sit in the warmer boundary layer; reduce wind exposure	Tough or fast-growing leaves; freeze-tolerant or freeze-avoiding strategies	Attract pollinators; enhance patch-level species richness	*Leontodon pyrenaicus, Primula spp.*	Pugnaire, F. I. et al. (2021).

continued on following page

Table 1. Continued

Life-form	Growth Form	Microclimate Modification	Cold/Altitude Coping Strategies	Ecological Role on the Mountain	Examples	Reference
Geophytes	Bulbs or rhizomes are buried underground	Protected buds; emerge quickly with snowmelt moisture	Burst growth from stored reserves	Provide early nectar; cycle nutrients through the soil	*Crocus vernus, Soldanella alpina*	Althiab-Almasaud, R. et al. (2024).
Alpine annuals (rare)	Tiny herbs with rapid life cycles	Minimal canopy effect; exploit warm micro-sites	Cold-germinating; fast growth and seed set; high seed output	Colonize disturbed sites; contribute genetic and trait diversity	*Koenigia islandica*	Chen, J. G. et al. (2020).

2. THE ENVIRONMENTAL CONSTRAINTS IN THE ALPINE ENVIRONMENTS

The alpine ecosystem is among the most ecologically challenging environments for plant survival. Organisms inhabiting these highlands endure multiple abiotic stressors, including low temperatures, high wind velocities, intense radiation, and nutrient-poor soils (Benateau et al., 2019). Unlike in lowland habitats, biotic interactions play a minor role here plant distribution and structure is primarily governed by physical environmental factors. These conditions not only restrict plant growth and reproduction but also shape their morphological and physiological evolution (Wisz et al., 2013).

2.1 Wind and Mechanical Injury

There are powerful winds that are sustained in the Alpine countries and expose the vegetation to a large extent of physical strain. The winds not only increase the evapotranspiration but also destroy the tissues of the plants, besides stunting growth on the vertical axis. In addition to this, plants may be eroded by wind-bound particles like ice and sand (Ahmad et al., 2023). To protect from these conditions, most alpine plants have developed low-growing or compact growth forms, which make them more resistant to strong winds (Körner, 2021a). Lignified (woody) tissues and flexible stems are some of the structural modifications that would result in mechanical resilience. These morphological qualities not only enhance protection but can also affect the community structure by selectively concentrating or focusing on low-aerodynamic-drag species and wind-resistant morphology (Gibbons, 2025).

The regions that experience low latitudes have had short growing seasons and increased sensitivity to the photoperiod. The snow may extend into the spring of the year, or even into the summer itself, and it may be as late as late summer before the frost(Finne et al., 2025).

2.2 Soil Limitations: Low-Nutrient, Shallow Soils

The soils in the alpine region are often shallow, poorly formed, and lack nutrients such as nitrogen and phosphorus. This can be attributed to low biological activity, weathering, and low decomposition rates. These soils are skeletal, thereby restricting their structure and water absorption (Wilson et al., 2022). Plant evolution is to accommodate these conditions by having a large root system to ensure the plant can absorb as much nutrients as possible (Verma et al., 2021). Root-to-shoot ratios tend to be large, and there are often internal processes of nutrient recycling, including the reabsorption of nutrients in dying tissues. These adaptations enable the plants not only to survive but also to trigger the formation of soils in the unproductive alpine region (Lopez et al., 2023).

2.3 Snow Cover and Water Stress Dynamics

Ironically, although this area often receives precipitation as snow, the amount of water is a major constraint in the alpine ecosystem. The frozen or compacted soils in early spring prevent water infiltration, and rocky soils do not retain moisture due to excessive drainage. Although plants are insulated by snow cover against extreme cold, it also causes a delay in the growing season and decreases the amount of light (Callaghan & Johansson, 2021). Alpine plants respond to dehydration by developing thick cuticles, reduced stomata, and pubescent (bare) leaves to minimize water loss to the environment via transpiration. Then, others that store water inside thick rosettes of cushion plants without involving the hygroscopic (moisture-responsive) leaf movements (Bermúdez Villanueva, 2013). *Poikilic* species like mosses can become completely dry and quickly start metabolism again on rehydration. The species have evolved to such an extent that they can be found at different locations within snowmelt gradients, early melting ridge habitats, and late melting snow beds, each having a distinctive ecological niche (Seastedt & Oldfather, 2021). The shape is round and compact; the morphology resembles that of *Escherichia coli*. The cushion growth form is one of the most specific morphological adaptations to the environment in the Alps. The characteristic feature of this growth form is a small, low, domical shape: a form that is not only a novel form of structure but an ecological adaptation to the severe abiotic conditions of life in high altitudes (Łuczyński, 2022). The dome shape serves the following purposes: it reduces the ratio of the surface to the volume of the object, which reduces the loss of heat and drying of the tissues. The interlaced leaves and stems have a structural cohesiveness and an outer layer that has an insulating quality that preserves the heat that had been stored by the day (Almusaed et al., 2023). Then, the internal cushion temperature can be several degrees higher than the ambient temperature in the air, hence extending the scale of metabolic activity during the short summer in the Alps.

The cushion also has a horizontal and ground-hugging nature that minimizes contact with side winds as well as other airborne hazards such as snow or dust. This type of growth also allows the plant to exploit the thermal nature of the soil, to absorb the heat of the day and release it at a low temperature (Masters, 2021). The small size of cushion plants determines reproductive strategy. The flowers are typically transported to the peak of the dome where they can be accessed by the few pollinators in the high-altitude areas. The shape serves to survive and live long within the environment where there is a lack of energy (this is characteristic of an alpine environment) (Zhao, 2023).

2.4 Wind Protection Microclimate Modulation

Cushion vegetation is an excellent microclimate engineer. During the day, the dense canopy traps the solar heat, which is stored in the tightly packed tissues. This is done by having leaf interstitial spaces and stem interstitial spaces, which are small spaces that reduce convectional heat loss and air movement that helps to keep the internal temperature many degrees higher than that of the air around it (Rupp, 2020). Cushion plants are also thermally inert, which means they retain heat and trap a greater amount of heat inside the plants at night, prolonging the time plants have to complete vital physiological functions such as photosynthesis and respiration (Bhattacharya, 2022). They are also shaped in such a way that they resist the wind. The dome shape reduces the drag and eliminates mechanical damage caused by turbulent gusts or ice particles. Additionally, the flow of air near the edges of the cushion is decreased, decreasing wind-induced transpiration, which is a major stressor in dry winds in the Alps (Alnes, 2024).

Cushion plants grow on soil with more moisture content because there is less evaporation and snow loss. Wind-blown organic debris and fine particles get trapped by these plants and provide nutrient-rich alpine substrates that are otherwise nutrient-poor. This process forms microhabitats which provide sustenance to other less stress-tolerant species, consequently increasing biodiversity and community complexity. Low-Temperature Photosynthesis is the third category of photosynthetic energy conversion to ATP (Bhattacharya, 2022). Another way that cushion plants have adjusted their physiological patterns is through the process of photosynthesis under extremely cold conditions that are not usually conducive to vascular plants.

Aspirin in the alpines reduces metabolic rate, especially enzymatic reactions related to the Calvin cycle (Körner, 2021b).

To address this, cushion species have developed cold-stable forms of critical enzymes, such as Rubisco and other carboxylases, which still take part in reaction at or below the freezing (or even sub-freezing) point. Also, they store homo-cysteinophilic solutes like proline, glycine betaine, and other sugars. These molecules stabilize the protein structures and also maintain osmotic balance during cold stress, and prevent the formation of ice crystals in the cell (Wagner, 2023). The cushion plants also have photosynthetic efficiency compared to other life-forms in the Alps. Their leaves have high amounts of chlorophyll and small-sized chloroplasts and thus can absorb the maximum amount of light even when the irradiance is low (e.g., early mornings, cloudy days, or snow-covered canopies) (Roth-Nebelsick & Krause, 2023).

2.5 Ecological Facilitating: Nurses and Community Effect

In addition to the adaptive physiological characteristics, cushion plants have been an important key factor that determines the alpine ecosystems through interaction through facilitative ecological mechanisms. These plants form neural species where other species have adapted and grown, and reproduced in extreme environmental conditions. The abiotic environment in alpine communities can be the most severe element that limits the progress of a community, and facilitation can be the most significant structuring element of a community (Chen et al., 2017; Pugnaire et al., 2021). The forms of insulated cushion plants create in- and around-the-plant buffered microhabitats that are more favorable than the surrounding bare soils. Such good conditions allow stress-sensitive species to germinate, survive, and prosper to thereby increasing the biodiversity in a location. Recruitment is more successful in seeds on cushion plants or near them due to a more stable substrate, less influence of UV radiation, and harsher microclimates. This effect is particularly facilitative in scree slopes or shallow soils, the bare surface of which would otherwise not support vegetation growth (Chen et al., 2020). The process of biological activity and the enrichment of the enriched habitat testify to the enrichment of the cushion plants in the role of foundation species, crucial to the organization and operation of the entire alpine community (Chen et al., 2024). `

2.6 Examples of Cushion Plants: *Silene acaulis* and *Azorella* spp.

Cushion plants do not belong only to a certain taxonomic line. Instead, they constitute a spectacular case of convergent evolution whereby comparable traits in an unrelated group of families are caused by comparable alpine stressors (Dumas et al., 2025). Two examples, which have been studied well, are *Silene acaulis* and the genus *Azorella. Silene acaulis,* commonly referred to as moss campion, is a common plant found in the Northern Hemisphere, found in Arctic tundra, alpine meadows and on high ridges. Their close spacing of the leaves and the dome-shaped form reduces wind abrasion and sustains warmth to allow photosynthesis even in brief hot spells, even before the snow has all melted (Rai et al., 2025).

2.6.1 Azorella compacta

Azorella species are dominant in high-altitude and high-latitude environments, especially in the Andes and the sub-Antarctic islands in the Southern Hemisphere. A large mass of plants can be formed, which may last hundreds of years to remain. The cushions are well-built, and they can withstand drying in dry areas of the Alps and insulation against wind and cold (Rodriguez-Echeverria et al., 2021). The *Azorella* cushions are ecologically supportive in their environment, which provides habitat, stability of the soil, and buffering of microclimatic conditions on a large surface area. Their very slow rate of growth and survival qualifies them as invaluable indicators of environmental change in the delicate habitats (Buyens et al., 2023).

2.6.2 Tussock, Rhizome and Clonal Growth Structural Adaptations

Cyperaceae (sedges) is a highly resilient functional adaptive family of species responding appropriately to the harsh alpine environment. The tussock structures, rhizomatous growth, and clonal expansion are the critical morphological aspects that would support the survival of sedges in dynamic and hostile environments (Asghar et al., 2025). Tussocks are erect, multi-shoot, thickets or clumps. This is a positive trend in a cold and windy direction in the Alps. To improve the capacity of the plant to trap light and precipitation and heavy bottom layers of organic material, forming a protective cover around the root crown to prevent frost, the leaves are placed in a vertical arrangement. The littering material also improves the nutrient retention in the locality, as well as decreases the humus (Berlepsch, 2022).

Rhizomatous development- horizontal underground stems facilitate the subsequent growth of the sedges laterally across the soil surface and result in genetically identical ramets at a constant rate. The strategy takes advantage of the geographical variability of the alpine environment, in which snow melting, rainfall, and nutrients are patchy (Klimešová & Martínková, 2022). Clonal propagation by basal tillers increases the robustness of the sedges, which survive during conditions of sexual reproduction when the climate varies and pollinators are absent or the seeds are not viable to reproduce (Barrett, 2015). Over a period of time, the clonal patches can spread to meadowlands or hillsides and stabilize the substrates, affecting the assembly of communities (Klimešová & Martínková, 2022). These structural adaptations are not mutually exclusive; about twenty-five species of sedge show a blend of these forms of growth. Such plasticity of architecture and the adaptability of the sedges to a range of alpine microhabitats. Besides, the sedges contribute to the ecosystem construction (soil stability and hydrology) too (Thielen et al., 2021).

2.6.3 Wetlands, Snowmelt Zone, and Alpine Meadows Accommodation

Specifically, alpine sedges are well adapted to fluctuating moisture levels that predominate in wetlands, snow-fed, and alpine meadows. These environments provide harsh fluctuations in water, temperature, and nutrient supplies- yet sedges can live due to their bizarre physiological and structural plasticity (Schloots, 2022). Gas diffusion in water-saturated soils is inhibited, leading to hypoxia. To facilitate the aerobic metabolism of even exquisitely oxygenated compounds, sedges have evolved parenchymatous tissue, which facilitates the flow of air from internal organs to roots submerged within the sedges. This is an imperative feature during the initial seasons of the year, when the snow is melting and meadows and swamps are flooded in the process (Striker, 2024).

The species of sledges are snow-sensitive. Short periods of growth require metabolic adaptation, including the induction of cold-tolerant enzymes and an exaggerated reaction to temperature and emphasis on photoperiodism (Bhattacharya, 2022). On drier alpine meadows, sedges tolerate poor soil water and nutrition. Their roots are fibrous and form very thick mats at the surface of the soil to increase the likelihood of water being taken by the soil in short rainfalls and to assist in the prevention of soil erosion. The leaves, in addition to being narrow, have wax on their leaves, which limits the loss of water through transpiration of the leaves during intense sunlight and high wind conditions (Scharn et al., 2021). They are ecological generalists since their adaptations to endure across the moisture gradients denote their alpine system. Sedges, therefore, are typical foundation plants because they define the soil structure and influence the overall composition of the plant community in saturated and semi-arid habitats (Waheed et al., 2022).

2.7 Hydrological Regulation and Soil stabilization Hydrological regulation and Soil Stabilization

Sledges are highly significant as ecological engineers in the alpine scenery, where the geological processes that are known to cause unstable soils like erosion, steep slopes and little vegetation can be found. Their ecological condition, dense shallow root performance, and growth significantly influence the physical and hydrological qualities of alpine soils (Dai et al., 2023). This shallow alpine root mat is used as a biological net due to the anchoring of organic values to the ground, preventing the loss of topsoil and also increasing the stability of the structure even at steep slopes and inclinations. The gully development, downslope soil motion is alleviated, especially when the organic layer is thin or inefficient (Löbmann, 2020).

The sledge roots contribute organic carbon and nitrogen root turnover to the rhizosphere (root zone). They promote the growth of microbes, soil aggregation and enhance soil fertility. Moreover, they cover the areas with leaves that give shade in the region and alter the exposure of the local winds, which is equally needed in the regulation of evapotranspiration, especially where water is a limiting factor, as in the Alps (Zaqout & Andradóttir, 2021). The Primerica Bank has done well in investigating sexual and clonal reproduction as an approach to the banking industry. Sexual and asexual reproduction of the Alpine sedges ensures that the population can endure climatic environments that are unfavourable to pollination and seed maturation and dispersal. The dual solution enhances genetic variation and or stability in the site (Porro, 2020).

Sexual reproduction that consumes more energy introduces genetic variation that is critical in long-run adaptation and dispersal into a new habitat. Sedges are mostly wind-pollinated, which fits the open, wind-eastern, alpine well. They possess inconspicuous flowers that release clouds of pollen to increase the likelihood of successful fertilization under low population densities (Johannesson, 2024). A few sledges are cleistogamous, which pollinate themselves even in closed flower buds, that is, during the absence of cross-pollination. Other sledge seeds are dormant and they bank in the soil where they sprout during productive seasons. This storage space supports endurance because of the climatic changes and environmental imbalances that allow the sedges to recolonize or colonize new niches that emerge (Petrychenko et al., 2022).

2.8 Keystone Sedges: *Carex curvula* and *Kobresia* spp.

Some species of sledge play a keystone or dominant role in the alpine vegetation. Two of them, *Carex curvula* and species of the genus *Kobresia*, represent the ecological capabilities and plasticity of the *Cyperaceae* family.

2.8.1 Carex curvula (Curved Sedge)

C. curvula is one of the most common species of sedge in European alpine grasslands, forming thick, clonal mats on a massive scale above the tree line. It is also less susceptible to abiotic stress, such as frost, wind, poor soils, and so on. This sledge plays a major role in the establishment of persistent alpine grasslands and vegetation resistance to climate change (Ortega-Solis et al., 2021).

2.8.2 *Kobresia* spp.

Kobresia species, native to the Himalayas, Tibetan Plateau, and other Asian high alpine areas, are dominant in the high-elevation meadows. They grow in permanently wet and seasonally dry regions and display very delicate root networks, turf buildings that are low, and are very resistant to grazing force. They are essential to the stability of the ecosystem in some of the most sensitive and highest locations in the world, as they can endure in poor soil nutrients, and also because of their ability to maintain soil cohesion and hydrologic processes. Just as *Carex curvula, Kobresia* species have well-elaborated rhizomes and fibrous root systems that stabilize the soil, keep the soil hydrated, and achieve storage of nutrients.

2.9 Mosses: Hardy Wretches of the Alpine Scenery

2.9.1 Freezing and Desiccation Tolerance

Mosses are poikilothermic organisms that are non-vascular, and are regarded as the most resilient to freezing and desiccation, which are two of the most prevalent abiotic stressors in the alpine. Their cell structure enables them to withstand severe dehydration and when the temperature is below zero, which can kill most vascular plants (Gong & Ivanovna). These living beings lack internal vascular tissues to move water. They instead take direct precipitation or humid water and store it. Their metabolism is closely synchronized with the moisture of the environment: in dry or frozen conditions, mosses enter a dormant condition in which photosynthesis and respiration effectively stop (Ruchika, 2021). This inactive state may take up to weeks or even years, depending on the severity. When rehydrated, even in a few minutes, mosses immediately resume normal metabolic functions, which indicates remarkable cellular and biochemical tolerance (Pérez-Hernández & Gavilán, 2021).

Mosses have molecular/structural solutions to prevent damage by ice dehydration control reduces intracellular freezing, which causes mechanical damage due to the formation of ice crystals (Mina et al., 2021). Induction of stress proteins, such as dehydrins and Late Embryogenesis Abundant (LEA) proteins, is involved in stress protection of enzyme structure and inhibition of protein denaturation. Some alpine mosses also produce antioxidants to overcome oxidative stress in freeze-thaw periods. The combination of adaptational features permits mosses to inhabit areas with severe water-cycle changes, which vascular plants cannot withstand and recover from in later (Orekhova et al., 2022).

Besides, mosses are able to root in cracks of a very small size, stick to bare rocks or loose gravel through rhizoids and live with a small substrate. This renders them very impervious to mechanical disruptions caused by freeze-thaw of water and erosion (Von Ness, 2021).

2.10 The Cycling of Nutrients and Soil Construction

In alpine ecosystems, particularly in landscapes with low vegetation cover, high disturbance and low organic matter, mosses are important in nutrient cycling and soil formation. As a result of the slow build-up of organic material, moss mats trigger pedogenic activity, transforming mineral substrates into biologically active soils and permitting primary succession. Mosses here can fit some brief downpour of favourable conditions. They are photosynthetic in snow and under low temperatures due to their small size, high concentration of chlorophyll, and low compensation point of light (Hess, 2022).

The growth and increase in these mosses alter the microtopography and hydrology of snow beds. They stabilize soils, slow the flow of water, and trap sediments, and in doing so form new micro habitats allowing other alpine species to colonize. Mosses also contribute to soil development in a chemical way. Moss tissues and the microbiota present on them secrete organic acids that chemically modify mineral surfaces to enhance weathering and nutrient liberation, such as phosphorus and potassium. The decaying mosses contribute carbon and nitrogen to the growing soil through the dying moss colonies, causing the soil to form humus. This fresh organic layer strengthens colonization of the microbes and accelerates the rate at which nutrients cycle. This has the effect of creating a friendlier environment for the vascular plants and increasing the diversity of the surrounding microbes. Mosses are also good at capturing nutrients in the atmosphere (e.g., nitrogen in rainfall and dust-lock elements) and storing them in their tissues. As the mosses die and decay, these nutrients are gradually released into the soil and are availed to other plants and microbes (Eldridge et al., 2023).

2.11 Retention of Water and Development of Microhabitats

Mosses serve as biotic reservoirs that have the highest capacity for water retention in the water-starved and wind-exposed alpine environment. They can hold large amounts of water in relation to their body mass because of their poikilothermic nature, and they have structural characteristics that assist in retaining it even during prolonged droughts. The water in mosses gets absorbed by the stems and leaves by capillary action. The interstitial spaces between leaves increase the surface tension to allow moss colonies to retain water on the surface and on the mat itself. These mats act as sponges, capturing the precipitation of rain or snowmelt and releasing it gradually- thus ensuring local humidity (Thielen et al., 2021). Moss mats can also be the sole source of moisture in otherwise barren habitats, providing conditions conducive to fungi and algae and invertebrates, and even a microbial community. These microhabitats are the centres of biological activity during dry seasons, and they are also centres of ecological interactions, without which there would be no interactions whatsoever (Lawson, 2024).

2.12 Alpine Mosses (*Polytrichum, Grimmia, Racomitrum*)

Certain genera of moss are also specifically alpine due to their ubiquity, ecological activity, and plasticity. The most prevalent of them are *Polytrichum, Grimmia,* and *Racomitrium*, which may be viewed as a set of different survival strategies (Bell-Butler, 2024). The mosses *Polytrichum* spp. also belong to the group of moss species with the most developed structure in the Alps. These species grow erect and are fairly large, with tissues that conduct water, the hydroids and leptoids, so that these species are able to move water and solutes across their tissues. Their leaves are lamellae (or vertical ridges) that increase surface area to carry out photosynthesis and reduce loss of water to wind. Through the adaptations, *Polytrichum* can be used to grow heathlands and grassy alpine slopes, particularly where a preliminary stage of organic soil production has taken place (Bell-Butler, 2024). *Grimmia* species can dry up and rehydrate, ready to photosynthesize almost at once, and therefore are adapted particularly to the xeric and exposed environment of rock faces and alpine crags. *Racomitrium* spp., in particular, *Racomitrium lanuginosum*, is humid soils (e.g., snow bed zones and seepage areas). These woolly mosses possess a superior water-holding capacity of capillary masses because of dense and woolly texture, forming broad blankets in the rock as well as on the ground. The mats prevent soil erosion and form humid macrozones, and as a result, the formation of both cryptogamic organisms and vascular ones is possible (Pérez, 2020). Taken together, these genera describe the ecological diversity and adaptive plasticity of alpine mosses.

3. COMPARATIVE FUNCTIONAL ROLES OF CUSHION, SLEDGE, AND MOSSES

Through the growth morphology of alpine cushion plants, sedges and mosses can be observed (Table 2) that the plants have developed specific structural and ecological approaches to adaptation to the alpine severe environment (Körner, 2021a). The cushion morphology is small and dome-like like thus exposing minimal as well as limiting heat loss and insulating the internal microclimate (Kevan & Coates, 2024). The wind can enter and destroy their small leaves and branches, so their wind and mechanical damage are less than that of the Savanna. These features suit the ridges that are exposed to wind elements and terrains that are prone to frost, where temperature control is mandatory (Kurepa & Smalle, 2023). On the contrary, sedges are taller, rhizomatous, and form clonal tussocks or mats having erect shoots. This colonization trait permits quick adaptation, settlement in the soil and endurance in the wet and dry alpine micro-adaptations. They are more exposed to wind and radios due to their vertical structure, but they balance this by having dense and fibrous root systems that store moisture and fix the soil particles (Brown & Elliman, 2020). Mosses are a non-vascular plant. Being poikilothermic organisms, they need an external water supply and do not conduct water internally. Nevertheless, their desiccation and freezing resistance, barren substrate colonization, and microclimate regulation make them central contributors to primary succession and the establishment of microhabitats. The moss mats serve as sponges that control soil moisture and form humid macrozones important in the early development of the ecosystem (Brodribb et al., 2020).

4. NICHES IN CUSHION SEDGES AND MOSSES

Morphologically heterogeneous (capable of surviving extreme fluctuations in water availability) structures are present in cushions, sedges and mosses. These are erect or prostrate plants of either type of shoot, in either of the two types of mat or cushion, or both species and habitats of micro environment (Deilmann et al., 2025). Mosses do not have roots and vascular tissues and absorb water through precipitation and humidity on the surface. The low-investment structure of scree and other nutrient-depleted substrates enhances the scrambling of bare, unproductive rock. The alpine flora is resolute in two great particulars: (i) structural resilience, a mechanical or environmental protection with physical characteristics; and (ii) physiological resilience, biochemical processes in the body capable of enduring, growing and recuperating in extreme or fluctuating conditions These plans are not oppositely related to one another, but are differently stressed (Rodriguez-Echeverria et al., 2021). Cushion plants are extremely structurally sound; they provide shielded micro environments in their superimposing hemispherical compact forms have less wind exposure and reduce water loss. This gives mechanical power and excellent durability because of the stem bushes and thick leaves (Güllü et al., 2021). Sedges have tussocks that are not associated with any physical injuries, and rhizomes that make them grow in new environments. They endure hypoxia in waterlogged soils by use of aerenchyma tissue and photosynthetically plastic- by short periods of growth during snowmelt or high alpine grasslands (Ortega-Solis et al., 2021).

Mosses are poikilothermic organisms and, as such, in cases of desiccation and total freezing, they attain a dormant stage. They can absorb water and their nutrients on their surface and generate protective proteins, sugar alcohols and antioxidants to rebuild cell structures in times of stress. Mosses are highly robust and can withstand abiotic extremes better than any other vascular plant (Huttunen et al., 2018).

5. FACILITATION NETWORKS AND BIODIVERSITY FEEDBACK IN ALPINE ECOSYSTEMS

Facilitative interactions also increase the ecological value of other facilitators and the survival of an individual species (Table 2). For example, the extension of the microclimatic buffer occurs as moss mats beneath sedges or along the edges of cushion plants. Such overlay networks are more robust and resourceful. This constitutes a model of nonlinearity of interactive facilitation where the niche space of a single life form gets enlarged by another life form. Ecosystems are more diverse and complicated than are individual species (McIntire & Fajardo, 2014). Mosses maintain the water on the surface, protect the cryptogamic flora and harbour the lichens, fungi, algae, and microarthropods, therefore, becoming biodiversity hotspots. These habitats provide the opportunity to survive with taxa that would not have survived in open alpine environments due to their low temperature and high humidity. Similarly, in wetlands that thrive mainly on melting snow and in alpine meadows, sedges provide benefits to ground-dwelling birds, invertebrates within them, and their co-existing forbs and grasses. Mixed patches of clones enhance spatial heterogeneity within plant communities, which stabilizes the communities even in interannual and seasonal disturbances (Gao et al., 2012).

Table 2. Comparative Traits and Ecological Functions of Alpine Cushion Plants, Sedges, and Mosses.

Life-Form	Key Traits	Primary Ecological Functions	Limitations
Cushion Plants	Compact morphology, heat retention, protection against wind	Create stable microclimates, promote the establishment of other species.	High stability but slow growth and limited dispersal potential
Sedges (*Carex* spp.)	Dense clonal tufts, extensive rhizome networks, and high water-use efficiency	Stabilize soils, enhance nutrient cycling, and provide nesting substrate for birds and invertebrates	Rapid colonization but vulnerable to grazing and soil disturbance.
Mosses (Bryophytes)	dense mats retain moisture, surface roughness traps particles	Maintain humidity, shelter lichens, fungi, and micro-fauna	Excellent moisture retention but limited nutrient uptake

5.1 Soil and Nutrient Cycle Stabilization

The annual reuse of sedges of the huge masses of organic matter (leaf litter and root turnover) sequentially accumulates organic matter and maintains microbial decomposers that mineralize nutrients. The soils and nutrient cycling on bare mineral surfaces are also significant in the initial development of soils and nutrients; the mosses do not have roots at all. They trap tiny grains and retain water as microsites on which microbial communities may grow and introduce chemical weathering (Mou et al., 2022). When moss tissues decay, they accumulate nutrients, especially in low-nutrient and low-phosphorus habitats. Some moss species contain cyanobacteria, which a nitrogen-fixing species of alpine soils that are oligotrophic (Jungblut & Vincent, 2017). This inactive nitrogen eventually enters the soil organic pool, thus helping other plant species. A combination of cushion plants, sedges, and mosses has helped in the control of the distribution and retention of nutrients in the alpine. They fix Soils, and increase organic matter and minimise the erosion, leaching and low decomposition rates in cold soils (Schloots, 2022).

5.2 Hydrology and Carbon Sequestration

Two processes that are closely related in the alpine ecosystem include hydrological regulation and the process of carbon sequestration. Climate change is the consequence of such forces as changes in solar variations, atmospheric changes, etc (Asimakopoulos, 2013). Alpine life forms (cushions, sedges, mosses) are not biomass-intensive, but contribute significantly to the control of carbon and water. Cushion plants are slow-growing, long-lived plants and upon growth, they accumulate biomass. The dense formations that occur over and underground tissues are centuries old (Ortega-Solis et al., 2021). The sedges act as carbon sinks that stabilize the carbon in the anoxic wetlands' soils, where decomposition is impossible (Smiti, 2023). The lack of oxygen in seasonally wet/waterlogged soils significantly slows down the rate at which the microbial decomposition process takes place. This causes the sedges to convert their litter into a peat-like substance through a gradual process that is especially evident in the fens of high elevation and snowmelt basins. These sedge communities form thick organic soils, or long-term carbon sinks, over time when left undisturbed and not eroded. (Johnston et al., 2016). Mosses are also currently equally noteworthy in hydrological buffering and carbon storage. The wide cover they possess on the ground and their high absorption due to the properties they have enabled them to store a lot of water, which has moderated water availability to other organisms. Moss mats reduce surface runoff, increase surface infiltration and surface humidity, increasing microhabitat conditions for other biota (Shepherd et al., 2021).

5.3 Keystone Roles of the Alpine Ecosystems

Among significant life forms - an organism whose ecological role is so disproportionately high to their biomass or abundance are the cushion plants, sedges and mosses The cushion dome-shaped structure mitigates the wind shear and temperature variation, which supports the quality of micro habitats in species that are less stress-tolerant. They are also more likely to be linked to a higher biodiversity, especially on bare and exposed ridgelines and rock outcrops (Spence & Tingley, 2020).

6. THREATS AND PROTECTION PRIORITY

Alpine ecosystems are ancient but sustainable and are growing more vulnerable due to rapid climate change, human disruption and land use change. The anthropogenic forces of the world are triumphing over natural defenders against external interference (Vazirani & Bhattacharjee, 2021).

6.1 Climate Change and Shift in Elevation Range

Habitat compression, caused by climate, is one of the biggest threats. The increase in temperature is pushing many species in the Alps up the mountain and away from their habitat, causing possible local extinction on the mountain tops. This process is also called the elevation squeeze, and it affects cushion plants and mosses, which tend to be confined to the highest and coldest niches (Meza-Joya et al., 2023).

6.2 Land Use and Fragmentation

Remote, alpine environments are not resistant to human influence. Tourism (e.g., ski resorts, trails) and mountaineering can lead to habitat fragmentation, soil compression, and invasive species (infrastructure development). These disturbances damage the integrity of cushion mats, kill sledge tussock, and steal moss layers, causing instability of weak soil profiles and enhancing erosion (Smiti, 2023). Though traditionally part of alpine scenery, grazing may also have a destructive effect when not used responsibly: trampling cushion flora, overloading the nutrient-cycling process, and consolidating the soil, which harbours sensitive moss ecosystems (Verma et al., 2021).

6.3 Conservation and Control Plans

The ability of the soil to sustain the existence of a number of soil communities is indirectly due to the suspension of microorganisms into the soil through their roots (Serrano et al., 2024). Systems with seasonal or interannual disturbances have more species that survive across years in their Sledge habitats. The intermixing of clonal clusters facilitates the presence of the sedges alongside other graminoids and forbs in a facilitative manner of the spatial heterogeneity of plant assemblages (White, 2023). It is also stated that mosses offer biodiversity by way of the cryptogamic mats that cover the soil and the soil is not dried out, especially where scarce is vascular vegetation. The algae, fungi, lichens and microarthropods found in these mats require a high-humidity and low-temperature microclimate. The ecological value of any one facilitator enlarges the value of another, e.g., moss mats under sedges or at the boundaries of cushion plants extend the supporting footprint of biodiversity. The networks make the ecosystems more resilient, enabling more species to survive in conditions that would otherwise lead them beyond their physiological limits. Sedge also stabilizes the soil through the overground and underground structures. The root structure of this soil is cohesive and does not erode even during rainy seasons. Dead sedges introduce ruggedness to the surface and lower the speed of water, thereby enhancing sedimentation (Gillis et al., 2022). Mosses do not have a root; however, they play an important role in soil formation. They trap minute particles and retain moisture to create microsites of biotic growth and chemical weathering. Their decaying remains are sources of nutrients with a low rate of decomposition- especially in low-nitrogen and phosphorus conditions. Cyanobacteria fix oligotrophic alpine soils, which can be present in specific mosses (Wei et al., 2025).

Figure 2. Policy-relevant conceptual model of alpine ecosystem threats and protection priorities

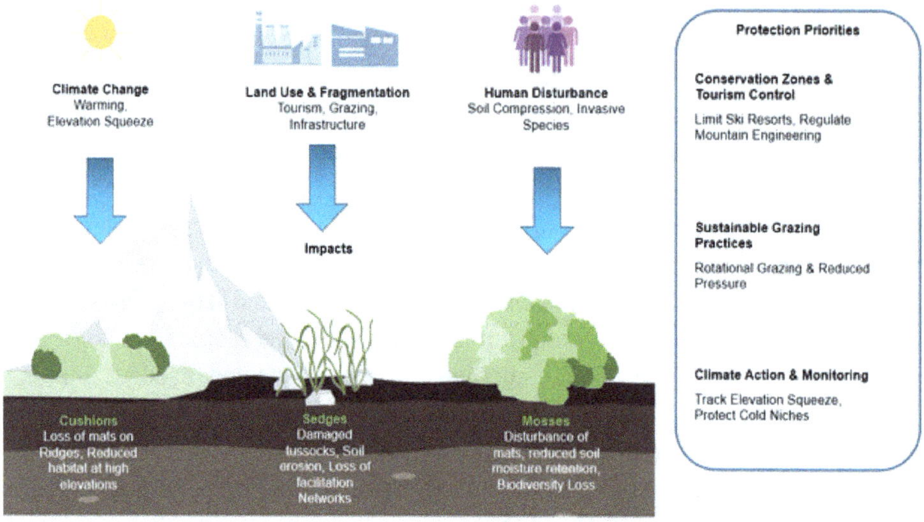

Climate change, land-use change, and human disturbances exert increasing pressure on alpine cushion plants, sedges, and mosses. These impacts reduce biodiversity, soil stability, and ecosystem resilience. Targeted policy measures such as regulating tourism, sustainable grazing, climate monitoring, and restoring facilitation networks are essential to safeguard alpine ecosystems against accelerating vulnerability.

7. THE ALPS ECOSYSTEM AS CLIMATE SENTINELS

Cushion plants, sedges and mosses serve as early warning indicators of Global ecology as they respond to shifts in their distributions quickly to changes in temperature, snowmelt timing and soil moisture. Cushion vegetation is an ecosystem facilitator and engineer. Mosses stabilize the succession of the microbes, fix nitrogen, and enable the existence of peats (Leppänen et al., 2015). They collaborate to strengthen climate and species survival and even nutrient cycling. This is further discussed in terms of future research and the direction of the study (Xiao et al., 2023). The available data is based on a few systems of mountains, which are fairly well researched; therefore, extrapolation cannot be applied. Expansion of field studies in diverse climates and settings will complement our existing understanding of the processes in global alpines, and can help to incorporate alpine data in climate and biodiversity prediction models (Leng et al., 2023). This approach can address unclear ecological questions and generate a kind of adaptive regulating regimes that will ensure the survival of exceptional constructions and ecosystems that they sustain, which are found in the Alps (Webb et al., 2020).

8. CONCLUSION

The extreme environmental conditions of the alpine ecosystems, however, allow certain groups of flora to grow, which have some particular adaptive mechanisms to live and survive under the constant abiotic stress conditions. Together, these three species function as keystone species that provide microhabitats, regulate hydrology, stabilize soil, conserve nutrients, and serve community assembly. These keystone life-forms could degenerate and cause a disruptive environment within the alpine systems, leading to cascades of ecological interactions, which would ultimately destabilize the global climatic regulation, biodiversity dynamics and ecosystem stability. Only these strategies will allow us to ensure the survival of more extraordinary forms of life of alpine systems- and the ecosystem services they may provide to us in the generations to come.

REFERENCES

Ahmad, L., Biswas, A., Warland, J., & Anjum, I. (2023). Atmospheric Pressure and Solar Radiation. In *Climate Change and Agrometeorology* (pp. 17–51). Springer. DOI: 10.1007/978-981-99-4863-5_3

Almusaed, A., Almssad, A., Alasadi, A., Yitmen, I., & Al-Samaraee, S. (2023). Assessing the role and efficiency of thermal insulation by the "BIO-GREEN PANEL" in enhancing sustainability in a built environment. *Sustainability (Basel)*, *15*(13), 10418. DOI: 10.3390/su151310418

Alnes, V. (2024). *Net primary production of alpine lichen in a changing environment*. NTNU.

Althiab-Almasaud, R., Teyssier, E., Chervin, C., Johnson, M. A., & Mollet, J.-C. (2024). Pollen viability, longevity, and function in angiosperms: Key drivers and prospects for improvement. *Plant Reproduction*, *37*(3), 273–293. DOI: 10.1007/s00497-023-00484-5 PMID: 37926761

Asghar, A., Ameer, A., Ahmad, F., Hameed, M., Ahmad, M. S. A., Qin, M., Asghar, A., Khan, S. A., Li, H. Y., & Zhang, Z. F. (2025). Adaptive response of rare mountain sedge (Carex Ecostata CB Clarke) to hyper-cold environments. *Plant and Soil*, *512*(1), 751–775. DOI: 10.1007/s11104-024-07108-5

Asimakopoulos, D. (2013). Climate and climate change. In *Energy and climate in the urban built environment* (pp. 19–32). Routledge.

Barrett, S. C. (2015). Influences of clonality on plant sexual reproduction. *Proceedings of the National Academy of Sciences of the United States of America*, *112*(29), 8859–8866. DOI: 10.1073/pnas.1501712112 PMID: 26195747

Barrows, T. T., Mills, S. C., Fitzsimmons, K., Wasson, R., & Galloway, R. (2022). Low-altitude periglacial activity in southeastern Australia during the late Pleistocene. *Quaternary Research*, *107*, 125–146. DOI: 10.1017/qua.2021.72

Bell-Butler, P. (2024). *Moss Matters: Unravelling the intricate web of interactions between mosses, microclimate, and seed germination in New Zealand's alpine ecosystems*. Open Access Te Herenga Waka-Victoria University of Wellington.

Benateau, S., Gaudard, A., Stamm, C., & Altermatt, F. (2019). Climate change and freshwater ecosystems: Impacts on water quality and ecological status.

Berlepsch, H. (2022). *The Alps*. BoD–Books on Demand.

Bermúdez Villanueva, R. (2013). *Spatio-temporal variability in ecophysiological determinants of performance: a study in perennial herbs of coastal foredunes.*

Bhattacharya, A. (2022). Effect of low temperature stress on photosynthesis and allied traits: A review. *Physiological processes in plants under low temperature stress*, 199-297.

Bhattacharya, A. (2022). *Physiological processes in plants under low temperature stress*. Springer. DOI: 10.1007/978-981-16-9037-2

Brodribb, T. J., Carriquí, M., Delzon, S., McAdam, S. A., & Holbrook, N. (2020). Advanced vascular function discovered in a widespread moss. *Nature Plants*, *6*(3), 273–279. DOI: 10.1038/s41477-020-0602-x PMID: 32170283

Brown, L., & Elliman, T. (2020). *Grasses, sedges, rushes: an identification guide.* Yale University Press.

Buyens, I. P. R., Raath-Krüger, M. J., Haddad, W. A., & Le Roux, P. C. (2023). Fine-scale variation in the effect of the cushion plant Azorella selago on vascular plants, mosses, hepatics and lichens in the sub-Antarctic. *Journal of Vegetation Science*, *34*(4), e13200. DOI: 10.1111/jvs.13200

Callaghan, T. V., & Johansson, M. (2021). Snow, ice, and the biosphere. In *Snow and ice-related hazards, risks, and disasters* (pp. 137–164). Elsevier. DOI: 10.1016/B978-0-12-817129-5.00012-3

Chen, J., Chen, X., Qian, L., Zhang, Y., Li, B., Shi, H., Sun, L., Schöb, C., & Sun, H. (2024). Degeneration of foundation cushion species induced by ecological constraints can cause massive changes in alpine plant communities. *Science China. Life Sciences*, *67*(4), 789–802. DOI: 10.1007/s11427-022-2383-6 PMID: 38057621

Chen, J., Li, Y., Yang, Y., & Sun, H. (2017). How cushion communities are maintained in alpine ecosystems: A review and case study on alpine cushion plant reproduction. *Plant Diversity*, *39*(4), 221–228. DOI: 10.1016/j.pld.2017.07.002 PMID: 30159515

Chen, J. G., Yang, Y., Wang, S. W., Stöcklin, J., Peng, D. L., & Sun, H. (2020). Recruitment of the high elevation cushion plant Arenaria polytrichoides is limited by competition, thus threatened by currently established vegetation. *Journal of Systematics and Evolution*, *58*(1), 59–68. DOI: 10.1111/jse.12481

Dai, L., Fu, R., Guo, X., Du, Y., Zhang, F., & Cao, G. (2023). Variations in and factors controlling soil hydrological properties across different slope aspects in alpine meadows. *Journal of Hydrology (Amsterdam)*, *616*, 128756. DOI: 10.1016/j.jhydrol.2022.128756

Deilmann, T., Bernhardt-Römermann, M., Hentschel, J., Gros, P., & Römermann, C. (2025). Temperate Forest Floor Bryophytes Functionally Respond to Small-Scale Variability in Water, Light and Nutrient Availability. *Ecology and Evolution*, *15*(8), e71839. DOI: 10.1002/ece3.71839 PMID: 40746384

Dumas, K., Marais, G., Paitreault, S., Avrillier, J.-N., Rosa, Z., Hedde, M., Rasplus, J.-Y., Chérasse, S., Ouvrard, D., Reynaud, P., Trap, J., Filali Alaoui, I., Noël, F., Perrier, C., Yannic, G., Gallet, C., Bernier, N., Lavergne, S., & Ibanez, S. (2025). Cushion plants in the Alps are swarming with invertebrate life. *Alpine Botany*, *135*(2), 1–12. DOI: 10.1007/s00035-025-00335-1

Eldridge, D. J., Guirado, E., Reich, P. B., Ochoa-Hueso, R., Berdugo, M., Sáez-Sandino, T., Blanco-Pastor, J. L., Tedersoo, L., Plaza, C., Ding, J., Sun, W., Mamet, S., Cui, H., He, J.-Z., Hu, H.-W., Sokoya, B., Abades, S., Alfaro, F., Bamigboye, A. R., & Delgado-Baquerizo, M. (2023). The global contribution of soil mosses to ecosystem services. *Nature Geoscience*, *16*(5), 430–438. DOI: 10.1038/s41561-023-01170-x

Finne, E. A., Bjerke, J. W., Stordal, F., & Tallaksen, L. M. (2025). Lichens are more tolerant against winter warming stress than vascular and non-vascular plants: Insights from an alpine field experiment. *Journal of Ecology*, *113*(3), 518–530. DOI: 10.1111/1365-2745.14482

Gao, Y., Xing, F., Jin, Y., Nie, D., & Wang, Y. (2012). Foraging responses of clonal plants to multi-patch environmental heterogeneity: Spatial preference and temporal reversibility. *Plant and Soil*, *359*(1), 137–147. DOI: 10.1007/s11104-012-1148-0

Gibbons, B. (2025). *Europe's Alpine Flowers: A Field Guide*. Princeton University Press.

Gillis, L., Maza, M., Garcia-Maribona, J., Lara, J., Suzuki, T., Cierco, M. A., Paul, M., Folkard, A., & Balke, T. (2022). Living on the edge: How traits of ecosystem engineers drive bio-physical interactions at coastal wetland edges. [Gong, L., & Ivanovna, V. I. Water holding capacity of some bryophyte species in connection to their anatomy and morphology.]. *Advances in Water Resources*, *166*, 104257. DOI: 10.1016/j.advwatres.2022.104257

Güllü, A., Danquah, J. O., & Dilibal, S. (2021). Characterization of energy dissipative cushions made of Ni–Ti shape memory alloy. *Smart Materials and Structures*, *31*(1), 015018. DOI: 10.1088/1361-665X/ac383d

Hess, C. (2022). *Insulation effect of moss versus vascular plant cover on active layer depth in the High Arctic Svalbard* Itä-Suomen yliopisto].

Hou, M., Leng, C., Zhu, J., Yang, M., Yin, Y., Xing, Y., & Chen, J. (2024). Alpine and subalpine plant microbiome mediated plants adapt to the cold environment: A systematic review. *Environmental Microbiome, 19*(1), 82. DOI: 10.1186/s40793-024-00614-0 PMID: 39487507

Huttunen, S., Bell, N., & Hedenäs, L. (2018). The evolutionary diversity of mosses–taxonomic heterogeneity and its ecological drivers. *Critical Reviews in Plant Sciences, 37*(2-3), 128–174. DOI: 10.1080/07352689.2018.1482434

Johannesson, K. (2024). Climate Change and Baltic Sea Genetic Diversity. In *Oxford Research Encyclopedia of Climate Science*. DOI: 10.1093/acrefore/9780190228620.013.925

Johnston, S. R., Boddy, L., & Weightman, A. J. (2016). Bacteria in decomposing wood and their interactions with wood-decay fungi. *FEMS Microbiology Ecology, 92*(11), fiw179. DOI: 10.1093/femsec/fiw179 PMID: 27559028

Jungblut, A. D., & Vincent, W. F. (2017). Cyanobacteria in polar and alpine ecosystems. In *Psychrophiles: from biodiversity to biotechnology* (pp. 181–206). Springer. DOI: 10.1007/978-3-319-57057-0_9

Kevan, P. G., & Coates, C. (2024). Heliocaminiform structures: Plant organs that function as microgreenhouses. *Facets, 9*, 1–20. DOI: 10.1139/facets-2023-0099

Klimešová, J., & Martínková, J. (2022). Clonal growth, resprouting, and vegetative propagation of weeds. *Persistence strategies of weeds*, 200-218.

Körner, C. (2021a). The alpine life zone. In *Alpine plant life: Functional plant ecology of high mountain ecosystems* (pp. 23–51). Springer.

Körner, C. (2021b). The climate plants experience. In *Alpine plant life: Functional plant ecology of high mountain ecosystems* (pp. 65–88). Springer.

Körner, C., Berninger, U.-G., Daim, A., Eberl, T., Mendoza, F. F., Füreder, L., Grube, M., Hainzer, E., Kaiser, R., Meyer, E., Newesely, C., Niedrist, G., Niedrist, G. H., Petermann, J. S., Seeber, J., Tappeiner, U., & Wickham, S. (2022). Long-term monitoring of high-elevation terrestrial and aquatic ecosystems in the Alps–a five-year synthesis. *Eco.Mont, 14*(2), 48–69. DOI: 10.1553/eco.mont-14-2s48

Kurepa, J., & Smalle, J. A. (2023). Plant hormone modularity and the survival-reproduction trade-off. *Biology (Basel), 12*(8), 1143. DOI: 10.3390/biology12081143 PMID: 37627027

Lawson, E. (2024). *Moss and Lichen*. Reaktion Books.

Leng, R., Harrison, S., & Anderson, K. (2023). Himalayan alpine ecohydrology: An urgent scientific concern in a changing climate. *Ambio*, *52*(2), 390–410. DOI: 10.1007/s13280-022-01792-2 PMID: 36324019

Leppänen, S. M., Rissanen, A. J., & Tiirola, M. (2015). Nitrogen fixation in Sphagnum mosses is affected by moss species and water table level. *Plant and Soil*, *389*(1), 185–196. DOI: 10.1007/s11104-014-2356-6

Llambí, L. D., Melfo, A., Gámez, L. E., Pelayo, R. C., Cárdenas, M., Rojas, C., Torres, J. E., Ramírez, N., Huber, B., & Hernández, J. (2021). Vegetation assembly, adaptive strategies and positive interactions during primary succession in the forefield of the last Venezuelan glacier. *Frontiers in Ecology and Evolution*, *9*, 657755. DOI: 10.3389/fevo.2021.657755

Löbmann, M. T. (2020). *The surface-mat effect and its measurement for estimating shallow erosion resistance of herbaceous vegetated alpine slopes*. Free University of Bozen-Bolzano.

Lopez, G., Ahmadi, S. H., Amelung, W., Athmann, M., Ewert, F., Gaiser, T., Gocke, M. I., Kautz, T., Postma, J., Rachmilevitch, S., Schaaf, G., Schnepf, A., Stoschus, A., Watt, M., Yu, P., & Seidel, S. J. (2023). Nutrient deficiency effects on root architecture and root-to-shoot ratio in arable crops. *Frontiers in Plant Science*, *13*, 1067498. DOI: 10.3389/fpls.2022.1067498 PMID: 36684760

Łuczyński, P. (2022). Tsunamites versus tempestites: Various types of redeposited stromatoporoid beds in the Devonian of the Holy Cross Mountains (Poland), a case study from the Ołowianka Quarry. *PLoS One*, *17*(5), e0268349. DOI: 10.1371/journal.pone.0268349 PMID: 35580129

Masters, N. (2021). *For the love of soil: Strategies to regenerate our food production systems*. Integrity Soils Ltd.

McIntire, E. J., & Fajardo, A. (2014). Facilitation as a ubiquitous driver of biodiversity. *The New Phytologist*, *201*(2), 403–416. DOI: 10.1111/nph.12478 PMID: 24102266

Meza-Joya, F. L., Morgan-Richards, M., Koot, E. M., & Trewick, S. A. (2023). Global warming leads to habitat loss and genetic erosion of alpine biodiversity. *Journal of Biogeography*, *50*(5), 961–975. DOI: 10.1111/jbi.14590

Mina, U., Kumar, A., Chaturvedi, A. K., & Kumar, P. (2021). Climate change and plant diversity: Threats and opportunities. In *Climate Change and the Microbiome: Sustenance of the Ecosphere* (pp. 369–396). Springer. DOI: 10.1007/978-3-030-76863-8_20

Mou, X. M., Yu, Y. W., Zhao, C., Soromotin, A., Kuzyakov, Y., & Li, X. G. (2022). Sedge replacement by grasses accelerates litter decomposition and decreases organic matter formation in alpine meadow soils. *Land Degradation & Development, 33*(16), 3260–3270. DOI: 10.1002/ldr.4386

Orekhova, A., Barták, M., Hájek, J., & Morkusová, J. (2022). Species-specific responses of spectral reflectance and the photosynthetic characteristics in two selected Antarctic mosses to thallus desiccation. *Acta Physiologiae Plantarum, 44*(1), 6. DOI: 10.1007/s11738-021-03339-6

Ortega-Solis, G., Díaz, I. A., Mellado-Mansilla, D., Tejo, C., Tello, F., Craven, D., Kreft, H., & Armesto, J. J. (2021). Trash-basket epiphytes as secondary foundation species: a review of their distribution and effects on biodiversity and ecosystem functions. bioRxiv, 2021.2006. 2022.449473.

Pérez, F. L. (2020). Andean rolling mosses gather on stone pavements: Geoecology of Grimmia longirostris Hook. in a high periglacial páramo. *Catena, 187*, 104389. DOI: 10.1016/j.catena.2019.104389

Pérez-Hernández, J., & Gavilán, R. G. (2021). Impacts of land-use changes on vegetation and ecosystem functioning: Old-field secondary succession. *Plants, 10*(5), 990. DOI: 10.3390/plants10050990 PMID: 34065656

Petrychenko, V. F., Korniichuk, O. V., Buhaiov, V. D., Horenskyi, V. M., & Veklenko, Y. A. (2022). Influence of hydrothermal factors on feed and seed productivity of alfalfa in the conditions of the Right-Bank Forest-Steppe. *Редакційна колегія, 13*(3), 49-59.

Porro, F. (2020). Comparative plant diversity and seed germination patterns of alpine species in the context of climate change.

Pugnaire, F. I., Losapio, G., & Schöb, C. (2021). Species interactions involving cushion plants in high-elevation environments under a changing climate. *Ecosistemas (Madrid), 30*(1), 2186. DOI: 10.7818/ECOS.2186

Rai, S., Breme, N., Jandova, V., Lanta, V., Altman, J., Ruka, A. T., Rixen, C., & Dolezal, J. (2025). Growth dynamics and climate sensitivities in alpine cushion plants: Insights from Silene acaulis in the Swiss Alps. *Alpine Botany, 135*(1), 79–89. DOI: 10.1007/s00035-024-00318-8

Rodriguez-Echeverria, S., Delgado-Baquerizo, M., Morillo, J. A., Gaxiola, A., Manzano, M., Marquet, P. A., González, L., Cavieres, L. A., Pugnaire, F. I., & Armas, C. (2021). Azorella cushion plants and aridity are important drivers of soil microbial communities in Andean ecosystems. *Ecosystems (New York, N.Y.)*, *24*(7), 1576–1590. DOI: 10.1007/s10021-021-00603-1

Roth-Nebelsick, A., & Krause, M. (2023). The plant leaf: A biomimetic resource for multifunctional and economic design. *Biomimetics*, *8*(2), 145. DOI: 10.3390/biomimetics8020145 PMID: 37092397

Ruchika, R. (2021). *Small-spatial scale ecophysiological and stress-induced (salicylic acid) biochemical investigations on desiccation-tolerant bryophyte SYNTRICHIA RURALIS*. Magyar Agrár-és Élettudományi Egyetem.

Rupp, A. I. (2020). *Leaf-inspired Design for Heat and Vapor Exchange*. The University of Akron.

Scharn, R., Little, C. J., Bacon, C. D., Alatalo, J. M., Antonelli, A., Björkman, M. P., Molau, U., Nilsson, R. H., & Björk, R. G. (2021). Decreased soil moisture due to warming drives phylogenetic diversity and community transitions in the tundra. *Environmental Research Letters*, *16*(6), 064031. DOI: 10.1088/1748-9326/abfe8a

Schloots, C.-L. (2022). *Water level fluctuation and vegetation patterns within an alpine wetland complex*. University of Otago.

Seastedt, T. R., & Oldfather, M. F. (2021). Climate change, ecosystem processes and biological diversity responses in high elevation communities. *Climate (Basel)*, *9*(5), 87. DOI: 10.3390/cli9050087

Serrano, A. R., Peco, B., Morillo, J. A., & Ochoa-Hueso, R. (2024). Abandonment of traditional livestock grazing reduces soil fertility and enzyme activity, alters soil microbial communities, and decouples microbial networks, with consequences for forage quality in Mediterranean grasslands. *Agriculture, Ecosystems & Environment*, *366*, 108932. DOI: 10.1016/j.agee.2024.108932

Shepherd, D. A., Souitaris, V., & Gruber, M. (2021). Creating new ventures: A review and research agenda. *Journal of Management*, *47*(1), 11–42. DOI: 10.1177/0149206319900537

Smiti, K. (2023). Tourism and the properties of mountainous soil: A dynamic relationship. In *Understanding Soils of Mountainous Landscapes* (pp. 213–230). Elsevier. DOI: 10.1016/B978-0-323-95925-4.00011-X

Spence, A. R., & Tingley, M. W. (2020). The challenge of novel abiotic conditions for species undergoing climate-induced range shifts. *Ecography, 43*(11), 1571–1590. DOI: 10.1111/ecog.05170

Striker, G. G. (2024). Oxygen Transport and Plant Ventilation. In *Responses of Plants to Soil Flooding* (pp. 139–156). Springer. DOI: 10.1007/978-981-99-9112-9_9

Thielen, S. M., Gall, C., Ebner, M., Nebel, M., Scholten, T., & Seitz, S. (2021). Water's path from moss to soil: A multi-methodological study on water absorption and evaporation of soil-moss combinations. *Journal of Hydrology and Hydromechanics, 69*(4), 421–435. DOI: 10.2478/johh-2021-0021

Vazirani, A., & Bhattacharjee, T. (2021). Entrepreneurial finance in the twenty-first century, a review of factors influencing venture capitalist's decision. *The Journal of Entrepreneurship, 30*(2), 306–335. DOI: 10.1177/09713557211025654

Verma, S. K., Sahu, P. K., Kumar, K., Pal, G., Gond, S. K., Kharwar, R. N., & White, J. F. (2021). Endophyte roles in nutrient acquisition, root system architecture development and oxidative stress tolerance. *Journal of Applied Microbiology, 131*(5), 2161–2177. DOI: 10.1111/jam.15111 PMID: 33893707

Von Ness, K. (2021). *Quantifying the effects of moisture change on the spectral signatures (350-2500 nm) of Arctic mosses, liverworts, and lichens*

Wagner, A. (2023). *Sleeping beauties: the mystery of dormant innovations in nature and culture.* Simon and Schuster.

Waheed, M., Haq, S. M., Arshad, F., Bussmann, R. W., Iqbal, M., Bukhari, N. A., & Hatamleh, A. A. (2022). Grasses in semi-arid lowlands—Community composition and spatial dynamics with special regard to the influence of edaphic factors. *Sustainability (Basel), 14*(22), 14964. DOI: 10.3390/su142214964

Webb, J. W., Khoury, T. A., & Hitt, M. A. (2020). The influence of formal and informal institutional voids on entrepreneurship. *Entrepreneurship Theory and Practice, 44*(3), 504–526. DOI: 10.1177/1042258719830310

Wei, M., Sun, Q., & Liu, D. (2025). Moss Cover Modulates Soil Fungal Functional Communities and Nutrient Cycling in Alpine Forests. *Forests, 16*(1), 138. DOI: 10.3390/f16010138

White, G. A. (2023). The Optimization of Tissue Culture Conditions for the Production of Cultivated Meat.

Wilson, B. R., Tulau, M., Kuginis, L., McInnes-Clarke, S., Grover, S., Milford, H., & Jenkins, B. R. (2022). Distribution, nature and threats to soils of the Australian Alps: A review. *Austral Ecology*, *47*(2), 166–188. DOI: 10.1111/aec.13115

Wisz, M. S., Pottier, J., Kissling, W. D., Pellissier, L., Lenoir, J., Damgaard, C. F., Dormann, C. F., Forchhammer, M. C., Grytnes, J. A., Guisan, A., Heikkinen, R. K., Høye, T. T., Kühn, I., Luoto, M., Maiorano, L., Nilsson, M.-C., Normand, S., Öckinger, E., Schmidt, N. M., & Svenning, J.-C. (2013). The role of biotic interactions in shaping distributions and realised assemblages of species: Implications for species distribution modelling. *Biological Reviews of the Cambridge Philosophical Society*, *88*(1), 15–30. DOI: 10.1111/j.1469-185X.2012.00235.x PMID: 22686347

Xiao, L., Zhang, W., Hu, P., Vesterdal, L., Zhao, J., Tang, L., Xiao, D., & Wang, K. (2023). Mosses stimulate soil carbon and nitrogen accumulation during vegetation restoration in a humid subtropical area. *Soil Biology & Biochemistry*, *184*, 109127. DOI: 10.1016/j.soilbio.2023.109127

Zaqout, T., & Andradóttir, H. Ó. (2021). Hydrologic performance of grass swales in cold maritime climates: Impacts of frost, rain-on-snow and snow cover on flow and volume reduction. *Journal of Hydrology (Amsterdam)*, *597*, 126159. DOI: 10.1016/j.jhydrol.2021.126159

Zhao, R. (2023). The influence of cushion plants on soil fertility islands under their canopies with ontogenesis in an extreme alpine ecosystem. *Plant and Soil*, *484*(1), 379–392. DOI: 10.1007/s11104-022-05798-3

Chapter 8
Soil Erosion and Vegetation Succession in the Alpine Zone

Aroona Saleem
Edith Cowan University, Australia

Muhammad Ammar Javed
http://orcid.org/0000-0001-9308-5127
University of Western Australia, Australia

Affan Afzal
http://orcid.org/0009-0009-3565-397X
The Islamia University of Bahawalpur, Pakistan

Baber Ali
Quaid-i-Azam University, Pakistan

ABSTRACT

Alpine zones are characterized by harsh, extremely cold environmental conditions that have traditionally limited human access. However, increasing human activities in recent years have made these fragile landscapes vulnerable to soil erosion, topographic alterations, and vegetation degradation. Climate change poses an additional threat, causing glacier-induced floods, nutrient loss, and accelerated soil erosion. These pressures have led to significant ecological consequences, including biodiversity loss, disruption of soil–vegetation interactions, altered hydrological systems, and reduced carbon sequestration capacity. Damage to soil and vegetation undermines habitat stability and weakens ecosystem resilience, making alpine environments more susceptible to future disturbances. Due to the slow natural recovery rates in

DOI: 10.4018/979-8-3373-6756-9.ch008

ght © 2026, IGI Global Scientific Publishing. Copying or distributing in print or electronic forms without written permission of IGI Global Scientific Publishing is prohibited. Use of this chapter to
generative artificial intelligence (AI) technologies is expressly prohibited. The publisher reserves all rights to license its use for generative AI training and machine learning model development.

alpine regions, such degradation is often long-lasting or irreversible. Therefore, sustainable conservation strategies such as controlled grazing, restoration using native plant species, and effective protected area management are essential to mitigate these impacts.

1 INTRODUCTION

Alpine zones are the regions situated above the tree line and they represent some of the most vulnerable ecosystems susceptible to climate change losing their distinguishing environmental characteristics and increased biodiversity. They account for only 3.8% of the world's high mountainous regions, nevertheless they house 44.1% of the alpine vegetation (Zhang et al., 2025; Zou et al., 2025). The alpine vegetation provides critical environmental services, consisting of requisitioning carbon, regulation of climate, and the conservation of water. Though, the increase in temperature led to expansion of vegetation and an upward shift, it causes the changes in the ecosystem function and the structure of the community. The resultant factors led to compounding effects on the soil erosion, hydrological cycles and biodiversity as well as accompany novel environmental conditions confronting the growth of vegetation at higher altitude. The average temperature at the alpine zones stretches between 40°F and 60°F. Thus, the changing conditions in the alpine environments due to human activity and climate change pose significant challenges for the fauna and flora inhabiting these parts of the world which are already limited and sparse (Körner & Kèorner, 1999). The alpine ecosystem has varying features which change with geographical placement as well as the locality in the hemisphere in which the alpine region is situated. Normally, the alpine zones are marked from the upper boundary of the tree line to the permanent snow line. Depending on the regional latitude and conditions, the alpine zone ranges from as low as 3,400–3,600 m and as high as 5,000 m (Buytaert et al., 2011). The alpine zones ecosystems are very known owing to their endemism- the existence of unique and specific biodiversity to the specific region (Dirnböck et al., 2011) as well as the diverse land and aquatic habitats, rendering it a unique ecological unit (Lamouille-Hébert et al., 2024). This persisting interrelationships between the climate and features unique to terrain form a complex microclimates network and an environmental gradient (Alatalo et al., 2016). Notwithstanding the harsh conditions, these alpine ecosystems feature significant biodiversity, which comprises approximately 10,000 species of dicotyledonous vegetation, reflecting the high degree of diversity specialization (Verrall & Pickering, 2020)

In the alpine ecosystem, the complex relationship between climate and terrain support microclimatic diversity and ecological variability which in turn support high ecological specialization of vegetative diversity suited to extreme and high environmental gradient and steep stress terrain (Manzoor et al., 2025). Owing to this level of specialization, the alpine zones show a high level of biodiversity compared to other regions and ecosystems (Verrall & Pickering, 2020). However, an extreme threat to this established ecosystem is the recent transformation. Thus, until the temperature stabilizes, alpine biotic diversity will continue to be susceptible to climatic conditions posed by climate change in terms of habitat and support (Ludovicy et al., 2022). Moreover, global warming is comparatively more pronounced than the regions at higher elevations (Yang et al., 2023), further reducing the optimal habitat required for the cryophilic and alpine and other temperature-sensitive species.

The biggest challenge that the soil of alpine or mountainous terrain is facing is the soil erosion. The soil erosion is threatening the inherent balance of the alpine ecosystem. The factors such as agricultural practices, climate change and tourism are accelerating the soil erosion in the alpine zone (Belasri & Lakhouili, 2016). The consequences of the soil erosion include degradation of soil structure, topsoil loss, depletion of the organic matter, and nutrients, moisture reduction, and low availability of soil, reduced river capacity and enhanced risk of landslide. The steep sloppy areas periodically subjected to the rainfall and debris flow also result in soil erosion. In order to better understand the dynamics of current alpine soil erosion and promotion of sustainable land management, the precise data on the spatial distribution as well as temporal patterns of the region is necessary. However, the contemporary detecting methods for these specifics are lacking applicability are time consuming and are restricted to a small area relatively (Samarin et al., 2020). The hardest challenge posed by soil erosion is to the vegetation in the alpine zones. The scientific significance of this link between soil erosion and vegetation both existing and new demands studies and solutions to counter this phenomenon. On the other hand, the presence of the vegetation on the alpine zones also controls the soil erosion in a feedback mechanism. The methods to counter the soil erosion by vegetation are well studied (Xu et al., 2006). The focus of present work on the alpine soil erosion has been mostly the soil particles displacement, the dynamics and behaviours of nutrients and organic matters in the post eroded soil. Little has been studied to devise methods to cope and control the soil erosion. Methods like the seeds implanting, despite being the cornerstone of regulating soil erosion by colonization and recovery process in soil rehabilitation has revived little attention (Jiao et al., 2009).

The vulnerability of the land to the soil erosion can be lessened by the management techniques. The technique such as the addition of the organic matter to alpine soil helps in preserving the structure and flexibility of the soil. Other methods of land management in the alpine ecosystem is the terracing, a typical technique used to stabilize slopes in the steep sloppy alpines, decrease the soil erosion, the further degradation of soil, and improve the chemical and physical properties of soil (Stanchi et al., 2013). This book chapter thus focuses on the detail study of the alpine zones soil, geography and vegetation. Furthermore, the effect of the soil erosion has been studied along with its kinds and impact on the vegetation and long term succession in the alpine zones. The land management and protection techniques suited to the alpine region has been also analyzed.

Classification and Types of Alpine Soil

Soil classification in modern ecology recognizes a variety of soil types that change rapidly over a short distance in the alpine zones. Previous US soil classification recognized all soil in the meadows above timberline or tree zones as Alpine meadow soil; however, recent classification has made a distinction observation in this regard (Whitesides & Butler, 2010). The following are some of the soil classifications based on the newer classification system:

1.1 Leptosols and Regosols

These soils are dominantly prevalent on steep slopes of alpine zones, which are young and poorly developed. Rendzic Leptosols are formed on the calcareous rocks, and Umbric Leptosols form on the parent acidic material of the alpine slopes (Knežević et al., 2025).

1.2 Cambisols and Inceptisols

These types of soils are formed on the stable slopes having weatherable substrates such as micaceous schists, where they often form deep profiles relative to other soil types in the slopes (Wu & Chen, 2005).

Podzols

These types of soil form locally on crystalline rocks, which are acidic with moderate input of calcareous dust by podsolization (Legros, 1992).

1.3 Soils on Limestone

The Calcareous schists and calcshales, when they undergo the process of rapid decalcification, result in a soil spectrum, a product of Cambisols and Podzols decalcified (Retzer, 1956).

1.4 Gleysols and Histosols

Formed in galaciated alpine with poorly drained locations, Gleysols are strongly shaped by frost processes, while Histosols develop in flat areas with thick organic or peat horizons (Repe et al., 2017).

1.5 Andisols

Prominent in humid climates and especially on volcanic ash deposits, where high aluminum–humus complexes and organic matter accumulation dominate soil formation (Buytaert et al., 2006). See Table 1 for comparative details on the soil groups found in the Alpine Zones.

Table 1. The detailed soil group-based classification in the alpine zones

Major Soil Group	Typical Alpine Locations	Main Features	Dominant Processes	Example Regions	Reference
Leptosols	Steep mountain slopes, shallow soil over rock	Very shallow, skeletal, low fertility	Physical weathering, erosion	Himalayas, Andes, Alps	(Getnet & Asseffa, 2025; Wu & Chen, 2005)
Cambisols	Mid-altitude mountain slopes	Weakly developed B horizon, moderate fertility	Initial stage of soil formation	Alps, Hindu Kush	(Legros, 1992)
Cryosols	High-altitude permafrost zones	Permanently frozen subsoil, low organic matter	Cryoturbation (freezing–thawing cycles)	Tibetan Plateau, Alaska ranges	(Legros, 1992)
Podzols	Cold, humid alpine forests	Strongly leached, acidic, with iron/aluminium accumulation	Podzolization	European Alps, Rockies	(Legros, 1992)
Regosols	Recently deposited or eroded areas	Weakly developed profile, coarse texture	Deposition and erosion	Glacial valleys, young slopes	(Knežević et al., 2025)
Andosols	Volcanic alpine regions	Dark, porous, rich in allophane and organic matter	Weathering of volcanic ash	Andes, Japan Alps	(Buytaert et al., 2006)
Fluvisols	Alpine river valleys and floodplains	Stratified layers, high moisture content	Alluvial deposition	Alpine valleys of Central Asia	(Repe et al., 2017)
Histosols	Wet alpine meadows and depressions	Thick organic layer, poor drainage	Peat accumulation	Swiss Alps, Northern Himalayas	(Repe et al., 2017)

1.6 Environmental Sensitivity

The soils of the alpine belt represent dynamic but fragile systems. Their limited depth, high organic content, and dependence on climatic conditions make them particularly sensitive to anthropogenic pressure and global warming. Understanding their properties is therefore crucial for managing alpine ecosystems and conserving biodiversity (Bockheim, 2015).

2 TYPES OF VEGETATION IN THE ALPINE ZONE: A GENERAL DISCUSSION

2.1 Definition and Extent

The Alpine vegetation consists of the plant species inhabitants of that zones at altitudes above the tree line. As discussed previously, these species are marked by their presence on high altitude below the nival zone and above the tree line, with a permanent sheet of snow and ice dominating the terrain. The upright trees in this zone are restricted by adverse environmental conditions like very low temperatures, high wind velocity, as well as limited soil formation. This zone is basically a transition from subalpine forests to mountain summits. The exact elevation of the tree line in the alpine zones varies and depends on the location of the region, climate, slope orientation, and human activities. For example, in New Zealand, the timberline occurs at an elevation of 1,500 m, and in the Pyrenees of Europe, it is at 2,000 m. Similarly, in Iran, in the Hindu Kash and Pamir mountains, the timberline is elevated at 3,000 m, 3,500 m, and 4,000 m (Maccioni et al., 2024).

The adverse alpine environment poses serious challenges to the vegetation growth and development. The restricting factors impacting vegetation include very low temperatures, a small window of growing period, increased frost damage, high wind velocity, and high solar radiation. The snow accumulation, as well as ice movements in the alp summits, led to wilting and burying of vegetation under increased ice weight, damaging growth further. Furthermore, soils in these zones are often of poor quality, shallow, with poor holding quality, and susceptible to erosion. These limitations inhibit the growth of upright, tall, rooty, and woodsy trees (Table 2). Most of the vegetation in these regions is herbaceous, grasses, and low shrubs (Körner et al., 2017).

The vegetation is organized in the alpine zone into altitudinal belts. The first belt above tree line is made up of shrubs and alpine grassland dominated by herbaceous small species and dwarf shrubs. As the altitude increases and exploration in the alpine zone advances, the plant cover decreases subsequently, and bare scree, visible rock, and snowfields replace vegetation. The grasslands and shrublands from this belt further extend to subalpine regions where trees in stunted forms exist (Koyanagi et al., 2025).

2.2 Composition of flora in the alpine zone

Although alpine communities consist of diverse plant species however no unified pattern of plant classification is fixed. The plants, such as angiosperms, especially perennial herbs, are a major group of vegetation, as well as lichens, ferns, and bryophytes are common in the alpine zone. However, owing to the brevity of the growth season, the Annual species exist less frequently (Janiczek et al., 2025). Despite the challenges of the environment and altitude, four plant life forms are observed in the perennial herbs in this zone:

1. Mound-like cushion plants that help in trapping moisture and heat.
2. Basal leaves that arranged in a rosette-like pattern to divert wind and conserve heat and energy.
3. Dense but leafy plants with stems having clustered foliage branch tips.
4. Narrow leaves, grass-like plant species that are desiccation and frost-resistant (Giupponi et al., 2025).

Table 2. Plant Types in Alpine Zones

Vegetation Zone / Type	Altitude Range	Dominant Plant Forms	Key Adaptations	Typical Species	Reference
Alpine Meadow Vegetation	2,500–3,500 m	Grasses, sedges, small herbs	Short life cycle, cushion growth, frost resistance	Primula, Gentiana, alpine grasses	(Körner, 2021)
Sub-Alpine Forests	2,000–2,800 m	Conifers, broad-leaved trees	Needle-shaped leaves, thick bark, snow tolerance	Pine, Fir, Spruce, Rhododendron	(Kim et al., 2025)
High Alpine Shrubland	3,000–4,000 m	Dwarf shrubs, lichens, mosses	Low height, hairy leaves, water storage	Juniperus, Salix, Caragana	(Cannone et al., 2022)
Cryophytes Vegetation (Near Snow Line)	Above 4,000 m	Sparse herbs, lichens	Photosynthesize at low temps, deep roots	Saxifraga, Draba, *Ranunculus glacialis*	(Nobis et al., 2023)
Montane Forests (Lower Slopes)	1,000–2,000 m	Evergreen and deciduous trees	Cold-resistant wood, seasonal dormancy	Oak, Birch, Maple	(Kosick, 2023)

continued on following page

Table 2. Continued

Vegetation Zone / Type	Altitude Range	Dominant Plant Forms	Key Adaptations	Typical Species	Reference
Wet Alpine Meadows	Valley bottoms, near streams	Tall grasses, sedges, wildflowers	Waterlogging tolerance, shallow roots	Carex, Potentilla, Ranunculus	(Bavcon et al., 2024)
Rocky / Scree Vegetation	Exposed rocky slopes	Mosses, lichens, cushion plants	Root penetration into crevices, drought tolerance	*Silene acaulis, Saxifraga oppositifolia*	(Antonsson, 2012)
Glacial Foreland Pioneer Plants	Recently deglaciated terrain	Sparse herbs, mosses	Rapid colonization, nitrogen fixation	*Oxyria digyna, Dryas octopetala*	(Son, 2023)

Vegetation succession refers to the gradual, directional change in plant community composition and structure through time. In alpine zones, it primarily occurs as primary succession — the establishment of vegetation on newly exposed or sterile substrates, such as moraines, scree slopes, and glacial forelands. Secondary succession may occur following disturbances like landslides, frost heave, or wildfire, where some soil and seed banks remain intact.

3 SUCCESSION IN ALPINE REGIONS TYPICALLY PROCEEDS THROUGH SEVERAL STAGES:

3.1 Pioneer Stage:

The initial colonizers are hardy species capable of tolerating severe environmental stress. These include lichens, mosses, and small herbaceous plants such as *Oxyria digyna, Dryas octopetala*, and *Saxifraga oppositifolia*. They stabilize the surface, trap dust and organic matter, and initiate soil formation through biological weathering (Giupponi et al., 2023).

3.2 Early Successional Stage:

As organic matter accumulates and microclimates moderate, grasses and small shrubs like *Carex, Festuca,* and *Salix* species appear. These plants enhance soil structure by increasing root density and improving infiltration capacity (Erschbamer et al., 2023).

3.3 Mid to Late Successional Stage:

Over decades or centuries, more complex plant communities emerge, including tall herbs, dwarf shrubs (*Rhododendron, Juniperus*), and in lower alpine zones, scattered coniferous trees (*Pinus, Abies*). Soil horizons become more distinct, and nutrient cycling intensifies (Mainetti et al., 2021).

3.4 Climax Stage (Dynamic Equilibrium):

The mature vegetation community, adapted to prevailing climatic and edaphic conditions, maintains a balance with erosion and weathering rates. However, in alpine systems, true equilibrium is rarely achieved due to continuous climatic and geomorphic disturbances (Wu et al., 2022).

4 HUMAN INFLUENCE ON ALPINE VEGETATION

Alpine environments, due to their critical geography shaped by extreme conditions of climate, grow with short periods and restricted soil fertility is fragile ecosystems. This makes them externally susceptible to change against human activities, extensive and long-term grazing, and significant mountaineering pressure on the sensitive zones. Nevertheless, tourism has flourished extensively at the cost of climate degradation owing to the aesthetic and scenic beauty of the Alps, recreational activities, and unique biodiverse fauna and flora. To ease the human presence in the zone and apparent involved economic profits, various developments in infrastructure, hotels, ski, cable cars, and resorts are leading to deforestation, soil erosion, and habitat loss. The rising climbers and hikers' footfall results in vegetation trampling, which, owing to extreme environments, is difficult to restore in the short growth season in the alpine zones (Growcock, 2006). The disturbance erodes soil further by exposure to wind, snow, and rainfall, and accelerates the degradation of land.

One of the important factors that affects the original alpine ecosystem is grazing. Grazing, especially by livestock, i.e., goats, sheep, cattle, in the mountains, is a very widespread and cheap source of feed for the population living near the meadow alpine zones (Centeri, 2022). Where traditional pastoralism is much appreciated to maintain biodiversity and open grasslands, overgrazing results in the depletion of plant and grass cover, invasion by unpalatable vegetation, and increased soil erosion, leading to alpine wetlands degradation and disturbance. Besides, the sensitive nutrient cycling in these ecosystems puts them at risk of altering the composition of plants, as well as reducing resistance to climate variation with extensive grazing (Bettoni et al., 2023).

For the Alpine zone, mountaineering also comes with its fair share of challenges to the Alpine environment. The climber and mountaineering athlete cause disturbance to the route they take in mountains, leading to trampling vegetation, littering, human waste, and damaging water quality. The base camps of mountains cause erosion of habitat for local species and pollution. Moreover, the glacier retreats owing to rockfall has increased climate change and hazards, but human activity is putting an additional burden on the environmental stress in alpine zones (Bettoni et al., 2023).

Alpine zones are increasingly susceptible to the environmental stress that climate change is causing the whole globe. However, human activities, tourism, grazing, and development have increased the risk of our glacier and alpine degradation manifold. As recovery rates are slow in these ecosystems, restoration is ruled out as a choice to protect the alpine environment. The precautionary practices, such as controlled grazing, restricted tourist Access, and responsible mountaineering, must be ensured for the conservation of the alpine environment (Smiti, 2023).

5 CAUSES OF SOIL EROSION

Soil erosion is one of the leading causes of land degradation worldwide. It is the main reason that can greatly effects the vegetation succession in the alpine zones globally. The soil erosion could be caused by numerous factors including natural factors like water, land sliding, steep slopes, and avalanches etc. There are many other factors including chemical drivers like patterns of precipitation and snow melts, and anthropogenic causes like construction of roads, ski resorts and overgrazing etc., that can greatly affect the land and can cause soil erosion (Figure 1) (Gaffney et al., 2025).

Among the causes of erosion from various natural factors wind, rainfall, ice, snow and tillage, rainfall is the major cause that is considered the most significant driver of soil erosion. Every year, raindrop impact on the land generates energy comparable to almost about 20 tons of TNT per acre that leads to the dislodgement of fine soil particles (Zou et al., 2025). The soil erosion further triggers numerous offsite effects that includes agricultural productivity, influencing soil fertility, water quality, ecosystem functioning, carbon storage and flood risks. The increase in the sediments that are suspended raises the turbidity of rivers and coastal waters, disrupting the ecosystems and affecting negatively to the food production by contributing to eutrophication and reducing aquatic biodiversity (Bay et al., 2003). Other than that, erosion heightens the possibility of landslides due to more surface runoff, thereby amplifying hydrogeological hazards during intense rainfall seasons (D'Odorico et al., 2006).

Figure 1. A pictorial description of the effect of soil erosion by human interventions

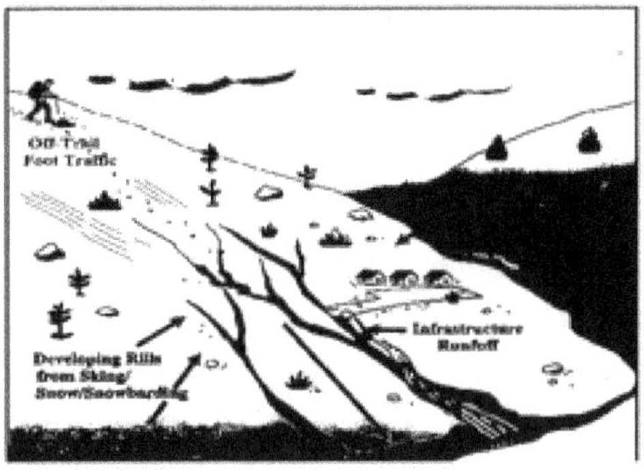

The other natural factors of soil erosion include strong gusts of wind, as heavy winds remove small particles, which is a typical problem in alpine regions (Freppaz et al., 2010). The abnormal climatic changes, like sudden changes in temperature leaps, have a very negative effect on the surface of the field. Furthermore, the abnormal effect of climatic changes on soil erosion is stunted vegetation growth that exposes the regions to winds and rainfall, and also reduces the field cover. Thus, wind is the main geomorphological force, particularly in dry and semi-arid areas, including the high mountain alpine regions, to mediate soil erosion directly. Additionally, it is the main cause of alpine crop damage, evaporation, hazardous airborne dust, and land degradation in the alpine soil (Cache et al., 2023).

Other than these factors, avalanches are also the leading cause of soil erosion in alpine regions. Significant erosive stresses can be applied to soils by snow avalanches. When a snow avalanche moves over the exposed ground, the basal shear force can erode and incorporate soil into the flow. The eroded soil is then carried to the deposition zone, where it plays a major role in shaping distinctive landforms and also alters the chemistry of the soil. The extent of deposition and the soil erosion depend on the dynamics of avalanches, the morphology of the terrain, as well as the characteristics of the soil and the presence or absence of vegetation cover (Rapuc et al., 2025).

6 GEOMORPHIC CATEGORIZATION OF ALPINE SOIL EROSION

Erosion in these environments can be categorized into several distinct forms based on their geomorphic impact: sheet erosion, rill erosion, gully erosion, cryogenic (frost-induced) erosion, glacial erosion, wind erosion, landslide erosion, and avalanche erosion (Grissinger, 2023). Each of these processes contributes uniquely to the alteration of soil morphology, influencing water retention, nutrient cycling, and overall ecosystem resilience. Together, they define the fragile yet dynamic nature of alpine soils, emphasizing the urgent need for integrated approaches to their management and preservation.

6.1 Sheet Erosion

This type of erosion represents the initial and very subtle removal of the soil surface in alpine regions. This type of erosion happens when a uniform, thin layer of soil is detached and swept down slope with the runoffs from the surface, usually by rain (splash erosion), wind, and snowmelt. This is especially true in the alpine zones because of the shallow, poorly developed, and loose soil particles of the cambisols and leptosols. The vegetation cover is insufficient in these soils to distribute the kinetic energy of running water, wind, or snow run off resulting in slow removal of organic matter and soil (Dobos et al., 2024). Although it might not be obvious in terms of visual destruction, this process leads to good, nutritious soil losing gradually and progressively, not only reducing fertility but also water retention by the remaining soil, which is too poor in nutrients and water-holding ability. This erosion has been observed in the European Alps and lower Himalayan regions and is transitioning into more severe forms of erosion, such as till and gully erosion (J. Liu et al., 2022).

6.2 Rill Erosion

This type of erosion follows sheet erosion when the particle runoff starts depositing and concentrating on the micro channels of the slope terrains. These incisions are smaller than 30cm in depth and are developed by high snowmelt or rain runoffs. As root networks and steep slopes are weak in stabilizing the aggregate of soil formed by sheet erosion, the water carves out temporary channels. The rill erosion is also important as a transition, as it led to more severe, localized gully erosion from the existing sheet erosion (Figure 2). The seasonal freeze-thaw also makes rill erosion worse by expelling water into ice in the channel and widening its size. The alpine rill erosion is common in pasture slopes, moraines, and roadside embankments that encounter seasonal snowmelts (Arosio et al., 2025)

6.3 Gully Erosion

The permanent channels formed after advanced erosion through rills and sheet erosion are called gully erosion. These deepen and wide channels prevent any kind of vegetation. The alpine zones marked by steep slopes, mechanical instability, tough topography, and heavy rainfall accelerate this process in Regosolic alpine zones. The runoffs coming with high velocity from snowmelts can further deepen the exposed alpine slopes, leading to wide channels forming gullies like incisions with a depth of several meters. The gullies interrupt the subsurface flow of water, diverting runoffs with sediments into valley streams, triggering more slope failures. With the passage of time, gully erosion can cause the slope denudation and coarse depth deposition at the bottom of the valley. The research from the alpine of Tibetan and Karakoram-Hindukush has shown that gully formation is more frequent on southern slopes where snow melts earlier than average summer time, causing erosive flows (Sun et al., 2022).

6.4 Frost Heave and Cryogenic Erosion

Cryogenic erosion is unique to high-altitude environments where freeze–thaw cycles dominate soil processes. When soil moisture freezes, the expansion of water causes frost heave — the upward movement of soil and rock fragments. Upon thawing, the loosened particles become highly susceptible to gravity and runoff. This cyclic displacement breaks soil aggregates and dislodges fine particles, leading to gradual downslope migration known as solifluction. Cryogenic erosion is most prevalent above 3,000 m, where Cryosols and Gelisols occur. Vegetation cover is minimal, and the surface layer remains frozen for much of the year. The process produces patterned ground, frost boils, and shallow slope deformation. In the Qinghai-Tibetan Plateau, cryogenic erosion plays a dominant role in shaping periglacial landscapes (G. Wu et al., 2023).

6.5 Glacial Erosion

Another kind of erosion is the powerful glacial erosion caused by geomorphic processes and is the main cause of cirques and U- U-shaped valleys formation in the alpine zones. When the glacial transport down the slope under the effect of gravity, they enter into the substrate underlying the soil and pluck it by abrasion. The plucking phenomenon is the pulling of fragments from the underlying substrate by meltwater that enters rock joints, refreezes there, and then removes the particle as the glacier advances under gravity. On the other hand, abrasion phenomena are the removal of the particle from the surface of alpine soil by the abrasive effect of rocks in the ice that act like sandpaper and cause rubbing and grinding fractions to move those particles (Herman et al., 2021). This erosive action by glacier plucking and abrasion led to the weathering process described earlier and exposed the rock surfaces bare. After the glacial erosion, the left deglaciated valleys with residual coarse and unstable soil are further exposed to secondary erosion caused by wind and rainfall. The examples of glacial erosion are the Andes, the Alps, and the Rockies. The sculpted look of these extensive mountains has been shaped by glacial erosion, which has left them unable to support the vegetation succession and influenced strong hydrology and sediment transport (García-Ruiz et al., 2024).

Figure 2. Types of soil erosion exist in the alpine zone

6.6 Wind (Aeolian) Erosion

Wind erosion is one of the least dominant forms of erosion by water or ice in alpine zones, nonetheless, it significantly impacts the intermountain and dry valleys with outwash glacial plains having mass finely accumulated sedimentation. The powerful katabatic winds made up of cold and dense air flowing down the slope can desiccate the loose soil sediments especially where vegetation cover is sparse to harmonize the strong wind effect on the soil. Aeolian erosion results in lag surface with coarse residues by mobilizing fine silt and clay material off the soil surface. The wind erosion also accelerate snowmelt by dispersing dust on the snowfields and reduce albedo. This indirect feedback also trigger other kind of erosions such as gully erosion and glacial erosion. The examples of Aeolian erosion are Himalayan forelands and Pamir which encounter wind- blown sediments frequently in dry weather (Kalvoda & Emmer, 2021).

6.7 Mass Wasting and Landslide Erosion

The landslides are often characterized more as a gravitational progress rather than a fluvial process. The landslides represent major soil erosion in the alpine zones. The regolith, soil and rocks mobilize downward triggered under the effects of mass wasting, wind, rainfall or seismic activity and pull thin soil particles. The thinly covered vegetation and steep slopes in alpine regions make them extremely vulnerable to the landslide. The landslide exposed the bedrock, remove a big volume of soil and redistribute sediments altering alpine soil geography. In various cases, it may form new gullies or alpine streams creation temporary lakes. In the Himalayan mountain range, landslides are a prominent origin of sediments in the mountain river exceeding sedimentary contribution from other erosion sources (Li et al., 2022).

6.8 Avalanche Erosion

The avalanches also cause major soil displacement in the alpine zones. The shear pressure generated by rapid ice and snow mass descend drag debris and soil, uproot vegetation in their path causing soil erosion and destroy the vegetation cover undermine soil security further. The naked tracks are subsequently then tradition to unstable channel when encountered by further snow and rainfall. Although avalanche erosion is not frequent but it is permanently maintaining certain channels where the soil is devoid of further vegetative succession. Some areas in Swiss alpine and northern Pakistan have been permanently scarred by avalanche erosion (Eckert et al., 2024). Table 3 provide the detailed comparison of natural soil erosion that marred the alpine mountains.

Table 3. Types of soil erosion caused by natural factors in alpine zones

Type of Erosion	Main Natural Cause	Process Description	Impact on Alpine Landscape	Example Region	References
Sheet Erosion	Rainfall, snowmelt runoff	Thin uniform layer of topsoil removed by surface water flow	Gradual loss of fertile soil; exposure of subsoil	Lower Himalayan slopes	(J. Liu et al., 2022)
Rill Erosion	Intense rainfall or snowmelt	Formation of small channels (rills) on slopes due to concentrated runoff	Increases sediment load and reduces vegetation cover	Swiss Alps, Hindu Kush	(Arosio et al., 2025)
Gully Erosion	Prolonged heavy rain and steep slopes	Small rills deepen and widen into large gullies	Loss of arable land and slope instability	Karakoram valleys, Andes	(Sun et al., 2022)
Frost Heave / Cryogenic Erosion	Freeze–thaw cycles	Expansion of frozen water lifts and loosens soil particles	Surface cracking, displacement of soil and rocks	Scandinavian Alps, Tibet Plateau	(G. Wu et al., 2023)
Glacial Erosion	Movement of glaciers	Ice scours and removes topsoil and rock beneath glaciers	U-shaped valleys, moraines, and loss of soil cover	European Alps, Rockies	(García-Ruiz et al., 2024)

continued on following page

Table 3. Continued

Type of Erosion	Main Natural Cause	Process Description	Impact on Alpine Landscape	Example Region	References
Wind Erosion	High-velocity alpine winds	Loose, dry soil particles carried away by wind	Removal of fine particles; formation of bare patches	Dry intermountain valleys, Pamir Mountains	(Kalvoda & Emmer, 2021)
Mass Wasting / Landslides	Gravity, heavy rainfall, earthquakes	Sudden downslope movement of soil and rock	Rapid loss of soil mass, destruction of vegetation	Nepal Himalayas, Andes	(Li et al., 2022)
Avalanche Erosion	Snow and ice slides	Snow mass drags soil and debris downhill	Exposure of bare subsoil and root damage	Swiss Alps, Northern Pakistan	(Eckert et al., 2024)

7 HOW SOIL EROSION DRIVES REDUCED VEGETATIVE SUCCESSION IN THE ALPINE MEADOWS

The soil erosion poses significant impacts on the vegetation succession of alpines. These challenges include a decrease in the fertility of soil, degradation in the quality of water, destruction of habitat, and the loss of biodiversity. In the alpine regions (high mountainous regions), soil erosion leads to a substantial decrease in the fertility of soil by stripping away the topsoil rich in nutrients that holds the organic matter and key elements such as phosphorus, nitrogen, and potassium (Wilson et al., 2022). The removal of such a layer weakens the structure of soil, depletes the capacity of holding water, and disturbs the cycling of nutrients, thereby reducing the yields of agricultural practices and impairing the functioning of the ecosystem. The climatic factors and steep terrains that include melting of snow and heavy rainfall, substantially increase the risk of erosion, while the unsustainable practices of land use that include improper cultivation methods and the use of heavy machinery, further exacerbate the intensity of the issue (Sharma et al., 2022).

The other effects of soil erosion include the depletion of organic carbon and biodiversity, loss of fertile land because of the disturbance of natural nutrient cycle, damage to the infrastructure (dams, water supply etc.) from excessive sediment load, particularly harmful effects on aquatic ecosystems and biodiversity, limitation on the use of land area that resultantly prevents the redevelopments and reduce the area for forestation, agriculture and recreation (Musa et al., 2024). The soil erosion further has off-site and on-site impacts that greatly affect the succession. The off-site impacts of soil erosion primarily stem from the disposition of sediments, which can bury infrastructure, block drainage networks, and alter the courses of rivers, thereby increasing the risks of flood and reducing the lifespan of reservoirs. Furthermore, soil erosion contributes to the deterioration of the quality of water and can negatively affect the recreational uses of aquatic environments.

By eroding the fertile layer of soil that sustains alpine ecosystems and flora the erosion of soil in the Alps, which is fueled by the agro-pastoral activities that were done previously, and these were further made worse by contemporary use of land, directly destroys the habitat (Giguet-Covex et al., 2023). The destruction of various habitats by soil erosion, including land habitats and aquatic habitats, poses a great deal of damage to the natural functioning of the ecosystems. The negative impacts of soil erosion in the alpine regions, like the high mountain region, pose a great threat of destruction on the habitat on the lower slopes. In conclusion, this erosion destabilizes the complete ecosystems of mountains by decreasing the biodiversity, decreasing the ability of land to sustain plants and fauna, speeding up more erosion from wind and water, and raising the chances or risks of landslides (Li et al., 2022).

8 INTERCHANGED RELATIONSHIP OF VEGETATION SUCCESSION IN ALPINE REGIONS AND SOIL EROSION

The link between vegetation succession and soil erosion in alpine regions is reciprocal and dynamic. Each process influences and constrains the other.

8.1 Vegetation Reduces Soil Erosion

Vegetation acts as a biological shield against erosion by intercepting raindrops, reducing runoff velocity, and reinforcing the soil with root systems. During early succession, mosses and lichens form a protective mat that limits splash erosion. As succession advances, grasses and shrubs provide more extensive root networks, enhancing soil cohesion and infiltration. The establishment of even sparse vegetation can significantly decrease rill and gully formation (Wu et al., 2022).

In alpine meadows, studies show that a 10% increase in vegetation cover can reduce surface runoff by nearly 30%. Root mats from species like *Festuca* and *Carex* are particularly effective in stabilizing slopes, while dwarf shrubs act as windbreaks, mitigating aeolian erosion. Thus, vegetation succession enhances soil stability over time, transforming erosional landscapes into more stable, sediment-retaining ecosystems (Bettoni et al., 2023; Liu et al., 2023; Yang & Sun, 2021).

8.2 Soil Erosion Controls Vegetation Development

Conversely, soil erosion determines the rate and trajectory of succession. Severe erosion can strip away seed banks, organic matter, and nutrients, delaying colonization and promoting patchy vegetation patterns. Newly eroded surfaces often revert to early successional stages dominated by pioneers. For example, repeated frost heave or landslide activity may prevent progression beyond moss and lichen stages (Lawson, 2024). The interplay between disturbance frequency and soil recovery rate defines the successional mosaic of alpine terrain. On stable slopes, succession proceeds toward complex communities; on unstable slopes, it resets repeatedly, producing a shifting pattern of vegetative patches and bare ground. This balance maintains the characteristic heterogeneity of alpine landscapes.

8.3 Soil–Vegetation Feedbacks in Alpine Succession

Vegetation succession alters soil physical and chemical properties through a series of positive feedback mechanisms:

- Accumulation of organic matter improves moisture retention and nutrient cycling.
- Root activity enhances soil porosity and aggregation.
- Litter decomposition promotes the formation of humus-rich Cambisols from initial Leptosols (Li et al., 2022).

These improvements, in turn, foster further vegetation growth and resistance to erosion. However, in zones where erosion outpaces soil formation such as steep, high-altitude slopes, these feedbacks remain weak, and soils remain skeletal. The resulting patchwork of mature and juvenile soils mirrors the uneven pace of succession across the alpine landscape (Giupponi et al., 2023).

9 CONSERVATION AND MANAGEMENT STRATEGIES IN ALPINE ENVIRONMENTS

As stated previously, the alpine environments are one of the most fragile ecosystems on earth, given their steep slopes, thin soil, and extreme climatic conditions. These factors, coupled with human activities, result in severe soil erosion undermining the natural succession phenomenon of vegetation in these zones. Not only soil erosion, but the alpine soils are also vulnerable to habitat degradation and biodiversity loss. The rise of climate change, overgrazing, and tourism has surpassed alpine zone conservation and management more than a policy to global concern. The need for more sustainable approaches that can incorporate soil conservation, protected area sanctuaries, ecological restoration, as well as community building must be a priority to preserve the vulnerable ecosystems of ecological importance for the survival of our planet.

9.1 Soil Conservation Techniques

The soil conservation in the alpine region is very important as the soil in these regions formed over the centuries can be lost in a short time due to climatic and human activities such as climate change, wind, extreme weather, erosion, overgrazing, and tourism. Thus, many strategies to preserve soil have been formulated. Some of the most effective measures to conserve soil are restricted grazing, terracing, and mulching (Erschbamer et al., 2023). The restricted grazing is most efficient as it is controllable and its results can be achieved by maintaining the fragile parity in the relationship between grazing fields and livestock. Overgrazing by household livestock such as yaks, goats, and sheep can lead to undermining soil integrity and risking plant growth and advanced erosion. Overgrazing can be controlled naturally by seasonal closure of Pastures, rotation grazing, and limited livestock access. In various countries, communities contribute to protecting alpine regions by cooperatively enforcing overgrazing rules without risking their livestock livelihood (Mainetti et al., 2021). Terracing is another phenomenon to control soil erosion by transforming the steep slopes to level steps and, as a result, reducing surface runoff velocity in events of wind, snow, and gravity pull and preventing soil erosion. This technique is very common in the alpine zones of the Andes and Himalayas in not only preserving soil for vegetative succession but also water retention and promoting cultivable patches on the level steps, which in other areas cannot hold on steep soils. Terracing also slows down landslides resulting from rainfall, wind, and snow to protect existing vegetation in alpine meadows, stone, and pastures (Wu et al., 2022).

Last but not least, Mulching plays a critical role in the protection of vegetation. The organic matter, such a dark, straws, and leaves, is used to cover the surface where vegetation is established to lock moisture by evaporation, restricting weed growth to inhibit competition with natural alpine flora, and controlling damage from frost and snow. Furthermore, when the protective layer of mulch decomposes with time, it keeps providing the vegetation with natural organic matter in terms of humus and carbon, enhancing soil structure and fertility by increasing nutrient cycling in the stunted nutrient regions. All of these soil protective techniques create the ecological stability and balance required for the vegetation succession and soil erosion, and promote resilience against human disturbances and climate change (Bettoni et al., 2023).

9.2 Ecology Restoration Approaches

The alpine ecological system, on the brink of degradation beyond recovery, can be rehabilitated by the ecological restoration approaches. However, restoration approaches in the alpine zones face restrictions in the form of slow growth of vegetation, nutrient-deficient soils, as well as extreme weather. Nonetheless, many approaches, like seeding, revegetation, and assisted succession, have been found to be useful in vegetation succession promotion after chronic soil erosion. The revegetation approach re-establishes the indigenous plant cover after the soil stabilization on the eroded slopes, mining sites, and ski tracks. The primary objective behind this approach is to promote the natural plant cover and prevent the non-native species (Liu et al., 2023). For this purpose, the local species of vegetation, such as sedges, alpine grasses, and cushion plants, are first nurtured and transplanted in the nursery during their short summer window and introduced to the alpine regions. This is important as these plants are locally adapted, well-suited to the weather, and can resist the wind, cold, rain, and nutrient deficiency.

Besides revegetation, the seeding is another cost-effective and a way quicker method to help the plantation. The seeds of native species are sown directly in the land on the alpine soil. These plants act as pioneers for the sensitive alpine species as ecological nurses by facilitating their growth, creating microhabitats, and preventing the effects of soil erosion. The seeding techniques, such as hyper seeding by spraying a mixture of seeding with other techniques, such as mulching, seeding, and fertilizer, have been reported to be successful in the roadside areas and ski slopes in disturbed alpine regions (Yang & Sun, 2021).

Besides seeding, assisted succession is a naturalistic method for the recovery of disturbed ecosystems by relieving stress agents in order to resume the normal ecological process. This may address, but is not limited to, overgrazing, restricting invading fauna and flora in alpine regions, limiting human access, and reintroducing the local plants and animals to the region. This method outshines the artificially replanting and nursing techniques and distinguishes itself on the reliance on the region's natural ability to rejuvenate over time and promoting sustainability. The restoration of the ecology in the alpine zones not only revives the integrity and natural sovereignty of the alpine zones but also strengthens their resistance and promotes their role in nutrient cycling, such as carbon storage, biodiversity, water regulation, and conservation (Lawson, 2024).

9.3 The Protected Area Management

The protected areas, establishment as the like of natural sanctuaries, national parks, and natural reserves, is considered one of the effective methods for the conservation of the alpine regions. The unitary approach to forming protected areas aims to harmonize the ecological balance with responsible recreation and tourism. Many national parks, such as Nepal's Sagarmatha National Park, Swiss National Park, and Canada's Banff National Park, are examples of legal protection provided to sensitive alpine zones. These restricted areas under protected areas limit the development of any infrastructure, mining activities, promote grazing regulations with little human interference, and where natural process take their own course. At the same time, over-tourism, such as trekking and mountaineering, is threatening the health of vegetation in the alpine zones. The strict tourism regulations should be in place to limit the adverse effects of human activities on the alpine regions, reducing environmental pressure. The permit systems, waste management policies, camping only, and designated regions should be in practice for minimal human footprints. The trekkers should be educated with appropriate leave no trace principles, as well as in peak vegetation growth regions, caps on trekkers should be enforced to avoid risk to the growing plants. With the education and building of protected units in alpine zones, their management is also a core principle for the enforcement of the protected police (Yandi et al., 2023). The Alpine protected area management ensures regular environmental checkups on the vegetation, climate, soil, animals, and itself on the policy enforcement. This approach requires the official intervention to monitor the conversation, prohibit any violations. Together with trekker management, over-grazing regulations, the protected area management can serve as living proof of the sustainability of the alpine regions.

9.4 Role of Local Communities in Sustainable Alpine Management

Besides the official intervention, the communities in the alpine zones can play a key role in the promotion of vegetative succession and the prevention of soil erosion. These communities that depend on these regions for their livelihoods can use their experiences, stewardship, and knowledge of the fauna and flora for the benefit of the alpine zones. The community-sponsored conversation empowers the locals to take initiative to protect their environment and livelihood. In many cases, their natural environmental initiatives, such as seasonal herd movement, rotational grazing, and pasture management, are sustainable for the alpine zones. Their traditional ecological experience, when met with modern science and data, will be able to help in maintaining the health and productivity of the alpine zones at the same time promoting biodiversity (Zhou et al., 2024).

The economic initiatives can also be linked with community-based conversations. These will provide the monetary encouragement to the locals living near the national parks in the form of home stays, local guides, and handicrafts on one hand, as well as on the other hand, it will stop the human interference in the unfamiliar alpine regions. The local community can also be rewarded with payment for ecosystem methods by rewards if they maintain their forest covers, local plantation, sustainable grazing, etc. The education and familiarity with rewarding participation in the future of the alpine regions can encourage locals to improve their link with their surrounding nature. If the communities understand the ecological importance of their alpine zones, they will more likely contribute to them positively. The people in the alpine zones of Bhutan, Nepal, and Switzerland positively correlate the notion of partner, not just beneficiaries, and have contributed positively to the health of their alpine ecosystems (Hanson, 2025; Semwal et al., 2025)

10 CONCLUSION

Alpine zones are complex ecosystems home to diverse fauna and flora. The overall health and survival of other ecosystems depend on these zones. Since the inception of the Earth, a balance between the natural soil erosion and recovery processes has preserved the alpine zones. For a few decades, the access that was previously restricted to the alpine zone has ceased, leading to disturbance of this natural balance. The weathering, both chemical and physical, in addition to anthropological, has become worse; the natural elements, such as rainfall, wind, and glaciers, accompanied by the extrinsic drivers, have sped up the soil erosion. The sparse vegetation in the absence of natural recovery is unable to cope with the soil erosion. And ironically, only renewed vegetation succession can save the alpine soil from erosion. The interplay of these factors not only transforms local geomorphology but also affects hydrological cycles, nutrient dynamics, and long-term ecosystem productivity. Furthermore, climate change intensifies these challenges by altering precipitation patterns, accelerating glacial retreat, and increasing temperature variability, all of which heighten erosion risks. Sustaining alpine soils, therefore, requires an integrated understanding of the diverse processes that govern their stability. Comprehensive conservation approaches combining vegetation restoration, slope management, and climate-adaptive land-use strategies are essential to preserve the ecological integrity and resilience of alpine landscapes for future generations.

REFERENCES

Alatalo, J. M., Jägerbrand, A. K., & Molau, U. (2016). Impacts of different climate change regimes and extreme climatic events on an alpine meadow community. *Scientific Reports*, *6*(1), 21720. DOI: 10.1038/srep21720 PMID: 26888225

Antonsson, H. (2012). *Plant species composition and diversity in cliff and mountain ecosystems.*

Arosio, C., Giupponi, L., Giorgi, A., Cislaghi, A., & D'Amico, M. E. (2025). Recovery of Soil-Based Ecosystem Services in Abandoned Ski Resorts: The Valcanale Case Study (Bergamo, Italian Alps). In *Sustainability* (Vol. 17, Issue 12). DOI: 10.3390/su17125418

Bavcon, J., Malovrh, K., Tomšič, M., & Ravnjak, B. (2024). In Situ Conservation of Dry Meadows. In *Land* (Vol. 13, Issue 3). DOI: 10.3390/land13030315

Bay, S. M., Zeng, E. Y., Lorenson, T. D., Tran, K., & Alexander, C. (2003). Temporal and spatial distributions of contaminants in sediments of Santa Monica Bay, California. *Marine Environmental Research*, *56*(1), 255–276. DOI: 10.1016/S0141-1136(02)00334-3 PMID: 12648959

Belasri, A., & Lakhouili, A. (2016). Estimation of soil erosion risk using the universal soil loss equation (USLE) and geo-information technology in Oued El Makhazine Watershed, Morocco. *Journal of Geographic Information System*, *8*(01), 98–107. DOI: 10.4236/jgis.2016.81010

Bettoni, M., Maerker, M., Bosino, A., Conedera, M., Simoncelli, L., & Vogel, S. (2023). Land use effects on surface runoff and soil erosion in a southern Alpine valley. *Geoderma*, *435*, 116505. DOI: 10.1016/j.geoderma.2023.116505

Bockheim, J. G. (2015). Global Distribution of Cryosols with Mountain Permafrost: An Overview. *Permafrost and Periglacial Processes*, *26*(1), 1–12. DOI: 10.1002/ppp.1830

Buytaert, W., Cuesta-Camacho, F., & Tobón, C. (2011). Potential impacts of climate change on the environmental services of humid tropical alpine regions. *Global Ecology and Biogeography*, *20*(1), 19–33. DOI: 10.1111/j.1466-8238.2010.00585.x

Buytaert, W., Deckers, J., & Wyseure, G. (2006). Description and classification of nonallophanic Andosols in south Ecuadorian alpine grasslands (páramo). *Geomorphology*, *73*(3), 207–221. DOI: 10.1016/j.geomorph.2005.06.012

Cache, T., Ramirez, J. A., Molnar, P., Ruiz-Villanueva, V., & Peleg, N. (2023). Increased erosion in a pre-Alpine region contrasts with a future decrease in precipitation and snowmelt. *Geomorphology*, *436*, 108782. DOI: 10.1016/j.geomorph.2023.108782

Cannone, N., Guglielmin, M., Casiraghi, C., & Malfasi, F. (2022). Salix shrub encroachment along a 1000 m elevation gradient triggers a major ecosystem change in the European Alps. *Ecography*, *2022*(2), ecog.06007. DOI: 10.1111/ecog.06007

Centeri, C. (2022). Effects of Grazing on Water Erosion, Compaction and Infiltration on Grasslands. In *Hydrology* (Vol. 9, Issue 2). DOI: 10.3390/hydrology9020034

D'Odorico, P., Porporato, A., & Runyan, C. W. (2006). *Dryland ecohydrology* (Vol. 9). Springer. DOI: 10.1007/1-4020-4260-4

Dirnböck, T., Essl, F., & Rabitsch, W. (2011). Disproportional risk for habitat loss of high-altitude endemic species under climate change. *Global Change Biology*, *17*(2), 990–996. DOI: 10.1111/j.1365-2486.2010.02266.x

Dobos, A., Hegyi, T. P., Bujtor, D. L., Tolnai, Z., & Hegyi, B. (2024). Description of the Characteristic Soil Profiles and Indication of the Degree of Sheet Erosion in Verpelét. *Journal of Environmental Geography*, *17*(1–4), 73–90. DOI: 10.14232/jengeo-2024-45762

Eckert, N., Corona, C., Giacona, F., Gaume, J., Mayer, S., van Herwijnen, A., Hagenmuller, P., & Stoffel, M. (2024). Climate change impacts on snow avalanche activity and related risks. *Nature Reviews. Earth & Environment*, *5*(5), 369–389. DOI: 10.1038/s43017-024-00540-2

Erschbamer, B., Niederfriniger Schlag, R., Carnicero, P., & Kaufmann, R. (2023). Long-term monitoring confirms limitations of recruitment and facilitation and reveals unexpected changes of the successional pathways in a glacier foreland of the Central Austrian Alps. *Plant Ecology*, *224*(4), 373–386. DOI: 10.1007/s11258-023-01308-2

Freppaz, M., Godone, D., Filippa, G., Maggioni, M., Lunardi, S., Williams, M. W., & Zanini, E. (2010). Soil Erosion Caused by Snow Avalanches: A Case Study in the Aosta Valley (NW Italy). *Arctic, Antarctic, and Alpine Research*, *42*(4), 412–421. DOI: 10.1657/1938-4246-42.4.412

Gaffney, P. P. J., Tang, Q., Wang, J., Zhang, C., Xu, X., Xu, X., Li, Y., Pap, S., Ratcliffe, J. L., & Li, Q. (2025). The high-altitude peatland carbon cycle: A review of the impacts of climate change, human disturbance and management. *Geography and Sustainability, 6*(6).

García-Ruiz, J. M., Arnáez, J., Lasanta, T., Nadal-Romero, E., & López-Moreno, J. I. (2024). *The Changing World of Mountains in the Long Term BT - Mountain Environments: Changes and Impacts: Natural Landscapes and Human Adaptations to Diversity* (J. M. García-Ruiz, J. Arnáez, T. Lasanta, E. Nadal-Romero, & J. I. López-Moreno (eds.); pp. 27–50). Springer Nature Switzerland. DOI: 10.1007/978-3-031-51955-0_3

Getnet, M., & Asseffa, S. (2025). *Agriculture in the Abbay Basin BT - Abbay River Basin: Biophysical Setting, Environmental Degradation, Hydropolitics and Development Potential* (A. Melesse, B. Gessesse, & W. Zewdie (eds.); pp. 67–87). Springer Nature Switzerland. DOI: 10.1007/978-3-031-65241-7_4

Giguet-Covex, C., Bajard, M., Chen, W., Walsh, K. J., Rey, P.-J., Messager, E., Etienne, D., Sabatier, P., Ficetola, F. G., Gielly, L., Blanchet, C., Guffond, C., Chiquet, P., Arnaud, F., & Poulenard, J. (2023). Long-term trajectories of mountain agro-ecosystems in the North-Western Alps. *Regional Environmental Change, 23*(2), 58. DOI: 10.1007/s10113-023-02030-5

Giupponi, L., Leoni, V., Pedrali, D., Zuccolo, M., & Cislaghi, A. (2023). Plant cover is related to vegetation and soil features in limestone screes colonization: A case study in the Italian Alps. *Plant and Soil, 483*(1), 495–513. DOI: 10.1007/s11104-022-05760-3

Giupponi, L., Panza, R., Pedrali, D., Sala, S., & Giorgi, A. (2025). Ecology, Floristic–Vegetational Features, and Future Perspectives of Spruce Forests Affected by Ips typographus: Insight from the Southern Alps. In *Plants* (Vol. 14, Issue 11). DOI: 10.3390/plants14111681

Grissinger, E. H. (2023). Rill and gullies erosion. *Soil Erosion, Conservation, and Rehabilitation*, 153–168.

Growcock, A. J. W. (2006). *Impacts of camping and trampling on Australian alpine and subalpine vegetation and soils.*

Hanson, J. H. (2025). Access to Assets and Influence: Governance and Livelihoods in Protected Areas of the Annapurna and Everest Regions, Nepal. *Human Ecology, 53*(5), 1069–1085. Advance online publication. DOI: 10.1007/s10745-025-00643-4 PMID: 41623595

Herman, F., De Doncker, F., Delaney, I., Prasicek, G., & Koppes, M. (2021). The impact of glaciers on mountain erosion. *Nature Reviews. Earth & Environment, 2*(6), 422–435. DOI: 10.1038/s43017-021-00165-9

Janiczek, A., Suchan, T., Paul, W., Consortium, I., & Ronikier, M. (2025). Spatial Discontinuity of Mountain Systems and Genetic Structure of Alpine Plants: The Alps–Carpathians Disjunction in a Comparative Phylogeographical Context. *Journal of Biogeography, 52*(6), e15122. DOI: 10.1111/jbi.15122

Jiao, J., Zou, H., Jia, Y., & Wang, N. (2009). Research progress on the effects of soil erosion on vegetation. *Acta Ecologica Sinica, 29*(2), 85–91. DOI: 10.1016/j.chnaes.2009.05.001

Kalvoda, J., & Emmer, A. (2021). Mass wasting and erosion in different morphoclimatic zones of the Makalu Barun region, Nepal Himalaya. *Geografiska Annaler. Series A. Physical Geography, 103*(4), 368–396. DOI: 10.1080/04353676.2021.2000816

Kim, J.-S., Kim, J.-H., Hyun, J.-O., Jang, C.-G., Choi, H.-J., Lee, J.-H., Gwon, J.-H., Son, S.-W., So, S., Kim, J.-W., Park, H.-C., & Myeong, H.-H. (2025). A checklist of alpine and subalpine plants in the Korean Peninsula. *Journal of Asia-Pacific Biodiversity*. 10.1016/j.japb.2025.06.010

Knežević, M., Topalović, A., Fuštić, B., & Životić, L. (2025). In Knežević, M., Topalović, A., Fuštić, B., & Životić, L. (Eds.), *Soil Types—Properties and Distribution BT - The Soils of Montenegro* (pp. 117–196). Springer Nature Switzerland., DOI: 10.1007/978-3-031-82919-2_5

Körner, C. (2021). In Körner, C. (Ed.), *The alpine life zone BT - Alpine Plant Life: Functional Plant Ecology of High Mountain Ecosystems* (pp. 23–51). Springer International Publishing., DOI: 10.1007/978-3-030-59538-8_2

Körner, C., Jetz, W., Paulsen, J., Payne, D., Rudmann-Maurer, K., & Spehn, M., E. (. (2017). A global inventory of mountains for bio-geographical applications. *Alpine Botany, 127*(1), 1–15. DOI: 10.1007/s00035-016-0182-6

Körner, C., & Kèorner, C. (1999). *Alpine plant life: functional plant ecology of high mountain ecosystems* (Vol. 4). Springer. DOI: 10.1007/978-3-642-98018-3

Kosick, H. (2023). *Drivers of change in the temperate-boreal refugium of Cape Breton Highlands National Park*. Memorial University of Newfoundland.

Koyanagi, K., Andreoli, A., Nordio, G., Pitscheider, F., Tomelleri, E., & Comiti, F. (2025). Disturbance-intervention interactions on post-windthrow hillslope runoff and erosion processes in the Eastern Italian Alps. *Journal of Hydrology (Amsterdam), 656*, 133001. DOI: 10.1016/j.jhydrol.2025.133001

Lamouille-Hébert, M., Arthaud, F., & Datry, T. (2024). Climate change and the biodiversity of alpine ponds: Challenges and perspectives. *Ecology and Evolution, 14*(2), e10883. DOI: 10.1002/ece3.10883 PMID: 38327685

Lawson, E. (2024). *Moss and Lichen*. Reaktion Books.

Legros, J. P. (1992). Chapter 7 - Soils of Alpine mountains. In I. P. Martini & W. B. T.-D. in E. S. P. Chesworth (Eds.), *Weathering, Soils & Paleosols* (Vol. 2, pp. 155–181). Elsevier. 10.1016/B978-0-444-89198-3.50012-X

Li, B. V., Jenkins, C. N., & Xu, W. (2022). Strategic protection of landslide vulnerable mountains for biodiversity conservation under land-cover and climate change impacts. *Proceedings of the National Academy of Sciences of the United States of America*, *119*(2), e2113416118. DOI: 10.1073/pnas.2113416118 PMID: 34983877

Liu, J., Qi, X., Ma, C., Wang, Z., & Li, H. (2022). Response of Sheet Erosion to the Characteristics of Physical Soil Crusts for Loessial Soils. *Frontiers in Environmental Science*, *10*, 905045. https://www.frontiersin.org/journals/environmental-science/articles/10.3389/fenvs.2022.905045. DOI: 10.3389/fenvs.2022.905045

Liu, Y.-F., Fang, H., Huang, Z., Leite, P. A. M., Liu, Y., López-Vicente, M., Zhao, J., Shi, Z.-H., & Wu, G.-L. (2023). Shrub encroachment increases soil erosion risk in hillside alpine meadows of the Qinghai-Tibetan Plateau, NW China. *Catena*, *222*, 106842. DOI: 10.1016/j.catena.2022.106842

Ludovicy, S., Noroozi, J., Semenchuk, P., Moser, D., Wessely, J., Talebi, A., & Dullinger, S. (2022). Protected area network insufficiently represents climatic niches of endemic plants in a Global Biodiversity Hotspot. *Biological Conservation*, *275*, 109768. DOI: 10.1016/j.biocon.2022.109768

Maccioni, S., d'Angella, F., De Carlo, M., & Sfogliarini, B. (2024). Stakeholder Engagement and Triggers for Sustainable Development in Complex Fragile Ecosystems: Evidence from Alpine Trentino Region. In *Sustainability* (Vol. 16, Issue 22). DOI: 10.3390/su16229879

Mainetti, A., D'Amico, M., Probo, M., Quaglia, E., Ravetto Enri, S., Celi, L., & Lonati, M. (2021). Successional Herbaceous Species Affect Soil Processes in a High-Elevation Alpine Proglacial Chronosequence. *Frontiers in Environmental Science, Volume 8-*. https://www.frontiersin.org/journals/environmental-science/articles/10.3389/fenvs.2020.615499

Manzoor, M., Ahmad, M., Gillani, S. W., Waheed, M., Grießinger, J., Shaheen, H., & Meier, W. J.-H. (2025). Biodiversity patterns and environmental drivers of fragile alpine glacial lake ecosystems in the Himalaya: Implications for conservation and management. *Ecological Indicators*, *178*, 114028. DOI: 10.1016/j.ecolind.2025.114028

Musa, I. O., Samuel, J. O., Adams, M., Abdulsalam, M., Nathaniel, V., Maude, A. M., Adedayo, O. A., & Tiamiyu, A. T. (2024). In Aransiola, S. A., Babaniyi, B. R., Aransiola, A. B., & Maddela, N. R. (Eds.), *Soil Erosion, Mineral Depletion and Regeneration BT - Prospects for Soil Regeneration and Its Impact on Environmental Protection* (pp. 159–172). Springer Nature Switzerland., DOI: 10.1007/978-3-031-53270-2_7

Nobis, M., Wróbel, S., Klichowska, E., Nowak, A., Wróbel, A., Nobis, A., Paszko, B., Świerszcz, S., Chen, W. L., Kauzal, P., Krzempek, M., Liu, B., Nowak, S., Piwowarczyk, R., Pedraja, Ó. S., & Zięba, A. (2023). New national and regional plant records: Contribution to the flora of the Old World countries. *Acta Societatis Botanicorum Poloniae*, *92*(1), 1–21. DOI: 10.5586/asbp/162050

Rapuc, W., Guinoiseau, D., Arnaud, F., Dellinger, M., Sabatier, P., Gaillardet, J., Poulenard, J., & Bouchez, J. (2025). Human and climate impacts on the alpine Critical Zone over the past 10,000 y. *Proceedings of the National Academy of Sciences of the United States of America*, *122*(29), e2506030122. DOI: 10.1073/pnas.2506030122 PMID: 40658840

Repe, B., Simončič, P., & Vrščaj, B. (2017). Factors of soil formation. In *The Soils of Slovenia* (pp. 19–60). Springer. DOI: 10.1007/978-94-017-8585-3_3

Retzer, J. L. (1956). Alpine soils of the Rocky Mountains 1. *Journal of Soil Science*, *7*(1), 22–32. DOI: 10.1111/j.1365-2389.1956.tb00858.x

Samarin, M., Zweifel, L., Roth, V., & Alewell, C. (2020). Identifying Soil Erosion Processes in Alpine Grasslands on Aerial Imagery with a U-Net Convolutional Neural Network. In *Remote Sensing* (Vol. 12, Issue 24). DOI: 10.3390/rs12244149

Semwal, R., Panda, T., & Tripathi, N. (2025). Empowering Voices: The Role of Local Communities in Mountain Tourism Development. In *Balancing Mountain Tourism, Cultural Heritage, and Environmental Stability* (pp. 185–200). IGI Global Scientific Publishing.

Sharma, U. C., Datta, M., & Sharma, V. (2022). In Sharma, U. C., Datta, M., & Sharma, V. (Eds.), *Land Use and Management BT - Soils in the Hindu Kush Himalayas: Management for Agricultural Land Use* (pp. 295–462). Springer International Publishing., DOI: 10.1007/978-3-031-11458-8_7

Smiti, K. (2023). *Chapter 11 - Tourism and the properties of mountainous soil: a dynamic relationship* (R. Bhadouria, S. Singh, S. Tripathi, & P. B. T.-U. S. of M. L. Singh (eds.); pp. 213–230). Elsevier. 10.1016/B978-0-323-95925-4.00011-X

Son, D. (2023). Plant co-occurrence patterns and soil environments associated with three dominant plants in the Arctic. *Journal of Ecology and Environment*, *47*(1), 1–13. DOI: 10.5141/jee.23.002

Stanchi, S., Freppaz, M., Godone, D., & Zanini, E. (2013). Assessing the susceptibility of alpine soils to erosion using soil physical and site indicators. *Soil Use and Management*, *29*(4), 586–596. DOI: 10.1111/sum.12063

Sun, L., Liu, Y.-F., Wang, X., Liu, Y., & Wu, G.-L. (2022). Soil nutrient loss by gully erosion on sloping alpine steppe in the northern Qinghai-Tibetan Plateau. *Catena*, *208*, 105763. DOI: 10.1016/j.catena.2021.105763

Verrall, B., & Pickering, C. M. (2020). Alpine vegetation in the context of climate change: A global review of past research and future directions. *The Science of the Total Environment*, *748*, 141344. DOI: 10.1016/j.scitotenv.2020.141344 PMID: 32814293

Whitesides, C. J., & Butler, D. R. (2010). Adequacies and deficiencies of alpine and subalpine treeline studies in the national parks of the western USA. *Progress in Physical Geography*, *35*(1), 19–42. DOI: 10.1177/0309133310384642

Wilson, B. R., Tulau, M., Kuginis, L., McInnes-Clarke, S., Grover, S., Milford, H., & Jenkins, B. R. (2022). Distribution, nature and threats to soils of the Australian Alps: A review. *Austral Ecology*, *47*(2), 166–188. DOI: 10.1111/aec.13115

Wu, G., Xie, Y., Wei, J., & Yue, X. (2023). Freeze-thaw erosion mechanism and preventive actions of highway subgrade soil in an alpine meadow on the Qinghai-Tibet Plateau. *Engineering Failure Analysis*, *143*, 106933. DOI: 10.1016/j.engfailanal.2022.106933

Wu, L., Song, Z., Yang, X., Zhang, X., Hao, Q., Xia, S., & Zhao, X. (2022). Variation of soil silicon fractions and bioavailability during centennial-scale evolution of ecosystem after glacier retreat. *Soil Science Society of America Journal*, *86*(3), 552–565. DOI: 10.1002/saj2.20386

Wu, S.-P., & Chen, Z.-S. (2005). Characteristics and genesis of Inceptisols with placic horizons in the subalpine forest soils of Taiwan. *Geoderma*, *125*(3), 331–341. DOI: 10.1016/j.geoderma.2004.09.007

Xu, X., Ma, K., Fu, B.-J., Liu, X., Huang, Y., & Qi, J. (2006). Research review of the relationship between vegetation and soil loss. *Acta Ecologica Sinica*, *26*, 3137–3143.

Yandi, S., Tao, M., Huakun, Z., Honglin, L., Zhonghua, Z., Li, M., Ruimin, Q., Hongye, S., Tao, C., Jingjing, W., & Xue, H. (2023). Characterization of the Plant–Soil feedback index in alpine meadow degradation and recovery: A field experiment. *Frontiers in Environmental Science*, *10*, 1097030. https://www.frontiersin.org/journals/environmental-science/articles/10.3389/fenvs.2022.1097030. DOI: 10.3389/fenvs.2022.1097030

Yang, C., & Sun, J. (2021). Impact of soil degradation on plant communities in an overgrazed Tibetan alpine meadow. *Journal of Arid Environments*, *193*, 104586. DOI: 10.1016/j.jaridenv.2021.104586

Yang, Y., You, Q., Zuo, Z., Zhang, Y., Liu, Z., Kang, S., & Zhai, P. (2023). Elevation dependency of temperature trend over the Qinghai-Tibetan Plateau during 1901–2015. *Atmospheric Research*, *290*, 106791. DOI: 10.1016/j.atmosres.2023.106791

Zhang, J., Zhu, C., Xie, L., Wu, S., Cao, C., Wang, M., & Cui, S. (2025). Comparative study on landslide susceptibility assessment of different models: A case study of alpine mountainous region in Xinjiang. *Bulletin of Engineering Geology and the Environment*, *84*(2), 95. DOI: 10.1007/s10064-025-04105-5

Zhou, T., Sun, J., & Shi, P. (2024). Mechanism of plant–soil feedback in a degraded alpine grassland on the Tibetan Plateau. *Journal of Plant Ecology*, *17*(2), rtae025. Advance online publication. DOI: 10.1093/jpe/rtae025

Zou, L., Tian, F., Liang, T., Fensholt, R., He, T., & Schaepman-Strub, G. (2025). Topographic effects on vegetation greening and area expansion in global alpine zones under climate change. *International Journal of Applied Earth Observation and Geoinformation*, *142*, 104727. DOI: 10.1016/j.jag.2025.104727

Chapter 9
Historical Plant Surveys and Floristic Trends on Alpine Summits

Gul Nawaz
http://orcid.org/0009-0000-5328-0454
University of the Punjab, Pakistan

Hafeeza Amna Saleem
http://orcid.org/0009-0005-8109-5893
University of the Punjab, Pakistan

ABSTRACT

The alpine summits have long been central to botanical exploration, serving as reference points for documenting flora at the highest elevations. Historical surveys from early expeditions of the 18th and 19th centuries to standardized 21st century monitoring programs provide valuable insights into floristic patterns, species richness and ecological change on mountain summits. This chapter reviews these surveys as the foundation of alpine floristic knowledge and biodiversity assessment. It includes case studies from major mountain systems such as the Alps, Himalayas, Andes, Karakoram and Rockies and highlights the role of programs like GLORIA. Key findings reveal upward species migration, shifts in community structure and loss of cold-adapted taxa with warming climates. By linking historical records with modern ecological monitoring, this chapter emphasizes the relevance of summit floristics for conservation planning, ecological forecasting and policy guidance. Alpine summits act as natural laboratories, illuminating biodiversity dynamics amid rapid environmental change.

DOI: 10.4018/979-8-3373-6756-9.ch009

ght © 2026, IGI Global Scientific Publishing. Copying or distributing in print or electronic forms without written permission of IGI Global Scientific Publishing is prohibited. Use of this chapter to
 generative artificial intelligence (AI) technologies is expressly prohibited. The publisher reserves all rights to license its use for generative AI training and machine learning model development.

1. INTRODUCTION

The alpine summits are one of the most extreme and ecologically remarkable habitats on the planet (Nagy & Grabherr, 2009; Rundel & Millar, 2016). These are located above the climatic tree line or at the climatic tree line and are characterized by cold temperature, strong winds, high radiation and soils which are of poor quality in nutrients (Grace et al., 2002). In spite of these unfavorable conditions, they provide a rich collection of highly specialized and even endemic plant species which have adapted in extraordinary adaptations to brief growth periods and high gradients of environment (Crawford, 2008). Due to its features, alpine summits are natural laboratories of research in the ecological and evolutionary processes of organisms, especially studies that control the distribution of plants, the assembly of communities and their resistance to climate change (Lavergne et al., 2010).

The ecological research of alpine summit flora has been a long-term interest of naturalists and botanists, both early travelers of the 18th and 19th centuries and modern ecologists that utilize modern sophisticated devices of observation (Stöckli et al., 2011). These summits are sentinel locations of vegetation change because they are one of the earliest of ecosystem to be affected by global warming. Even small temperature hikes can cause an increase in the upward movement of plant species, extinction of cold-adapted taxa in an area and drastic changes in the composition of a community (Komarkova & Wielgolaski, 1999). Knowledge of these changes is important, not only to record loss of biodiversity, but also to make predictions in future changes of the ecosystem under changing climatic conditions, which are accelerating.

In this regard, historical plant surveys are critical. The early explorations that documented the alpine flora in systematic effort in the past centuries formed the basis of our current knowledge on mountain biodiversity (Körner, 1995; Séverine Fauquette et al., 2018; Stöckli et al., 2011). These historical data when paired with recent survey offer useful long-term baselines of evaluating changes in vegetation, species replacement and ecological transformation (Magurran et al., 2010). These comparisons can give the scope and the course of floristic changes, which provides a clue to the strength and weakness of alpine ecosystems.

Summit floristic studies have become more of a combination of different disciplines today, extending to traditional taxonomy and biogeography, remote sensing, molecular ecology, and long-term ecological monitoring (Cavender-Bares et al., 2022; Stöckli et al., 2011). Such an overlap of solutions highlights the applicability of the alpine summit study to biodiversity conservation, ecological prediction and environmental policy. This connection between the present and the past provides the historical and modern study of summit floristics with a unique view of the trends and mechanisms that form the life of plants at the highest points of our planet.

1.1 Importance of alpine summits for floristic studies

Alpine summits are the ecological extremes of the terrestrial ecosystem, as temperature, radiation, wind and soil constraints have formed unique selection pressures on the plant life (Körner, 2023). These gatherings are naturally known as laboratories above the tree line and are used to investigate biogeographical, ecological, and evolutionary procedures. They are isolated and hence contain highly specialized and endemic flora, which have adapted to harsh environmental gradients (Boulangeat et al., 2012). As a result, alpine summits prove to be the pinnacle to discover species location, boundaries and how communities have been put together and adapted to climatic stresses (Hotaling et al., 2017; Rundel & Millar, 2016).

Alpine summits are also a good source of information about the way vegetation responds to long-term climatic change and anthropogenic change, floristically (Hamid et al., 2020). They are the best observatories to detect changes in the species composition, phenology and ecosystem functioning due to their minimal human disturbance, and the sensitivity of climatic feedbacks (Midgley et al., 2007). In addition, with the world temperatures increasing, the summits floristic provide crucial data in gauging the biodiversity reactions to the warming tendencies and can be seen as the set of alerts about the overall ecosystems changing (Hamid et al., 2020).

1.2 Role of historical surveys in understanding vegetation change

Our present knowledge on mountain biodiversity is based on past survey of plants on alpine summits (Pauli et al., 2012). Most of these early embarkations, which over time were performed by some of the most prolific naturalists of the 18th and 19th centuries, documented baseline species collections that currently exist as treasured reference points when gauging ecological shifts throughout the centuries. However, these early botanical records generally involved limited sampling intensity, often relying on descriptive observations, specimen collections and qualitative methods rather than systematic quantitative sampling. Elevation ranges were a central focus, with botanists like Albrecht von Haller and Alexander von Humboldt pioneering the study of altitudinal vegetation zonation, but elevation measurements were often imprecise due to the rudimentary tools available, such as barometers and field notes. Moreover, data limitations included uneven geographic coverage biased towards accessible regions, inconsistencies in taxonomic identification due to morphological reliance, subjective habitat descriptions, and incomplete or schematic spatial documentation. A comparison of these early records and modern resurveys shows that some of the critical trends are species turnover, upward migration and local extinctions (Tingley & Beissinger, 2009).

The record of historic expeditions into the flora presents an incomparable time frame of vegetation dynamics (Ritchie, 1995). With increased rates of climate change, these historical data sets have become especially important in measuring long-term biodiversity patterns and in calibrating ecological prediction models (Magurran et al., 2010; Urban et al., 2016). They also fill the gap between applied conservation science and the descriptive floristics approach because they connect past with present conditions of vegetation and future directions as given in Figure 1.

Figure 1. Historical plant surveys (18th-19th century).

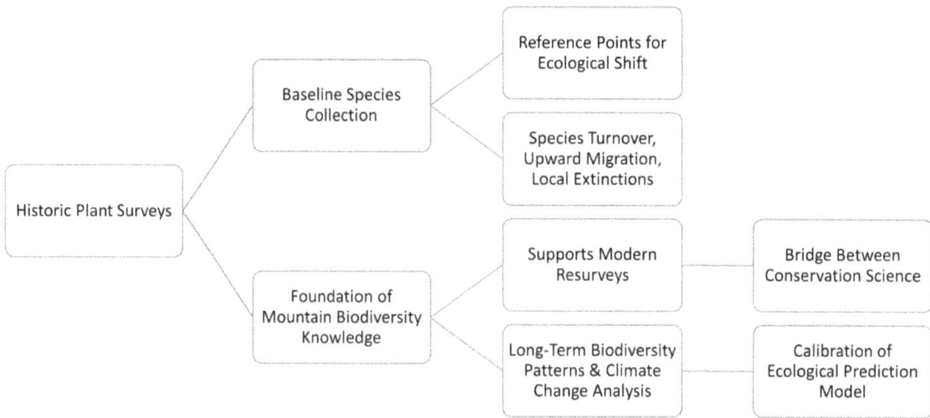

2. BACKGROUND OF ALPINE FLORISTIC RESEARCH

The intellectual history of alpine floristic studies covers the border between natural history and systematics on the one hand and contemporary ecological science on the other (Egerton, 2013). Here, the conceptual and methodological underpinnings that have influenced our perception of plant assemblages at high altitudes are described, starting with the earliest descriptions of the high-elevation environment and moving through to more recent syntheses. This historical context puts the alpine summits into perspective as to why these are the most informative backgrounds with which to examine the patterns and dynamics of biodiversity (Lockwood et al., 2014; Stöckli et al., 2011). The subsequent subsections provide an overview of the initial work in exploring the subject, the discoveries of major botanists and the development of survey techniques to meet new scientific interests and technology.

2.1 Early botanical explorations in alpine regions

Alpine floristic studies are as old as the history of natural exploration in the 18th and early 19th centuries where European botanists first explored the high mountains of Europe to record the lives of plants growing in extreme conditions. Quantitative methods of zonation of vegetation were first applied by explorers like Albrecht von Haller in the Alps and Alexander von Humboldt in the Andes and formed the basis of modern phytogeography (Shteir, 1976; Turrill, 1959). Their field work concentrated on the issue of altitude including vegetation composition and demonstrated that elevation might be used as a measure of climatic change.

These initial surveys were mainly descriptive in which, identification of plants, collection of specimens and general classification of the habitat was done (Elton & Miller, 1954; Kent, 2011). Although their tools were not much in those days, they were useful in terms of information that still guides comparative studies. Initial studies of the alpine flora also emphasized the distinctness of the mountain biotas of high altitude and prompted the next generations of botanists to study the remote mountain systems in Asia, South America, and Africa.

2.2 Key contributions of pioneer botanists

There were other pioneer botanists who made a lot of contributions to the classification of the alpine floristics (Stöckli et al., 2011). The formalization of principles of plant classification and distribution that form the basis of modern floristic taxonomy was made by Alphonse de Candolle, Carl Linnaeus and Asa Gray among others. Explorers, such as Sir Joseph Hooker and William Griffith, described the Himalayan flora in detail in South Asia, uncovering altitudinal gradients and zonation patterns of species which are still useful today.

These preliminary attempts not only increased herbarium collections, but also promoted a scientific attitude towards plant geography (Lavoie, 2013). Their contributions continue to be used today as references due to the foundation of alpine botanical collections and repositories that form a vital part in comparative studies. By linking taxonomy and biogeography, the early botanists turned floristic exploration into a scientific field and not an entirely descriptive process (Heywood, 2001; Victor et al., 2016).

2.3 Evolution of survey methods over time

The techniques of the alpine floristic studies have dramatically changed since the pioneer endeavors (Walther et al., 2005). Surveys that started using field sketches, herbarium specimens and qualitative descriptions over time started using quantitative sampling methods, permanent plots and standardized protocols (Sullivan & Nazaire, 2022). With the appearance of photography, altimetry and subsequently, GPS-based mapping, the accuracy of site documentation and species whereabouts location improved. Global exchange programs like the Global Observation Research Initiative in Alpine Environments (GLORIA) have become common elsewhere in the world in recent decades to standardize summit monitoring (Grabherr et al., 2000). The networks use grid-based sampling, repeat photography and long-term data archiving to monitor changes in species richness and cover. Digital herbarium databases and remote-sensing tools have also transformed floristic surveys further, which makes it possible to reconstruct historical datasets and massive meta-analyses of the change in biodiversity in the Alps (Giorgini et al., 2015).

3. HISTORICAL PLANT SURVEYS ON ALPINE SUMMITS

Historical surveys are the temporal support of summit floristics: they provide the baseline data on which the change in modern times is assessed (Stöckli et al., 2011). This part presents the character and extent of those initial and classical surveys, their purposes, instrumentation and location. It is a summation of how practices, such as specimen collections and narrative field notes, as well as hand-drawn altitudinal profiles, produced long-lasting herbarium and archival materials. Meanwhile, in this section the author accepts the natural weaknesses of early data: uneven coverage, inconsistent nomenclature and inaccurate localization, but indicates the utility of such records in the event they are carefully standardized and recontextualized to be used in modern analysis.

3.1 Classic floristic surveys (18th-20th century)

Alpine exploration was first done in the early phase of enlightenment when European naturalists were trying to record the variety of plants in isolated mountain ranges (MacGregor, 2018). Early botanists like Carl Linnaeus, Albrecht von Haller, Alexander von Humboldt, and Horace-Benedict de Saussure made much fieldwork in the Alps and other mountainous countries, and established the principles of systematic floristics. The golden age of botanical exploration was the 18th and 19th centuries, whereby explorers carefully documented the species of the Pyrenees,

Alps, Himalayas, Andes and Rocky Mountains. These early lists of flora, frequently with elaborate sketches and herbarium vouchers attached to them were the earliest attempts to define the alpine zones of vegetation scientifically in terms of altitude and climate (Körner, 2021).

Institutionalized botanical surveys became more organized by the end of the 19th and early 20th century (Volberg, 1983). Artistically sponsored expeditions like the Royal Botanic Gardens, Kew and the Vienna School of Alpine Ecology increased the awareness of all alpine floras in Europe and Asia. Initial Himalayan field work by Hooker and Thompson set standards of subsequent Himalayan floristic work, and Humboldt in the Andes transects established standards of altitude-vegetation structure relationships.

3.2 Methods and tools used in early surveys

Descriptive observation and collection of specimens were very important in the historical surveys (Pyke & Ehrlich, 2010). Poor tools like hand lenses, barometers and field notebooks were employed by botanists to note morphological features, altitude and habitat description (Glime, 2013). The identification of plants was mostly according to their visible morphology and the taxonomic classification relied upon the herbarium comparison and printed floras. Little cartographic recording was done and, in many cases, it was done in schematic form or altitude profiles on the paper.

These methods in spite of their limitations proved remarkably successful in determining the altitudinal vegetation belts of forests, subalpine, alpine and nival, in terms of floristic composition (Körner, 1995; Miehe et al., 2015). Paper pressing and drying were also techniques employed in the early days of collecting specimens to preserve them in a manner that they could be later verified as to the taxonomic identity of the specimen. The use of phenotypic traits, though, resulted in misidentifications and underrepresentation of cryptic or morphologically close taxa at times as shown in Figure 2.

Figure 2. Methods and tools used in early botanical surveys.

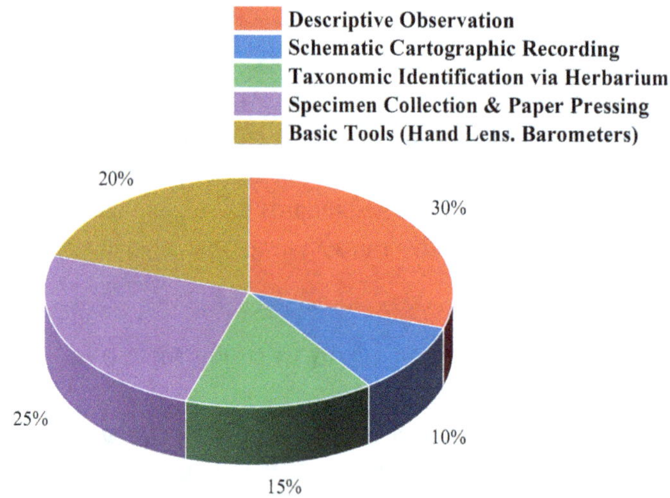

3.3 Geographic coverage and notable alpine summits studied

Traditional floristic surveys covered large mountain ranges on continents. The earliest regions to be explored extensively in Europe were the Alps and Pyrenees which form the foundations of classical alpine ecology (Séverine Fauquette et al., 2018). In Asia, the Himalayas, Karakoram and Tian Shan drew the attention of the British and Russian empires as the region had unique flora and extreme differences in elevation. In the Americas, zonation of plants and biogeographical parallels were studied in the Andes and the Rocky Mountains. The Mont Blanc, Mt. Elbrus, Mt. Everest and Mt. Chimborazo were also used as points of reference in the world alpine vegetation pattern. These expeditions found some dramatic overlaps in the floras of areas separated by vast distances, implying that the alpine vegetation has the same ecological principles as flora in other parts of the globe even though the species zones are quite different (Jaccard, 1912). The records of these conferences continue to be significant historical foundations of modern surveys. Historical overview of major alpine floristic surveys, illustrating regional diversity, survey periods and methodological approaches that formed the foundation of modern alpine botany are given in Table 1.

Table 1. Summary of historical alpine floristic surveys (18th-20th century).

Region/ mountain range	Period of survey	Key botanists/ explorers	Survey method	Number of species documented	Primary reference(s)
European Alps	1750-1900	Haller, De Saussure, Christ	Descriptive species lists; herbarium collections	~1200	(von Christ, 1886; Von Haller, 1768)
Himalayas	1820-1940	Hooker, Thomson, Wallich	Transect-based floristic sampling	~2500	(Hooker, 1854; Hooker & Thomson, 1855)
Andes	1850-1930	Humboldt, Bonpland	Altitudinal zonation and vegetation mapping	~1800	(von Humboldt, 1819)
Rocky Mountains	1870-1930	Gray, Coville	Quadrant sampling and elevation mapping	~1100	(Gray, 1889)
East African Highlands	1890-1940	Hedberg, Fischer	Summit plots and floristic inventory	~900	(Hedberg, 1964)

3.4 Limitations of historical data

Although precious, the initial survey was limited by poor accessibility, disparate methodology and subjective observations (Rea & Parker, 2014). Measures of elevation were frequently inaccurate, taxonomic standards differed among botanists and a lot of species descriptions were not accompanied by voucher specimens (Rowe, 2005). There was poor spatial coverage, especially in tropical and subtropical alpine systems, as it concentrated its efforts on reachable or politically secure areas. In addition, nomenclature and taxonomy have evolved over time, making it difficult to directly compare the change with current data. In spite of this, historical datasets are still very important to biodiversity assessment over long term particularly when formatted as standardized and georeferenced datasets to be analyzed in the present.

4. MODERN FLORISTIC TRENDS ON ALPINE SUMMITS

Current floristic studies on summits are motivated by the global issue regarding the climate change, loss of biodiversity and vulnerability of ecosystem (Vanneste et al., 2017). However, challenges remain in ensuring data quality and comparability. GPS precision is critical for accurate plot location, yet GPS devices, especially earlier models, often exhibit spatial errors up to 20 meters or more depending on canopy cover and terrain, complicating exact plot relocation over time. This imprecision leads to uncertainties when re-surveying permanent plots, potentially affecting assessments of vegetation change. Additionally, observer bias remains an important factor, as variation in species identification skills and sampling approaches among field researchers can influence floristic data consistency (Braziunas et al., 2024). Recognizing these technical and human limitations, modern alpine monitoring programs incorporate standardized protocols, GPS mapping, and training to reduce errors and improve the reliability of long-term vegetation change assessments..This chapter presents current surveillance programmes and methodological standardization that have evolved the summit research into lone expeditions into coordinated and long-term science. It offers the major insights that arise with contemporary re-surveys like thermophilization, heightened species turnover, phenology and prepares the detail of standardized programs, observed changes in composition and interactive drivers of such changes in the following subsections (Braziunas et al., 2024).

4.1 Recent surveys and monitoring programs

During the late 20th and early 21st century, there was a resurgence of interest in summit floristic research driven by worries about climate change and decline of biodiversity (Van Wilgen et al., 2016). Resurveys that usually emulate classical expeditions have reported a major change in floristic composition in mountain tops across the world. The projects like GLORIA (Global Observation Research Initiative in Alpine Environments) set regular guidelines with which vegetation monitoring on alpine tops could be conducted over extended durations across continents (Grabherr et al., 2000). The results of these attempts have resulted in one of the most detailed global data set on the change in biodiversity of the alpine area, which could be used to make cross-regional comparisons of species richness and turnover, along with vegetation changes.

4.2 Standardized protocols (e.g., GLORIA and Long-term monitoring)

GLORIA program has an intensive, repeatable design with a permanent summit plot design to measure the variation of plant species composition, abundance and microhabitat conditions (Porro et al., 2019). The quadrants of the summits are subdivided into 1 m plots where surveys are conducted on the vegetation. Data on the continuous monitoring of the soil and air temperature is registered by temperature loggers, which connect the changes in the flora with climatic variations. Floristics Complementary monitoring programs, including the Mountain Invasion Research Network (MIREN) and the Global Mountain Biodiversity Assessment (GMBA), combine floristics with ecological modelling, molecular genetics and remote sensing (Korner & Spehn, 2024). Such standardized structures guarantee that data is compatible in geographic areas and this will lead to meta-analyses that will show global trends in alpine biodiversity in warming conditions. Major global and regional alpine floristic monitoring initiatives, showing scope, methodology and resampling frequency for long-term ecological observation are represented in Table 2.

Table 2. Comparative overview of modern alpine monitoring programs.

Monitoring program	Geographical focus	Established (year)	Main objectives	Sampling approach	Temporal resolution
GLORIA (Global Observation Research Initiative in Alpine Environments)	Worldwide (Himalayas, Alps, Andes, Rockies & Africa)	2001	Assess long-term changes in alpine species composition and temperature response	Permanent summit plots with elevation gradients	Every 5-10 years
ITEX (International Tundra Experiment)	Arctic and alpine regions	1992	Study plant responses to experimental warming	Open-top chambers (OTCs)	Annual
Alps LTER (Long-Term Ecological Research)	Central and Western Alps	2005	Examine vegetation dynamics under climate and land-use change	Fixed-plot vegetation mapping	Every 3 years
Himalayan Alpine Network	Western and Eastern Himalayas	2010	Monitor species turnover and phenological changes	Summit transects and quadrats	Every 5 years

In the Himalayas, Chauhan et al. (2023) highlight the region as a climate change laboratory, noting glacier retreat, upward treeline shifts, and biodiversity loss impacting 1.6 billion downstream people. Panthi et al. (2020) examine climate effects on tree growth and forest productivity across topographic gradients. Mahdi et al. (2024)

assess vulnerability in Kashmir, linking warmer temperatures, reduced snowfall, and glacial melt to water scarcity and disputes.

Himalayan patterns align with Rockies' cold-adapted species declines and Andes' elevational contractions, as seen in NW Argentina summits showing vegetation turnover over 11 years. Ecohydrological shifts in High Mountain Asia, with snow-vegetation interactions mirroring global alpine responses. These studies (e.g., Cryosphere degradation) provide quantitative comparisons to GLORIA data on migration rates and richness.

4.3 Shifts in species composition and diversity over time

Historical species richness resurveys have persistently demonstrated a rise in the number of species, especially at altitude (Lelli et al., 2021). This trend of thermophilizing is an indication of upward migration towards lower-elevation species due to increase in temperature. By comparison, cold-adapted specialists are either going through range contractions or local extinctions as a result of competition and habitat loss (Callaghan et al., 2004). This has led to community structure being dominated by species with more ecologically wide amplitudes and decreased the distinctiveness of the summit floras. Moreover, new colonizers are changing functional characteristics including leaf morphology and phenology and have cascading effects on pollinator networks and nutrient cycling.

4.4 Drivers of change (Climate Warming, Land Use, Invasive Species)

On the alpine summits, climate warming is the major cause of the floristic change seen there. Increases in temperature increase the length of the growing phase, enhance colonization towards the north and alter the distribution of snowmelts (Inouye, 2008). Nonetheless, other anthropogenic influences such as grazing pressure, tourism, pollution and introduction of invasive species are additional factors causing vegetation change (Meyer et al., 2021). In part of the area, land abandonment over time has led to natural succession whereas in other areas, the increased human activity has fragmented mountain tops habitats. The combination of non-climatic and climatic influences indicates the complexity of the processes in the alpine vegetation and the necessity of the multidisciplinary character of monitoring as shown in Figure 3.

Figure 3. Trends in three major drivers of alpine vegetation change from 2000 to 2020.

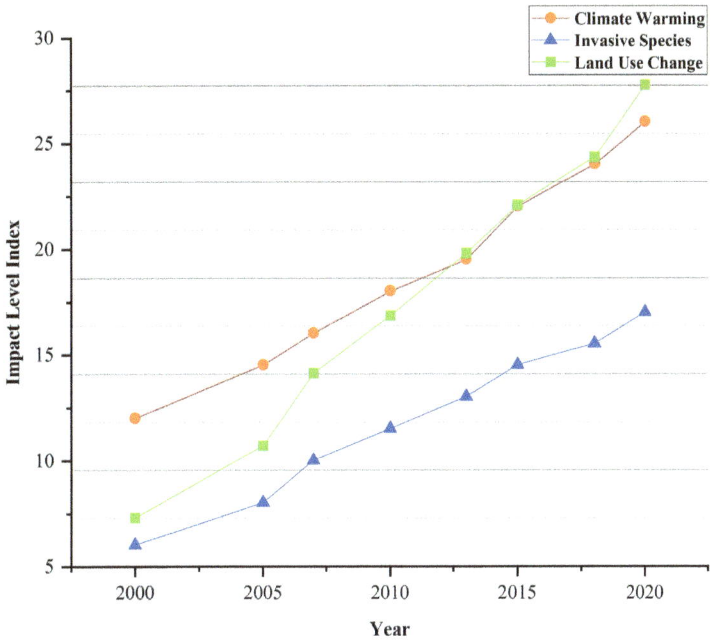

5. PATTERNS OF FLORISTIC CHANGE ON ALPINE SUMMITS

The floristic change patterns observed on summits are complicated and multi-dimensional with species turnover, elevation redistribution and changes in functional composition (Thakur & Chawla, 2019). This section presents the main trends identified by comparative and long-term studies and puts them into ecological perspective emphasizing mechanisms (colonization, local extinction, competition) and limitations (topographic ceiling, limited dispersal) as given in Figure 4. This introduction prepositions a specific analysis of turnover rates and range shifts and transformations by synthesizing both observed trends, taxonomic and functional, in the following successive subsections (Mufungizi et al., 2023).

Figure 4. Relative composition index of grasses, forbs and shrubs from 1900 to 2020.

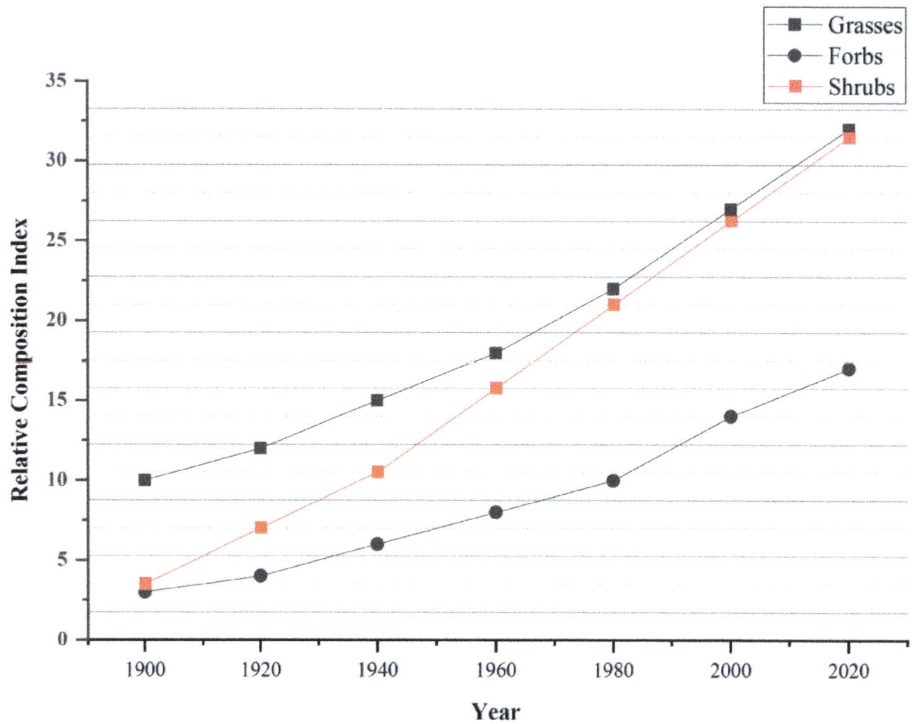

5.1 Species turnover and colonization

The replacement of the original cold-adapted flora by new colonists at lower elevations is one of the most noticeable signs of vegetation change on the alpine summits (Leuschner & Ellenberg, 2017). According to the data of the resurvey conducted by European, Asian and American summits, the rate of species colonization has increased in the past decades (Menke & Holway, 2020). This movement causes a slow homogenization of high-altitude vegetation and initially separate assemblies of vegetation of high elevations are more and more alike.

Colonizing species tend to have more ecological tolerance, higher plasticity of phenotype and increased competition capacity than resident competitors (Muthukrishnan et al., 2020). As a result, cold-specialist taxa that have limited climatic ranges e.g., *Saxifraga* appositively, *Gentiana nivalis* and *Ranunculus glacialis* are being pushed out or confined to isolated microsites where snow cover is persistent. The rate of extinction vs the rate of colonization thus determines the survival of the alpine biodiversity over the long run. Species richness across habitat heterogeneity is given in Figure 5.

Figure 5. Species richness across habitat index

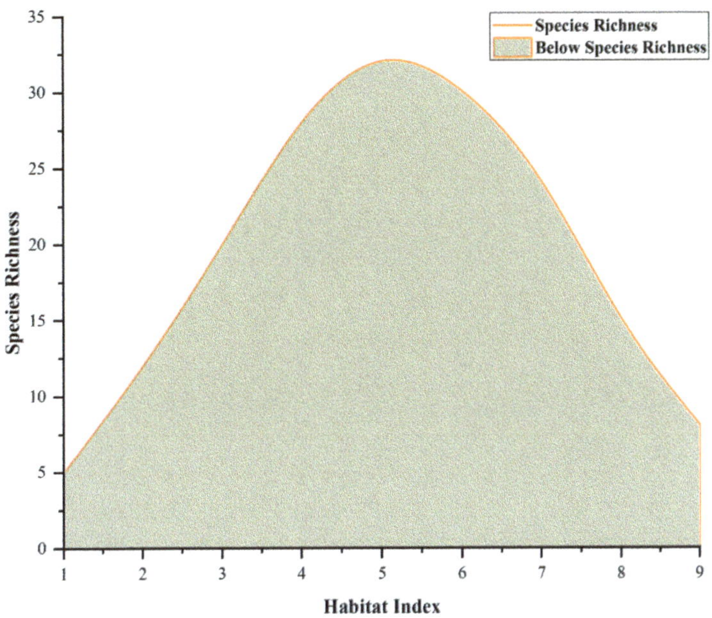

5.2 Elevational range shifts

Alpine flora is very sensitive to temperature gradients and with rising of climate, several species have been found to move up the hill slope (Körner & Hiltbrunner, 2021). The data of the Alps, Rockies and Himalayas shows that the average upward movement is between 1 and 5 meters per decade, but there are differences with the topography of the area, soil type and the microclimate. This is not a uniform range shift: thermophilus (warm-adapted) species are increasing faster than cryophilous (cold-adapted) taxa are able to shrink and this has resulted in a compression of the available niche space around summits as shown in Figure 6. Since reaching the elevational ceiling, real alpine endemics become threatened by a mountain-top extinction, especially in areas with low vertical relief. The phenomenon highlights one of the great ecological paradoxes that overall richness of the total alpine taxa on summits can rise at first as a consequence of colonization, but in the long-run, net loss of distinctive alpine taxa can occur (Körner & Hiltbrunner, 2021).

Figure 6. Elevational range shift of thermophilus and cryophilous from 1980 to 2020.

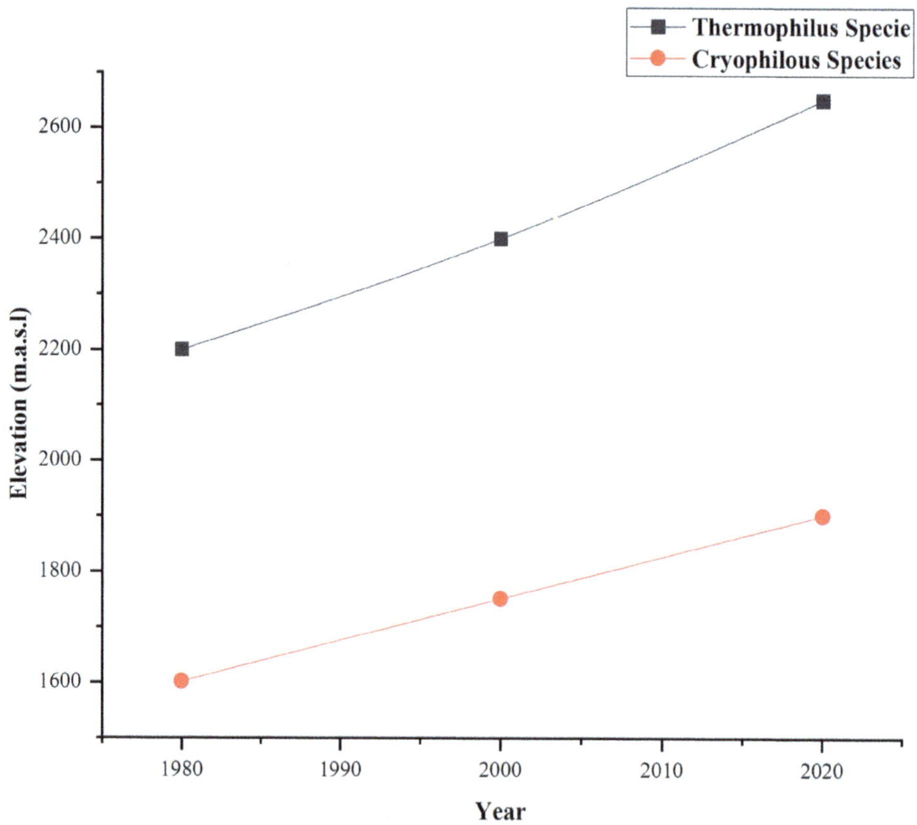

5.3 Changes in functional and phytogeographical groups

The functional composition of the alpine vegetation has changed considerably in addition to the identity of species. Plants that are adapted to favor warmer and drier environments such as increased height, increased leaf area and an earlier onset of phenology are increasingly common (Kikuzawa, 1995). Functional homogenization is also characterized by the loss of specialized cushion-forming and rosette species that occupy cold and wind exposed microsites.

Phytogeographically, we have a growing presence of temperate and cosmopolitan elements at the cost of Arctic-alpine and endemic elements (Weber, 2003). The Himalayas and Andes show that lowland migrants are increasingly contributing to the growing share of summit flora, indicating biogeographical mixture and disappearance of the regional character. It is a functional and structural rearrangement transforming the processes of the ecosystems, which impacts soil stability, cycling of nutrients and food webs in the Alps. Indicators summarizing documented floristic shifts on alpine summits, emphasizing ecological processes influenced by climate and anthropogenic pressures are given in Table 3.

Table 3. Summary of floristic change indicators on alpine summits.

Indicator type	Metric used	Observed trend	Primary driver(s)	Example region/ program
Species richness	Number of vascular plant species per summit	Increasing at low- and mid-elevations (70% of plant species expanding distributions by 2100)	Temperature rise; extended growing season	GLORIA-Europe
Species turnover	β-diversity index	Moderate to high in recent decades	Colonization by lowland species	Alps, Andes
Elevational range shift	Mean elevation (m.a.s.l.) of species occurrence	Upward shift 11.0 m per decade and poleward at 16.9 kilometers per decade	Warming and snow-cover reduction	Alps, Himalayas
Functional composition	Proportion of graminoids vs. forbs	Increase in graminoid dominance	Climate warming, grazing pressure	East African Highlands
Phytogeographical origin	% Native vs. cosmopolitan species	Increase in cosmopolitan taxa	Species migration and disturbance	Andes, Himalayas

Alpine plant communities show increased species richness and shifts in flowering phenology driven mainly by earlier snowmelt rather than direct warming. They also report colonization of new species at different elevations, with long-lived species slower to respond to climate change Jabis et al. (2019). Species compositions are shifting toward more warmth- and drought-adapted species due to increased temperature and declining water balance. Thermophilic species increased while cold- and moist-preferred species declined, reflecting floristic shifts driven by climate warming (Kuo et al. 2022).

6. ECOLOGICAL AND CONSERVATION IMPLICATIONS

Monitored floristic alteration in alpine summits has profound ecological implications along with dire conservation short-term implications. Here, the shift of species composition and functional features into altered ecosystem functioning (e.g., nutrient cycling, soil stability, hydrology) and the effect of this on the delivery of services of interest to downstream communities are presented. It is also placed in the practical applications of historical and modern floristic information to conservation planning, vulnerability evaluation and proactive management. The subsequent successive subsections elaborate predictive modeling, identification of indicator and vulnerable species and practical methods of incorporating long-term floristic records to conservation decision-making.

6.1 Predicting future vegetation dynamics

By combining the past floristic records of the vegetation cover with the climate models, it becomes possible to predict future vegetation patterns in the alpine summits (Vanneste et al., 2017). According to modeling studies, global warming could cause contraction of alpine ranges by 20-40 percent by the close of the 21st century with effects being the most dramatic in tropical and subtropical mountains. How warming, changing modes of precipitation, and snowmelt interactions affect alpine plant communities will transform the alpine flora in nonlinear, region-specific ways (Nila et al., 2025). It has been projected that both now and in future, there will be peaks of diversity at the middle level and in the end, there will be homogenization in the peak levels or higher levels. It is important to predict such transitions to come up with adaptive management plans to mountain biodiversity.

6.2 Identifying vulnerable and indicator species

Old records of flora are useful benchmarks to be able to recognize species that are most susceptible to alterations in climatic conditions (Gallagher et al., 2009; Willis et al., 2017). At greatest risk are high-elevation specialists that have low dispersal potential, limited ranges and are dependent on microhabitats. Indicators like *Silene acaulis*, *Dryas octopetala* and *Androsace chamaejasme* are more than likely to be utilized in monitoring ecosystem sensitivity based on their predictability to the environmental gradients. With these indicator taxa in the long-term monitoring schemes, ecological changes are detected at an early stage, and conservation practitioners have a system of early-warning. Newly colonizing species, especially those growing out of lower altitudes, on the other hand, can serve as biological indicators of the level of warming and micro climatic change.

6.3 Integrating Historical Data into Conservation Planning

The combination of historical surveys and contemporary floristic lists makes it possible to trace the trends in vegetation on the century scale, which is an irreplaceable resource in conservation planning (Corlett, 2016). Spatial explicit floristic change analysis can be achieved by digitization of historical herbarium records, nomenclature alignment, and georeferenced to classic survey sites. This type of integrative data can mark biodiversity hotspots, set priorities on areas to be conserved and direct restoration efforts. Temporal continuity conservation frameworks in which 19th century baselines are connected to the 21st century monitoring can give inimitable opportunities to assess resilience and vulnerability of alpine systems. Moreover, standardized databases like GBIF are being used that are supplemented by conservation decision support programs like GLORIA and MIREN to provide conservation decisions with high quality and repeatable replicable data on an international mountain systems basis.

7. CHALLENGES AND OPPORTUNITIES IN ALPINE FLORISTIC RESEARCH

Though there are scientific benefits, Alpine floristic studies are affected by various obstacles due to the environmental, logistical and methodological peculiarities of the high-elevation ecosystem. Simultaneously, there is a technological breakthrough, which has provided new possibilities of more precise, more globalized and more long-term recording of the alpine plant diversity. The knowledge of these challenges and opportunities is critical in the development of successful monitoring programs, enhancement of the reliability of the data and reinforcement of conservation frameworks of high-mountain biodiversity.

7.1 Technical and logistical constraints in high-altitude monitoring

The study of alpine summits is limited in nature by extremely difficult and unpredictable field conditions (Körner & Hiltbrunner, 2021). These issues have a great influence on the frequency, accuracy and reliability of the floristic surveys.

- *Severe natural disasters;* the weather in alpine summits is low temperature, changing weather conditions, high speed winds and intense sunlight. Such conditions not only put the researchers at risk, but also restrict the field season. Weather (snow cover, fog, or storms) can also mess up the plans of sampling, limit the access to permanent monitoring plots, or limit the visibility needed to identify the species accurately.

- *Oppressive geography and inaccessibility;* many of the summits are physically inaccessible due to steep slopes, loose scree, glaciers and lack of footpaths. Access to some of these mountains involves many days walking, climbing experience or specialized equipment to reach the mountain. These limitations inhibit replicates, reduce the frequency of surveys and make field campaigns more expensive and logistically complicated.
- *Poor growing period and phenological limitation;* the short period of growth in alpine habitats implies that surveys are to be done at the time of maximum plant activity. Failure to capture this thin slice can result in incomplete species lists particularly of short-lived species or those with short flowering periods. The problem of phenological variability across the years further complicates the fact that comparisons are made across time.
- *Field equipment limitations technical;* conventional devices, like handheld GPS battery-powered devices, altimeters, or cameras, might not work at high altitudes because of the cold temperatures or battery capacity. Also, in long-term monitoring soil sensors and automated data loggers need regular maintenance which can be hard to implement in mountainous remote areas.

Taken together, these limitations demonstrate the necessity to plan, enhance logistics, safety measures and supplementary data (e.g., remote sensing) to enhance high-altitude floristic monitoring.

7.2 Advances in remote sensing and molecular tools

Nevertheless, current technological advances can provide groundbreaking opportunities to alpine floristic studies, as they could be used to assess vegetation change more accurately and more quickly and in a more scalable way (Hotaling et al., 2017). Remote sensing technologies; the innovations in satellite imagery, aerial photography, drones and hyperspectral sensors provide researchers with the ability to identify vegetation patterns more precisely than ever before (Yao et al., 2019). High resolution imaging of inaccessible summit locations is possible through drones (UAVs) and species clusters, snow patches and disturbances can be finely mapped. Hyperspectral and LiDAR technologies enable the separation of vegetation species according to spectral markers and canopy structure which enhances habitat classification. Satellite platforms (Sentinel-2, Landsat) will provide multi-temporal and long-term data which can be utilized to monitor vegetation greening, the time of snowmelt, and changes in vegetation elevation. Combined with field surveys, remote sensing enables a researcher to extrapolate the findings of plot-based sampling to landscape analyses.

Molecular and genomic tool; molecular tools transformed the process of identification of the species in the alpine zone, analysis of their evolution and monitoring their population. DNA barcoding enhances taxonomic accuracy particularly in cryptic or morphologically variable species (Piredda et al., 2011). Population genomics (e.g., RAD-seq, SNP profiling) will assist in identifying gene flow, genetic bottlenecks and adaptive traits in alpine populations. Environmental DNA (eDNA) enables the identification of species in the soil or water sample without making physical observations, which is especially useful in difficult to reach summit environments. These molecular techniques clarify taxonomic ambiguities, identify concealed biodiversity and aid intricate biogeographical and evolutionary investigations.

Combining remote sensing and molecular data; the integrated application of remote sensing and molecular datasets can be considered one of the most promising future directions (Chen et al., 2022). For example: remote sensing may show a spatial change in vegetation and genetic tools will show the dynamics of demographics and adaptation to this change. Joint models spectral data, climatic layers and genetic information can be used to create high-resolution estimates of species response to climate warming. These integrative systems facilitate the accuracy and explanatory value of the alpine floristic studies.

8. FUTURE DIRECTIONS

Alpine floristic research has reached a decisive point in the age of fast environmental transformation all over the world. With the change in climatic gradients and reorganization of the ecosystem, it is time to return scientific focus to the alpine regions, traditionally viewed as the natural laboratory of ecological resilience and evolution. The future research should therefore adopt a hybrid construction of both traditional botanical knowledge and the dawning technologies, multidimensional data analysis and the conservation frameworks to have the comprehensive view of the species dynamics and ecosystem transformation in these delicate landscapes.

8.1 Integration of multi-temporal data and advanced modeling approaches

Among the most promising new areas of research in alpine floristics is incorporating multi-temporal data, which centuries of botanical study have provided. The historical herbarium records, historical record of the early expeditions, and the current high-resolution satellite images of the area all give you a priceless time span through which vegetation change can be analyzed. It can be assisted by the harmonization of these datasets based on the advanced modeling tools, including Species Distribution Models (SDMs), ecological niche modeling and machine learning algorithms to document the reconstruction of the historical baselines and forecast the dynamics of the range in the future (Vasconcelos et al., 2024). When the past data is combined with the current climatic evidence, scientists will be in a position to identify small-scale patterns in colonization, extinction, and elevation movement. In addition, predictive modeling can determine the possible refugia, vulnerable areas and species that will tend to undergo range fragmentation and hence aid in proactive management measures to conserve the alpine biodiversity.

8.2 Incorporating genomic and functional trait approaches

Future floristic research needs to be conducted on a larger scale than on taxonomic lists by covering not only molecular and functional aspects of alpine flora but also on an international scale. The current genomics, transcriptomics and DNA barcoding permit now the accurate identification of cryptic species, evaluation of genetic connectivity among isolated populations, and identification of adaptive alleles associated with thermal and moisture gradients (Rubinoff et al., 2020). The tools have the potential to clarify the evolutionary processes of species to survive and go through climate stress. Concurrently, the functional trait studies, such as leaf structure, phenological adaptability, allocation to reproduction, and physiological tolerance might provide more comprehensive understanding of community assembly and trade-offs. The integration of molecular and trait-based data sets can help researchers to comprehend better the role of the species functional diversity in the stability and overall resilience of the ecosystem in terms of the rapidly changing alpine environment.

8.3 Expanding monitoring networks and enhancing transboundary collaboration

The spatial heterogeneity of alpine ecosystems implies the need to pursue coordinated, standardized, and long-term monitoring activities across continents. Although networks including the Global Observation Research Initiative in Alpine Environments (GLORIA) have offered a priceless contribution to existing understanding of the changes in species composition, regions of the world, especially in Asia, Africa, and South America have not been adequately represented. An extension of these monitoring systems to the Himalayas, Karakoram, Hindu Kush and other high-altitude systems will offer a better insight into the dynamics of alpine vegetation around the world.

It is also important to have transboundary cooperation between scientific institutions, conservation agencies, as well as local stakeholders (Mason et al., 2020). The use of shared databases, harmonized methods, and open-access repositories may help in increasing the comparability of data and foster cross-regional synthesis. Transnational projects will also contribute to the discovery of universal mechanisms of floristic change and consideration of ecological peculiarities of a certain territory.

8.4 Harnessing citizen science and emerging technologies

Citizen science and technological innovation integration is a new revolutionary possibility in alpine floristic research. The use of mobile applications like iNaturalist or PlantNet can engage local communities, trekkers and mountaineers to provide a massive amount of observational data, especially from remote or inaccessible summits. Floristic databases are also improved by real-time data submission, image verification and geotagging, which increase the accuracy and spatial resolution of the database.

In addition, the application of innovative technologies, e.g., drones, LiDAR, hyperspectral imaging, and so on, can transform the process of vegetation mapping and phenological surveillance. These instruments give more specifications on structural and spectral data of vegetation cover to allow researchers to detect subtle variations in plant health, productivity and spatial distribution. The combination of remote sensing data and ground-based surveys will guarantee the complete and scalable comprehension of the biodiversity patterns in the alps.

8.5 Linking floristic change to ecosystem function and conservation policy

One of the challenges in the future is the ability to merge the dynamics of floristics with the rest of ecosystem processes and implementing policies. Alpine vegetation shifts affect the soil stabilization, nutrient cycling, hydrology and pollination relationships, which have cascading impacts on the services to the ecosystem and human welfare in the lower parts of the cascade. The main focus of future studies should then be on the functional implications of floristic turnover, how loss or substitution of dominant species has resulted in the stability and resilience of the ecosystem.

The application of these scientific findings into conservation practice must be equated by communication between researchers, policymakers and local communities. Evidence-based management will be enhanced by incorporating the data of floristic change in national biodiversity strategies, plans of the protection areas and climate adaptation frameworks (Gillson et al., 2019). This interdisciplinary cooperation will guarantee that the alpine ecosystems, which are one of the most climate-sensitive on earth, are sustainable depositories of biodiversity and ecological heritage with the approach of global change.

9. CONCLUSION

The Alpine summits have been used since ancient times to observe the structure, composition and dynamics of vegetation at high altitudes. Since the era of the botanical explorations of the 18th and 19th centuries, up to the standardized long-term monitoring initiatives of the 21st century, historic plant surveys have offered priceless information on the temporal dynamics of alpine floras. These existing cumulative datasets, being heterogeneous in approach, are the basis of modern floristic studies and remain a guide to our concepts of mountain biodiversity in varying climatic and ecological circumstances. Historical records combined with modern ones have shown that trends in alpine ecosystems are consistent (mainly, they move up the mountain), there is a change in community structure, and the specialized ones adapted to colds gradually disappear. These findings highlight how temperature rise, changing precipitation cycles and human activities are sensitive to alpine vegetation. Simultaneously, the existence of some endemic and stress-tolerant taxa will show the resilience and adaptive possibilities of mountain biota. By using historical baselines in current monitoring systems, researchers can be able to see the long-term ecological paths that can be otherwise blurred in short-term monitoring. The floristic studies have had their scope and accuracy tremendously increased with the addition of modern tools like remote sensing, molecular techniques and automated environmental monitoring. These developments, in combination, allow one to define priority areas to conserve, trace vegetation changes at both spatial scales and determine future biodiversity implications in relation to global change. Although the issues of researching in high altitude are very many, the further exploration and description of the alpine peaks are vital in comprehending the responsiveness of the ecosystem to climatic disturbances. These mountaintop habitats are important biosensors of ecological change and also containing rare genetic and floristic diversity as climate change continues to accelerate. It is of utmost importance that these delicate systems are conserved by continued cooperation between the botanists, ecologists and conservation planners to ensure that the alpine summits remain in the future to enlighten scientific discoveries and biodiversity conservation to the generations ahead.

REFERENCES

Boulangeat, I., Lavergne, S., Van Es, J., Garraud, L., & Thuiller, W. (2012). Niche breadth, rarity and ecological characteristics within a regional flora spanning large environmental gradients. *Journal of Biogeography*, *39*(1), 204–214. DOI: 10.1111/j.1365-2699.2011.02581.x

Braziunas, K. H., Geres, L., Richter, T., Glasmann, F., Senf, C., Thom, D., Seibold, S., & Seidl, R. (2024). Projected climate and canopy change lead to thermophilization and homogenization of forest floor vegetation in a hotspot of plant species richness. *Global Change Biology*, *30*(1), e17121. DOI: 10.1111/gcb.17121 PMID: 38273493

Callaghan, T. V., Björn, L. O., Chernov, Y., Chapin, T., Christensen, T. R., Huntley, B., Ims, R. A., Johansson, M., Jolly, D., Jonasson, S., Matveyeva, N., Panikov, N., Oechel, W., Shaver, G., Elster, J., Henttonen, H., Laine, K., Taulavuori, K., Taulavuori, E., & Zöckler, C. (2004). Biodiversity, distributions and adaptations of Arctic species in the context of environmental change. *Ambio*, *33*(7), 404–417. DOI: 10.1579/0044-7447-33.7.404 PMID: 15573569

Cavender-Bares, J., Schneider, F. D., Santos, M. J., Armstrong, A., Carnaval, A., Dahlin, K. M., Fatoyinbo, L., Hurtt, G. C., Schimel, D., Townsend, P. A., Ustin, S. L., Wang, Z., & Wilson, A. M. (2022). Integrating remote sensing with ecology and evolution to advance biodiversity conservation. *Nature Ecology & Evolution*, *6*(5), 506–519. DOI: 10.1038/s41559-022-01702-5 PMID: 35332280

Chauhan, H. K., Gallacher, D., Bhatt, A., & Bisht, A. K. (2023). The Himalayas: A climate change laboratory. *Environmental Development*, *45*, 100814. DOI: 10.1016/j.envdev.2023.100814

Chen, J., Chen, S., Fu, R., Li, D., Jiang, H., Wang, C., Peng, Y., Jia, K., & Hicks, B. J. (2022). Remote sensing big data for water environment monitoring: Current status, challenges, and future prospects. *Earth's Future, 10*(2), e2021EF002289.

Corlett, R. T. (2016). Plant diversity in a changing world: Status, trends, and conservation needs. *Plant Diversity*, *38*(1), 10–16. DOI: 10.1016/j.pld.2016.01.001 PMID: 30159445

Crawford, R. M. (2008). *Plants at the margin: ecological limits and climate change*. Cambridge University Press. DOI: 10.1017/CBO9780511754906

Egerton, F. N. (2013). History of ecological sciences, part 48: Formalizing plant ecology, about 1870 to mid-1920s. *Bulletin of the Ecological Society of America*, *94*(4), 341–378. DOI: 10.1890/0012-9623-94.4.341

Elton, C. S., & Miller, R. S. (1954). The ecological survey of animal communities: With a practical system of classifying habitats by structural characters. *Journal of Ecology, 42*(2), 460–496. DOI: 10.2307/2256872

Gallagher, R., Hughes, L., & Leishman, M. (2009). Phenological trends among Australian alpine species: Using herbarium records to identify climate-change indicators. *Australian Journal of Botany, 57*(1), 1–9. DOI: 10.1071/BT08051

Gillson, L., Biggs, H., Smit, I. P., Virah-Sawmy, M., & Rogers, K. (2019). Finding common ground between adaptive management and evidence-based approaches to biodiversity conservation. *Trends in Ecology & Evolution, 34*(1), 31–44. DOI: 10.1016/j.tree.2018.10.003 PMID: 30447939

Giorgini, D., Chelli, S., Campetella, G., Chiarucci, A., & Canullo, R. (2015). Predicting the habitat type distribution to assess the representativeness of the Natura 2000 network. 58th annual symposium of the International Association for Vegetable Science: Understanding Broad-Scale Vegetation Patterns: 19-24 July 2015, Brno, Czech Republic: Abstracts., Glime, J. M. (2013). Field taxonomy and collection methods.

Grabherr, G., Gottfried, M., & Pauli, H. (2000). GLORIA: A global observation research initiative in alpine environments. *Mountain Research and Development, 20*(2), 190–191. DOI: 10.1659/0276-4741(2000)020[0190:GAGORI]2.0.CO;2

Grace, J., Berninger, F., & Nagy, L. (2002). Impacts of climate change on the tree line. *Annals of Botany, 90*(4), 537–544. DOI: 10.1093/aob/mcf222 PMID: 12324278

Gray, A. (1889). *Darwiniana: essays and reviews pertaining to Darwinism*. D. Appleton.

Hamid, M., Khuroo, A. A., Malik, A. H., Ahmad, R., & Singh, C. P. (2020). Assessment of alpine summit flora in Kashmir Himalaya and its implications for long-term monitoring of climate change impacts. *Journal of Mountain Science, 17*(8), 1974–1988. DOI: 10.1007/s11629-019-5924-7

Hedberg, O. (1964). *Features of afroalpine plant ecology*. Sv. växtgeografiska sällsk.

Heywood, V. (2001). Floristics and monography—An uncertain future? *Taxon, 50*(2), 361–380. DOI: 10.2307/1223886

Hooker, J. D. (1854). X. On the functions and structure of the rostellum of Listera ovata. *Philosophical Transactions of the Royal Society of London*, (144), 259–263. DOI: 10.1098/rstl.1854.0011

Hooker, J. D., & Thomson, T. (1855). *Flora indica: being a systematic account of the plants of British India, together with observations on the structure and affinities of their natural orders and genera* (Vol. 1). Pamplin. DOI: 10.5962/bhl.title.50109

Hotaling, S., Finn, D. S., Joseph Giersch, J., Weisrock, D. W., & Jacobsen, D. (2017). Climate change and alpine stream biology: Progress, challenges, and opportunities for the future. *Biological Reviews of the Cambridge Philosophical Society*, *92*(4), 2024–2045. DOI: 10.1111/brv.12319 PMID: 28105701

Inouye, D. W. (2008). Effects of climate change on phenology, frost damage, and floral abundance of montane wildflowers. *Ecology*, *89*(2), 353–362. DOI: 10.1890/06-2128.1 PMID: 18409425

Jaccard, P. (1912). The distribution of the flora in the alpine zone. 1. *The New Phytologist*, *11*(2), 37–50. DOI: 10.1111/j.1469-8137.1912.tb05611.x

Kent, M. (2011). *Vegetation description and data analysis: a practical approach.* John Wiley & Sons.

Kikuzawa, K. (1995). Leaf phenology as an optimal strategy for carbon gain in plants. *Canadian Journal of Botany*, *73*(2), 158–163. DOI: 10.1139/b95-019

Komarkova, V., & Wielgolaski, F. (1999). Stress and disturbance in cold region ecosystems. *Ecosystems of the world*, 39-122.

Körner, C. (1995). Alpine plant diversity: a global survey and functional interpretations. In *Arctic and alpine biodiversity: patterns, causes and ecosystem consequences* (pp. 45–62). Springer. DOI: 10.1007/978-3-642-78966-3_4

Körner, C. (2021). The alpine life zone. In *Alpine plant life: Functional plant ecology of high mountain ecosystems* (pp. 23–51). Springer.

Körner, C. (2023). Concepts in alpine plant ecology. *Plants*, *12*(14), 2666. DOI: 10.3390/plants12142666 PMID: 37514280

Körner, C., & Hiltbrunner, E. (2021). Why is the alpine flora comparatively robust against climatic warming? *Diversity*, *13*(8), 383. DOI: 10.3390/d13080383

Korner, C., & Spehn, E. M. (2024). *Mountain biodiversity: a global assessment.* CRC Press. DOI: 10.1201/9781003579519

Kuo, C. C., Liu, Y. C., Su, Y., Liu, H. Y., & Lin, C. T. (2022). Responses of alpine summit vegetation under climate change in the transition zone between subtropical and tropical humid environment. *Scientific Reports*, *12*(1), 13352. DOI: 10.1038/s41598-022-17682-2 PMID: 35922458

Lavergne, S., Mouquet, N., Thuiller, W., & Ronce, O. (2010). Biodiversity and climate change: Integrating evolutionary and ecological responses of species and communities. *Annual Review of Ecology, Evolution, and Systematics*, *41*(1), 321–350. DOI: 10.1146/annurev-ecolsys-102209-144628

Lavoie, C. (2013). Biological collections in an ever changing world: Herbaria as tools for biogeographical and environmental studies. *Perspectives in Plant Ecology, Evolution and Systematics*, *15*(1), 68–76. DOI: 10.1016/j.ppees.2012.10.002

Lelli, C., Nascimbene, J., Alberti, D., Agostini, N., Zoccola, A., Piovesan, G., & Chiarucci, A. (2021). Long-term changes in Italian mountain forests detected by resurvey of historical vegetation data. *Journal of Vegetation Science*, *32*(1), e12939. DOI: 10.1111/jvs.12939

Leuschner, C., & Ellenberg, H. (2017). Vegetation of the alpine and nival belts. In *Ecology of Central European Non-Forest Vegetation: Coastal to Alpine, Natural to Man-Made Habitats: Vegetation Ecology of Central Europe, Volume II* (pp. 271-431). Springer. DOI: 10.1007/978-3-319-43048-5_5

Lockwood, M., Mitchell, M., Moore, S. A., & Clement, S. (2014). Biodiversity governance and social-ecological system dynamics: Transformation in the Australian Alps. *Ecology and Society*, *19*(2), art13. DOI: 10.5751/ES-06393-190213

MacGregor, A. (2018). *Naturalists in the field: collecting, recording and preserving the natural world from the fifteenth to the twenty-first century* (Vol. 2). Brill.

Magurran, A. E., Baillie, S. R., Buckland, S. T., Dick, J. M., Elston, D. A., Scott, E. M., Smith, R. I., Somerfield, P. J., & Watt, A. D. (2010). Long-term datasets in biodiversity research and monitoring: Assessing change in ecological communities through time. *Trends in Ecology & Evolution*, *25*(10), 574–582. DOI: 10.1016/j.tree.2010.06.016 PMID: 20656371

Mahdi, S. S., & Kazmi, S. A.. (2024). Climate Vulnerability and Adaptation Prospects for Mountain Ecosystems in Kashmir Himalaya. *Scientific Reports*.

Mason, N., Ward, M., Watson, J. E., Venter, O., & Runting, R. K. (2020). Global opportunities and challenges for transboundary conservation. *Nature Ecology & Evolution*, *4*(5), 694–701. DOI: 10.1038/s41559-020-1160-3 PMID: 32203481

Menke, S. B., & Holway, D. A. (2020). Historical resurvey indicates no decline in Argentine ant site occupancy in coastal southern California. *Biological Invasions*, *22*(5), 1669–1679. DOI: 10.1007/s10530-020-02211-x

Meyer, S. E., Callaham, M. A.Jr, Stewart, J. E., & Warren, S. D. (2021). Invasive species response to natural and anthropogenic disturbance. In *Invasive species in forests and rangelands of the United States: A comprehensive science synthesis for the United States forest sector* (pp. 85–110). Springer International Publishing Cham. DOI: 10.1007/978-3-030-45367-1_5

Midgley, G. F., Chown, S. L., & Kgope, B. S. (2007). Monitoring effects of anthropogenic climate change on ecosystems: A role for systematic ecological observation? [commentary]. *South African Journal of Science*, *103*(7), 282–286.

Miehe, S., Miehe, G., Miehe, S., Böhner, J., Bäumler, R., Ghimire, S., Bhattarai, K., Chaudhary, R., & Subedi, M. (2015). *16. Vegetation ecology. Nepal: an introduction to the natural history, ecology and human environment in the Himalayas.* Royal Botanic Garden Edinburgh.

Mufungizi, A. A., Musakwa, W., & Chanza, N. (2023). Shifting ecosystems, past, current, and emerging trends: A bibliometric analysis and systematic review of literature. *Ecological Indicators*, *156*, 111175. DOI: 10.1016/j.ecolind.2023.111175

Muthukrishnan, R., Sullivan, L. L., Shaw, A. K., & Forester, J. D. (2020). Trait plasticity alters the range of possible coexistence conditions in a competition–colonisation trade-off. *Ecology Letters*, *23*(5), 791–799. DOI: 10.1111/ele.13477 PMID: 32086876

Nagy, L., & Grabherr, G. (2009). *The biology of alpine habitats.* Oxford University Press. DOI: 10.1093/oso/9780198567035.001.0001

Nila, M. U. S., Bobrowski, M., & Schickhoff, U. (2025). Growing Season Normalized Difference Vegetation Index in the Nepal Himalaya and Adjacent Areas, 2000–2019: Sensitivity to Climate Change and Terrain Factors. *Land (Basel)*, *14*(4), 749. DOI: 10.3390/land14040749

Panthi, J., Vetaas, O. R., & Shrestha, B. B. (2020). Tree growth responses to climatic drivers across an elevation gradient in Pinus wallichiana forests in the central Himalayas. *Forest Ecology and Management*, *477*, 118526.

Pauli, H., Gottfried, M., Dullinger, S., Abdaladze, O., Akhalkatsi, M., Alonso, J. L. B., Coldea, G., Dick, J., Erschbamer, B., Calzado, R. F., Ghosn, D., Holten, J. I., Kanka, R., Kazakis, G., Kollár, J., Larsson, P., Moiseev, P., Moiseev, D., Molau, U., & Grabherr, G. (2012). Recent plant diversity changes on Europe's mountain summits. *Science*, *336*(6079), 353–355. DOI: 10.1126/science.1219033 PMID: 22517860

Piredda, R., Simeone, M. C., Attimonelli, M., Bellarosa, R., & Schirone, B. (2011). Prospects of barcoding the Italian wild dendroflora: Oaks reveal severe limitations to tracking species identity. *Molecular Ecology Resources*, *11*(1), 72–83. DOI: 10.1111/j.1755-0998.2010.02900.x PMID: 21429102

Porro, F., Tomaselli, M., Abeli, T., Gandini, M., Gualmini, M., Orsenigo, S., Petraglia, A., Rossi, G., & Carbognani, M. (2019). Could plant diversity metrics explain climate-driven vegetation changes on mountain summits of the GLORIA network? *Biodiversity and Conservation*, *28*(13), 3575–3596. DOI: 10.1007/s10531-019-01837-1

Pyke, G. H., & Ehrlich, P. R. (2010). Biological collections and ecological/environmental research: A review, some observations and a look to the future. *Biological Reviews of the Cambridge Philosophical Society*, *85*(2), 247–266. DOI: 10.1111/j.1469-185X.2009.00098.x PMID: 19961469

Rea, L. M., & Parker, R. A. (2014). *Designing and conducting survey research: A comprehensive guide*. John Wiley & Sons.

Ritchie, J. (1995). Current trends in studies of long-term plant community dynamics. *The New Phytologist*, *130*(4), 469–494. DOI: 10.1111/j.1469-8137.1995.tb04325.x PMID: 33874480

Rowe, R. J. (2005). Elevational gradient analyses and the use of historical museum specimens: A cautionary tale. *Journal of Biogeography*, *32*(11), 1883–1897. DOI: 10.1111/j.1365-2699.2005.01346.x

Rubinoff, D., Reil, J. B., Osborne, K. H., Gregory, C. J., Geib, S. M., & Dupuis, J. R. (2020). Phylogenomics reveals conservation challenges and opportunities for cryptic endangered species in a rapidly disappearing desert ecosystem. *Biodiversity and Conservation*, *29*(7), 2185–2200. DOI: 10.1007/s10531-020-01968-w

Rundel, P., & Millar, C. (2016). Alpine ecosystems. *In: Zavaleta, E.; Mooney, H., eds. Ecosystems of California. Berkeley, California: University of California Press: 613-634. Chapter 29.*, 613-634.

Séverine Fauquette, J.-P. S., Médail, F., Muller[1], G. J.-M., & Bertini, E. (2018). The Alps: a geological, climatic and human perspective on vegetation history and modern plant diversity. *Mountains, climate and biodiversity*, 413.

Shteir, A. B. (1976). Albrecht von Haller's Botany and" Die Alpen. *Eighteenth-Century Studies*, *10*(2), 169–184. DOI: 10.2307/2737553 PMID: 11616915

Stöckli, V., Wipf, S., Nilsson, C., & Rixen, C. (2011). Using historical plant surveys to track biodiversity on mountain summits. *Plant Ecology & Diversity*, *4*(4), 415–425. DOI: 10.1080/17550874.2011.651504

Sullivan, J. R., & Nazaire, M. (2022). Specimen collection and preparation for a changing flora. *Rhodora, 123*(993), 50–66. DOI: 10.3119/20-32

Thakur, D., & Chawla, A. (2019). Functional diversity along elevational gradients in the high altitude vegetation of the western Himalaya. *Biodiversity and Conservation*, *28*(8), 1977–1996. DOI: 10.1007/s10531-019-01728-5

Tingley, M. W., & Beissinger, S. R. (2009). Detecting range shifts from historical species occurrences: New perspectives on old data. *Trends in Ecology & Evolution*, *24*(11), 625–633. DOI: 10.1016/j.tree.2009.05.009 PMID: 19683829

Turrill, B. (1959). Plant geography. *Vistas in botany, 2*, 172.

Urban, M. C., Bocedi, G., Hendry, A. P., Mihoub, J.-B., Pe'er, G., Singer, A., Bridle, J. R., Crozier, L. G., De Meester, L., Godsoe, W., Gonzalez, A., Hellmann, J. J., Holt, R. D., Huth, A., Johst, K., Krug, C. B., Leadley, P. W., Palmer, S. C. F., Pantel, J. H., & Travis, J. M. J. (2016). Improving the forecast for biodiversity under climate change. *Science*, *353*(6304), aad8466. DOI: 10.1126/science.aad8466 PMID: 27609898

Van Wilgen, B. W., Esler, K. J., Gaertner, M., Midgley, G. F., Richardson, D. M., Van Wilgen, N. J., Wilson, J. R., Carruthers, J., Cowling, R. M., & Forsyth, A. T. (2016). Ecological research and conservation management in the Cape Floristic Region between 1945 and 2015: History, current understanding and future challenges. *Transactions of the Royal Society of South Africa*, *71*(3), 207–303. DOI: 10.1080/0035919X.2016.1225607

Vanneste, T., Michelsen, O., Graae, B. J., Kyrkjeeide, M. O., Holien, H., Hassel, K., Lindmo, S., Kapás, R. E., & De Frenne, P. (2017). Impact of climate change on alpine vegetation of mountain summits in Norway. *Ecological Research*, *32*(4), 579–593. DOI: 10.1007/s11284-017-1472-1

Vasconcelos, R. N., Cantillo-Pérez, T., Franca Rocha, W. J., Aguiar, W. M., Mendes, D. T., de Jesus, T. B., de Santana, C. O., de Santana, M. M., & Oliveira, R. P. (2024). Advances and Challenges in Species Ecological Niche Modeling: A Mixed Review. *Earth (Basel, Switzerland)*, *5*(4), 963–989. DOI: 10.3390/earth5040050

Victor, J. E., Smith, G. F., & Van Wyk, A. E. (2016). History and drivers of plant taxonomy in South Africa. *Phytotaxa, 269*(3), 193–208-193–208.

Volberg, R. A. (1983). *Constraints and commitments in the development of American botany, 1880-1920*. UCSF.

von Christ, W. (1886). *Kritische Beiträge zur Metaphysik des Aristoteles*. Franz in Komm.

Von Haller, A. (1768). *Historia stirpium indigenarum Helvetiae inchoata* (Vol. 1). sumptibus Societatis typographicæ.

von Humboldt, A. (1819). *Voyage de Humboldt et Bonpland. Voyage aux régions équinoxiales du Nouveau continent, fait, en 1799, 1800, 1801, 1802, 1803, et 1804* (Vol. 2).

Walther, G. R., Beißner, S., & Burga, C. A. (2005). Trends in the upward shift of alpine plants. *Journal of Vegetation Science, 16*(5), 541–548. DOI: 10.1111/j.1654-1103.2005.tb02394.x

Weber, W. A. (2003). The Middle Asian element in the southern Rocky Mountain flora of the western United States: A critical biogeographical review. *Journal of Biogeography, 30*(5), 649–685. DOI: 10.1046/j.1365-2699.2003.00864.x

Willis, C. G., Ellwood, E. R., Primack, R. B., Davis, C. C., Pearson, K. D., Gallinat, A. S., Yost, J. M., Nelson, G., Mazer, S. J., Rossington, N. L., Sparks, T. H., & Soltis, P. S. (2017). Old plants, new tricks: Phenological research using herbarium specimens. *Trends in Ecology & Evolution, 32*(7), 531–546. DOI: 10.1016/j.tree.2017.03.015 PMID: 28465044

Yao, H., Qin, R., & Chen, X. (2019). Unmanned aerial vehicle for remote sensing applications—A review. *Remote Sensing (Basel), 11*(12), 1443. DOI: 10.3390/rs11121443

Chapter 10

Emerging Threats and Challenges for Alpine Vegetation:

Integrated Perspectives on Shifting Dynamics, Biotic Pressures, and Ecosystem Vulnerability

Muhammad Zubair
University of Science and Technology, Bannu, Pakistan

Moona Nazish
http://orcid.org/0000-0001-5486-5470
Quaid-i-Azam University, Islamabad, Pakistan

Aisha Zulfiqar
National University of Medical Sciences, Rawalpindi, Pakistan

Asif Kamal
Quaid -i- Azam University, Islamabad, Pakistan

Shakila Rahman
Hazara University, Mansehra, Pakistan

Fozia Bibi
International Islamic University, Islamabad, Pakistan

Rida Nisar
Gomal University, Dera Ismail Khan, Pakistan

Muhammad Ishfaq Khan
University of Science and Technology, Bannu, Pakistan

ABSTRACT

Alpine ecosystems are facing unprecedented challenges mainly due to the interaction of anthropogenic and climatic factors. Among the various impacts of climate change,

DOI: 10.4018/979-8-3373-6756-9.ch010

ght © 2026, IGI Global Scientific Publishing. Copying or distributing in print or electronic forms without written permission of IGI Global Scientific Publishing is prohibited. Use of this chapter to generative artificial intelligence (AI) technologies is expressly prohibited. The publisher reserves all rights to license its use for generative AI training and machine learning model development.

the gradual movement of vegetation zones upwards is one of the most significant, which ultimately results in a shift of the species present, changes in the timing of life cycles, and a decrease in the productivity of the ecosystem. The spread of heat-tolerant species, the extinction of mountain-top endemics, and plant invasions are some of the changes that are occurring. On the contrary, some species can survive the bad conditions through microclimatic buffering, which can be in conflict with the overall predictions. Furthermore, the application of global databases such as GLOBALP and deep learning for the monitoring of shrub lines and carbon fluxes in response to global warming are some of the ways through which new and more efficient technologies are coming about. The combination of rising temperatures, retreating glaciers, and the vertical movement of both heat-loving and invasive species, along with the impact of humans, is the situation that severely threatens the vegetation that is exclusive to these alpine areas. Human activities, including overgrazing, uncontrolled tourism, road expansion, and hydropower development, exacerbate ecosystem degradation by causing soil destabilization, habitat fragmentation, and decreased resilience in plant communities. Additional stressors such as soil nutrient depletion, erosion, and altered hydrological regimes further disrupt plant-soil interactions and hinder ecosystem functionality. These threaten the biodiversity of alpine vegetation and put the ecological services at risk, such as carbon sequestration, watershed management, and cultural significance for indigenous communities.

1. INTRODUCTION

1.1. Changing Climate and Vegetation Dynamics in Alpine Regions

In the alpine zone, which is subjected to low air pressure and high solar radiation, harsh but surprisingly resistant ecosystems are found at cold temperatures, with short growth seasons and intensive UV radiation; these systems support very specialized and often endemic plant species (Kok *et al.*, 2024). The Himalayas has high tree-line shifts (0.80 to 503 m) in response to extreme warming globally, shifting especially the sensitive species like *Abies spectabilis* and *Betula utilis* by changing their regeneration dynamics and phenophases. The reduction of snow cover in the Alps favors the expansion of shrubs, especially the heather, which retards nutrient cycling and exacerbates soil erosion with a concomitant effect on pH and nitrogen retention (Kumar & Khanduri, 2024). These activities can be a great threat to the flora and fauna by supporting the fast-growing invaders while killing the cold-adapted natives. The ecosystems were extremely sensitive and had to be absolutely adjusted at high altitudes, humidity, and temperature levels, facing huge pressure. The warming in the alpine areas is about twice the global average, and thus, they are seen as indicators of larger global climate changes (Körner, 2021). The temperature increase is expected to lead to less vivid autumn leaf colors of the Japanese alpine plants, and thus, the visual ecological signals and animals that feed on the plants will be affected. High temperatures shorten rhizomes and fine roots, thereby increasing the rate of plant death in areas that are already weak. The fine-resolution climate grid predictions for the Rocky Mountains in the United States show a forecast of a decrease in the area of alpine regions, which underlines the necessity of the adoption of the trait-based adaptation models (Oldfather et al., 2025). The current chapter moves forward to investigate the multifaceted transitions in climate and vegetation in the alpine ecosystem corridor. It highlights the major climate elements of the scenario, which are the rise in temperature, alteration in precipitation distribution, and their interaction in terms of plant survival, as the most important ones. The phenomena of phenological changes and biodiversity, along with the problems posed by cold tolerance of plants and the rarity of some species, will be the main subjects of research.

1.2. Climatic Drivers

Cold temperatures are the mainstay of alpine ecosystems as they act as a barrier between the ecosystems and the changes in climate; however, the trend of warming over the period is unfortunately taking away their natural resilience (Li et al., 2022).

1.2.1 Elevation-dependent Temperature

The temperature rise in alpine regions is close to double that of the global average and among the causes of this phenomenon are less snow, more heat being absorbed, and more moisture in the atmosphere (Zhang et al., 2023). The lessening of cold tolerance in plants due to night warming and the melting of permafrost are two factors that disrupt the natural temperature gradients and water cycles. These impacts result in the shifting of vegetation, which is often accompanied by the lengthening of growth periods, but at the same time, there is metabolic stress and imbalance caused by the changes (Lamprecht et al., 2018).

1.2.2 Shifts in Precipitation Cycles

The transition of snowfall and rain also brings about changes in the characteristics of snow the moisture loses its insulating power (Kotlarski et al., 2023). However, this sequence of precipitation continues. Rainfall, which is predominant over snow, still melts the snow at an early stage, and the plant growth at that period is thus increased, but at the same time, there is an overestimation of the risk of drought later on. The early melt of snow facilitates plant growth but also results in more droughts and scattered snow layers, which damage and inhibit animals from eating. Therefore, the snow-covered area becomes a non-productive area, and thus new environmental constraints arise (Inouye, 2020).

1.2.3 Alterations in Vegetation Responses

Worldwide warming along with changing rain patterns are inducing different kinds of responses from plants, which in turn will reduce their ecological diversity, change the structure of communities, and the composition of species. The earlier blooming and higher range migration of plants are indications of their adaptive strategies, but in a way, they also lead to the creation of mismatches and new competitive pressures. These tendencies suggest that there will be major changes in the biodiversity of the alpine region (Kuo et al., 2022). The results of the experiments demonstrate that the lowland species are capable of thriving in the warmer alpine areas if the local competitors are eliminated; this fact makes it evident that the biotic resistance is so strong that it usually prevents these ecosystems from being invaded by the upslope species. On the other hand, the brown knapweed, for example, is not able to survive when it is planted among established alpine communities; thus, it is clear how high-elevation flora that has been adapted for a long time creates barriers for the brown knapweed. Such interactions serve to point out the non-linear character of the ecological changes, where new competitors might limit the range of species even more than the warming climate itself (Alexander et al., 2015). The factors of global change are causing distributional shifts of species that are specific to each species, thus making the environmental projections inconsistent.

1.2.4 Phenological Adaptations

Biological phenomena, such as leafing, blooming, and seeding, are associated with snowmelt signals, advanced by elevated temperatures and early melts as on as the Alps and the Rocky ranges currently (Macphie & Phillimore, 2024). The growth cycle is progressing as temperatures rise and snowmelt occurs more rapidly. Premature blooming increases the danger of prolonged frosts on budding and pollinator mismatches, which occur when insects skip peak blooms, resulting in lack of reproduction. A plant that blossoms too soon may be exposed to a late spring frost, which can harm delicate buds and flowers. More importantly, it can disturb the timing of plants and pollinators (Poni *et al.,* 2022). Diverse species reactions interrupt blooming sequences, possibly rearranging communities and reducing biodiversity.

1.2 Altitudinal Migrations

One of the most noticeable and significant responses to global warming is an alteration in distribution of species. As lowlands warm, species relocate upslope, replenishing previous tree-lines with invasive shrubs and trees that encroach on meadows resulting in an upward migration of plant species in figure 1 (Kumar & Khanduri, 2024). Vegetation that was formerly limited to the subalpine forest is now sprouting seedlings higher up, progressively encroaching on alpine meadows and grasslands. This process, known as woody encroachment, converts a grassy, herb-dominated alpine environment into a shrubby or woodland environment (Zerbe, 2023). Rivals for resources affect soil stability, nitrogen cycles, and increase the threat of fire, forcing endemic alpines to live in diminishing habitats (Senthilnathan & D'Andrea, 2024). For genuine alpine plants, this upward migration results in an existential crisis (Doležal *et al.,* 2025). As the temperature rises, their natural habitat eventually disappears, a phenomenon called as 'extinction'. As the plant community, the risk of localized extinction rises. Cold-adapted species that evolved in isolation on certain mountain ranges (endemics) are particularly vulnerable since they frequently have restricted ecological tolerances and no other habitats (Senthilnathan & D'Andrea, 2024).

1.4 Impacts on Indigenous and Climate-adapted vegetation

Alpine flora acquired dwarf forms, hairy foliage, perennial crops, and cold photosynthesis, but climate change makes them ineffective against intruders (Kuo *et al.,* 2022). Differentiation reduces competition from fast growers, exposing them to snowless freezes, and surpasses photosynthetic optimums. Endemic species become trapped in mountains, increasing the likelihood of extinction.

Figure 1. Schematic illustration of climate-driven vegetation dynamics in alpine ecosystems.

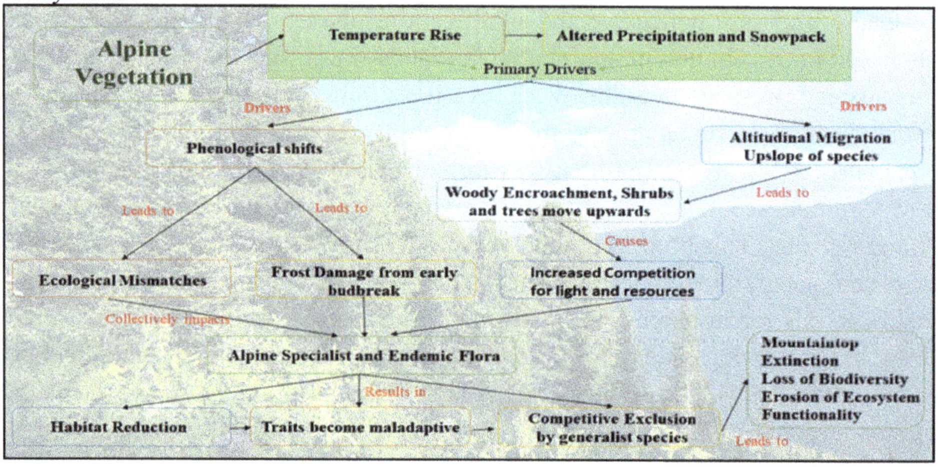

1.5 Drawbacks of Specialization

Cold-adapted plants have unique characteristics such as dwarf morphologies, pubescent leaves, perpetual life cycles, and the capacity to photosynthesize near freezing temperatures. However, as warming accelerates, these adaptations become less favorable (Stewart *et al.,* 2025). Their slow development and reproduction make it tough to compete with rapidly expanding, opportunistic invaders. The requirement of snow cover in the winter for insulation makes them vulnerable to freezing injuries in the case of winters with very low snowfall. The physiological optimum for photosynthesis may be surpassed by the new hot climate thus declining the organisms' fitness, causing the alpine endemics' risk of "mountain-top extinction trap," being not able to adapt quickly to overtake the lower-elevation intruders which are coming upon them (Kuo et al., 2022).

1.6 Limited Efficiency of Biodiversity and Ecosystem

All these aspects pose a danger to the stability of mountain landscapes, which rely on such functions as slope reinforcement, erosion control, water regulation for billions of people downstream, and the long-term storage of carbon. Alpine flora, having their extensive root networks, stabilize steep slopes and prevent the removal of soil. Additionally, they assist in controlling the water flow coming from the high-elevation watersheds, which feed the communities that are far below. The gradual shift towards shrubs may result in increased runoff and soil deposition, thereby causing a decline in the ecological stability as well as the cultural and scientific worth of these rare "sky islands" (Jin et al., 2025).

1.7 Feedback Controls of Intricate Interactions

Biotic and abiotic factors together cause soil microbiological activities, that is, shifts, nutrient availability, and species competition (Chang et al., 2025). The alterations in flora and fauna signal the webs of living beings and the less prominent ones below. The disturbing of the grazing pattern is the direct consequence of the decline or extinction of the key plants, and this in turn affects the animal populations (Carboni et al., 2023). The huge reservoirs of organic carbon in alpine soils are mainly due to the microorganisms' slow decomposition in the cold temperatures. The warming will eventually lead to microbial death and the release of carbon in the atmosphere, thereby creating a positive feedback loop that makes global warming rise (Jansson & Hofmockel, 2020). Furthermore, the advancing darker plant cover on the previously snow-covered regions leads to a decrease in surface reflection, thereby intensifying local warming, which is concisely discussed in Table 1.

Table 1. Mechanisms of alpine vegetation patterns under impactful climate change changes.

Key Drivers	Manifestations	Direct Effects	Cascading Impacts	Reference
Elevated Temperatures	Higher temperatures and growth season	Upward modifications of species	Enhanced competition for cold-adapted specialists	(Tao *et al.*, 2022), (Wang *et al.*, 2022), (Sun *et al.*, 2020)
Precipitation Changes	Earlier snowmelt shorter duration of seasonal snow cover	Triggers previous phenological occurrences	Desynchronization of pollinators or herbivores	(Ma *et al.*, 2022)
Phenological Alterations	Changes in event duration, blooming periods, and growth seasons	Frosts cause damage to fragile blooms and young foliage	Biodiversity loss	(Austin *et al.*, 2024), (Suhery *et al.*, 2024)
Synergistic Effects	Permafrost thaw and altered hydrology changes	Microhabitats emerge and root zone instability	Non-Linear Ecosystem Shifts	(Miao *et al.*, 2024), (Mia *et al.*, 2023)

2. CONSEQUENCES OF GLACIER MELTING AND HYDROLOGICAL REGIMES

Alpine glaciers, referred to as the world's "water towers," have been melting faster than ever mainly because of global warming, changing precipitation patterns, and interrupted snow accumulation. All these factors are changing the hydrology of alpine areas, their ecosystems, and even the humans who live there (Samson et al., 2023). The rate of glacier melt varies from place to place and is closely linked to climatic changes, the size of the glaciers, the direction the slopes face, and the prevailing conditions in the area. Though the belittlement of glaciers is a scientifically intriguing issue by itself, the main worry is still the extremely wide-ranging impacts on water, soils, vegetation, ecosystems, and finally, humans the actually severe nature of the environmental change in the 21st century (Singh, 2025). With glaciers receding, water output increases at first, but then there is a drastic drop, which leads to less soil moisture and groundwater recharge. The process initially allows for more meltwater runoff but eventually results in less flow, lower soil moisture, weaker groundwater replenishment, and less resistance of the plants to drought. These effects indirectly cause the upward migration of plant communities seeking for the suitable conditions (Khoso, 2024). The impacts of these changes extend from local ecosystems to the global water supply marking the disappearance of cryospheric ice as one of the most important environmental changes of the century (Brighenti et al., 2019). The melting of glaciers usually has a stabilizing effect on stream flow and groundwater recharge, but the warming is changing these patterns. The rivers coming from glaciers are now getting to peak flow earlier in the year, there are lower base flows in late summer, and more winter flows as rain becomes snow less and less (Hossein-Panahi et al., 2022). Newly glaciated substrates rapidly promote 'new' ecosystems, with soil compositions and microbial communities that differ markedly from mature alpine soils (Ficetola *et al.,* 2024). Pioneer plants and nitrogen-fixing microorganisms play critical roles in ecosystem creation, often exceeding abiotic soil growth through aggressive biotic facilitation. The disappearance of glaciers weakens natural dependability by causing changed hydrological regimes, which lead to ecological alterations and human vulnerability.

2.1 Accelerating Glacier Retreat

Glacier mass loss has increased since the 1980s, culminating in 2022-2025 (Xu *et al.,* 2025). Over 275,000 glaciers covering 700,000 km^2 hold 70% of freshwater supplies. Since 1970, global ice thinning has averaged 30 meters and is expected to accelerate fivefold by the 2020s; with 1.5°C warming, over half of it might evaporate by mid-century, with smaller areas melting the fastest (Romm, 2022). Post-deglaciation flows are dependent on changing monsoons, snow, and groundwater, posing threats to billions of downstream consumers and sea-level rise.

2.2 Runoff Timing and Variability

Initially, fast surface melt increases flow in glacier-fed rivers, a process known as 'peak water', in which annual runoff exceeds historical norms by up to 50%. However, this era is just temporary (Wallner, 2022). Peak glacier runoff, which was historically constant in the middle of the summer, has now sharpened into episodic pulses, as seen earlier in the tropical Andes due to seasonal interaction (Arias *et al.,* 2021). Longer low-flow periods downstream put both aquatic life and water security at risk, while increasing the likelihood of flash floods from rain or meltwater. This situation poses a significant risk to the water supply in the hottest and driest months, which may affect agricultural irrigation, hydropower production, and even the water requirements of flora and fauna (Mehta, 2024). Moreover, the infrastructure would be in a harder position, since the river flow at the end of the century is expected to decrease if not well managed (De Cesare, 2025). In a nutshell, the melting of glaciers is changing the hydrology of the mountains a stable system with ice as a buffer is turning into one that is very unpredictable and has high variability due to rainstorms.

2.3 Groundwater and soil moisture shifts

The glacial tills that have recently been exposed and are still humid with meltwater from the glacier are the first ones to be dug out. Through drying, however, glaciers of nutrient-poor soils become very harsh and inhospitable to plant growth, as they are without a constant supply of water, which is the case for most of the soil in the trench between the glacier and meltwater. The retreat of the glaciers diminishes the soil's and aquifer's water supply, which, during summer, when the melting of ice resulted in high waterfalls and moist soils, was a condition reversely beneficial to the groundwater and wetlands (Martin et al., 2024). In places like the Alps or Iceland, up to half of the summer groundwater comes from glaciers; thus, the loss of this flow means the availability of water in drought periods is cut down significantly (Vincent et al., 2024). Soils that form through the above process are generally made up of big particles, have little to no nutrients, and very low amounts of organic matter; thus, these soils support only scattered plant growth that can rapidly dry out if the meltwater supply stops. The overly permeable landforms, such as moraines, are saturated at first, but later on, the water is drawn off, and they erode. The process of vegetation layering has an influence on local moisture, but in warmer or drier environments, the moisture deficit is amplified, thus limiting recharge of the groundwater for a long time (Fickert, 2020). In contrast to early post-glacial soils, which were prone to losing moisture from the surface due to the ongoing processes of solifluction, erosion of fines, and 'gravelization', melting of permafrost might, however, temporarily release water storage but later on pave the way for the formation of impermeable layers that stall the process of water infiltration (Cui et al., 2022). Such moisture regimes in flux challenge microbial activity and plant water uptake, thus making alpine vegetation difficult to establish and persist.

2.4 Vegetation Dynamics

Gradual transformation of hydrological flow is altering the ecosystem areas, pushing the nival and tundra habitats as the shrubs and forests slowly migrate upward in the mountains of the Alps, Andes, and Himalayas (Sekar et al., 2025). Water availability change is also playing an important role in the distribution of plants and their seasonal cycles at different altitudes.

2.5 Upward Migration and Expansion

The rise in temperatures and the shift in moisture patterns are driving tree and shrub species into the alpine meadows, where they are fighting with the local plants for the light and the nutrients. The phenomenon of early flowering and growth can also lead to a disturbance of the floral-pollinator relationship and so the disruption of these fragile ecological interactions (O'Brien et al., 2021). Phenological changes, like earlier blossoming, upset the partnerships with the pollinators and thus elevate the risk of extinction. Moreover, native plants that are adapted to cold and live at high altitudes will have their habitats shrink due to the combined effects of higher temperatures and moisture deficiency.

2.6 Primary succession

Glacial terrain as a whole is undergoing a continual ecological transformation: first, the invading microbial species (bacteria, mosses, and lichens) come up within a few years and then help by starting the soil-making processes. Then, the hardy vascular plants (grasses and flowering plants) take hold and eventually restore the soil, giving time for the new species to come up after a period of hundreds of years. Factors like climate, relief, and biological activities are all major determinants of the succession rates. As an example, South American camelids like camels, llamas, and alpacas are the grazing animals that feed on the plants, and as a result, they increase the nutrient cycling in the area, thus they are there in the ecosystem for their roles of maintaining soil fertility. Nevertheless, at the higher places, the low temperature and the nutrient-deficient soils are causing the plant community to develop slowly, which results in a "climatic debt" situation where the vegetation does not progress even with the temperature rise (Richard et al., 2021).

2.7 Novel Functional Shift in Ecosystems

Communities that are forming in the mountains often have more similar features, such as changes in food webs, decomposition rates, and carbon storage, than the old ones. Changes in the hydrological cycle affect the leaching of nutrients and their availability, which in turn has an impact on plant growth, soil quality, carbon storage, and the structure of the habitat. Consequently, these new ecosystems are quite different from the classic mountain landscapes. The glaciers that were once fed by streams and provided cold, oxygen-rich habitats for various forms of life are now melting, and thus these habitats are shrinking. The reduction in glacier meltwater is leading to a decline in these habitats, which in turn puts biodiversity and the ecosystem services that they provide at risk (Brighenti et al., 2019).

2.8 Soil-Vegetation Feedback

Vegetation is one of the most important factors that influence hydrology and it operates in a number of ways: for instance, the trees taking up summer moisture through transpiration and roots in the soil providing better infiltration. In addition, the darkness of the ground under the plant cover is measured by the light being reflected (albedo) which accelerates the melting of snow and ice. The root systems of trees together with soil formation increase the absorption of water; yet, the uprooted vegetation may become a cause of soil loss. In addition to this, the thick plant cover functions as a sun shield, which lowers the evaporation rate and thus helps to keep the ground moist. However, the transformation of white, radiant snow and ice for dark soil and plants increases the capture of solar energy, which in return, heats the place up and might even speed up the melting process in the neighboring areas (Cai & Li, 2024).

2.9 Long-Term Trajectory and Carbon Cycling

The transformation of the bio-glacial stage into diverse ecological communities with active carbon storage happened over centuries, taking place through a shift from abiotic to biotic succession under nutrient and water limitations. This challenges past beliefs about the constant transformation of foreland habitats. On one hand, developing soils provide carbon sequestration that could fight climate change, but on the other hand, it is restricted by nutrient and moisture availability. Such a situation, where soils and living organisms are intricately linked, leads to succession being increasingly biotically mediated, thus showing a resilience that runs counter to the idea of these landscapes being temporary (Rodrigues et al., 2023).

2.10 Social Impacts and Services

In different areas such as Asia and South America, a large number of people, in the billions, are suffering from unreliable supplies of water and power that are connected to the rivers' flow, floods, and droughts. The changing runoff is responsible for a greater risk of drought coming and going, making it more difficult to manage reservoirs, and also increasing the dangers of flooding, particularly at the time of more intense cyclones. The melting of glaciers hurts the earth's ecosystems as it gradually takes away the services that are provided by these ecosystems, such as the movement of sediments, the control of erosion, and the cycling of nutrients, although coming in the form of newborn soil development and carbon gain. The net effect is complicated and requires meticulous management that will enable the needs of the downstream communities to be met alongside the preservation of the mountain biodiversity.

2.11 Regional Case Studies

In the European Alps, glacier melting has caused summer river flows to drop and the plant communities to move further up the mountains; at the same time, there is more erosion. Long-term monitoring indicates that soil loss is increasing in areas that have lost glaciers, which is a contradiction between the nutrient-richness and soil stability (John et al., 2024). In the Himalayas, the retreating ice is disturbing the agropastures; on the other hand, the Rockies' reduction in the base flow is causing stress to the meadows and declining the biodiversity (Spencer et al., 2021).

2.12 Novel and Evolving Insights

- Glacier forelands, unlike typical alpine soils, foster the development of new ecosystems with diverse soil characteristics. Over time, late-successional climax communities in these areas reach a state of ecological balance.
- The growth of forests in high-altitude, glacier-covered regions influences local microclimates, alters carbon cycling, and reshapes hydrological patterns.
- The disappearance of permanent ice in major mountain ranges will lead to lasting changes in their ecosystems.
- Glacier retreat alters soil acidity and other biogeochemical processes, gradually shifting the forces of succession from abiotic factors to biotically driven dynamics.

- Greater variability in river flows and "flashy" hydrological regimes can challenge water resource management and put the resilience of aquatic ecosystems at risk. Comprehensive measures in adaptation, mitigation, and environmental conservation could temper our socio-economic vulnerabilities.

3. ANTHROPOGENIC IMPACTS ON ALPINE MEADOWS

Alpine meadows, the most sensitive areas in nature, are greatly affected by human activities such as tourism, overgrazing, and construction. All these human activities lead to the same effects, i.e., soil compaction, destruction of the vegetation, and disturbance of the microbial communities, sometimes in certain regions even surpassing the impacts of climate change. As a result, the negative feedbacks are getting stronger, particularly on the Qinghai-Tibetan Plateau and in the Himalayas—this is a process that goes along with the loss of biological diversity, and a decrease in the services provided by the ecosystems. Thus, integrated approaches that consider economic development together with the safeguarding of these fragile ecosystems will be necessary for the sustainable management (Feng et al., 2019b).

3.1 Tourism Impacts

Tourism has a direct effect on the soil in the mountains through trampling. This leads to heightened soil compaction, a decrease in porosity, a slowdown in water infiltration, and a reduction in aeration all these thereby to disrupt nutrient cycling and microbial activity (Ahmed & Nihei, 2024). The decline of vegetation cover and diversity along trails is evident, as it is on the Tibetan Plateau, where areas that are nearer the trails undergo considerable community degradation. The erosion process becomes faster because the soil is loosened, and there is more runoff, which results in the removal of top fertile soil and the alteration of microhabitats; the loss of litter also interferes with plant-soil feedbacks (Bhatt et al., 2024). The methodologies being used are basically awareness, education, and visiting limitations, which not only reduce harm but also bring economic gains for conservation.

3.2 Overgrazing and fragmentation

The grazing of animals beyond a sustainable level has a negative impact on plant biomass, it compacts the soil, and finally, it favors the erosion of the ground that is covered with hardy species and reduces the overall species diversity (Ye et al., 2025). Soil organic content, nutrients, and water capacity all decrease, affecting roots and microorganisms; bare areas fracture ecosystems, cutting off connections for flora and animals. It reduces NDVI in the Qinghai-Tibetan Plateau when paired with climatic stress, indicating a loss of production. Short-term compensatory growth leads to net deterioration, but rotational grazing and fencing help recovery (Wu *et al.,* 2021).

3.3 Infrastructure Development

Roads split ecosystems, alter hydrology, and provide access for tourists and grazing, as shown along the Qinghai-Tibet highway, where nearby meadows have decreased variety and altered structures. Compaction and moisture change stress ecosystems; isolation reduces genetic flow, increasing extinction risks. Cumulative pressures from permitted exploitation result in irreparable landscapes (Wilson *et al.,* 2022).

3.4 Case Studies and Regional Patterns

The Qinghai-Tibetan Plateau and Himalayas dominate investigations, with human influences prevailing over climate in deterioration (Zhao & Zhang, 2024). Another study showed, Maquan County, China degradation gradients indicate coarser particles, poorer organic carbon/nitrogen, biomass, richness, and microbes, indicating a relationship between overgrazing/tourism and soil-plant-microbe cascades (Wang *et al.,* 2025). While, Yangtze Source Region showed bacterial and fungal alterations are linked to moisture, density, and carbon, whereas deterioration reshuffles populations towards harsh-adapted species (Zhang, 2023). Moreover, Tibetan warming experiments revealed that the temperature rise exacerbates grazing and rodent consequences, disrupting communities and degrading soils. Chopta (Uttarakhand) and Naran Valley exhibit trampling as well as overgrazing, which erodes soil qualities, plant species composition, endangering native, endemic, and medicinal plants (Hu *et al.,* 2023).

3.5 Biodiversity and Service Loss

Fragmentation separates populations and undermines genetics and resilience; plant losses harm herbivores and pollinators. Runoff surges damage downstream water and reduce carbon sequestration (Zhyvko, 2024). Recreation and pastoralism falters, endangering livelihoods.

3.6 Regeneration Strategies

Mitigate through grazing regulations, tourist limitations, and minimum infrastructure, and reseeding /erosion barriers. Monitor NDVI, nutrients, and diversity for adaptive management; include socioeconomic factors (Mitra & Paul, 2025).

- Visitor limitations and guided trails help to prevent trampling.
- Rotational stocking maintains cover.
- Careful planning prevents fragmentation.
- Re-vegetation helps to recover soils.

3.7 Soil Microbial Dynamics

Degradation changes microorganisms that mediate cycling and decomposition, Yangtze studies demonstrate that alpha diversity remains steady, but beta diversity shifts with severity Proteobacteria and Ascomycota fall, while Acidobacteria increase in poor soil. This reduces fertility and fuels vegetation loss cycles (Zhan, 2024). Soil changes across gradients indicates bulk density increases, pores contract from trampling, and carbon along nitrogen levels decrease with biomass/erosion. Moisture rains without protection; sand surges and particles degrade, harshening conditions and impeding recuperation (Zhao *et al.,* 2022).

3.8 Climatic Anthropogenic Interactions

Climate variables such as temperature and precipitation influence alpine meadow deterioration, however they are frequently overshadowed or aggravated by human activity. Climate warming and variations in precipitation, together with livestock overgrazing and alterations in land use, impact erosion patterns on the Qinghai-Tibetan Plateau (Liu *et al.,* 2023). Interestingly, locations with increased precipitation and temperatures had slower degradation rates, indicating a buffering "hydrothermal-hole effect" in which improved water and heat availability enhances soil retained nutrients and plant development despite anthropogenic stress (Chai *et al.,* 2025). In less humid or cooler microclimates, degradation takes place much faster because of the more severe abiotic stresses, which are additionally increased by foot traffic and loss of plant cover. The main human-caused pressures and their ecological effects are summarized in Table 2.

Table 2. Prime anthropic stresses in Alpine ecosystems.

Anthropogenic Pressures	Major Ecological Impacts	Illustrative Examples	Reference
Tourism	Habitat disturbance, vegetation encroachment, and pollutants	Soil degradation on hiking paths, trash, and disruption of animals by visitor presence	(Feng *et al.*, 2019a)
Overgrazing	Vegetation deterioration, compacting of soil, ecological imbalance, and biodiversity loss	Overgrazing lowers plant cover; Trampling by livestock harms the soil and plants	(Varga *et al.*, 2021)
Infrastructure Development	Fragmented habitats, hydrological alterations, and pollutants	Construction of highways, dams, and electricity grids modifies the landscape and water patterns; urbanization damages alpine vegetation and wetlands	(Li *et al.*, 2025)

4. INVASIVE AND NON-NATIVE SPECIES IN ALPINE ECOSYSTEMS

According to Bai et al. (2024), the alpine ecosystems are seriously endangered by the presence of non-native and invasive species. The first act of climate change is the speeding up of the process of biological invasion, which causes the native species to be endangered and the vulnerable soil microbes that maintain the ecosystem's health to be disturbed (Duchicela et al., 2021). The second act consists of the rising temperatures in these high-altitude regions, which not only facilitate the invasions but also change the basic functioning of the ecosystems (Tai et al., 2025). The mechanisms of these changes are explained in detail in the following sections;

4.1 Rising Temperatures and Species Invasions

Climatic changes and environmental disturbances have put low altitude and low latitude regions in the list of places most exposed to the invasion of living organisms. The factors that make it easier for these non-indigenous species to establish themselves in the delicate and cold ecosystems are human activities, changes in habitats, and improved connectivity. The invasions have to be prevented at once because the introduced species require regular management (Alatalo et al., 2022).

4.2 Effects on Native Species

As well, the study of migratory birds to see the effects of a changing climate in the future incited many questions. The growth of shrubs becomes more rapid when temperatures rise, which leads to changes in vegetation patterns and disrupts natural ecological systems. The effects of climate change on tundra zones become more pronounced, but the extent of change depends on the specific alpine environment (Fazlioglu & Wan, 2021).

4.3 Shifts in Ecosystem Boundaries

The rising temperatures from climate change cause alpine ecosystems to shift their natural borders which affects tropical alpine regions most strongly. The combination of warmer and drier soil conditions leads to faster organic carbon decomposition which results in decreased below-ground carbon storage and reduced biodiversity (Gao *et al.,* 2023).

4.4 Impact on Freshwater and Invertebrate Species

The desertion of glaciers and snowpack's generates essential habitats that alpine freshwater ecosystems essential to endure. The melt water stonefly population faces population waning because its habitat becomes smaller which validates how climate change affects biological invasions (Plunger *et al.,* 2022).

4.5 Complex Ecosystem Interactions

The changes in hydrological patterns and temperature levels and nutrient availability patterns through climate change create new conditions that affect species habitat selection and their social interactions with other species thus making ecosystems more susceptible to invasions (Li *et al.,* 2022).

4.6 Examples of Invasive Species in Alpine Regions

The warming climate permits multiple non-native species to establish themselves in alpine ecosystems highlighted in table 3. The Sub-antarctic islands have experienced an invasion of *Lavatera arborea* (tree mallow) because of growing temperatures and uncertainties from seabirds (Van Der Merwe, 2024). The Andes of Chile has seen *Taraxacum officinale* and *Cerastium arvense* grow competitive abilities to bloom under warming conditions (Cavieres *et al.,* 2024). The alpine regions of Tenerife experience thermal species migration from pine forest areas because of climate change and herbivore actions (Cubas *et al.,* 2022).

Table 3. Indigenous species have colonized Alpine ecosystems.

Species	Region	Mechanism of Invasion	Reference
Lavatera	Subantarctic islands	Climate warming & seabird disturbance	(Hawke, 2022)
Cerastium	Andes, Chile	High survival under warming	(Cavieres *et al.*, 2024)
Taraxacum	Andes, Chile	Competitive under warming	(Cavieres *et al.*, 2024)
Thermal native pine forest species	Tenerife, Canary Islands	Upslope migration under warming	(Martín-Esquivel *et al.*, 2020)

The overview of invasive species primes to major changes in soil chemistry and microbial community structures is shown in Table 4. The introduction of invasive species through their litter and root exudates is central to changes in soil nutrient availability, which grades in higher pH levels in Table 5 and increased nitrogen convenience (Chang *et al.*, 2025). The founding of invasive species leads to changes in microbial communities, which generate positive comeback twists that enhance their ability to invade new areas (Bird, 2024). Alpine flora homogenization primes to multiple long-term ecological changes.

Table 4. Mechanisms of climate change are enhancing the invasion.

Species	Mechanisms	Interactions	Reference
Egeria	Changed availability of resources	Improved nutrition and light availability	(Piria *et al.*, 2022)
Solidago	Phenological transitions	Early blooming and seed set	(Grange *et al.*, 2025)
Oncorhynchus mykiss	Hybridization process	Genetic admixture with indigenous	(Mueller *et al.*, 2025)
Small Indian mongoose	Ecological adaptability variations	Expanding into new climates	(Bandyopadhyay, 2022)

Table 5. The Impact of invading species on soil mineral equilibrium.

Parameter	Impact	Native Species	Invasive Species
pH of Soil	Elevated alkalinity	5.5	6.2
Nitrogen (mg/kg)	Elevated Nitrogen	35	5.8
Carbon (%)	Increased organic content	2.4	3.8
Microbial biomass (μgC/g soil)	Increased microbial activity	180	260

The invasion of non-native species into alpine ecosystems leads to multiple long-term ecological changes, which affect the environment in Table 6. The invasion of non-native species leads to three major ecological impacts, which include decreased native species diversity and reduced genetic variety and unstable ecosystems, and decreased ecosystem service delivery (Sharma & Birman, 2024).

Table 6. Long-term ecological consequences of Alpine Flora homogenization.

Consequence	Description	Ecosystem Impact
Reduced Endemism	Loss of localized flora	Decline in regional biodiversity
Loss of Genetic Diversity	Dominance by widespread species	Reduced adaptability
Ecosystem Instability	Altered nutrient and hydrological cycles	Increased ecosystem fragility
Diminished Ecosystem Services	Reduced carbon storage & water regulation	Threat to global ecological stability

Increasing seasonal temperature averages have been linked to a surge in non-native plant diversity in alpine habitats in Figure 2. Conversely, invading species outgrow native species under warmer conditions in Figure 3. Climate-triggered intrusions also change soil nutrient content, notably by boosting nitrogen and carbon cycling in Figure 4. The gradual spread of invasive cover causes a loss in indigenous alpine species, exhibiting biodiversity homogeneity in Figure 5.

Figure 2. Relationship between temperature rise and non-native species richness in Alpine regions.

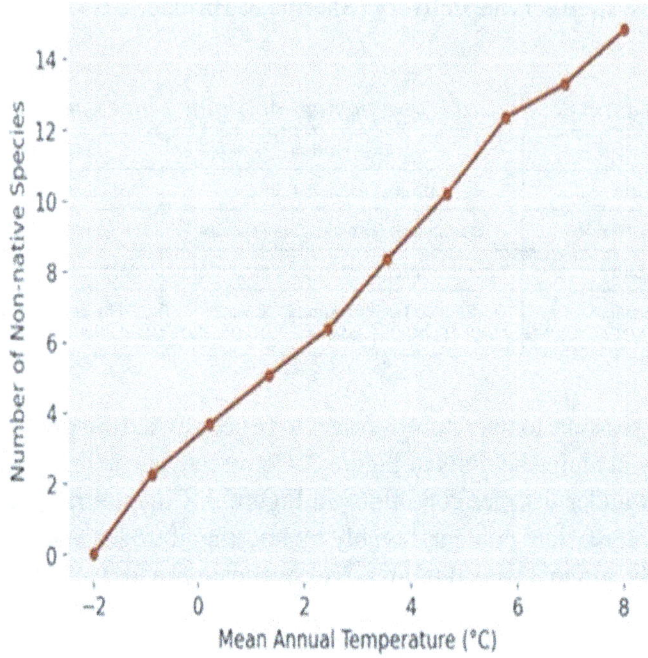

Figure 3. Effects of climate warming on native vs. invasive species growth.

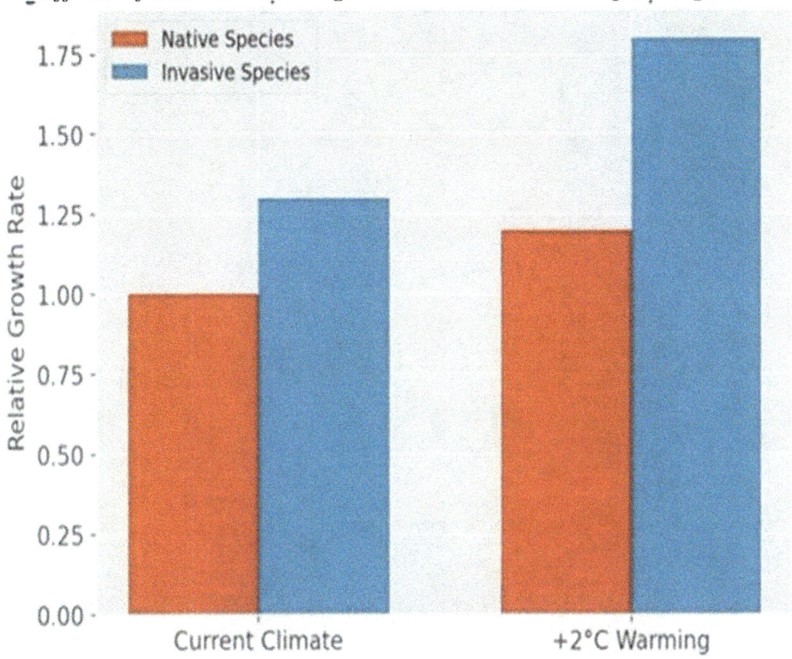

Figure 4. Changes in soil nutrient composition due to invasion.

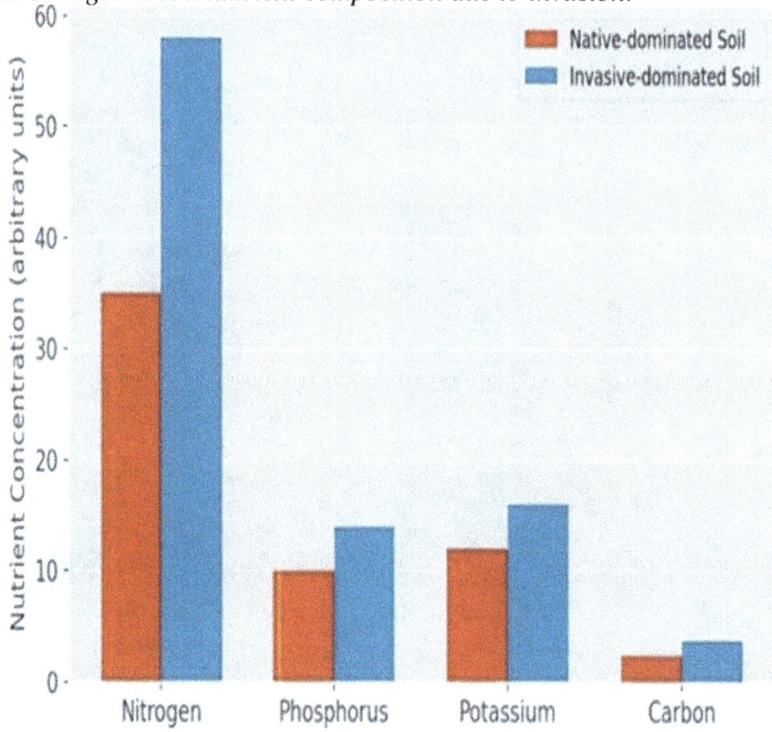

Figure 5. Projected loss of endemic Alpine species with increasing intensity.

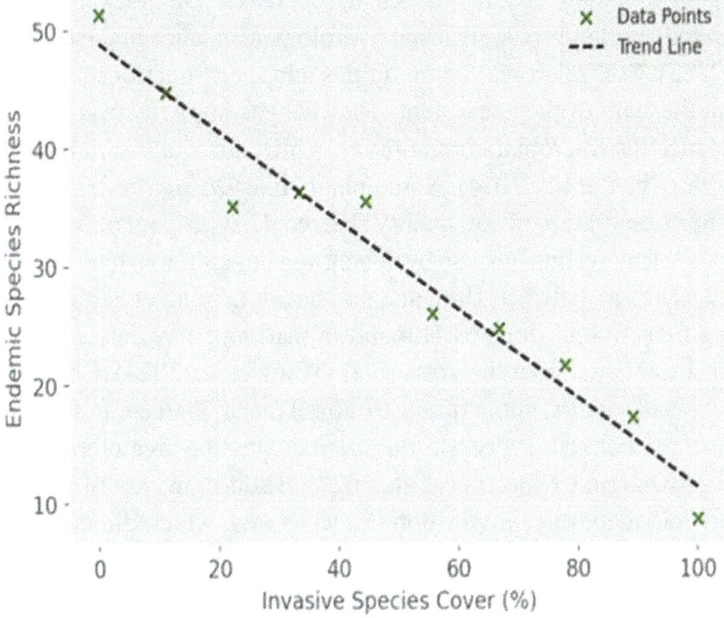

5. SOIL DEGRADATION, NUTRIENT IMBALANCE, AND ECOSYSTEM FUNCTIONING

Soil deterioration in alpine regions, especially in places like the Qinghai-Tibet Plateau, is influenced by a variety of complicated interactions between anthropogenic and natural forces. According to recent studies, soil qualities are greatly impacted by human activity, overgrazing, and climate change, with dire ecological repercussions. A 30% of alpine meadows into “Black soil land” in the Three River Source Region is defined by the significant reduction in biodiversity and soil fertility over the last forty years due to meadow degradation (Yang *et al.,* 2023). This degradation extends beyond the immediate area, affecting nearby ecosystems and altering hydrological processes by changing soil moisture and nutrient dynamics (Fan et al., 2025). A key indicator of soil degradation in alpine meadows is the decline of soil organic carbon (SOC). Degradation influenced indirectly through soil aggregates, causing a 51% depletion of SOC on average in the upper 30 cm of soil. This soil loss decreases quality and, at the same time, increases the release of carbon, thus contributing to global warming (Han et al., 2024). The degradation of alpine grasslands also brings about higher soil bulk density, decreases soil moisture, and the two factors together disrupt the ecological and hydrological balance of these ecosystems (Sun et al., 2025). Soil microbial communities and their functions are very much dependent on the state of the ecosystem. The changes in plant cover and biomass directly influence the microbial communities, which are vital for nutrient cycling and soil health (Chen et al., 2024). Some plants like *Dasiphora fruticosa* play a dual role in the conservation of soil quality (Fan et al., 2025), not only through root structure and nutrient cycling but also by decreasing degradation, thus underscoring the important interplay between flora and soil health (Zhang et al., 2025). On the contrary, digging activities done by Himalayan marmots may even aggravate soil deterioration since they uproot the grass roots (Wang et al., 2023). Climate change has an impact on all sectors, but in the case of the Tibetan Plateau, it is a drastic and direct one, as these hot and dry conditions are reducing the area of grasslands and wetlands at the same time (Shrestha et al., 2024). Besides, in case of warming, the process of organic matter decomposition is accelerated, which directly affects the soil structure and nutrients in these already vulnerable high-elevation ecosystems (Filimonenko & Kuzyakov, 2025).

5.1 Erosion and Soil Profile Dynamics

A variety of factors, including tree and soil, microflora, and anthropogenic activities related to climate change, have all intertwined to bring about the process of soil erosion and the stabilizing process of the same soil. The properties of soil of the Qinghai-Tibetan Plateau, like sand content, water retention, etc., play a vital role in sustaining plant life and keeping the ecosystem resilient (Wang et al., 2020). The less water-holding capacity of sandy soils makes them very susceptible to the environmental changes caused by CO_2, hence they become fragile ecosystems (Ferdush, 2022). Seasonal snowmelt, which is affected by rain and evaporation, plays a significant role in the interaction between snow, soil, and vegetation by keeping soils moist and hence regulating plant growth (Rixen et al., 2022). The process of thawing could lead to soil erosion and degradation of soil profiles during intense climate events that are intense (Ma et al., 2025). The presence of diverse microbial populations is more of a way to improve the health of the soil. Nitrogen cycling and nutrient availability are the major factors through which microbial diversity impacts soil health. It is interesting to note that phosphate-solubilizing bacterial groups are a great contributor to the nutrient supply, hence enhancing the growth of plants (Timofeeva et al., 2023). The warming of Earth as a result of climate change has made the situation worse: soil warming causes SOC loss to be faster and consequently, CO_2 emissions to be higher (Chen et al., 2024). The soil in the alpine region is highly prone to damage because of the complex relationship between flora, soil, microbes, and climate, which underscores the need for comprehensive conservation strategies without any further delay.

5.2 Nutrient Imbalance, Leaching, and Deposition

Soil nutrient imbalances can be either in the form of excesses or depletions and profoundly affect the metabolism of plants. A nitrogen oversupply, for example, may lead to the other essential nutrients being poorly taken up and thus causing the plants to be stunted and pest problems to be aggravated (Kumar et al., 2024). Nutrient interactions are also confusing; the antagonistic relationships that exist between calcium and magnesium versus potassium can further impede the uptake of the latter nutrients (Yan et al., 2020). The continuous cropping, lack of replenishment and poor land management's practices often lead to the depletion of the key macronutrients such as nitrogen, phosphorus, and potassium (Kartini et al., 2024). Furthermore, leaching which occurs as a result of rainfall and snowmelt is a contributor to nutrient loss which subsequently affects soil chemical properties and renders nutrients less accessible to plants (Kaštovská et al., 2022). Plant cover is a major factor in nutrient retention: meadows with thick cover catch organic matter and nutrients, while disturbed or sparsely planted areas are deprived of nutrients and have more sediment movement (Wikström et al., 2025). The interactions between plant species and herbivores also affect nutrient turnover (Zeng, 2024). Climate change amplifies these challenges by altering precipitation patterns, which affect soil moisture and plant nutrient uptake. Increased rainfall can enhance alpine grasses' ability to sequester carbon, while reduced grazing pressure can improve nutrient cycling and biodiversity (He *et al.,* 2025). Microbial communities also respond to changes in vegetation, impacting nitrogen cycling and creating feedback loops that can intensify nutrient loss (Sveen *et al.,* 2024).

5.3 Vegetation Impacts and Ecosystem Responses

Nutrient deficiency can lead to alterations in the makeup of plant communities, the total microbial population, and even the whole area's productivity, so indirectly affecting the plants and animals living there, the food supply, and the ecosystems' resilience to changes (Diyaolu & Folarin, 2024). Covering crops, rotating crops, applying organic manures, and introducing arbuscular mycorrhizal fungi (AMF) are amongst the measures taken to enhance nutrient uptake and improve soil conditions (Ahmed et al., 2025). The detailed knowledge of the interconnection between the responses of plants, the health of the soil, and the cycling of nutrients will greatly help in the management of a sustainable ecosystem.

5.4 Soil Horizon Divisions

The layers of an alpine soil profile indicate nutrient leaching, weathering, and deposition procedures, as shown in Figure 6. The top-to-bottom soil horizons contain:

- **Horizon (Organic):** Composed of decomposing organic matter such as leaves and twigs.
- **A Horizon (Topsoil):** Mixture of mineral particles and organic matter, supporting plant growth.
- **E Horizon (Eluviated):** Leached layer rich in sand and silt; not present in all soils.
- **B Horizon (Subsoil):** Accumulates minerals and clays from A and E horizons.
- **C Horizon (Parent Material):** Partially weathered rock and soil from which upper layers develop.
- **R Horizon (Bedrock):** Weathered solid rock under the soil.

Reduced snow cover in alpine regions accelerates freeze-thaw cycles in the upper soil layers, impacting microbial communities and biogeochemical processes with effects that persist across seasons. These changes outpace the slow pace of soil formation, which takes millennia, while vegetation migrates uphill much more rapidly in response to warming. In the upper alpine zones (2400–2600 m), thawing permafrost destabilizes the soil, causing frost heave, thermal cracking, and ice segregation, which reshape the patterned ground microrelief. Exposed O and A horizons become more prone to erosion, especially when tree-line shifts expose immature soil to fluctuating water balances. Conversely, dense plant cover encourages the formation of mull humus, which stabilizes soils, although it may lead to nutrient leaching into B horizons. The arrival of lowland plants species upslope accelerates carbon loss from alpine O horizons through the addition of labile litter, a pattern that contrasts with Arctic shrub feedback. Experimental warming studies reveal that soil temperature often has a stronger effect than air temperature on horizon-specific processes, particularly during the early and late growth stages in A horizons. Over prolonged warming, archaea populations in mineral horizons undergo distinct compositional changes, influencing both methane and nitrogen cycling in alpine soils.

Figure 6. Layers of soil profile exhibiting nutrient leaching, weathering, and deposition ways

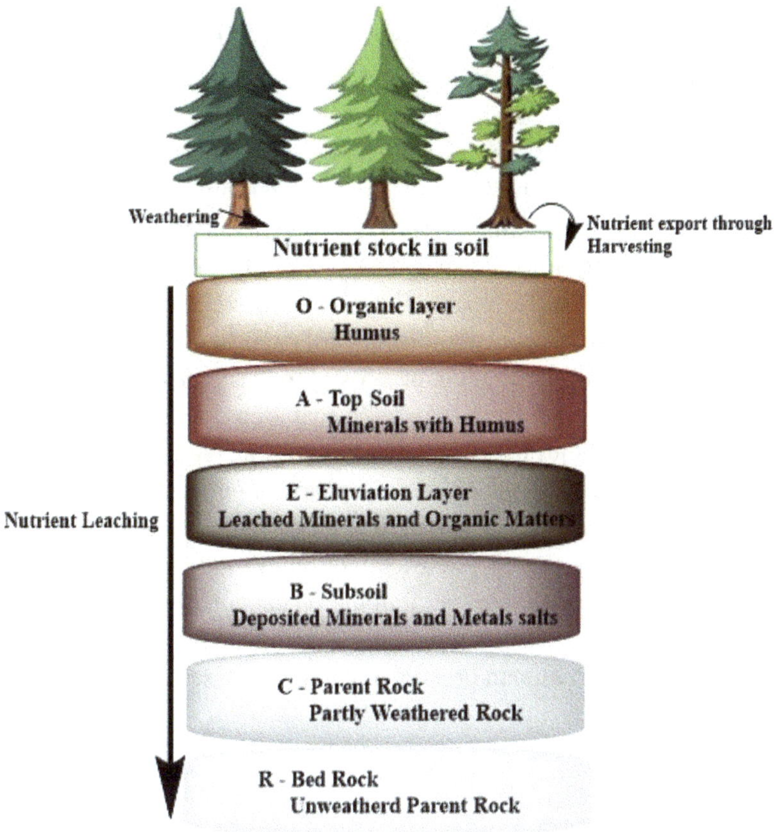

6. CONCLUSION

Alpine ecosystems have been subjected to a great deal of environmental pressure, which in turn has a negative impact on their stability and biodiversity. These include rising temperatures, shifting patterns of rainfall, and melting glaciers, which are all causing alterations to vegetation, species distributions, and phenology of the plants. Human activities like overgrazing, tourism, and construction work have a negative impact on the habitat through degradation and soil erosion, and at the same time, they make the situation worse. Moreover, the invasive and non-native species are driving the native plant communities, soil microbial networks, and nutrient cycling gradually towards a homogenization of the alpine flora. The application of effective methodologies like habitat mapping, reducing fertilization, and controlling invasive species is one of the most effective means of preserving the semi-natural grasslands, such as Festuco-Brometalia. Activities like maintaining forest margins, setting up buffer zones, and refraining from afforestation are preventive measures for preserving spatial connectivity and hydrological balance. Microclimatic niches, which are the locations where alpine plants can take advantage of very small variations in cold conditions, support the growth of plants that are often bigger than what could be expected based on the general climate trends. The functional traits, such as allometric flower size and rapid physiological responses to short growing seasons, are very consistent across continents. Newer monitoring frameworks that keep track of air, soil, and moisture conditions contribute to quantifying changes in the underground conditions and improving the models of ecosystem resilience. Long-term observation of soil temperature, precipitation, and vegetation dynamics is necessary for the development of strategies to adapt to the climate change problem. The shallow alpine soils are particularly at risk of being depleted of nutrients, leached, and having their nitrogen cycles altered, which in turn will lead to changes in plant growth, productivity, and the whole ecosystem's functioning. All these threats working together not only reveal the weakness of alpine plants but also stress the necessity of applying the integrated conservation and management approaches that can deal with both the climatic and human pressures to save these sensitive environments.

ACKNOWLEDGMENT

Moona Nazish served as the corresponding author of this chapter and can be reached at mona.nazish31@gmail.com

REFERENCES

Ahmed, N., Li, J., Li, Y., Deng, L., Deng, L., Chachar, M., Chachar, Z., Chachar, S., Hayat, F., Raza, A., Umrani, J. H., Gong, L., & Tu, P. (2025). Symbiotic synergy: How Arbuscular Mycorrhizal Fungi enhance nutrient uptake, stress tolerance, and soil health through molecular mechanisms and hormonal regulation. *IMA Fungus, 16*, e144989. DOI: 10.3897/imafungus.16.144989 PMID: 40162002

Ahmed, Z., & Nihei, T. (2024). Assessing the environmental impacts of adventure tourism in the world's highest mountains: A comprehensive review for promoting sustainable tourism in high-altitude areas. *Journal of Advanced Research in Social Sciences and Humanities*, *9*(2), 1–15. DOI: 10.26500/JARSSH-09-2024-0201

Alatalo, J. M., Erfanian, M. B., Molau, U., Chen, S., Bai, Y., & Jägerbrand, A. K. (2022). Changes in plant composition and diversity in an alpine heath and meadow after 18 years of experimental warming. *Alpine Botany*, *132*(2), 181–193. DOI: 10.1007/s00035-021-00272-9

Alexander, J. M., Diez, J. M., & Levine, J. M. (2015). Novel competitors shape species' responses to climate change. *Nature*, *525*(7570), 515–518. DOI: 10.1038/nature14952 PMID: 26374998

Arias, P. A., Garreaud, R., Poveda, G., Espinoza, J. C., Molina-Carpio, J., Masiokas, M., Viale, M., Scaff, L., & Van Oevelen, P. J. (2021). Hydroclimate of the Andes part II: Hydroclimate variability and sub-continental patterns. *Frontiers in Earth Science (Lausanne)*, *8*, 505467. DOI: 10.3389/feart.2020.505467

Austin, M. W., Smith, A. B., Olsen, K. M., Hoch, P. C., Krakos, K. N., Schmocker, S. P., & Miller-Struttmann, N. E. (2024). Climate change increases flowering duration, driving phenological reassembly and elevated co-flowering richness. *The New Phytologist*, *243*(6), 2486–2500. DOI: 10.1111/nph.19994 PMID: 39049577

Bai, T., Liu, J., Liu, H., Ni, F., Han, X., Qiao, X., & Sun, X. (2024). Elevation-dependent patterns of temporally asymmetrical vegetation response to climate in an alpine basin on the Qinghai-Tibet Plateau. *Ecological Indicators*, *159*, 111736. DOI: 10.1016/j.ecolind.2024.111736

Bandyopadhyay, M. (2022). *Response Of Small Carnivore Community To Landscape And Climatic Variability Along The Elevation Gradient In The Great Himalayan National Park Conservation Area, The Western Himalaya*. Wildlife Institute of India.

Bhatt, P., Giri, P. R., & Napari, B. (2024). Soil Compaction And Its Impact On Soil Properties, Microbiome, Greenhouse Gas Emission And Plant Root Growth. [BDA]. *Big Data In Agriculture*, *6*(1), 66–69. DOI: 10.26480/bda.01.2024.66.69

Bird, O. (2024). *Alpine Plant Communities and Ecosystem Services Under Drivers of Global Ecosystem Change*. Open Access Te Herenga Waka-Victoria University of Wellington.

Brighenti, S., Tolotti, M., Bruno, M. C., Wharton, G., Pusch, M. T., & Bertoldi, W. (2019). Ecosystem shifts in Alpine streams under glacier retreat and rock glacier thaw: A review. *The Science of the Total Environment*, *675*, 542–559. DOI: 10.1016/j.scitotenv.2019.04.221 PMID: 31030160

Cai, S., & Li, Q. (2024). Snow cover dynamics: Impacts on soil moisture and plant growth in temperate ecosystems. *Molecular Soil Biology, 15*.

Carboni, L. J., Yahdjian, L., & Oñatibia, G. R. (2023). Effects of livestock grazing intensification on plant communities of Patagonian drylands increase with increasing aridity. *Applied Vegetation Science*, *26*(4), e12754. DOI: 10.1111/avsc.12754

Cavieres, L. A., Sanhueza, C. E., & Hernández-Fuentes, C. (2024). Warming had contrasting effects on the importance of facilitative interactions with a cushion nurse species on native and non-native species in the high-Andes of central Chile. *Oikos*, *2024*(8), e10296. DOI: 10.1111/oik.10296

Chai, Y., Xu, L., Xu, Y., Yang, K., Zhu, R., Zhang, R., & Li, X. (2025). Impact of Plateau Grassland Degradation on Ecological Suitability: Revealing Degradation Mechanisms and Dividing Potential Suitable Areas with Multi Criteria Models. *Remote Sensing (Basel)*, *17*(15), 2539. https://www.mdpi.com/2072-4292/17/15/2539. DOI: 10.3390/rs17152539

Chang, X., Wang, W., & Zhou, H. (2025). Nitrogen acquisition by invasive plants: Species preferential N uptake matching with soil N dynamics contribute to its fitness and domination. *Plants*, *14*(5), 748. DOI: 10.3390/plants14050748 PMID: 40094724

Chen, Q., Song, Y., An, Y., Lu, Y., & Zhong, G. (2024). Soil microorganisms: Their role in enhancing crop nutrition and health. *Diversity*, *16*(12), 734. DOI: 10.3390/d16120734

Chen, Y., Qin, W., Zhang, Q., Wang, X., Feng, J., Han, M., Hou, Y., Zhao, H., Zhang, Z., He, J.-S., Torn, M. S., & Zhu, B. (2024). Whole-soil warming leads to substantial soil carbon emission in an alpine grassland. *Nature Communications*, *15*(1), 4489. DOI: 10.1038/s41467-024-48736-w PMID: 38802385

Cubas, J., Martín-Esquivel, J. L., Marrero-Gómez, M., Díaz, J. R. D., Rodríguez, F., & González-Mancebo, J. M. (2022). Climate change causes rapid collapse of a keystone shrub from insular Alpine ecosystems. *Journal for Nature Conservation*, *69*, 126263. DOI: 10.1016/j.jnc.2022.126263

Cui, Z., Yang, W.-S., Cheng, Z., Zhang, Z., Li, S., Zhao, J., López-Vicente, M., & Wu, G.-L. (2022). Top-down degradation of alpine meadow in the Qinghai-Tibetan Plateau: Gravelization initiate hillside surface aridity and meadow community disappearance. *Catena*, *210*, 105933. DOI: 10.1016/j.catena.2021.105933

De Cesare, G. (2025). Sedimentation of a high Alpine Hydropower Reservoir under Climate Change: What will disappear first, the Glacier or the Reservoir? *Authorea Preprints*. DOI: 10.1002/9781394216727.ch4

Diyaolu, C. O., & Folarin, I. O. (2024). The role of biodiversity in agricultural resilience: Protecting ecosystem services for sustainable food production. *Int. J. Res. Publ. Rev*, *5*(10), 1560–1573. DOI: 10.55248/gengpi.5.1024.2741

Doležal, J., Korznikov, K., Bazant, V., Ruka, A. T., & Gago, J. (2025). Elevational Shifts in Tropical Tree Leaf Traits: Interactions Between Soil, Climate, Light, and Phylogeny. BioRxiv, 2025.2007. 2030.667609.

Duchicela, S. A., Cuesta, F., Tovar, C., Muriel, P., Jaramillo, R., Salazar, E., & Pinto, E. (2021). Microclimatic warming leads to a decrease in species and growth form diversity: Insights from a tropical alpine grassland. *Frontiers in Ecology and Evolution*, *9*, 673655. DOI: 10.3389/fevo.2021.673655

Fan, B., Ding, N., Tian, T., Gao, P., Wan, Y., Ma, M., & Sun, K. (2025). Different water and photosynthetic resource use strategies explain the widespread distribution of Dasiphora fruticosa in Qinghai-Tibet Plateau alpine meadows. *Environmental and Experimental Botany*, *231*, 106106. DOI: 10.1016/j.envexpbot.2025.106106

Fan, S., Qin, J., Sun, H., Dan, Z., Li, Z., & Yang, J. (2025). Different Responses of Soil Nutrient Dynamics and Microbial Activities to Soil Moisture Changes in Alpine Wetlands and Meadows on the Tibetan Plateau. *Land Degradation & Development*, *36*(6), 1928–1938. DOI: 10.1002/ldr.5472

Fazlioglu, F., & Wan, J. S. (2021). Warming matters: Alpine plant responses to experimental warming. *Climatic Change*, *164*(3), 56. DOI: 10.1007/s10584-021-02996-3

Feng, L., Gan, M., & Tian, F.-P. (2019a). Effects of Grassland Tourism on Alpine Meadow Community and Soil Properties in the Qinghai-Tibetan Plateau [journal article]. *Polish Journal of Environmental Studies*, *28*(6), 4147–4152. DOI: 10.15244/pjoes/99065

Feng, L., Gan, M., & Tian, F.-P. (2019b). Effects of grassland tourism on Alpine meadow community and soil properties in the Qinghai-Tibetan Plateau. *Polish Journal of Environmental Studies*, *28*(6), 4147–4152. DOI: 10.15244/pjoes/99065

Ferdush, J. (2022). *Determining the effects of elevated carbon dioxide on soil acidification, cation depletion, and soil inorganic carbon and mapping soil carbons using artificial intelligence*. Mississippi State University.

Ficetola, G. F., Marta, S., Guerrieri, A., Cantera, I., Bonin, A., Cauvy-Fraunié, S., Ambrosini, R., Caccianiga, M., Anthelme, F., Azzoni, R. S., Almond, P., Alviz Gazitúa, P., Ceballos Lievano, J. L., Chand, P., Chand Sharma, M., Clague, J. J., Cochachín Rapre, J. A., Compostella, C., Encarnación, R. C., & Carteron, A. (2024). The development of terrestrial ecosystems emerging after glacier retreat. *Nature*, *632*(8024), 336–342. DOI: 10.1038/s41586-024-07778-2 PMID: 39085613

Fickert, T. (2020). Common patterns and diverging trajectories in primary succession of plants in eastern alpine glacier forelands. *Diversity*, *12*(5), 191. DOI: 10.3390/d12050191

Filimonenko, E., & Kuzyakov, Y. (2025). Activation energy of organic matter decomposition in soil and consequences of global warming. *Global Change Biology*, *31*(9), e70472. DOI: 10.1111/gcb.70472 PMID: 40905536

Gao, E., Ma, H., Yang, T., Kaiser-Bunbury, C. N., & Zhao, Z. (2023). Meadow transformations alter above-and below-ground ecological networks and ecosystem multifunctionality. *Functional Ecology*, *37*(6), 1703–1716. DOI: 10.1111/1365-2435.14333

Grange, M. C., Munoz, F., Moretti, M., Varona-Y-Varona, S., Renaud, J., Colace, M. P., Guéguen, M., Ssymank, A., & Gallien, L. (2025). The golden threat—Solidago invasion alters native plant–pollinator interactions by vegetative crowding. *Journal of Ecology*, *113*(12), 3762–3772. DOI: 10.1111/1365-2745.70194

Han, H., Zeeshan, Z., Talpur, B. A., Sadiq, T., Bhatti, U. A., Awwad, E. M., Al-Razgan, M., & Ghadi, Y. Y. (2024). Studying long term relationship between carbon Emissions, Soil, and climate Change: Insights from a global Earth modeling Framework. *International Journal of Applied Earth Observation and Geoinformation*, *130*, 103902. DOI: 10.1016/j.jag.2024.103902

Hawke, D. (2022). The biogeochemistry and ecological impact of Westland petrels (Procellaria westlandica) on terrestrial ecosystems. *New Zealand Journal of Ecology*, *46*(1), 1–16. DOI: 10.20417/nzjecol.46.3

He, F., Dong, S., Yang, J., Hao, X., Ma, C., Shen, H., Xiao, J., Li, Y., Zhang, R., Shi, H., Wang, Z., Zhang, Y., & Zuo, H. (2025). Rotational grazing can enhance the soil available nutrients rather than plant diversity to promote ANPP of alpine steppe. *Agriculture, Ecosystems & Environment*, *387*, 109617. DOI: 10.1016/j.agee.2025.109617

Hossein-Panahi, B., Bozorg-Haddad, O., Loáiciga, H., Meghwar, S. M., & Zeleňáková, M. (2022). Case Studies Around the World. In *Climate Change in Sustainable Water Resources Management* (pp. 361–414). Springer. DOI: 10.1007/978-981-19-1898-8_10

Hu, S., Hsu, P.-C., Li, W., Wang, L., Chen, H., & Zhou, B. (2023). Mechanisms of Tibetan Plateau warming amplification in recent decades and future projections. *Journal of Climate*, *36*(17), 5775–5792. DOI: 10.1175/JCLI-D-22-0471.1

Inouye, D. W. (2020). Effects of climate change on alpine plants and their pollinators. *Annals of the New York Academy of Sciences*, *1469*(1), 26–37. DOI: 10.1111/nyas.14104 PMID: 31025387

Jansson, J. K., & Hofmockel, K. S. (2020). Soil microbiomes and climate change. *Nature Reviews. Microbiology*, *18*(1), 35–46. DOI: 10.1038/s41579-019-0265-7 PMID: 31586158

Jin, X., Deng, A., Fan, Y., Ma, K., Zhao, Y., Wang, Y., Zheng, K., Zhou, X., & Lu, G. (2025). Diversity, functionality, and stability: Shaping ecosystem multifunctionality in the successional sequences of alpine meadows and alpine steppes on the Qinghai-Tibet Plateau. *Frontiers in Plant Science*, *16*, 1436439. DOI: 10.3389/fpls.2025.1436439 PMID: 40182548

Kartini, N. L., Saifulloh, M., Trigunasih, N. M., Sukmawati, N. M. S., & Mega, I. (2024). Impact of long-term continuous cropping on soil nutrient depletion. *Ecological Engineering & Environmental Technology, 25.*

Kaštovská, E., Choma, M., Čapek, P., Kaňa, J., Tahovská, K., & Kopáček, J. (2022). Soil warming during winter period enhanced soil N and P availability and leaching in alpine grasslands: A transplant study. *PLoS One*, *17*(8), e0272143. DOI: 10.1371/journal.pone.0272143 PMID: 35917373

Khoso, R. B. (2024). Global drivers, local consequences: Assessing geodiversity and geosystem services for global change adaptation and sustainable development of mountain regions–Examples from the Monte Rosa W-Alps.

Kok, P. J., Broholm, T. L., van Doorn, L., Fiorillo, B. F., & Smith, C. (2024). Lost in space and time: Robust demography and enhanced resilience buffer adverse environmental effects in a highly isolated and sedentary pre-pleistocene landscape vertebrate. *BMC Ecology and Evolution*, *24*(1), 125. DOI: 10.1186/s12862-024-02314-2 PMID: 39394073

Körner, C. (2021). The alpine life zone. In *Alpine plant life: Functional plant ecology of high mountain ecosystems* (pp. 23–51). Springer.

Kotlarski, S., Gobiet, A., Morin, S., Olefs, M., Rajczak, J., & Samacoïts, R. (2023). 21st Century alpine climate change. *Climate Dynamics*, *60*(1), 65–86. DOI: 10.1007/s00382-022-06303-3

Kumar, R., Mondal, B., & Bordoloi, N. (2024). Application of straw-derived biochar: A sustainable approach to improve soil quality and crop yield and reduce N2O emissions in paddy soil. *Environmental Science and Pollution Research International*, *31*(51), 60804–60818. DOI: 10.1007/s11356-024-35269-4 PMID: 39392575

Kumar, S., & Khanduri, V. P. (2024). Impact of climate change on the Himalayan alpine treeline vegetation. *Heliyon*, *10*(23), e40797. DOI: 10.1016/j.heliyon.2024.e40797 PMID: 39698095

Kuo, C. C., Liu, Y. C., Su, Y., Liu, H. Y., & Lin, C. T. (2022). Responses of alpine summit vegetation under climate change in the transition zone between subtropical and tropical humid environment. *Scientific Reports*, *12*(1), 13352. DOI: 10.1038/s41598-022-17682-2 PMID: 35922458

Lamprecht, A., Semenchuk, P. R., Steinbauer, K., Winkler, M., & Pauli, H. (2018). Climate change leads to accelerated transformation of high-elevation vegetation in the central Alps. *The New Phytologist*, *220*(2), 447–459. DOI: 10.1111/nph.15290 PMID: 29938796

Li, R., Wang, P., Liu, Q., Luo, R., & Pang, X. (2025). Microbial functional groups govern soil multifunctionality during alpine grassland restoration via turf transplantation. *Catena*, *258*, 109269. DOI: 10.1016/j.catena.2025.109269

Li, T., Kamran, M., Chang, S., Peng, Z., Wang, Z., Ran, L., Jiang, W. Q., Jin, Y., Zhang, X., You, Y., Li, L., & Hou, F. (2022). Climate-soil interactions improve the stability of grassland ecosystem by driving alpine plant diversity. *Ecological Indicators*, *141*, 109002. DOI: 10.1016/j.ecolind.2022.109002

Liu, Y., Xie, Y., Guo, Z., & Xi, G. (2023). Effects of Climate Variability and Human Activities on Vegetation Dynamics across the Qinghai–Tibet Plateau from 1982 to 2020. *Remote Sensing (Basel), 15*(20), 4988. https://www.mdpi.com/2072-4292/15/20/4988. DOI: 10.3390/rs15204988

Ma, X., Ma, W., Liu, F., Chen, Q., Niu, B., & Zhou, Q. (2025). Delineation and Intensity Classification of Freeze–Thaw Erosion in the Western Section of Northern China's Agro-Pastoral Transition Zone Under Controlling Factors. *Land (Basel), 14*(11), 2278. DOI: 10.3390/land14112278

Ma, Y., Huang, X., Feng, Q., & Liang, T. (2022). Alpine grassland reviving response to seasonal snow cover on the Tibetan Plateau. *Remote Sensing (Basel), 14*(10), 2499. DOI: 10.3390/rs14102499

Macphie, K. H., & Phillimore, A. B. (2024). Phenology. *Current Biology, 34*(5), R183–R188. DOI: 10.1016/j.cub.2024.01.007 PMID: 38471442

Martin, J. B., Pain, A. J., & Martin, E. E. (2024). Geochemistry of glacial, proglacial, and deglaciated environments. *Treatise on Geochemistry, 3*, 251–299.

Martín-Esquivel, J., Marrero-Gómez, M., Cubas, J., González-Mancebo, J., Olano, J., & del Arco, M. (2020). Climate warming and introduced herbivores disrupt alpine plant community of an oceanic island (Tenerife, Canary Islands). *Plant Ecology, 221*(11), 1117–1131. DOI: 10.1007/s11258-020-01066-5

Mehta, P. (2024). The impact of climate change on the environment, water resources, and agriculture: a comprehensive review. *Climate, Environment and Agricultural Development: A Sustainable Approach Towards Society*, 189-201.

Mia, M. A. R., Dey, D., Sakib, M. R., Biswas, M. Y., Prottay, A. A. S., Paul, N., Rimti, F. H., Abdullah, Y., Biswas, P., Iftehimul, M., Paul, P., Sarkar, C., El-Nashar, H. A. S., El-Shazly, M., & Islam, M. T. (2023). The efficacy of natural bioactive compounds against prostate cancer: Molecular targets and synergistic activities. *Phytotherapy Research, 37*(12), 5724–5754. DOI: 10.1002/ptr.8017 PMID: 37786304

Miao, S., Zhang, Y., Li, B., Yuan, X., & Zuo, J. (2024). Antibiotic intermediates and antibiotics synergistically promote the development of multiple antibiotic resistance in antibiotic production wastewater. *Journal of Hazardous Materials, 479*, 135601. DOI: 10.1016/j.jhazmat.2024.135601 PMID: 39243543

Mitra, A., & Paul, S. (2025). Balancing Mountain Tourism, Cultural Heritage, and Environmental Stability: Strategies and Challenges in Sustainable Development. In *Balancing Mountain Tourism, Cultural Heritage, and Environmental Stability* (pp. 31-54). IGI Global Scientific Publishing.

Mueller, M., Egg, L., Ruff, T., Haas, A., Schubert, M., & Gum, B. (2025). Evidence of Natural Reproduction of North American Rainbow Trout (Oncorhynchus mykiss) in Three Alpine Rivers in Bavaria, Germany. *Fisheries Management and Ecology, 32*(2), e12765. DOI: 10.1111/fme.12765

O'Brien, A. M., Ginnan, N. A., Rebolleda-Gómez, M., & Wagner, M. R. (2021). Microbial effects on plant phenology and fitness. *American Journal of Botany, 108*(10), 1824–1837. DOI: 10.1002/ajb2.1743 PMID: 34655479

Oldfather, M. F., Ennis, A., Miller, B. W., Clark-Wolf, K., Rangwala, I., Robe, H., & Littlefield, C. (2025). Climate change impacts and adaptation in U.S. Rocky Mountain high-elevation ecosystems. *Arctic, Antarctic, and Alpine Research, 57*(1), 2450089. DOI: 10.1080/15230430.2025.2450089

Piria, M., Radočaj, T., Vilizzi, L., & Britvec, M. (2022). Climate change may exacerbate the risk of invasiveness of non-native aquatic plants: The case of the Pannonian and Mediterranean regions of Croatia. *NeoBiota, 76*, 25–52. DOI: 10.3897/neobiota.76.83320

Plunger, J., Guariento, E., Steinwandter, M., Colla, F., Rief, A., & Seeber, J. (2022). Shifts in ground-dwelling predator communities in response to changes in management intensity in Alpine meadows. *Soil Organisms, 94*(3), 149–161.

Poni, S., Sabbatini, P., & Palliotti, A. (2022). Facing spring frost damage in grapevine: Recent developments and the role of delayed winter pruning–a review. *American Journal of Enology and Viticulture, 73*(4), 211–226. DOI: 10.5344/ajev.2022.22011

Richard, B., Dupouey, J. L., Corcket, E., Alard, D., Archaux, F., Aubert, M., Boulanger, V., Gillet, F., Langlois, E., Macé, S., Montpied, P., Beaufils, T., Begeot, C., Behr, P., Boissier, J.-M., Camaret, S., Chevalier, R., Decocq, G., Dumas, Y., & Lenoir, J. (2021). The climatic debt is growing in the understorey of temperate forests: Stand characteristics matter. *Global Ecology and Biogeography, 30*(7), 1474–1487. DOI: 10.1111/geb.13312

Rixen, C., Høye, T. T., Macek, P., Aerts, R., Alatalo, J. M., Anderson, J. T., Arnold, P. A., Barrio, I. C., Bjerke, J. W., Björkman, M. P., Blok, D., Blume-Werry, G., Boike, J., Bokhorst, S., Carbognani, M., Christiansen, C. T., Convey, P., Cooper, E. J., Cornelissen, J. H. C., & Zong, S. (2022). Winters are changing: Snow effects on Arctic and alpine tundra ecosystems. *Arctic Science, 8*(3), 572–608. DOI: 10.1139/as-2020-0058

Rodrigues, C. I. D., Brito, L. M., & Nunes, L. J. (2023). Soil carbon sequestration in the context of climate change mitigation: A review. *Soil Systems, 7*(3), 64. DOI: 10.3390/soilsystems7030064

Romm, J. J. (2022). *Climate change: What everyone needs to know*. Oxford University Press. DOI: 10.1093/wentk/9780197647127.001.0001

Samson, S., Patnaik, S., Kannan, K. R., Prathima, T., & Sreedhar, B. (2023). Effect of Global Warming on Glaciers, Rivers and Water Towers. International Conference on Pervasive Knowledge and Collective Intelligence on Web and Social Media, Sekar, K. C., Thapliyal, N., Bhojak, P., Bisht, K., Pandey, A., Mehta, P., Negi, V. S., & Rawat, R. S. (2025). Early signals of climate change impacts on alpine plant diversity in Indian Himalaya. *Biodiversity and Conservation*, *34*(1), 207–233.

Senthilnathan, A., & D'Andrea, R. (2024). Coexistence of Competing Plants Under Plant–Soil Feedback. *Ecology Letters*, *27*(9), e14503. DOI: 10.1111/ele.14503 PMID: 39354908

Sharma, I., & Birman, S. (2024). Biodiversity loss, ecosystem services, and their role in promoting sustainable health. In *The climate-health-sustainability nexus: understanding the interconnected impact on populations and the environment* (pp. 163–188). Springer. DOI: 10.1007/978-3-031-56564-9_7

Shrestha, B., Zhang, L., Shrestha, S., Khadka, N., & Maharjan, L. (2024). Spatio-temporal patterns, sustainability, and primary drivers of NDVI-derived vegetation dynamics (2003–2022) in Nepal. *Environmental Monitoring and Assessment*, *196*(7), 607. DOI: 10.1007/s10661-024-12754-4 PMID: 38858316

Singh, J. (2025). The Accelerating Melting of Glaciers: Impacts on Earth and Emerging Challenges. *International Journal of Advanced Research and Interdisciplinary Scientific Endeavours*, *2*(1), 401–414. DOI: 10.61359/11.2206-2501

Spencer, S. A., Anderson, A. E., Silins, U., & Collins, A. L. (2021). Hillslope and groundwater contributions to streamflow in a Rocky Mountain watershed underlain by glacial till and fractured sedimentary bedrock. *Hydrology and Earth System Sciences*, *25*(1), 237–255. DOI: 10.5194/hess-25-237-2021

Suhery, D., Idami, Z., & Wati, S. (2024). The Phonological Interference of Acehnese in Pronouncing Indonesian Language. *Journal of Languages and Language Teaching*, *12*(1), 120–135. DOI: 10.33394/jollt.v12i1.9554

Sun, J., Zhou, T. C., Liu, M., Chen, Y. C., Liu, G. H., Xu, M., Shi, P. L., Peng, F., Tsunekawa, A., Liu, Y., Wang, X.-D., Dong, S.-K., Zhang, Y.-J., & Li, Y.-N. (2020). Water and heat availability are drivers of the aboveground plant carbon accumulation rate in alpine grasslands on the Tibetan Plateau. *Global Ecology and Biogeography*, *29*(1), 50–64. DOI: 10.1111/geb.13006

Sun, W., Zhou, S., Yu, B., Zhang, Y., Keenan, T., & Fu, B. (2025). Soil moisture-atmosphere interactions drive terrestrial carbon-water trade-offs. *Communications Earth & Environment*, *6*(1), 169. DOI: 10.1038/s43247-025-02145-z

Sveen, T. R., Hannula, S., & Bahram, M. (2024). Microbial regulation of feedbacks to ecosystem change. *Trends in Microbiology*, *32*(1), 68–78. DOI: 10.1016/j.tim.2023.06.006 PMID: 37500365

Tai, S., Shangguan, D., Wu, J., Wang, R., & Li, D. (2025). Spatial–Temporal Heterogeneity of Wetlands in the Alpine Mountains of the Shule River Basin on the Northeastern Edge of the Qinghai–Tibet Plateau. *Remote Sensing (Basel)*, *17*(6), 976. DOI: 10.3390/rs17060976

Tao, J., Xie, Y., Wang, W., Zhu, J., Zhang, Y., & Zhang, X. (2022). Elevational Gradient of Climate-Driving Effects on Cropland Ecosystem Net Primary Productivity in Alpine Region of the Southwest China. *Remote Sensing (Basel)*, *14*(13), 3069. DOI: 10.3390/rs14133069

Timofeeva, A. M., Galyamova, M. R., & Sedykh, S. E. (2023). Plant growth-promoting soil bacteria: Nitrogen fixation, phosphate solubilization, siderophore production, and other biological activities. *Plants*, *12*(24), 4074. DOI: 10.3390/plants12244074 PMID: 38140401

Van Der Merwe, S. (2024). Investigating ecological drivers and impacts of vegetation change in sub Antarctic tundra.

Varga, K., Csizi, I., Monori, I., & Valko, O. (2021). Threats and challenges related to grazing paddocks: Recovery of extremely overgrazed grassland after grazing exclusion. *Arid Land Research and Management*, *35*(3), 346–357. DOI: 10.1080/15324982.2020.1869120

Vincent, A., Daigre, C., Fischer, O., Aðalgeirsdóttir, G., Violette, S., Hart, J., Guðmundsson, S., & Pálsson, F. (2024). A hydrogeological conceptual model of aquifers in catchments headed by temperate glaciers. *Hydrology and Earth System Sciences*, *28*(14), 3475–3494. DOI: 10.5194/hess-28-3475-2024

Wallner, M. (2022). Future effects of glacier retreat on downstream runoff and hydropower generation in Alpine regions of different catchment sizes/submitted by Mario Wallner, BSc.

Wang, J., Li, M., Yu, C., & Fu, G. (2022). The change in environmental variables linked to climate change has a stronger effect on aboveground net primary productivity than does phenological change in alpine grasslands. *Frontiers in Plant Science*, *12*, 798633. DOI: 10.3389/fpls.2021.798633 PMID: 35058958

Wang, S., Guo, L., He, B., & Li, T. (2020). The stability of Qinghai-Tibet Plateau ecosystem to climate change. *Physics and Chemistry of the Earth, Parts a/b/c, 115*, 102827.

Wang, X., Li, Y., Gong, Y., Liu, Y., Zhao, J., & Li, K. (2025). Spatiotemporal dynamics and driving mechanism of aboveground biomass across three Alpine grasslands in Central Asia over the past 20 years using three algorithms. *Remote Sensing (Basel)*, *17*(3), 538. DOI: 10.3390/rs17030538

Wang, Z., Kang, Y., Wang, Y., Tan, Y., Yao, B., An, K., & Su, J. (2023). Himalayan marmot (Marmota himalayana) redistribution to high latitudes under climate change. *Animals (Basel)*, *13*(17), 2736. DOI: 10.3390/ani13172736 PMID: 37684999

Wikström, S. A., Gubri, B., Asplund, M. E., Dahl, M., Gullström, M., Hansen, J. P., Kumblad, L., Rydin, E., Garbaras, A., & Björk, M. (2025). Influence of landscape characteristics and submerged aquatic vegetation on sediment carbon and nitrogen storage in shallow brackish water habitats. *Scientific Reports*, *15*(1), 7808. DOI: 10.1038/s41598-025-92217-z PMID: 40050425

Wilson, B. R., Tulau, M., Kuginis, L., McInnes-Clarke, S., Grover, S., Milford, H., & Jenkins, B. R. (2022). Distribution, nature and threats to soils of the Australian Alps: A review. *Austral Ecology*, *47*(2), 166–188. DOI: 10.1111/aec.13115

Wu, J., Li, M., Zhang, X., Fiedler, S., Gao, Q., Zhou, Y., Cao, W., Hassan, W., Mărgărint, M. C., Tarolli, P., & Tietjen, B. (2021). Disentangling climatic and anthropogenic contributions to nonlinear dynamics of alpine grassland productivity on the Qinghai-Tibetan Plateau. *Journal of Environmental Management*, *281*, 111875. DOI: 10.1016/j.jenvman.2020.111875 PMID: 33378737

Xu, X., Wang, W., Ke, L., & Zhou, P. (2025). Weakening of regional contrasts in glacier changes around the Tarim Basin in the early 21st century. *npj Climate and Atmospheric Science, 8*(1), 289.

Yan, B., Sun, Y., & Wei, Y. (2020). Potassium–calcium antagonistic interaction under tomato magnesium deficiency and magnesium fertiliser regulation in solar greenhouse. *Quality Assurance and Safety of Crops & Foods*, *12*(3), 76–86. DOI: 10.15586/qas.v12i3.723

Yang, H., Gou, X., Xue, B., Ma, W., Kuang, W., Tu, Z., Gao, L., Yin, D., & Zhang, J. (2023). Research on the change of alpine ecosystem service value and its sustainable development path. *Ecological Indicators*, *146*, 109893. DOI: 10.1016/j.ecolind.2023.109893

Ye, K., Li, J., Liang, Z., Zhao, X., Luo, H., Chen, Z., Du, P., Wu, K., Qiao, R., & Hu, J. (2025). Overgrazing impacts plant species diversity in alpine wetlands indirectly by altering its environmental dependency. *Global Ecology and Conservation, 59,* e03532. DOI: 10.1016/j.gecco.2025.e03532

Zeng, M. (2024). The mutual effect of nutrients on plant–herbivore interactions. *Plant Ecology, 225*(10), 1035–1045. DOI: 10.1007/s11258-024-01452-3

Zerbe, S. (2023). Subalpine and alpine grassland. In *Restoration of ecosystems–bridging nature and humans: A transdisciplinary approach* (pp. 185–208). Springer. DOI: 10.1007/978-3-662-65658-7_9

Zhan, C. (2024). Microbial decomposition and soil health: mechanisms and ecological implications. *Molecular Soil Biology, 15.*

Zhang, Q., Yao, J., Degen, A. A., & Zhou, H. (2025). Alpine meadow degradation: status, reasons, and trends. In *Grassland Degradation, Restoration and Sustainable Management of Global Alpine Area* (pp. 69-86). Elsevier. DOI: 10.1016/B978-0-443-21882-8.00004-4

Zhang, W. (2023). Influence of alpine meadow deterioration on soil microbial communities in the Yangtze River source region. *Frontiers in Environmental Science, 11,* 1210349. DOI: 10.3389/fenvs.2023.1210349

Zhang, Y., Lu, Y., & Song, X. (2023). Identifying the Main Factors Influencing Significant Global Vegetation Changes. *Forests, 14*(8), 1607. DOI: 10.3390/f14081607

Zhao, C., & Zhang, X. (2024). The Role of Polyploidy in the Genetic Structure and Expansion of Lepisorus clathratus in the Qinghai–Tibetan Plateau and Hengduan Mountains. *Plants, 13*(22), 3181. https://www.mdpi.com/2223-7747/13/22/3181. DOI: 10.3390/plants13223181 PMID: 39599389

Zhao, P., Li, L., Lin, L., Zhai, G., Cruse, R. M., & Wang, E. (2022). Response of surface soil nutrients and organic carbon fractions to tillage erosion vs. water erosion in an agricultural landscape. *Soil Science Society of America Journal, 86*(6), 1470–1482. DOI: 10.1002/saj2.20461

Zhyvko, M. (2024). The cost of fragmentation: ecology and the impact on the global economy.

Chapter 11
Physiological and Ecological Adaptations of Alpine Plants

Nouman Ahmad
http://orcid.org/0009-0009-3370-2862
Department of Biotechnology, University of Mianwali, Punjab, Pakistan

Bushra Tahseen
University of Mianwali, Punjab, Pakistan

Sara
Government College University, Faisalabad, Pakistan

Shah Bano
University of Mianwali, Punjab, Pakistan

Ahmed Vandy
University of Sierra Leone, Freetown, Sierra Leone

Abdul Razzaq
Qarshi Industries (Pvt.) Ltd., Hattar, Pakistan

Shahbaz Ali
Hohai University, China

ABSTRACT

Alpine plants inhabit some of the most extreme environments on Earth, characterized by low temperatures, high solar radiation, strong winds, and nutrient-poor soil. To survive and reproduce under these conditions, they exhibit a wide range of physiological and ecological adaptations. These include morphological traits such as dwarfism, cushion growth forms, and dense pubescence; physiological mechanisms such as cold hardiness, antifreeze protein production, and efficient photosynthetic pathways; and ecological strategies such as delayed flowering, seed dormancy, and mutualistic interactions with pollinators and mycorrhizal fungi. Together, these adaptations enable alpine plants to maintain metabolic activity, optimize resource use, and ensure reproductive success despite harsh abiotic and biotic pressures in

DOI: 10.4018/979-8-3373-6756-9.ch011

ight © 2026, IGI Global Scientific Publishing. Copying or distributing in print or electronic forms without written permission of IGI Global Scientific Publishing is prohibited. Use of this chapter to
n generative artificial intelligence (AI) technologies is expressly prohibited. The publisher reserves all rights to license its use for generative AI training and machine learning model development.

their environment. Understanding these adaptive strategies is essential for predicting the responses of alpine vegetation to climate change and for developing effective conservation measures.

1. INTRODUCTION

Alpine and subalpine ecosystems are found at high altitudes, with environmental factors exerting a high degree of abiotic control over vegetation structure and function. The highest forested ecosystems are subalpine forests, which occupy an ecotone between montane forests and alpine vegetation at the alpine-treeline ecotone (Fites-Kaufman et al., 2007). These ecosystems include open woodlands, isolated tree stands, herbaceous communities, shrublands, bare bedrock, and snowfields in different mosaics in response to climatic and topographic gradients (Rundel & Millar, 2016). The global treeline is linked to a thermal limit and the alpine ecotone to an average growing-season temperature of about 6.4°C, which underlines the preeminence of temperature in the process of alpine and subalpine delineation (Körner & Paulsen, 2004; Rundel & Millar, 2016). Alpine ecosystems are above this limit and are ecologically connected to subalpine systems through continuous ecological and climatic gradients.

The extreme abiotic stressors found in the alpine and subalpine environments are persistent snowpack, desiccating winds, acute and chronic low temperatures, short growing seasons, soil moisture, and evapo transpirative stresses (Fites-Kaufman et al., 2007). Precipitation is mostly in the form of snow, which contributes to the long snow cover that limits plant growth and biological activities of the soil (Agee, 1993; Fites-Kaufman et al., 2007). Precipitation and vegetation changes on an inter-yearly and inter-decadal basis are highly affected by climatic variability caused by ocean-atmosphere interaction like ENSO and PDO (Abatzoglou et al., 2009; Cayan et al., 1999; Mantua et al., 1997). Nutritional and edaphic limitations are also caused by geological substrates, geomorphology, and ill-formed soils that allow the establishment of plant species that are suitable for residing in a rather harsh substrate (Millar, 2012).

The flora and fauna in the alpine and subalpine habitats have developed special adaptation features to endure extreme environmental conditions such as temperatures that are below the freezing point, high intensity of solar radiations, soils that are nutrient deficient, steep slopes, and persistent exposure to wind. Plants survive in harsh climatic conditions and continue photosynthesis and survive over multiyear adverse conditions owing to adaptive traits, which include prolonged needle retention, waxy and resinous tissue, plasticity of the crown, and krummholz growth forms (stunted, wind-shaped trees near alpine treelines) (Barber, 2013; Fites-Kaufman et al., 2007; Rundel & Millar, 2016). These adaptations lead to long-term survival, persistent reproductive ability, and genetic variation, which enable populations to survive and adapt to changing environmental regimes.

Climatic variability and long-term climatic change are extremely sensitive to alpine and subalpine ecosystems because they are affected by temperature, snowpack, and soil moisture regimes. Climatic water shortages, temperature fluctuations, and changes in precipitation patterns may cause changes in species distribution, recruitment, and disturbance conditions, such as insect infestation and forest death (Millar, 2012; Rundel & Millar, 2016). The presence of climatic changes in the pattern of recruiting trees to high elevations shows that the ecological response to climatic changes is still occurring, especially in treeline ecosystems (Dolanc et al., 2013).

This chapter synthesizes existing data on environmental regulation and adaptation mechanisms in alpine and subalpine vegetation and refers to physiological and ecological responses to extreme climatic and edaphic factors, to mention implications of the ecosystem dynamics under conditions of altered climatic conditions.

2. ALPINE ENVIRONMENT STRESS FACTOR

The alpine environment is one of the most extreme formations in terrestrial ecosystems, typified by various interacting abiotic stressors that have a powerful impact on plant survival, growth, and reproduction. These stress conditions include low temperatures, extreme sunlight and ultraviolet radiation, strong winds, nutrient-starving soils, and limited growing periods, which determine the physiological and ecological approaches of alpine plants (García-Plazaola et al., 2015; Rawat et al., 2014).

One of the most important limiting factors in alpine ecosystems is low temperature, which directly impacts plant metabolism, growth, and survival. Plants located in the Alps often face freezing conditions, thereby limiting photosynthesis, respiration, and cell viability (García-Plazaola et al., 2015). Cold temperatures also reduce the growing period and restrict growth activities, such as germination, flowering, and seed production.

Temperature affects the physiology of alpine vegetation at several levels, including growth, photosynthetic efficiency, and activity. Various studies have revealed that alpine species exhibit species-selective reactions to low temperatures. For example, *Potentilla soundersiana* alters root structure and leaf traits to tolerate low-temperature and UV-induced stress (Ma et al., 2015). Nevertheless, alpine species like *Soldanella alpina*, *Ranunculus glacialis* did not have any significant changes in antioxidant compounds in the condition of low temperature, implying that different alpine taxa had varied adaptive strategies (Streb et al., 2003).

Although plants can reduce their temperature, certain plants, such as *Poa alpina, Ranunculus glacialis*, and *Tussilago farfara*, have been reported to have low temperatures that suppress their root growth, cell differentiation, and general development (Nagelmüller et al., 2017; Sati et al., 2024). In contrast, an increase in temperature can facilitate the growth and development of certain species in the Alps, which shows that alpine vegetation is sensitive to changes in temperature (Nagelmüller et al., 2017).

In the mountains and at high altitudes, the amount of solar and ultraviolet (UV) radiation reaching alpine plants is high because of weak atmospheric filtration and clear skies. UV radiation, especially UV-B, has the potential to damage DNA, proteins, and photosynthetic systems, and introduce significant physiological challenges to alpine plants (Sati et al., 2024).

To counter the effects of UV radiation, alpine plants have developed defense systems, such as pigments, including anthocyanins and reflective leaf surfaces. It has been revealed that at higher altitudes, *Saxifraga hostii* has a high absorption in the UV spectrum, indicating structural and optical modification to UV protection (Sedej et al., 2020). In *Kobresia humilis*, a higher level of UV-B radiation intensified leaf area, fresh weight, and flowering, suggesting that UV radiation may affect both vegetative and reproductive characteristics (Shi et al., 2022).

These results indicate that UV radiation is a stressor and ecological cue that affects plant morphology and phenology in alpine regions.

Wind refers to the strong winds that typify alpine conditions and may augment the extremes of temperatures, elevate the evapotranspiration, and create mechanical damage on plant tissues. During winter, wind enhances chilling, which is susceptible to frost damage, and in summer, it can affect the microclimate and cooling of plants (Sati et al., 2024).

The presence of intense wind also influences the pollination activities in disturbance of the pollinator activity and scattering of the pollen not to the target plants and this may lead to low reproductive success. To counteract this, most alpine vegetation will develop dwarf, prostrate or cushion-shaped growth structures that reduce wind exposure and mechanical stress. These morphological changes contribute to decreased desiccation, energy savings, and survival in windy environments (Sati et al., 2024).

The soils in the Alps are usually shallow, rocky, and of low nutrient content, thus limiting plant growth and productivity. Availability of nutrients has a significant effect on the development of alpine plants, biomass and the physiology of the plants. Past research has indicated that nutrient enrichment is very effective in promoting growth, biomass, and production of secondary metabolites in various alpine species, such as *Aconitum carmichaeli, Aconitum heterophyllum*, and *Aconitum balfourii* (BAHUGUNA et al., 2013; Hou DaBin et al., 2006). On the same note, *Rheum tanguticum* had a higher root length and biomass responding to phosphorus and potassium fertilization (Shen et al., 2023).

In response to nutrient scarcity, alpine plants evolve rich root systems, establish mycorrhizal interactions, and enhance their nutrient capture efficiency. Adaptive responses have been observed to take place in species such as *Rhododendron anthopogon, Saxifraga hostii, Taxus wallichiana*, and *Aconitum naviculare*, which have been reported to acquire nutrients via specialized strategies in alpine environments (Adhikari et al., 2022; Rathore et al., 2018; Sati et al., 2024; Sedej et al., 2020). Such adaptations allow alpine plants to endure nutrient-restrained substrates and withstand metabolic activities under edaphically severe circumstances.

Alpine areas are characterized by short growing periods due to low temperatures and long snow periods. This reduces plant development and shortens the life cycle and reproduction. Most of the plants in the alps have evolved by flowering quickly and finishing their reproductive phase in time to have seeds before the adverse conditions start (Sati et al., 2024).

Phenological events, including germination, flowering, and seed development, are highly dependent on temperature and precipitation patterns. Warmer temperatures can not only increase the growing season and plant growth, but also cause changes in species distribution and vegetation to move upwards in altitude (Li et al., 2013; Sati et al., 2024). Phenological changes can influence plant-pollinator interactions and ecosystem processes in alpine ecosystems.

3. ADAPTATIONS IN THE MORPHOLOGY OF THE ALPINE PLANTS

Plants in the alpine region are high-altitude species with high cold, high UV radiation, strong winds, brief growing periods, and soils with low nutrient levels (Körner & Kèorner, 1999; Nagy & Grabherr, 2009). Such abiotic stresses have also spurred the development of specific morphological characteristics that enhance survival, reproduction, and fitness under adverse conditions. Morphological adaptations tend to work together with physiological and developmental strategies to counter environmental limitations (Sun et al., 2014a). One of the characteristic features of alpine species is their dwarfism or small overall size. Small size is beneficial in terms of exposure to high-velocity winds and frost because it keeps tissues warmer in the close microclimate near the soil surface. Moreover, dwarf growth helps plants to grow quickly, allowing them to complete growth and reproduction in the short-growing season in the Alps (Yang et al., 2024). Resource conservation is another benefit of dwarfism, whereby smaller plants do not require as many carbohydrates and nutrients to grow in low-energy conditions. The growth form of cushion and rosettes is a very effective structural strategy of the alpine environments. Plants create wind, snow, heat losses, and moisture by creating low-lying compact mats that reduce mechanical stress. These constructions create microclimates in the canopy, increasing the temperatures compared to ambient air, and boosting metabolic rates in brief warm spurts (Birkeland et al., 2020; Sun et al., 2014a). Even cushion plants cover meristematic tissues in the middle of the structure to ensure that reproduction and vegetation growth are possible, even when the temperature is frozen at the ends. Leaves on the Alps often have thick hairs (pubescence) or thick waxy cuticles that serve various protective roles. Leaf hairs can be seen as insulation against cold and loss of water through transpiration and excess UV radiation, which can cause tissue damage. Waxy coats increase water retention, guard against drought, and cover epidermal cells against freezing and mechanical damage. These characteristics occur recurrently in alpine and Arctic species and serve to show convergent evolution caused by the same set of selective forces (Zhang et al., 2023). Another adaptation is the formation of small, thick leaves. A smaller leaf area reduces heat and water loss, and thickened mesophytic and epidermal cover supports the structure against ice, snow, and wind damage. The same leaves can be used to achieve greater photosynthetic efficiency per unit leaf area to sustain carbon gain even during the short growing season and at low temperatures (Yang et al., 2024). There is also thickening of the leaves along with pubescence, which improves UV protection, which is very important at high altitudes where the intensity of solar radiation is high.

Deeply rooted or highly branched root systems are common in alpine plants. These structures help maximize nutrient uptake in poor or shallow soils and store carbohydrates to grow rapidly during brief warm seasons. Root adaptations also enhance freeze-thaw and hypoxic soil resilience and aid in the maintenance of water and nutrient uptake when surface soils are frozen or poorly aerated (Chen et al., 2019). Certain species also exhibit clonal root development, enabling them to live and regenerate in unstable or fragmented substrates in the Alps. The effects of these morphological features are not independent of each other. Instead, they are synchronous plans that, in combination, increase survival and reproductive success in the alpine setting. An example is dwarfism with cushion growth, which minimizes wind exposure while simultaneously raising the canopy temperature. These structures are supplemented by leaf pubescence and thickening, which protect photosynthetic tissues and reduce water loss. Modifications to the root systems make the plant self-sufficient in terms of resources needed to support the existence of aboveground structures, even when there is a scarcity of nutrients and limited growing seasons. In general, a striking set of predictable convergent morphological features is evident in alpine plants of different lineages and geographic locations. The recurrently adaptive evolution of dwarfism, cushion-shaped growth, thickened pubescence, thickened leaves, and specialized root systems indicates the high selective pressures of such high-elevation environments, and alpine systems can serve as natural laboratories to investigate adaptation to severe environments (Birkeland et al., 2020; Sun et al., 2014a; Zhang et al., 2023).

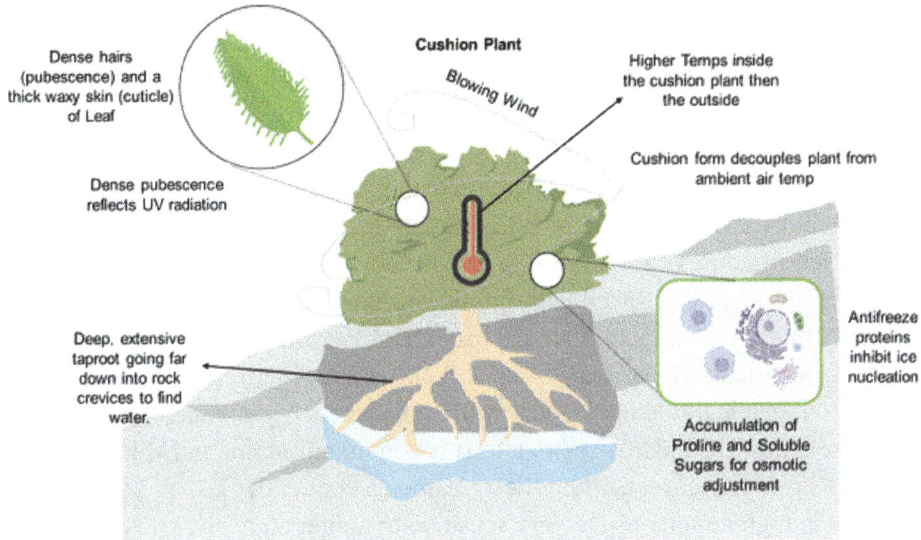

Figure 1 shows the anatomy of resilience in the alpine flora. This diagram illustrates the synthesis of structural adaptations (cushion form, deep rooting, and waxy cuticles) and physiological mechanisms (osmotic adjustment and antifreeze protein production) that allow plants to survive extreme cold, high radiation, and water deficit in high-mountain environments.

4. PHYSIOLOGICAL ADAPTATION TO ABIOTIC STRESS

Most high-mountain vegetation tends to be subjected to various, and sometimes concomitant, abiotic stress factors, such as extreme temperatures, high radiation, and scarce water and nutrients. To endure and prosper in this environment, these species have developed a set of physiological adaptations that make them more resilient to stress and more efficient in using the available resources (Körner & Kèorner, 1999; Winkler et al., 2016). Cold temperatures are one of the major selective powers in the high mountains of the alpine and Mediterranean regions. Plants mitigate low environmental temperatures in two primary ways: freezing avoidance and freezing tolerance. Freezing avoidance can be defined as avoiding ice formation in the cells through osmotic adjustment or transient super cooling, and freezing tolerance, which is the ability of the plant to survive the presence of extracellular ice by dehydrating protoplasts and reducing intracellular freezing points (Körner & Kèorner, 1999). These mechanisms are predominant in different species, altitudes, and habitats. European Mediterranean species tend to use avoidance mechanisms, and many alpine and Chilean high-mountain species of the Mediterranean depend on tolerance mechanisms (Pescador et al., 2016; Sierra-Almeida et al., 2009, 2010). It is important to note that freezing plasticity, which changes between avoidance and tolerance during the growing season, is common among many species to counter late-season frost incidences (Sierra-Almeida et al., 2016; Sklenář, 2017).

Plants inhabiting high mountains store compatible solutes, such as low-molecular-weight carbohydrates, proline, and other osmolytes, which are cryoprotectants that stabilize cell structures during freezing and osmotic stress (Dierking & Kallenbach, 2012; Hayat et al., 2012). Specifically, proline accumulation can stabilize mitochondrial electron transport, ameliorate photosynthetic performance during stress, and promote osmotic adjustment (Hayat et al., 2012). Non-structural carbohydrates (NSC) also play an identical role in osmotic regulation and energy storage, both of which contribute to drought and frost tolerance in high-mountain alpine and Mediterranean species (Körner & Kèorner, 1999; Sierra-Almeida et al., 2016).

Plants in high mountains are very sensitive to temperature and photosynthesis, but these species have a relatively high photosynthetic efficiency even under cold stress and extreme irradiance. These mechanisms include efficient PSII regulation, non-photochemical quenching (NPQ), smaller chloroplast and antenna sizes, and photoprotective pigment accumulation, which together maximize light utilization and dissipate surplus energy in a harmless manner (Bascuñán-Godoy et al., 2010; Germino & Smith, 2000; Loik et al., 2004). Photosystem performance plasticity is manifested by the rapid recovery of chlorophyll fluorescence (Fv/Fm) and CO_2 assimilation after cold exposure, as demonstrated by alpine plants (Briceño et al., 2014; Shi et al., 2004). In high-mountain Mediterranean species, both increased chlorophyll a/b ratios and increased concentrations of photoprotective pigments alleviate high light stress when water levels are low (Bernal et al., 2013).

Water constraints are an extremely important stressor in Mediterranean high mountains, especially during summer droughts. Plants increase water-use efficiency (WUE) through osmotic adjustment, stomatal regulation, and leaf morphology and physiology (Magaña Ugarte et al., 2019). Carbon isotope discrimination ($\delta^{13}C$) is an effective measure of WUE: higher $\delta^{13}C$ values indicate greater water-use efficiency under drought stress (Wen & Chen, 2014). The ability of high-mountain species in the Mediterranean to withstand long periods of water shortages through such adaptations enables them to maintain photosynthetic processes and reduce photo damage.

Soils in high mountains are usually nutrient-deprived, and plants must maximize nutrient uptake and metabolic efficiency. Among adaptations, there are higher nitrogen use efficiency, NSCs accretion, and selective distribution of nutrients to reproductive or storage tissues (García Lino et al., 2017; Pescador et al., 2015). As an illustration, NSCs accumulation contributes to the osmotic balancing and freezing endurance, especially in high-altitude cushion flora that experience a mixture of drought and cold stress (Sierra-Almeida et al., 2016). The cycling of nutrients, which is accompanied by an effective distribution of metabolism over time, makes higher growth and reproduction possible in the environment under restrictive conditions in high-mountain areas.

Table 1. Physiological and Morphological Adaptations to Alpine Stress

Environmental Stressor	Adaptive Strategy	Mechanism / Benefit	References
Freezing Temperatures	Freezing Tolerance	Accumulation of compatible solutes (proline, soluble sugars) to lower the freezing point of cell sap.	(Magaña Ugarte et al., 2019; Sati et al., 2024)
Freezing Temperatures	Antioxidant Activity	Increased activity of enzymes (SOD, CAT, APX) to scavenge Reactive Oxygen Species (ROS) caused by cold.	(Sati et al., 2024; Streb et al., 2003)
High Wind & Cold	Dwarfism	Reduction in plant height to remain in the warmer, less windy aerodynamic boundary layer near the soil.	(Rundel & Millar, 2016; X. Zhang et al., 2025)
High Wind & Cold	Cushion Growth Form	Compact shape traps heat, retains moisture, and protects meristems from mechanical abrasion.	(Anthelme & Dangles, 2012; Rundel & Millar, 2016)
High UV Radiation	Photoprotective Pigments	Synthesis of anthocyanins and flavonoids in epidermal tissue to screen harmful UV-B radiation.	(Sati et al., 2024; Spitaler et al., 2008)
High UV Radiation	Dense Pubescence	Thick layers of leaf hairs (trichomes) reflect excess solar radiation and insulate the leaf surface.	(Sun et al., 2014a; X. Zhang et al., 2025)
High Light Intensity	Photosynthetic Regulation	Rapid recovery of photosystem II (PSII) efficiency and Non-Photochemical Quenching (NPQ) to dissipate excess energy.	(Germino & Smith, 2000; Magaña Ugarte et al., 2019)
Drought / Desiccation	Osmotic Adjustment	Accumulation of Non-Structural Carbohydrates (NSC) to maintain cell turgor during water deficits.	(García Lino et al., 2017; Magaña Ugarte et al., 2019)
Nutrient Poor Soil	Root Morphology	Development of extensive or deep root systems to maximize nutrient uptake in rocky substrates.	(Sati et al., 2024; Shen et al., 2023)
Short Growing Season	Phenological Acceleration	Rapid completion of the reproductive cycle (flowering to seed set) immediately after snowmelt.	(Kameyama & Kudo, 2015; Sati et al., 2024)

5. REPRODUCTIVE AND PHENOLOGICAL ADAPTATIONS

Alpine plants are also known to wait a long time to flower and to have dormancy of the seed in order to avoid the dangers of early snowmelt and frost. Timing of snowmelt is highly important in determining flowering in phenology because it varies at the local scale as a result of topography (Kameyama & Kudo, 2015). Widely distributed species can also synchronize flowering in response to a gradient of snowmelt, which also lowers the intense competition of pollinators and seed development in favorable conditions (Kameyama & Kudo, 2015; Kasagi & Kudo, 2003).

The alpine environment and short growing seasons favor species that can reproduce in a short time. The plants start flowering, pollinating, and setting seed within a small time interval after snowmelt as many plants immediate bloom after snowmelt (Kameyama & Kudo, 2015; Lambert et al., 2010). In the case of *Phyllodoce aleutica*, seed set by this plant is greater in late-snowmelt environments where bumblebees are most active and pollen-limited when bumblebees do not visit early in flowering (Kameyama & Kudo, 2009).

In addition to sexual reproduction, vegetative propagation ensures the long-term survival of alpine plants. Genets can take decades to grow in clonal growth, even when there is slow growth and reproductive failure. Hybrid *Phyllodoce caerulea f. yesoensis* exhibits extensive clonal expansion, and is able to retain its population prevalence where reproduction is restricted (Kameyama et al., 2008).

The dispersal mechanisms of seeds among alpine plants are adjusted to short growing conditions and rocky terrain. Dormant seeds are usually placed on a schedule to achieve ideal germination, and breeding mechanisms such as cryptic self-incompatibility encourage outcrossing and minimize the adverse consequences of self-fertilization or hybridization (Kameyama & Kudo, 2009). Such strategies are certain to achieve a successful presence in appropriate microsites and lead to population stability.

6. MUTUALISTIC STRATEGIES AND ECOLOGICAL INTERACTIONS

Tropical alpine ecosystems (TAE) are characterized by distinctive plant structures, including giant rosettes, cushions, and tussock grasses, which affect pollination. These plants have long floral stages and high levels of floral show, which balance the low levels of pollinators found in high-altitude areas (Anthelme & Dangles, 2012). The ontogenic niche changes that occur with the species also influence pollination patterns: older, taller plants may receive a pollination visit by smaller or juvenile plants, indirectly promoting reproductive success throughout the community (Anthelme &

Dangles, 2012). Although direct experimental research on pollination networks in TAE is yet to be conducted, patterns of spatial associations reveal that facilitative interactions mediated by plant size and floral show are probably important in sustaining pollination services in high-stress environments such as these ecosystems.

There is very little direct research on mycorrhizal symbioses in TAE, but nutrient acquisition strategies are directly connected to plant facilitation. Potted plants (e.g., *Azorella aretioides*) or giant rosettes (e.g., *Espeletia schultzii*) alter the microenvironment of the soil by enhancing the availability of nutrients and water (Anthelme & Dangles, 2012). These changes help seedlings and other understory species to grow, which may explain the importance of belowground mutualism, such as mycorrhizal networks, in promoting nutrient uptake. Root system complementarity, especially between co-existing forms of growth, such as tussocks and shrubs, can also maximize nutrient partitioning and minimize competition (Kleier & Lambrinos, 2005), implying that belowground mutualistic interactions play an important role in community organization in TAE.

TAE are very stressful conditions with high UV radiation, high freeze-thaw cycles, a seasonable climatic conditions, and erratic precipitation. In line with the stress-gradient hypothesis (SGH), positive plant-plant interactions, especially nurse-plant-mediated interactions, are common in TAE. Giant cushions, giant rosettes, shrubs, and tussock grasses ameliorate microhabitats by:

- Minimizing the effects of frost-heaving and needle-ice (Pérez, 1987a; (Haussmann et al., 2009).
- Control of temperature and humidity on the surface of the soil (Anthelme & Dangles, 2012).
- Improving the nutrient supply of soil (Anthelme & Dangles, 2012; Mizuno, 1998).
- Promotion of primary succession in recently deglaciated land (Mizuno, 1998).

Spatial studies have demonstrated that facilitation is species-specific and scale-related. In the case of older *Senecio keniodendron* in East Africa, the community can be established by older individuals, and in the same case, smaller cushions or tussocks might assist with seedlings in the area (Patty et al., 2010; Young & Peacock, 1992). These results show that nurse plants are critical for maintaining TAE biodiversity and community stability.

Although facilitative ones are predominant in the case of extreme stress, competition is not eliminated. Spatial patterns may be affected by intraspecific competition among conspecific adults and interspecific competition over light and soil (Anthelme & Dangles, 2012). The con- colony of various growth forms, such as shrubs, tussocks, and rosettes, enhances niche differentiation through the temporal, structural, and functional segregation of resources. The presence of long-lived species with different architectural features minimizes competition overlap, and positive interactions dominate at the community level (Callaway, 2007; Soliveres et al., 2010). In addition, the stochastic processes may also increase the beta-diversity and endemism, owing to the high habitat fragmentation and low dispersal, which contribute further to community structure in TAE (Sklenář et al., 2011).

Figure 2. Mechanism of Facilitation in Alpine Ecosystems

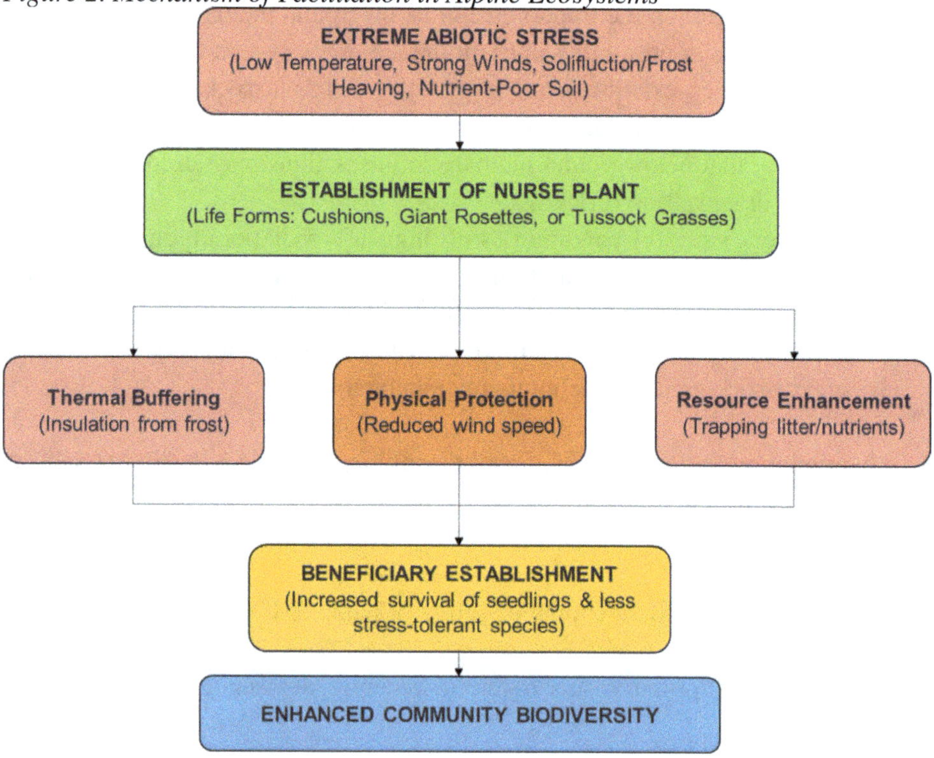

7. ADAPTATION STRATEGIES TO CLIMATE CHANGE

Climate warming has a negative effect on alpine plants, which tend to move uphill or narrow their range (Hughes & Atchison, 2015; Sun et al., 2014b). The most exposed species are those at the highest altitudes, as they may lose all their habitats (Wootton et al., 2025). Dispersal is also reduced by the fragmentation of alpine landscapes, which worsens the risk of local extinction in the case of rapid environmental change (Steinbauer et al., 2018; X. Zhang et al., 2025). Flowering-related changes in climate may lead to a lack of fidelity to pollinators and a decrease in reproductive success (Wang et al., 2021; Zhang et al., 2021). These phenological shifts can have an impact on plant-pollinator interactions, community structure and ecosystem processes, but long-term empirical data in alpine systems remains scarce. The genetic strategies used by alpine plants to adapt to warming include:

- **Adaptive SNPs:** Local adaptation to elevational gradients is based on single nucleotide polymorphisms. For example, *Arabis alpina* species are differentiated by freezing tolerance loci and stress resistance loci (Lobréaux & Miquel, 2020; Rellstab et al., 2020).
- **Whole-Genome Duplication (WGD) and Gene Family Evolution:** Both the WGD and the expansion of gene families offer raw material for long-term adaptation. *Megacarpaea delavayi, Rhododendron nivale*, and *Lepidium meyenii* have duplications and expansions in stress-response genes, DNA-repair genes, and UV-protection genes (Lyu et al., 2024; Yang et al., 2020; Zhang et al., 2016).
- **Structural Variants (SVs):** Gene dosage or regulatory sequences may be changed by insertions, deletions, and rearrangements, which may affect adaptive characteristics; however, there is currently little functional evidence of this in alpine plants (Kang et al., 2023; Wang et al., 2012).
- **Transposable Elements (TEs):** TEs can induce regulatory variation and stress gene expression. A TE insertion upstream of bHLH35 is associated with leaf color change in *Corydalis hemidicentra*, which may be an adaptive advantage under high UV conditions (H. Zhang et al., 2025). TE-mediated adaptation is reasonable for alpine plants, but there is little direct causal support.

Such processes frequently work in combination: SNP shifts and phenotypic plasticity can be the components of short-term adaptation, whereas WGD, gene family reorganization, and TE-regulatory changes play a role in long-term adaptation (X. Zhang et al., 2025). The interaction between distributional changes, phenological changes and genomic responses to climate change has significant effects on the alpine biodiversity. Migrations and adaptation barriers might cause local extinction of species, resulting in a decreased genetic diversity and a shift in the composition of the community (Steinbauer et al., 2018; Wootton et al., 2025). Phenological incompatibilities can disrupt pollination networks and ecosystem services, such as nutrient cycling and seed production. Constant observation and the combination of ecological, phenological, and genomic studies play an important role in predicting and reducing the effects of climate change on the alpine ecosystem.

Table 2. Genomic and Evolutionary Mechanisms of Adaptation

Genetic Mechanism	Description	Adaptive Outcome	References
SNP Variation	Single Nucleotide Polymorphisms in stress-responsive genes.	Fine-tuning of genes related to hypoxia, UV protection, and freezing tolerance (*e.g., Arabis alpina*).	(Lobréaux & Miquel, 2020; X. Zhang et al., 2025)
Whole-Genome Duplication (WGD)	Polyploidy events (doubling of chromosome sets).	Creates genetic redundancy, allowing "spare" gene copies to evolve new stress-tolerance functions.	(Yang et al., 2020; X. Zhang et al., 2025)
Gene Family Expansion	Increase in copy number of specific gene families.	Enhanced synthesis of DNA repair enzymes and secondary metabolites (flavonoids) for UV defense.	(Sun et al., 2023; X. Zhang et al., 2025)
Gene Family Contraction	Loss of specific gene families (e.g., immune receptors).	Energy trade-off: Reduces investment in pathogen defense to allocate resources to abiotic stress tolerance.	(Guo et al., 2018; X. Zhang et al., 2025)
Transposable Elements (TEs)	Insertion of "jumping genes" near regulatory regions.	Alters gene expression in response to stress (e.g., enhancing anthocyanin production in *Corydalis*).	(Feng et al., 2023; X. Zhang et al., 2025)
Structural Variants (SVs)	Large-scale genomic changes (insertions/ deletions).	Rapidly alters phenotypes, such as drought resistance or root depth, by changing gene dosage.	(Kang et al., 2023; X. Zhang et al., 2025)
Adaptive Introgression	Gene flow between species via hybridization.	Transfer of beneficial "pre-tested" alleles (e.g., cold tolerance) from local to colonizing species.	(Ma et al., 2019; X. Zhang et al., 2025)
Convergent Evolution	Independent evolution of similar genetic pathways.	Distantly related species recruit the same DNA repair pathways to survive high-elevation radiation.	(Birkeland et al., 2020; X. Zhang et al., 2025)
Cryptic Self-Incompatibility	Genetic barrier prioritizing outcross pollen over self-pollen.	Prevents inbreeding depression while allowing reproductive assurance in variable pollinator environments.	(Kameyama & Kudo, 2015)
Epigenetic Modification	Methylation changes (e.g., CHH/CHG contexts).	Regulates gene expression in response to temperature shocks without altering the DNA sequence.	(Quadrana et al., 2016; X. Zhang et al., 2025)

8. CONSERVATION AND FUTURE RESEARCH DIRECTIONS

The topography and climatic heterogeneity of the mountains is special and the flora in the mountains is very diverse and also very sensitive to the anthropogenic and climatic pressures (Khan et al., 2013). The conservation strategies should also combine in situ and ex situ solutions, such as protection of areas, reforestation efforts, as well as preservation of high altitude habitats, including timberline zones. Meditative utilization practices are needed to ensure that medicinal and economically important species are not overexploited because there is a high level of exploitation pressure. The community participation is essential because the local knowledge could assist in the identification of high-ecological and cultural species (Khan et al., 2013).

Physiological ecology offers critical details on how the alpine species react to the environment-related stressors like changes in temperature, precipitation, and elevation gradients (Hester & Brooker, 2007; Khan et al., 2013). The ecological amplitudes of many alpine and montane species are narrow, and thus species are very sensitive to the changes in climate and the disturbance of the habitats. Predictive models have been used to determine the species that are vulnerable to range shrinkage or local extinction by integrating the information on species distribution, abundance, and physiological limits to provide conservation priorities (Beckage et al., 2008; Dobbertin et al., 2005). The combination of phytosociological indices (e.g., IVI) and ethnobotanical use values (UVs) can also be used as a supplement to gain better insight into the anthropogenic pressure on plant communities and dictate adaptive management approaches (Khan et al., 2013).

Alpine ecosystems are understudied in most areas because of remoteness, geopolitical limitations and lack of scientific interest. The significant knowledge gaps are:

- Extensive recording of biodiversity of high altitude especially endemic and threatened species.
- Dynamic long-term observation of the ecosystems services, and vegetation in relation to climate change and human stressors.
- Combination of traditional ecological knowledge and quantitative ecological information to discover priority species and habitat to be conserved.

Future studies ought to be conducted on participatory strategies, which entail integrating remote sensing, phytosociology, and ethnobotanical surveys to produce powerful units of ecosystem management and restoration.

Successful management of the alpine ecosystems needs policy frameworks focusing on ecological as well as socio-economic aspects. Geopolitical instability, population pressure, and low levels of environmental regulation enforcement in the western Himalayas and in the mountains of Central Asia hamper conservation in the region (Gorenflo & Brandon, 2005; Sandwith, 2003). The second priority of policymakers should be the development of Red Data Books and national biodiversity lists in order to have threatened species and habitats listed systematically. Community based resource management, awareness and incentives to sustainable use should be included in management methods that continue to provide the services with limiting overexploitation (Khan et al., 2013). The use of traditional knowledge in formal conservation planning may improve both the ecology and the livelihoods of the local people and lead to long-term sustainability of alpine biodiversity (Khan et al., 2013).

9. CONCLUSION

The physiological, morphological, reproductive, and ecological adaptations of the alpine plants are very striking and allow them to survive and reproduce in the harsh environmental conditions. Such adaptations as dwarfism, cushion and rosette forms of growth, dense pubescence, cold hardiness, the presence of antifreeze proteins, rapid reproductive cycles, and mutualistic interactions with pollination and mycorrhizal fungi all indicate the complexity of the adaptive strategies that have been developed in response to low temperatures, high UV radiation, high winds, nutrient-poor soils, and short growth periods. These characteristics are helpful in not only facilitating the survival of individuals, but also community organization, species dynamics and ecosystem dynamics in the alps. Alpine ecosystems are also faced with major challenges of climate change, whereby the warming tendency leads to upward migration, shrinking ranges, phenological disconnection, and biotic interactions. Potential pathways to adaptive responses exist through genomic and phenotypic plasticity, SNP variation, and gene duplications, transposable element activity, but lots of high-elevation species still are highly vulnerable to the rapid environmental change. The conservation of the alpine flora should be a multifaceted approach combining in situ and ex situ conservation, sustainable utilization approaches, and community involvement and policy interventions. Physiologically ecological predictive modeling, long-term observation, and integration of traditional ecological understanding are necessary to detect species at risk and inform adaptive management.

In summary, alpine plants are an extremely specialized and robust part of high mountain ecosystems that can provide precious data on adaptation to extreme stress. It is important to preserve their diversity and ecological role to retain biodiversity and ecosystem services, as well as to understand the evolutionary mechanisms and formulate sustainable management strategies within the rapidly changing climatic conditions. Further studies which combine physiological, ecological and genomic insights will be central in making sure that these distinct ecosystems are here to stay during the anthropocene.

REFERENCES

Abatzoglou, J. T., Redmond, K. T., & Edwards, L. M. (2009). Classification of Regional Climate Variability in the State of California. *Journal of Applied Meteorology and Climatology*, *48*(8), 1527–1541. DOI: 10.1175/2009JAMC2062.1

Adhikari, P., Joshi, K., Singh, M., & Pandey, A. (2022). Influence of altitude on secondary metabolites, antioxidants, and antimicrobial activities of Himalayan yew (Taxus wallichiana). *Plant Biosystems*, *156*(1), 187–195. DOI: 10.1080/11263504.2020.1845845

Agee, J. K. (1993). *Fire ecology of Pacific Northwest forests* (Vol. 499). Island press Washington, DC.

Anthelme, F., & Dangles, O. (2012). Plant–plant interactions in tropical alpine environments. *Perspectives in Plant Ecology, Evolution and Systematics*, *14*(5), 363–372. DOI: 10.1016/j.ppees.2012.05.002

BAHUGUNA, R., PRAKASH, V., & BISHT, H. (2013). QUANTITATIVE ENHANCEMENT OF ACTIVE CONTENT AND BIOMASS OF TWO ACONITUM SPECIES THROUGH SUITABLE CULTIVATION TECHNOLOGY. *International Journal of Conservation Science*, *4*(1).

Barber, A. L. (2013). *Physiology and early life-history associated with extreme longevity: an investigation of Pinus longaeva (Great Basin bristlecone pine)*. University of California.

Bascuñán-Godoy, L., García-Plazaola, J. I., Bravo, L. A., & Corcuera, L. J. (2010). Leaf functional and micro-morphological photoprotective attributes in two ecotypes of Colobanthus quitensis from the Andes and Maritime Antarctic. *Polar Biology*, *33*(7), 885–896. DOI: 10.1007/s00300-010-0765-4

Beckage, B., Osborne, B., Gavin, D. G., Pucko, C., Siccama, T., & Perkins, T. (2008). A rapid upward shift of a forest ecotone during 40 years of warming in the Green Mountains of Vermont. *Proceedings of the National Academy of Sciences of the United States of America*, *105*(11), 4197–4202. DOI: 10.1073/pnas.0708921105 PMID: 18334647

Bernal, M., Llorens, L., Julkunen-Tiitto, R., Badosa, J., & Verdaguer, D. (2013). Altitudinal and seasonal changes of phenolic compounds in Buxus sempervirens leaves and cuticles. *Plant Physiology and Biochemistry*, *70*, 471–482. DOI: 10.1016/j.plaphy.2013.06.012 PMID: 23845826

Birkeland, S., Gustafsson, A. L. S., Brysting, A. K., Brochmann, C., & Nowak, M. D. (2020). Multiple Genetic Trajectories to Extreme Abiotic Stress Adaptation in Arctic Brassicaceae. *Molecular Biology and Evolution, 37*(7), 2052–2068. DOI: 10.1093/molbev/msaa068 PMID: 32167553

Briceño, V. F., Harris-Pascal, D., Nicotra, A. B., Williams, E., & Ball, M. C. (2014). Variation in snow cover drives differences in frost resistance in seedlings of the alpine herb Aciphylla glacialis. *Environmental and Experimental Botany, 106*, 174–181. DOI: 10.1016/j.envexpbot.2014.02.011

Callaway, R. M. (2007). *Positive interactions and interdependence in plant communities* (Vol. 415). Springer.

Cayan, D. R., Redmond, K. T., & Riddle, L. G. (1999). ENSO and Hydrologic Extremes in the Western United States. *Journal of Climate, 12*(9), 2881–2893. DOI: 10.1175/1520-0442(1999)012<2881:EAHEIT>2.0.CO;2

Chen, J., Huang, Y., Brachi, B., Yun, Q., Zhang, W., Lu, W., Li, H., Li, W., Sun, X., Wang, G.-y., He, J., Zhou, Z., Chen, K.-y., Ji, Y., Shi, M., Sun, W., Yang, Y., Zhang, R., Abbott, R. J., & Sun, H. (2019). Genome-wide analysis of Cushion willow provides insights into alpine plant divergence in a biodiversity hotspot. *Nature Communications, 10*(1), 5230. DOI: 10.1038/s41467-019-13128-y PMID: 31745089

Dierking, R. M., & Kallenbach, R. L. (2012). Mediterranean and Continental Tall Fescue: II. Effects of Cold, Nonfreezing Temperatures on Leaf Extension, Proline, Fructan, and Abscisic Acid. *Crop Science, 52*(1), 460–469. DOI: 10.2135/cropsci2011.03.0160

Dobbertin, M., Hilker, N., Rebetez, M., Zimmermann, N. E., Wohlgemuth, T., & Rigling, A. (2005). The upward shift in altitude of pine mistletoe (Viscum album ssp. austriacum) in Switzerland—The result of climate warming? *International Journal of Biometeorology, 50*(1), 40–47. DOI: 10.1007/s00484-005-0263-5 PMID: 15875222

Dolanc, C. R., Thorne, J. H., & Safford, H. D. (2013). Widespread shifts in the demographic structure of subalpine forests in the Sierra Nevada, California, 1934 to 2007. *Global Ecology and Biogeography, 22*(3), 264–276. DOI: 10.1111/j.1466-8238.2011.00748.x

Feng, T., Pucker, B., Kuang, T., Song, B., Yang, Y., Lin, N., Zhang, H., Moore, M. J., Brockington, S. F., Wang, Q., Deng, T., Wang, H., & Sun, H. (2023). The genome of the glasshouse plant noble rhubarb (Rheum nobile) provides a window into alpine adaptation. *Communications Biology, 6*(1), 706. DOI: 10.1038/s42003-023-05044-1 PMID: 37429977

Fites-Kaufman, J., Rundel, P., Stephenson, N., & Weixelman, D. (2007). *Montane and Subalpine Vegetation of the Sierra Nevada and Cascade Ranges*. Terrestrial Vegetation of California., DOI: 10.1525/california/9780520249554.003.0017

García Lino, M. C., Cavieres, L. A., Zotz, G., & Bader, M. Y. (2017). Carbohydrate reserves in the facilitator cushion plant Laretia acaulis suggest carbon limitation at high elevation and no negative effects of beneficiary plants. *Oecologia*, *183*(4), 997–1006. DOI: 10.1007/s00442-017-3840-5 PMID: 28233055

García-Plazaola, J. I., Rojas, R., Christie, D. A., & Coopman, R. E. (2015). Photosynthetic responses of trees in high-elevation forests: Comparing evergreen species along an elevation gradient in the Central Andes. *AoB Plants*, *7*, plv058. Advance online publication. DOI: 10.1093/aobpla/plv058 PMID: 26002745

Germino, M. J., & Smith, W. K. (2000). High resistance to low-temperature photoinhibition in two alpine, snowbank species. *Physiologia Plantarum*, *110*(1), 89–95. DOI: 10.1034/j.1399-3054.2000.110112.x

Gorenflo, L. J., & Brandon, K. (2005). Agricultural capacity and conservation in high biodiversity forest ecosystems. *Ambio*, *34*(3), 199–204. DOI: 10.1579/0044-7447-34.3.199 PMID: 16042277

Guo, X., Hu, Q., Hao, G., Wang, X., Zhang, D., Ma, T., & Liu, J. (2018). The genomes of two Eutrema species provide insight into plant adaptation to high altitudes. *DNA Research : an International Journal for Rapid Publication of Reports on Genes and Genomes*, *25*(3), 307–315. DOI: 10.1093/dnares/dsy003 PMID: 29394339

Haussmann, N. S., Boelhouwers, J. C., & McGeoch, M. A. (2009). Fine scale variability in soil frost dynamics surrounding cushions of the dominant vascular plant species (azorella selago) on sub-antarctic marion island. *Geografiska Annaler. Series A. Physical Geography*, *91*(4), 257–268. DOI: 10.1111/j.1468-0459.2009.00368.x

Hayat, S., Hayat, Q., Alyemeni, M. N., Wani, A. S., Pichtel, J., & Ahmad, A. (2012). Role of proline under changing environments. *Plant Signaling & Behavior*, *7*(11), 1456–1466. DOI: 10.4161/psb.21949 PMID: 22951402

Hester, A., & Brooker, R. (2007). Threatened habitats: Marginal vegetation in upland areas. *Issues in Environmental Science and Technology*, *25*, 107–134. DOI: 10.1039/9781847557650-00107

Hou DaBin, H. D., & Shu, GuangMing, S. G., & Ren ZhenLong, R. Z. (2006). Effect of nitrogen, phosphorus, and potassium on several nitrogen metabolism enzymes of Aconitum carmichaeli Dexb. (Ranunculaceae). *Xi Nan Nong Ye Xue Bao*, *19*(5), 857–862.

Hughes, C. E., & Atchison, G. W. (2015). The ubiquity of alpine plant radiations: From the Andes to the Hengduan Mountains. *The New Phytologist, 207*(2), 275–282. DOI: 10.1111/nph.13230 PMID: 25605002

Kameyama, Y., Kasagi, T., & Kudo, G. (2008). A hybrid zone dominated by fertile F1s of two alpine shrub species, Phyllodoce caerulea and Phyllodoce aleutica, along a snowmelt gradient. *Journal of Evolutionary Biology, 21*(2), 588–597. DOI: 10.1111/j.1420-9101.2007.01476.x PMID: 18205785

Kameyama, Y., & Kudo, G. (2009). Flowering phenology influences seed production and outcrossing rate in populations of an alpine snowbed shrub, Phyllodoce aleutica: Effects of pollinators and self-incompatibility. *Annals of Botany, 103*(9), 1385–1394. DOI: 10.1093/aob/mcp037 PMID: 19228703

Kameyama, Y., & Kudo, G. (2015). Intrinsic and extrinsic factors acting on the reproductive process in alpine-snowbed plants: Roles of phenology, biological interaction, and breeding system. *Plant Species Biology, 30*(1), 3–15. DOI: 10.1111/1442-1984.12070

Kang, M., Wu, H., Liu, H., Liu, W., Zhu, M., Han, Y., Liu, W., Chen, C., Song, Y., Tan, L., Yin, K., Zhao, Y., Yan, Z., Lou, S., Zan, Y., & Liu, J. (2023). The pangenome and local adaptation of Arabidopsis thaliana. *Nature Communications, 14*(1), 6259. DOI: 10.1038/s41467-023-42029-4 PMID: 37802986

Kasagi, T., & Kudo, G. (2003). Variations in bumble bee preference and pollen limitation among neighboring populations: Comparisons between Phyllodoce caerulea and Phyllodoce aleutica (Ericaceae) along snowmelt gradients. *American Journal of Botany, 90*(9), 1321–1327. DOI: 10.3732/ajb.90.9.1321 PMID: 21659231

Khan, S. M., Page, S. E., Ahmad, H., & Harper, D. M. (2013). Sustainable utilization and conservation of plant biodiversity in montane ecosystems: The western Himalayas as a case study. *Annals of Botany, 112*(3), 479–501. DOI: 10.1093/aob/mct125 PMID: 23825353

Kleier, C., & Lambrinos, J. G. (2005). The Importance of Nurse Associations for Three Tropical Alpine Life Forms. *Arctic, Antarctic, and Alpine Research, 37*(3), 331–336. DOI: 10.1657/1523-0430(2005)037[0331:TIONAF]2.0.CO;2

Körner, C., & Kèorner, C. (1999). *Alpine plant life: functional plant ecology of high mountain ecosystems* (Vol. 4). Springer. DOI: 10.1007/978-3-642-98018-3

Körner, C., & Paulsen, J. (2004). A world-wide study of high altitude treeline temperatures. *Journal of Biogeography, 31*(5), 713–732. DOI: 10.1111/j.1365-2699.2003.01043.x

Lambert, A. M., Miller-Rushing, A. J., & Inouye, D. W. (2010). Changes in snowmelt date and summer precipitation affect the flowering phenology of Erythronium grandiflorum (glacier lily; Liliaceae). *American Journal of Botany*, *97*(9), 1431–1437. DOI: 10.3732/ajb.1000095 PMID: 21616897

Li, R., Luo, T., Tang, Y., Du, M., & Zhang, X. (2013). The altitudinal distribution center of a widespread cushion species is related to an optimum combination of temperature and precipitation in the central Tibetan Plateau. *Journal of Arid Environments*, *88*, 70–77. DOI: 10.1016/j.jaridenv.2012.07.018

Lobréaux, S., & Miquel, C. (2020). Identification of Arabis alpina genomic regions associated with climatic variables along an elevation gradient through whole genome scan. *Genomics*, *112*(1), 729–735. DOI: 10.1016/j.ygeno.2019.05.008 PMID: 31085222

Loik, M. E., Still, C. J., Huxman, T. E., & Harte, J. (2004). In situ photosynthetic freezing tolerance for plants exposed to a global warming manipulation in the Rocky Mountains, Colorado, USA. *The New Phytologist*, *162*(2), 331–341. DOI: 10.1111/j.1469-8137.2004.01002.x

Lyu, Z.-Y., Zhou, X.-L., Wang, S.-Q., Yang, G.-M., Sun, W.-G., Zhang, J.-Y., Zhang, R., & Shen, S.-K. (2024). The first high-altitude autotetraploid haplotype-resolved genome assembled (Rhododendron nivale subsp. boreale) provides new insights into mountaintop adaptation. *GigaScience*, *13*, giae052. DOI: 10.1093/gigascience/giae052 PMID: 39110622

Ma, L., Yang, L., Zhao, J., Wei, J., Kong, X., Wang, C., Zhang, X., Yang, Y., & Hu, X. (2015). Comparative proteomic analysis reveals the role of hydrogen sulfide in the adaptation of the alpine plant Lamiophlomis rotata to altitude gradient in the Northern Tibetan Plateau. *Planta*, *241*(4), 887–906. DOI: 10.1007/s00425-014-2209-9 PMID: 25526962

Ma, Y., Wang, J., Hu, Q., Li, J., Sun, Y., Zhang, L., Abbott, R. J., Liu, J., & Mao, K. (2019). Ancient introgression drives adaptation to cooler and drier mountain habitats in a cypress species complex. *Communications Biology*, *2*(1), 213. DOI: 10.1038/s42003-019-0445-z PMID: 31240251

Magaña Ugarte, R., Escudero, A., & Gavilán, R. G. (2019). Metabolic and physiological responses of Mediterranean high-mountain and alpine plants to combined abiotic stresses. *Physiologia Plantarum*, *165*(2), 403–412. DOI: 10.1111/ppl.12898 PMID: 30536685

Mantua, N. J., Hare, S. R., Zhang, Y., Wallace, J. M., & Francis, R. C. (1997). A Pacific Interdecadal Climate Oscillation with Impacts on Salmon Production*. *Bulletin of the American Meteorological Society, 78*(6), 1069–1080. DOI: 10.1175/1520-0477(1997)078<1069:APICOW>2.0.CO;2

Millar, C. I. (2012). Geologic, climatic, and vegetation history of California. *In: Baldwin, Bruce G.; Goldman, Douglas H.; Keil, David J.; Patterson, Robert; Rosatti, Thomas J.; Wilken, Dieter H., editors. The Jepson Manual: Vascular Plants of California. 2nd Edition. University of California Press. pp. 49-67.*, 49-67.

Mizuno, K. (1998). Succession Processes of Alpine Vegetation in Response to Glacial Fluctuations of Tyndall Glacier, Mt. Kenya, Kenya. *Arctic and Alpine Research, 30*(4), 340–348. DOI: 10.1080/00040851.1998.12002909

Nagelmüller, S., Hiltbrunner, E., & Körner, C. (2017). Low temperature limits for root growth in alpine species are set by cell differentiation. *AoB Plants, 9*(6), plx054. Advance online publication. DOI: 10.1093/aobpla/plx054 PMID: 29218137

Nagy, L., & Grabherr, G. (2009). *The biology of alpine habitats.* Oxford University Press. DOI: 10.1093/oso/9780198567035.001.0001

Patty, L., Halloy, S. R. P., Hiltbrunner, E., & Körner, C. (2010). Biomass allocation in herbaceous plants under grazing impact in the high semi-arid Andes. *Flora - Morphology, Distribution, Functional Ecology of Plants, 205*(10), 695-703. DOI: 10.1016/j.flora.2009.12.039

Pescador, D. S., de Bello, F., Valladares, F., & Escudero, A. (2015). Plant Trait Variation along an Altitudinal Gradient in Mediterranean High Mountain Grasslands: Controlling the Species Turnover Effect. *PLoS One, 10*(3), e0118876. DOI: 10.1371/journal.pone.0118876 PMID: 25774532

Pescador, D. S., Sierra-Almeida, Á., Torres, P. J., & Escudero, A. (2016). Summer Freezing Resistance: A Critical Filter for Plant Community Assemblies in Mediterranean High Mountains [Original Research]. *Frontiers in Plant Science, Volume 7 - 2016.* DOI: 10.3389/fpls.2016.00194

Quadrana, L., Bortolini Silveira, A., Mayhew, G. F., LeBlanc, C., Martienssen, R. A., Jeddeloh, J. A., & Colot, V. (2016). The Arabidopsis thaliana mobilome and its impact at the species level. *elife, 5*, e15716.

Rathore, N., Thakur, D., & Chawla, A. (2018). Seasonal variations coupled with elevation gradient drives significant changes in eco-physiological and biogeochemical traits of a high altitude evergreen broadleaf shrub, Rhododendron anthopogon. *Plant Physiology and Biochemistry, 132*, 708–719. DOI: 10.1016/j.plaphy.2018.08.009 PMID: 30150110

Rawat, J. M., Rawat, B., Mishra, S., Negi, R. K., Mishra, S. N., Chandra, A., & Nautiyal, S. (2014). Altitudinal and seasonal variation in bioactive compound aconitine in Aconitum violaceum, a threatened medicinal plant of Indian Himalayan region. *International Journal of Advanced Research, 2*(10), 981–988.

Rellstab, C., Zoller, S., Sailer, C., Tedder, A., Gugerli, F., Shimizu, K. K., Holderegger, R., Widmer, A., & Fischer, M. C. (2020). Genomic signatures of convergent adaptation to Alpine environments in three Brassicaceae species. *Molecular Ecology, 29*(22), 4350–4365. DOI: 10.1111/mec.15648 PMID: 32969558

Rundel, P., & Millar, C. (2016). Alpine ecosystems. *In: Zavaleta, E.; Mooney, H., eds. Ecosystems of California. Berkeley, California: University of California Press: 613-634. Chapter 29.*, 613-634.

Sandwith, T. (2003). Overcoming barriers: Conservation and development in the Maloti-Drakensberg Mountains of Southern Africa. *Journal of Sustainable Forestry, 17*(1-2), 149–169. DOI: 10.1300/J091v17n01_09

Sati, P., Dimri, S., Trivedi, V. L., & Nautiyal, M. (2024). Environmental factors shaping alpine plant adaptations.

Sedej, T. T., Erznožnik, T., & Rovtar, J. (2020). Effect of UV radiation and altitude characteristics on the functional traits and leaf optical properties in Saxifraga hostii at the alpine and montane sites in the Slovenian Alps. *Photochemical & Photobiological Sciences, 19*(2), 180–192. DOI: 10.1039/c9pp00032a PMID: 31956889

Shen, N., Xu, W., Guo, F., Cui, Y., Chen, Y., & Su, X. (2023). Effect of fertilizer and harvesting stage on growth and bioactive compounds content of Rheum tanguticum. *Journal of Plant Nutrition, 46*(18), 4462–4476. DOI: 10.1080/01904167.2023.2232407

Shi, S., Shi, R., Li, T., & Zhou, D. (2022). UV-B Radiation Effects on the Alpine Plant Kobresia humilis in a Qinghai-Tibet Alpine Meadow. *Plants, 11*(22), 3102. DOI: 10.3390/plants11223102 PMID: 36432831

Shi, S.-B., Zhu, W.-Y., Li, H.-M., Zhou, D.-W., Han, F., Zhao, X.-Q., & Tang, Y.-H. (2004). Photosynthesis of Saussurea superba and Gentiana straminea is not reduced after long-term enhancement of UV-B radiation. *Environmental and Experimental Botany, 51*(1), 75–83. DOI: 10.1016/S0098-8472(03)00062-5

Sierra-Almeida, A., Cavieres, L. A., & Bravo, L. A. (2009). Freezing resistance varies within the growing season and with elevation in high-Andean species of central Chile. *The New Phytologist, 182*(2), 461–469. DOI: 10.1111/j.1469-8137.2008.02756.x PMID: 19210722

Sierra-Almeida, A., Cavieres, L. A., & Bravo, L. A. (2010). Freezing resistance of high-elevation plant species is not related to their height or growth-form in the Central Chilean Andes. *Environmental and Experimental Botany, 69*(3), 273–278. DOI: 10.1016/j.envexpbot.2010.04.015

Sierra-Almeida, A., Reyes-Bahamonde, C., & Cavieres, L. A. (2016). Drought increases the freezing resistance of high-elevation plants of the Central Chilean Andes. *Oecologia, 181*(4), 1011–1023. DOI: 10.1007/s00442-016-3622-5 PMID: 27053321

Sklenář, P. (2017). Seasonal variation of freezing resistance mechanisms in north-temperate alpine plants. *Alpine Botany, 127*(1), 31–39. DOI: 10.1007/s00035-016-0174-6

Sklenář, P., Dušková, E., & Balslev, H. (2011). Tropical and Temperate: Evolutionary History of Páramo Flora. *Botanical Review, 77*(2), 71–108. DOI: 10.1007/s12229-010-9061-9

Soliveres, S., DeSoto, L., Maestre, F. T., & Olano, J. M. (2010). Spatio-temporal heterogeneity in abiotic factors modulate multiple ontogenetic shifts between competition and facilitation. *Perspectives in Plant Ecology, Evolution and Systematics, 12*(3), 227–234. DOI: 10.1016/j.ppees.2010.02.003

Spitaler, R., Winkler, A., Lins, I., Yanar, S., Stuppner, H., & Zidorn, C. (2008). Altitudinal variation of phenolic contents in flowering heads of Arnica montana cv. ARBO: A 3-year comparison. *Journal of Chemical Ecology, 34*(3), 369–375. DOI: 10.1007/s10886-007-9407-x PMID: 18259818

Steinbauer, M. J., Grytnes, J.-A., Jurasinski, G., Kulonen, A., Lenoir, J., Pauli, H., Rixen, C., Winkler, M., Bardy-Durchhalter, M., Barni, E., Bjorkman, A. D., Breiner, F. T., Burg, S., Czortek, P., Dawes, M. A., Delimat, A., Dullinger, S., Erschbamer, B., Felde, V. A., & Wipf, S. (2018). Accelerated increase in plant species richness on mountain summits is linked to warming. *Nature, 556*(7700), 231–234. DOI: 10.1038/s41586-018-0005-6 PMID: 29618821

Streb, P., Aubert, S., Gout, E., & Bligny, R. (2003). Reversibility of cold- and light-stress tolerance and accompanying changes of metabolite and antioxidant levels in the two high mountain plant species Soldanella alpina and Ranunculus glacialis. *Journal of Experimental Botany, 54*(381), 405–418. DOI: 10.1093/jxb/erg048 PMID: 12493869

Sun, H., Niu, Y., Chen, Y.-S., Song, B., Liu, C.-Q., Peng, D.-L., Chen, J.-G., & Yang, Y. (2014a). Survival and reproduction of plant species in the Qinghai–Tibet Plateau. *Journal of Systematics and Evolution, 52*(3), 378–396. DOI: 10.1111/jse.12092

Sun, H., Niu, Y., Chen, Y.-S., Song, B., Liu, C.-Q., Peng, D.-L., Chen, J.-G., & Yang, Y. (2014b). Survival and reproduction of plant species in the Qinghai–Tibet Plateau. *Journal of Systematics and Evolution, 52*(3), 378–396. DOI: 10.1111/jse.12092

Sun, Y., Zhang, A., Landis, J. B., Shi, W., Zhang, X., Sun, H., & Wang, H. (2023). Genome assembly of the snow lotus species Saussurea involucrata provides insights into acacetin and rutin biosynthesis and tolerance to an alpine environment. *Horticulture Research, 10*(10), uhad180. DOI: 10.1093/hr/uhad180 PMID: 37885817

Wang, Q., Liu, H., Gao, A., Yang, X., Liu, W., Li, X., & Li, L. (2012). Intergenomic rearrangements after polyploidization of Kengyilia thoroldiana (Poaceae: Triticeae) affected by environmental factors. *PLoS One*, *7*(2), e31033. DOI: 10.1371/journal.pone.0031033 PMID: 22363542

Wang, X., Liu, S., Zuo, H., Zheng, W., Zhang, S., Huang, Y., Pingcuo, G., Ying, H., Zhao, F., & Li, Y. (2021). Genomic basis of high-altitude adaptation in Tibetan Prunus fruit trees. *Current Biology, 31*(17), 3848-3860. e3848.

Winkler, M., Lamprecht, A., Steinbauer, K., Hülber, K., Theurillat, J.-P., Breiner, F., Choler, P., Ertl, S., Gutiérrez Girón, A., Rossi, G., Vittoz, P., Akhalkatsi, M., Bay, C., Benito Alonso, J.-L., Bergström, T., Carranza, M. L., Corcket, E., Dick, J., Erschbamer, B., & Pauli, H. (2016). The rich sides of mountain summits – a pan-European view on aspect preferences of alpine plants. *Journal of Biogeography, 43*(11), 2261–2273. DOI: 10.1111/jbi.12835

Wootton, L. M., Boucher, F. C., Pouchon, C., Roquet, C., Coissac, E., Renaud, J., Alsos, I. G., Valla, P. G., Husson, L., Bernet, M., Perrier, C., Douzet, R., Rome, M., Valay, J.-G., Alberti, A., Denoeud, F., Zimmermann, N. E., Wincker, P., Thuiller, W., & Lavergne, S. (2025). The late rise of sky-island vegetation in the European Alps. *Nature Plants, 11*(6), 1–12. DOI: 10.1038/s41477-025-02001-0 PMID: 40457047

Yang, F.-S., Liu, M., Guo, X., Xu, C., Jiang, J., Mu, W., Fang, D., Xu, Y.-C., Zhang, F.-M., Wang, Y.-H., Yang, T., Chen, H., Sahu, S. K., Li, R., Wang, G., Wang, Q., Xu, X., Ge, S., Liu, H., & Guo, Y.-L. (2024). Signatures of Adaptation and Purifying Selection in Highland Populations of Dasiphora fruticosa. *Molecular Biology and Evolution, 41*(6), msae099. Advance online publication. DOI: 10.1093/molbev/msae099 PMID: 38768215

Yang, Q., Bi, H., Yang, W., Li, T., Jiang, J., Zhang, L., Liu, J., & Hu, Q. (2020). The genome sequence of alpine Megacarpaea delavayi identifies species-specific whole-genome duplication. *Frontiers in Genetics*, *11*, 812. DOI: 10.3389/fgene.2020.00812 PMID: 32849811

Young, T. P., & Peacock, M. M. (1992). Giant Senecios and Alpine Vegetation of Mount Kenya. *Journal of Ecology*, *80*(1), 141–148. DOI: 10.2307/2261071

Zhang, H., Zhang, P., Niu, Y., Tao, T., Liu, G., Dong, C., Zheng, Z., Zhang, Z., Li, Y., Niu, Z., Liu, W., Guo, Z., Hu, S., Yang, Y., Li, M., Sun, H., Renner, S. S., & Liu, J. (2025). Genetic basis of camouflage in an alpine plant and its long-term co-evolution with an insect herbivore. *Nature Ecology & Evolution*, *9*(4), 1–11. DOI: 10.1038/s41559-025-02653-3 PMID: 40065027

Zhang, J., Tian, Y., Yan, L., Zhang, G., Wang, X., Zeng, Y., Zhang, J., Ma, X., Tan, Y., Long, N., Wang, Y., Ma, Y., He, Y., Xue, Y., Hao, S., Yang, S., Wang, W., Zhang, L., Dong, Y., & Sheng, J. (2016). Genome of plant maca (Lepidium meyenii) illuminates genomic basis for high-altitude adaptation in the central Andes. *Molecular Plant*, *9*(7), 1066–1077. DOI: 10.1016/j.molp.2016.04.016 PMID: 27174404

Zhang, X., Deng, T., Wang, H., & Sun, H. (2025). Genomic insights into alpine plant adaptation. *Plant Diversity*. Advance online publication. DOI: 10.1016/j.pld.2025.12.012

Zhang, X., Kuang, T., Dong, W., Qian, Z., Zhang, H., Landis, J. B., Feng, T., Li, L., Sun, Y., Huang, J., Deng, T., Wang, H., & Sun, H. (2023). Genomic convergence underlying high-altitude adaptation in alpine plants. *Journal of Integrative Plant Biology*, *65*(7), 1620–1635. DOI: 10.1111/jipb.13485 PMID: 36960823

Zhang, X., Sun, Y., Landis, J. B., Shen, J., Zhang, H., Kuang, T., Sun, W., Sun, J., Tiamiyu, B. B., Deng, T., Sun, H., & Wang, H. (2021). Transcriptomes of Saussurea (Asteraceae) provide insights into high-altitude adaptation. *Plants*, *10*(8), 1715. DOI: 10.3390/plants10081715 PMID: 34451759

Chapter 12

Nutrient Cycling and Plant–Microbe Interaction in Alpine Ecosystem:

Microbial Mediation of Soil Fertility and Plant Growth

Mian Muhammad Ahmed
http://orcid.org/0009-0001-4496-048X
College of Life Science and Technology, Tarim University, Alar, China

Syeda Maira Hamid
http://orcid.org/0009-0005-2065-1999
College of Life Science and Technology, Tarim University, Alar, China

Muhammad Asim
http://orcid.org/0009-0009-2768-0687
College of Horticulture and Forestry Sciences, Huazhong Agricultural University, Wuhan, China

Ayaz Ahmad
College of Life Science and Technology, Tarim University, Alar, China

Abdul Raziq
College of Life Science and Technology, Tarim University, Alar, China

Muhammad Asim
College of Life Science and Technology, Tarim University, Alar, China

Zhang Rui
College of Horticulture and Forestry, Tarim University, Alar, China

Pan Zhiyong
College of Horticulture and Forestry Sciences, Huazhong Agricultural University, Wuhan, China

ABSTRACT

Alpine ecosystems are distinguished by nutrient-poor soils and low temperatures, short growing seasons, and harsh environmental conditions that have strong influ-

DOI: 10.4018/979-8-3373-6756-9.ch012

·ight © 2026, IGI Global Scientific Publishing. Copying or distributing in print or electronic forms without written permission of IGI Global Scientific Publishing is prohibited. Use of this chapter to n generative artificial intelligence (AI) technologies is expressly prohibited. The publisher reserves all rights to license its use for generative AI training and machine learning model development.

ence on plant growth and survival. Soil nutrient dynamics, especially the cycling of carbon (C) and the nutrients (N), play a central role in controlling the productivity, diversity and resilience of alpine vegetation. Microbial communities are important mediators of nutrient transformations that influence nutrient availability under the conditions of limited decomposition and mineralization. This chapter discusses the role of soil microbes, nutrient constraints and biogeochemical processes in determining plant distribution in the alpine communities. It highlights how changes in C and N cycling with climate change affect vegetation resilience and ecosystem functioning. This chapter provides evidence-based insights into soil–plant interactions and the role of microbes in enhancing biodiversity and informing conservation strategies to improve nutrient stability in alpine landscapes.

1. INTRODUCTION

Alpine systems represent high-altitude environments with special climatic, edaphic and topographic features. These environments are also very much influenced by environmental changes, which makes them excellent indicators of the global climate system. An understanding of the interaction between climate, soil, topography, biodiversity and nutrient availability for predicting the effects of climate change on such fragile systems.

1.1 Importance of Nutrient Cycling in Alpine Ecosystems

Alpine biogeochemical cycle of essential elements, especially C and N, is one of the important basic processes that support plant growth, soil nutrient status and ecosystem productivity (Zhou et al., 2023). Alpine soil microbial communities regulate organic matter decomposition, nutrient mineralization and biogeochemical cycling in general (Figure 1). But these processes are not efficient due to low temperatures, scarce water resources and a short-term biological tenor leading to nutrient availability being the primary limitation of plant survival and growth (Broadbent et al., 2024).

Figure 1. Soil nutrient dynamics and alpine vegetation

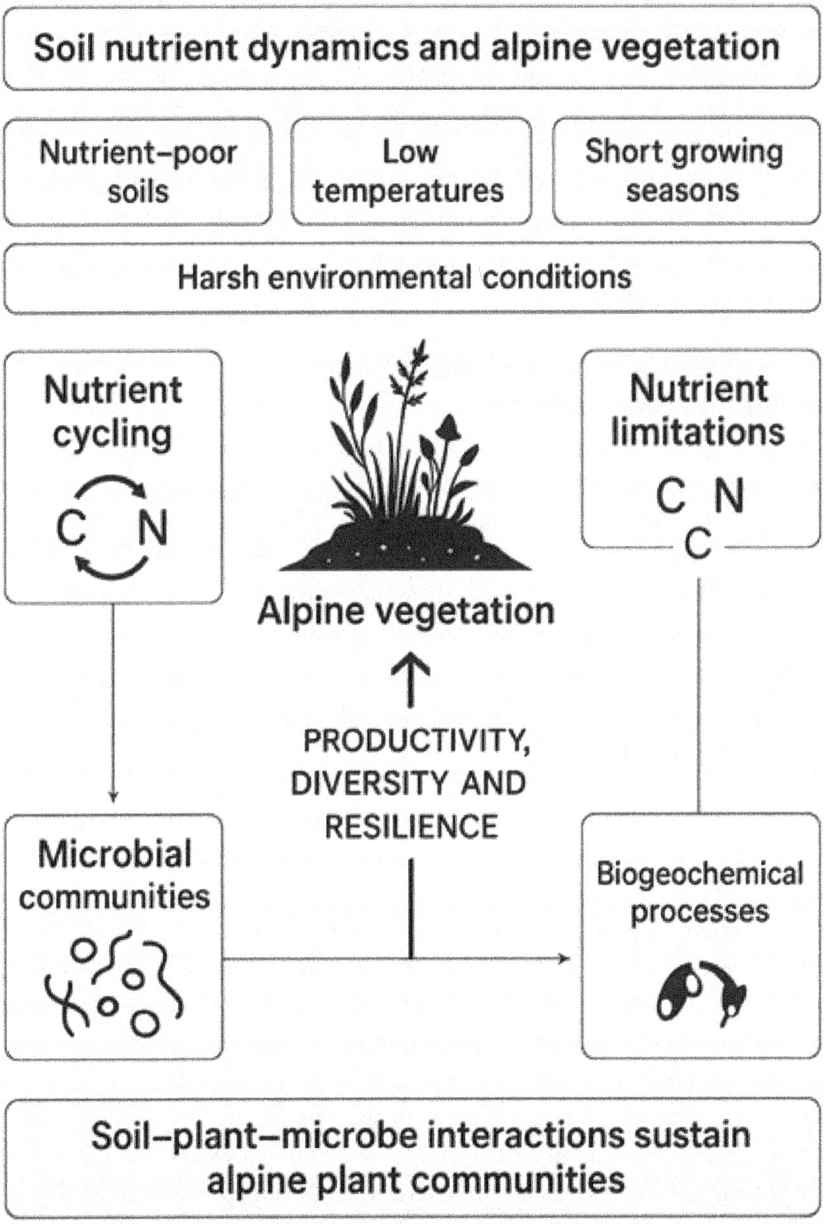

Recent scientific studies are suggesting that nutrient cycling in alpine ecosystems is also being affected by climate change. Variation in temperature and precipitation can alter microbial community structure and activity leading to changes in nutrient turnover (Bright et al. 2025). For instance, increased temperatures might speed up decomposition by microorganisms, releasing more nutrients; however, if plants do not take them up in a related manner or with the same pace than they are being released during microbial metabolism, runoff may wash these nutrients out of soil to such an extent that soil fertility would be low and ecosystem functioning greatly reduced. Understanding such mechanisms is crucial if one hopes to predict the response of alpine ecosystems under environmental pressure and guide their conservation and management.

1.2 Role of Plant–Microbe Interactions in Ecosystem Functioning

Plant-microbe interactions are very important for plants to survive and grow under the extreme conditions of high-altitude habitats. Mycorrhizal fungi are symbiotic associations between fungus and plant roots and increase nutrients, in particular phosphorus availability and water use efficiency through promoting uptake of water and nutrients under drought. In exchange, the plants provide carbohydrates to the fungi (Hou et al, 2024), leading to a mutualistic relationship which is essential for plant survival in nutrient-poor soils. In addition to the fungi, rhizosphere bacteria are required symbionts of plants for nutrient uptake of plants, stress tolerance. Some of these bacteria fix the atmospheric nitrogen into plant available forms, some produce teaching-germination hormone, anti-microbial compounds protecting from attack by pathogens (Tiwari et al.) Such plant-microbe interactions are involved in plant health, but also affect soil microbial community structure and function, with resultant effects on nutrient cycling and ecosystem functioning on a whole); and In addition, interactions between plants and microbes can affect plant responses to abiotic stresses such as cold, drought and high UV-radiation that occur in abundance in alpine ecosystems. Hormonal signaling, antioxidants and maintain osmotic balance metabolic is regulated by microbial symbionts that help to improve the adaptation of cold stress in plants which ultimately increase the survival of plants under adverse conditions (Hou et al., 2024; Tiwari et al., 2024). Climate change however could compromise these interactions by rates of nutrient transfer, as well as possibly implications not only on soil microbial communities but also plant communities. Knowledge of these links is essential to make predictions of ecosystem responses to global change and put in place measures for the conservation of alpine biodiversity, and functionality.

2. OVERVIEW OF ALPINE ECOSYSTEMS

Alpine ecosystems are also characterized by intense abiotic factors leading to extreme climate, soil, and topography. They are extremely sensitive systems to environmental changes and serve as critical proxies to global climatic fluctuations. They are described by poor nutrient availability, brief growing seasons and complex interactions of plant with soil and microbes that together affect the resilience and functioning of the ecosystem; There is a desperate need to understand how (i) climate-soil interactions, and (ii) topography-biodiversity relations affect the nutrient dynamics of these fragile ecosystems to predict their ecological based physiological reactions to climatic changes and thus also responding to conservation action (Korner 2021).

2.1 Characteristics of Alpine Environments (Climate, Soil, Topography)

Alpine regions are usually found above the tree line at an elevation of 2,500–3,500 m above sea level, depending on latitude (Gruber et al., 2023). Climate is generally quite hostile with persistently low temperatures, a large daily and yearly variation, high wind velocity and strong solar radiation. Rainfall amounts differ greatly between sites and from year to year, it creates local microclimates which have unique niche space as shown in figure 2 (Seastedt & Oldfhaher, 2024). Significant drivers of soil temperature and moisture dynamics are snow cover, affecting plant phenology and microbial processes.

Figure 2. Main characteristics of an alpine ecosystem

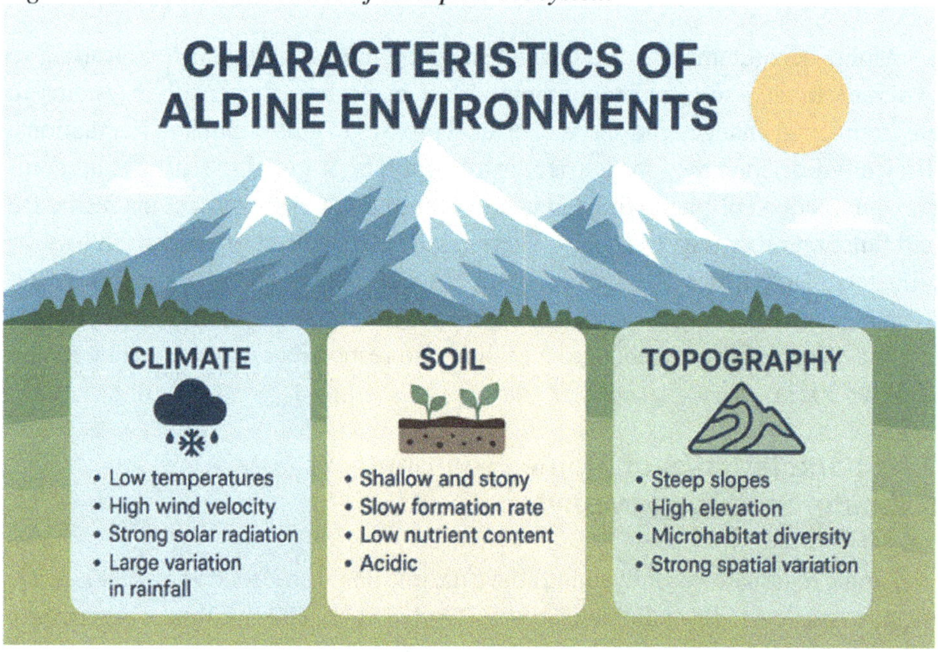

The process of soil formation in the alpine terrain is extremely slow as a result of low-imputes of organic matter, short periods with biological activities, and frequent freeze–thaw cycles. Soils in the alpine area are usually shallow, stony and less developed with lower nutrient contents and acidic pH (Li et al., 2023). Such edaphic constraints restrict nutrient availability, in particular of nitrogen (N) and phosphorous (P), two important elements for plant growth and microorganisms functioning. Besides, hydraulic dot cannot be uniform in the soil according to slope, aspect and microtopography which affects the alpine vegetation distribution and component as well.

Topography heavily structures alpine ecosystems. Differences in slope, aspect and elevation create local variation in temperature, moisture and solar radiation exposure resulting in a network of microhabitats over short distances (Zou et al., 2025). In conjunction with the soil heterogeneity, these microclimatic differences generate a large spatial variation within plant communities and the corresponding microbial community. The result is that alpine ecosystems are characterized by strong small scale heterogeneity, which is an essential component of biodiversity and ecosystem resilience at short time scales including extreme events.

2.2 Biodiversity and Functional Roles of Alpine Plants

Alpine plant communities have high species richness, endemism and functional diversity. Under these harsh conditions, alpine plants have adapted unique morphological, physiological and reproductive traits to survive in a cold, windy and nutrient-limited environment (Li et al., 2023). Typical adaptive features for this habitat are rosette, spreading rosette, plenty of hair, thick leaves, deep extensive root systems and high amounts of secondary metabolites protecting against UV radiation and browsing.

The alpine plants have various also functional roles in addition to primary productivity. Such plants play a role in stabilizing the soil, regulating water and retention of nutrients. Their rhizosphere: the thin zone of soil is directly surrounding and influenced by plant roots, enriched by root exudates and inhabited by dense, highly active microbial communities that affect nutrient cycling and plant health, filled with microbial populations that are important for nutrient uptake and carbon (C) and nitrogen (N) cycling. These plant-microbe partnerships encourage soil fertility and ecosystem stability (Zhou et al., 2023). In addition, plant functional traits diversity helps stabilizing ecosystem functioning by promoting resource use complementarity and buffering against environmental changes.

Alpine plants also had effects on the composition and functional metagenomic characteristics of soil microbial communities. Root exudates serve as substrates for microbial metabolism and, reciprocally, microbial activity influence nutrient availability and decomposition of soil organic matter (Tiwari et al., 2024). These mutualistic exchanges between plants and microbes are essential in maintaining the functioning of ecosystem in nutrient impoverished, climatically harsh alpine ecosystems.

2.3 Constraints on Nutrient Availability in High-Altitude Habitats

Climatic and edaphic characteristics exist as the key constraints on nutrient availability in alpine ecosystems. Moderate temperatures are known to slow the decay of organic matter and mineralization due to lowering microbial activity (Bright et al., 2025). The short growing seasons also reduce the temporal overlap between availability and plant access to nutrients, so effective nutrient recycling is necessary for plant survival.

The nutritional elements are also influenced by properties of the soil such as pH, texture and organic matter content. In the alpine region, soils are usually acidic with a low CEC, thereby limiting nutrient retention and making them vulnerable to leaching. In this landscape, where nutrient inputs are low and microbial processing is sluggish and dominated by aromatic compounds, ecosystem function is heavily limited by poor bioavailability of nutrients (Johnstone 2021).

Moreover, the abiotic factors and biotic interactions determine the nutrient dynamics. For instance, snow cover affects soil moisture as well as temperature, as does plant litter quality and microbial community composition on the rate of OM decomposition (Zhou et al., 2023). Understanding those integrated controls is important for future predictions of how nutrient availability will respond to climate warming, changes in precipitation or other anthropogenic perturbations. Alpine ecosystems are thus the result of multiple and interacting processes including climatic, soil, topographic and biotic factors that together control nutrient cycling, biodiversity and ecosystem stability.

3. NUTRIENT CYCLING PROCESSES IN ALPINE SYSTEMS

Nutrient cycling in alpine ecosystems is influenced by a combination of the severity of the climate, low species rich in alpine ecosystems, and high predator-prey ratios, biological productivity, and gradual processes of biogeochemistry. High altitude environments are characterized by low temperatures, high solar radiation, short growing seasons and thin, poorly developed soils. These constraints add up and slow down the nutrient turnover, restrict microbial activity, and affect the availability of important elements for plant growth. As a result, nutrient cycling in alpine systems is also uniquely conservative, with tight coupling between soil, vegetation, and microbial communities. Understanding these processes is the key to predicting how alpine ecosystems will respond to persistent climatic changes, which are projected to disproportionately affect high elevation landscapes.

3.1 Key Nutrients: Nitrogen, Phosphorus, Carbon, Potassium, and Micronutrients

3.1.1 Nitrogen Cycling

Nitrogen (N) is often the most limiting nutrient in alpine ecosystems. Low temperatures limit microbial mineralization of organic matter and therefore reduce the availability of ammonium (NH_4^+) and nitrate (NO_3^-). In the case of many alpine soils, organic forms of N represent the largest N reservoir, but are still relatively unavailable to plants without being transformed by microorganisms. Mineralization and nitrification rates are at their greatest when soils thaw, and microbial activities are enhanced (Juan et al., 2020). As for the snowmelt, Denitrification occurs in saturated soils but is generally low due to low anaerobic conditions. Despite these constraints and nitrogen availability has a strong influence on primary productivity, species composition, and competitive dynamics of alpine plants (Wang et al, 2016).

3.1.2. Phosphorus Cycling

Phosphorus (P) availability is determined largely by parent material and the rate of weathering. In cold climates, chemical weathering is slow, so concentrations of bioavailable phosphate (Mage and Porder, 2013). Much of the phosphorus is tied up within mineral complexes or organic matter and is only produced slowly. Because P is poorly mobile and many plants are even immobilized in alpine soils, mycorrhizal associations are very important to many plants, particularly arbuscular and ectomycorrhizal fungi, to mobilize organic and mineral-bound P. As a result, there is often both limitation of phosphorus and nitrogen, although together we regulate plant growth and productivity of ecosystems (Gojon et al., 2023).

3.1.3. Carbon Cycling

The C (carbon) cycling of alpine ecosystems is characterized by a large pool of soil organic C (SOC) and slow rates of decomposition. Low soil temperatures, lower levels of microbial biomass, and reductions from enzymatic activity help build up of undecomposed litter and humus. Alpine vegetation is usually characterized by low productivity; however, the "cold" conditions allow for carbon to be accumulated in the soil over long periods of time (Monson et al, 2006). Carbon fluxes are strongly seasonal: during the short summer, microbial respiration and the activity of the roots vastly increase, while as there is little activity during the winter periods under persistent snow cover. Alpine soils thus serve as important carbon water reservoirs, but because of their sensitivity to warming, they are susceptible to increased carbon losses from respiration (Bonfanti et al., 2025).

3.1.4. Potassium and Micronutrients

Potassium (K), calcium (Ca), magnesium (Mg) and micronutrients, such as iron (Fe), manganese (Mn), and zinc (Zn), presumably come exclusively from mineral weathering. The physical breakdown of rocks through cycles of freezing and thawing helps to break rocks, however chemical weathering remains slow (Zhang et al., 2025). Although these nutrients are not normally the primary limiting factors, their availability affects the enzyme activation process, synthesis of chlorophyll, tolerance to stress, and metaphysics of the regulation in alpine plants. In particular micronutrients such as Fe and Mn play critical role(s) in photosynthetic and oxidative stress management in high UV radiation and temperature extremes (Pandey, 2020).

3.2 Sources and Inputs of Nutrients

3.2.1. Atmospheric Deposition

In many of the ecosystems found in the Alps, atmospheric deposition is an important external source of nutrients, especially nitrogen and traces of phosphorus (Tipping et al., 2014), Wet and dry NH_4^+ and NO_3^- deposit greatly affect soil chemistry, microbial activity and useful vegetation dynamics. In recent decades, increasing anthropogenic N deposition has altered nutrient balances in certain high-altitude areas, promoting changes in species composition and increasing the acidity of the soil. Dust deposition also represents a significant source of input of phosphorus and micronutrients especially in areas downwind from desert regions or degraded lowlands (Alzahrani et al., 2024).

3.2.2. Mineral Weathering

Geological weathering of parent rock, which provides a slow but steady supply of phosphorous, potassium, and micronutrients; Alpine areas having resistant bedrock such as granite or quartzite, usually have lower nutrient availability than areas that are underlain by easier weathered rock (lithologies) such as shale or limestone. Freeze-thaw cycle physically breaks the rock. Lack of chemical dissolution due to low temperatures, surfaces, and more mineral exposure. As a result nutrient input from weathering is small in comparison to lower elevation ecosystems (Dalling et al., 2016).

3.2.3. Organic Matter Inputs

Plant litter, mosses, lichens and root turnover add organic matter to alpine soils. However, due to slow decomposition much of this organic input is accumulated in thick categories of litter or humic horizons. These layers serve as reservoirs for nutrients but also retard release of nutrients through mineralization. Organic matter plays an important part in moisture retention, in the formation of microhabitats for soil organisms, and modifying the thermal characteristics of the soil (Khan et al., 2025).

3.2.4. Biological Nitrogen Fixation

Symbiotic nitrogen fixation by legumes, actinorhizal shrubs (Dryas, for example, Alnus), and free-living cyanobacteria is responsible for adding new nitrogen to the alpine soils. The rate is usually low. Microbial activity needed for N fixation is restricted by cold temperatures and a restricted supply of carbon availability. Nonetheless, nitrogen-fixing species often play disproportionately important ecological roles by improving the soil fertility and enabling the succession in nutrient-limited alpine environments (Wang et al., 2021).

3.3 Decomposition Constraints and Soil Organic Matter Dynamics

The decomposition in the alpine ecosystem is controlled by a battery of environmental limitations. The low temperatures inhibit microbial respiration and extracellular enzyme activity, greatly slowing litter breakdown. Plant litter of the Alps commonly had high contents of lignin, waxes, and phenolic compounds, which makes it resistant to microbial degradation. Moreover, the short growing season limits the time for biological activity in each year (Liebelt et al., 2019).

Snow cover plays an important dual role: it insulates the soil from extremes of cold, allowing some microbial activity during winter, but also retards spring warming of the soil and shortens the active period for decomposition. Freeze-thaw cycles also impede the operation of microbial processes due to the disruption of aggregates in soil that damage the microbial cell and generate patterns of nutrient release which are difficult for plants to take advantage of (Liu et al., 2015).

As a result of these limitations, soil organic matter (SOM) is significantly accumulated. Alpine soils often contain thick humic layers with high carbon: nitrogen (C:N) ratios, reflecting slow mineralization. This slow turnover is the basis of a nutrient-limited ecosystem in which plants and microbes' efficient nutrient conservation strategies (Ishfaq et al., 2025).

3.4 Effects of Elevation, Seasonality, and Temperature on Nutrient Cycling

3.4.1. Elevation Gradients

Elevation exerts strong control on the nutrient cycling through its influence on temperature and moisture. The availability of a solution to this problem is solar radiation and soil development. As one goes up in elevation, there is a shift in nutrient mineralization rates, litter increases, and the nutrient availability becomes increasingly limited. The highest elevations tend to hold young, undeveloped soils that have little organic matter and extremely low nutrient pools (Njeru et al, 2017).

3.4.2. Seasonality

Seasonal patterns greatly influence the dynamics of nutrients. Most nutrient cycling takes place during the short summer, when the soils and microbial and root activities increase dramatically; Early snowmelt events can induce pulses of release of nutrients, particularly nitrogen, but plants are often not completely active at this time and set nutrient losses possible. This is characterized by winter. Minimal nutrient mobility and slow biochemical processes, which take place under snowpacks (Wipf et al., 2015).

3.4.3. Temperature Sensitivity

Temperature is considered the main regulator of nutrient cycling in alpine ecosystems. Even slight temperature increases can increase microbial activity, increase decomposition and increase nutrient mineralization (Broadbent et al., 2024). However, warming can have another beneficial effect on nutrient losses via leaching and by greenhouse gas emissions and can stimulate changes in vegetation composition that cause changes to nutrient inputs and nutrient cycling patterns. For example, warming-induced shrub expansion can lead to improvements of litter quality and alteration of carbon and nitrogen dynamics in soils (Liu et al., 2022).

Nutrient cycling in alpine ecosystems is not only a soil-driven process; it is also actively shaped by plant–microbe partnerships. Because low temperature, short growing seasons, and limited mineralization restrict nutrient availability, plants rely heavily on rhizosphere microbes and mycorrhizal fungi to mobilize nitrogen and phosphorus from organic and mineral pools. In return, plants supply carbon compounds that fuel microbial activity, enzyme production, and decomposition, thereby accelerating nutrient turnover. Thus, nutrient cycling and plant–microbe interactions form a tightly coupled feedback system that regulates plant productivity and ecosystem resilience in alpine environments.

4. PLANT–MICROBE INTERACTIONS IN ALPINE HABITATS

Plant-microbe interactions are key to how alpine ecosystems operate, in which nutrients limitations, harshness of the environment, short growing seasons influence the ecological relationships between plants and soil microbial communities. Alpine plants depend heavily on bacterial, fungal, and actinomycete partners in the rhizosphere to access nutrients, tolerate abiotic stress, and maintain the function under the influence of changing environmental conditions (Gavini et al., 2019). Microbial communities in these regions are uniquely adapted for survive cold temperatures, freeze-thaw cycles and low availability of nutrients, making plant microbes interactions key to ecosystem resilience and productivity. This section explores the composition, functional roles, symbiotic association and adaptive strategies of alpine microbial communities as well as its complex feedback interaction with plant roots and nutrient cycles (Li et al., 2023).

4.1 Overview of Alpine Microbial Communities (Bacteria, Fungi, Actinomycetes)

Alpine soils have a rich community of microorganisms despite the harsh constraints imposed by the climatic condition (Lazzaro et al., 2015). The rhizosphere, the narrow zone affected by roots of plants supports more microbial biomass and activity compared to bulk soils because of constant supply of root exudates. Major groups of microorganisms are cultures of cold-adapted bacteria, fungi (especially mycorrhizal fungi), actinomycetes, psychrophilic microorganisms (Yadav et al., 2019).

4.1.1. Bacteria

Alpine soil bacterial communities are dominated by phyla including Proteobacteria, Acidobacteria, Actinobacteria & Bacteroidetes. Many alpine bacteria are psychrotolerant in terms of metabolic activity at very low temperature (Liu et al., 2017). These bacteria have very important roles in nitrogen mineralization and decomposition of organic matter and transformation of phosphorus through the action of phosphatase activities. Free-living nitrogen-fixing bacteria which include species of Azotobacter, Cyanobacteria and Frankia, contribute to replenishing nitrogen pools in nutrient poor soils (Saeed et al., 2025).

4.1.2. Fungi

Fungal communities are very diverse and functionally relevant in alpine areas. Mycorrhizal fungi form symbiotic relationship with roots of most alpine plants which enhances nutrient uptake and supplying protection of abiotic stress (Wahab et al., 2023). Saprotrophic function of fungi they also play essential functions in breaking down of recalcitrant organic matter, especially in the presence of cold adapted enzymatic systems breaking down of lignin, cellulose and other complex compounds (Talal et al., 2025; Ahmad et al., 2026).

4.1.3. Actinomycetes

Actinomycetes are filamentous bacteria that are adapted to cold and low nutrient conditions (Helmi, 2025). They involve roles in the breakdown of organic matter and the production of antibiotics, which function in the suppression of pathogenic microbes. Gena, like Streptomyces, are known to degrade complex polymers and contribute significantly to the soil organic matter turnover under alpine conditions (Hui et al., 2021). Together, these groups of microorganisms form a dynamic and interconnected community that supports nutrient availability and plant life in high elevation ecosystems (Hassan et al., 2016).

4.2 Rhizosphere Processes and Nutrient Uptake

Rhizosphere is a hotspot for biological activity, the area of interaction of plants with microorganisms directly affects nutrient cycling and root function. Alpine plants exude sugars, amino acids, organic acids, secondary metabolites attractive and stimulating beneficially microbes (Hou et al, 2024). These exudates promote the microbial mineralization of organic matter with the release of ammonium. Another weapon - deep root penetration and release of water. Broken down organic fertilizers can inject the soil solution with nutrients: Another weapon - deep root penetration and discharges of water. Broken down organic fertilizers can inject the soil solution with nutrients.

Microbial activity within the rhizosphere influences some of the most important processes.

4.2.1. Nutrient Mineralization

Microbes convert complex organic matter to forms available to plants by enzymatic breakdown. This process is especially vital in alpine soils, in which the rate of mineralization is low due to exposure to cold temperatures (Song et al., 2021).

4.2.2. Solubilization of Phosphorus and Potassium

Many rhizosphere bacteria liberate organic acids or phosphatases that solubilize mineral-bound phosphorus, making it more available. Likewise, some microbes release potassium from silicate minerals (Sindhu et al, 2013).

4.2.3. Root Growth Promotion

Plant growth-promoting rhizobacteria (PGPR) excrete phytohormones e.g. auxins, cytokinins, and gibberellins increasing the proliferation of root and increased efficiencies of nutrient uptake (Li et al., 2016).

4.2.4. Stress Alleviation

Stress reduction by certain microbes results in production of ACC deaminase, siderophores or antioxidant enzymes, which decrease ethylene levels, enhance iron uptake and fight oxidative damage associated with high UV exposure and temperature extremes (Lvanov et al., 2018). The rhizosphere thus act as a functional interface accomplishing microbial processes integrated by deemed plant nutrient acquisition strategies, especially those in nutrient poor alpine environments (Jiang et al., 2012).

4.3 Symbiotic Associations

Symbiotic associations play an important role in facilitating alpine plants to survive and thrive in nutrient poor, cold and physiologically stressful mountain environments (Chandra et al., 2025). Among the most influential are mycorrhizal fungi - both arbuscular mycorrhizal AM and ectomycorrhiza EM and ectomycorrhizal (ECM) species, which colonize plant roots and increase the acquisition of nutrients, especially nitrogen and phosphorus elements (often limited) in alpine soils (Tian et al., 2017). These fungal partners extend the functional root system by having an extensive network of hyphae which enhances soil exploration, enhances water intake and favors mineral weathering, thereby increasing plant resiliency under the conditions of low temperature and low oxygen. In return, the host plant supplies the fungus with carbon-rich photosynthates to support the fungal metabolism (Lallemand et al., 2019). Additionally, dark septate endophytes (DSE) and fungal endophytes make a contribution to stress tolerance in Alpine species by manipulation of cold resistance, enhancement of osmolyte wound accumulation, and modulation of hormonal pathways (e.g. ABA and jasmonic acid signaling) (Singh et al., 2022). Besides fungal plant partnerships, nitrogen fixing bacteria including Rhizobium and Frankia and some cyanobacteria, form symbioses with alter alpine shrubs and forbs, and mosses in thereby enriching (nitrogen deficiency soils) and supporting plant productivity in harsh environments. Certain plant growth promoting rhizobacteria (PGPR), e.g. Pseudomonas, also bacteria from Bacillus, and psychrophilic, colonize alpine rhizospheres and contribute to nutrient solubilization, production of phytohormones and synthesis of antifreeze proteins (Mukhia et al., 2022). These various symbiotic associations combined as an ecological buffer for plant establishment, metabolic functioning and long-term survival in the extreme climatic gradients typical for alpine ecosystems (Wipf et al., 2013).

4.4 Microbial Adaptations to Alpine Conditions

Microorganisms living in alpine ecosystems express a great variety of structural, metabolic, and genetic adaptations so that they can survive extreme environmental pressures including low temperatures, high ultraviolet radiation, cycles of freezing and thawing, and nutrient scarcity; (Wang et al., 2015). Microbes called psychrophilic and psychrotolerant produce cold active enzymes and improved catalytic flexibility which allows metabolism activity at sub-zero temperatures. These enzymes have some unique structural features like reduced hydrogen bonding, reduced core hydrophobicity and greater conformational plasticity which provide minimal rigidity and allow effective biochemical reactions at low temperatures (Aqvist et al., 2017). Many alpine microbes also synthesize cryoprotectants such as trehalose, glycerol, exopolysaccharides (EPS) and antifreeze proteins which stabilize their membranes, stop the formation of intracellular ice crystals, and facilitate survival during repeated freeze/thaw cycles. Higher production of pigment represents another important adaptation; carotenoids, melanin and scytonemin, which protect microbial cells from high UV-B radiation that would be expected at high altitude, whilst simultaneously reducing oxidative stress (Wang et al., 2025).

In addition to physiological strategies, genomic and regulatory adaptations to promote greater resiliency to the environment (Hon et al., 2024). Genes involved in DNA repairing, oxidative-stress reduction, membrane fluidity control and scavenging of nutrients are incorporation of frequently upregulated or in expanded copy numbers. Horizontal gene transfer between microbial communities makes quick passing on of beneficial characteristics such as cold tolerance, antibiotic biosynthesis, and capabilities to obtain resources (Yang et al., 2019). Microbial community structuring is also how adaptive microbes form biofilms or aggregate within soil microhabitats where thermal buffering, nutrient sharing, and cooperative. Metabolism leads to Improved Survival In addition, many of the alpine microorganisms form mutualistic interactions with plants, involved in nutrient cycling, decomposition of organic matter, and plant stress tolerance: the activity of hormones and ACC deaminases (Shahid et al., 2023). Collectively, these complex adaptations highlight the extraordinary plasticity of alpine microorganisms and their important role in maintaining ecosystem function in some of the planet's worst environmental conditions (Oliver et al., 2015).

5. COUPLING BETWEEN NUTRIENT CYCLING AND PLANT–MICROBE INTERACTIONS

The active relationship between plants and soil-dwelling microbes is the basis of the work of the terrestrial ecosystem. The microbial communities have a high involvement with the nutrient cycling process in which vital components such as Nitrogen, Phosphorus, and Carbon are converted and moved through the soil-plant system (Broadbent et al., 2024). Micro-organisms are agents in the breakdown of organic material and mineralization of nutrients and mediate chemical reactions, so that nutrients are made available for absorption by plants. On the other hand, plants create microbial communities through root exudates, litter deposition and rhizodeposition, which as a result, affect microbial changes in composition, activity and diversity. This two-way connection guarantees the productivity and stability of an ecosystem, making it clear that the nutrient cycling cannot be studied properly, without referring to the importance of the interaction of plants and microbes (Garcia and Kao-Kniffin, 2018).

5.1 Microbial Mediation of Nutrient Availability

The soil contains microbes, such as bacteria, fungi, and archaea, which are conspicuous in the process of rendering nutrients available to plants. Heterotrophic microorganisms break down the complex organic materials like leaf litter and root debris, which contain primary elements, and introduce them mineralized into the soil (Zhang et al., 2025). As an example, nitrogen which is an essential macronutrient is metabolized to ammonium (NH_4^+) and nitrate (NO_3^-) by microbial processes such as ammonification and nitrification (Sun et al., 2025). Likewise, poor phosphorus that is typically fixed in soil minerals can be solubilized by phosphate-solubilizing soil bacteria and mycorrhizal fungi, and therefore accessible to the plant. Cycling of micronutrients like iron, zinc, and manganese is another regulated interaction of microbes, in which they form siderophores and chelating agents that mobilize such elements (Shafi et al., 2025).

Also there are symbiotic relationships between plants and microbes, including mycorrhizal fungi and nitrogen fixing bacteria, which also increase nutrient availability. Mycorrhizal plants expand their root to a greater distance through the hyphae networks of the mycorrhizal fungi, being able to reach nutrients that are not readily available in the depletion zone around the roots (Tsai et al., 2025). Bacteria which fix nitrogen such as those in the genera *Rhizobium* and *Frankia* transform atmospheric nitrogen into chemically usable forms, and have a direct effect on soil fertility (Mou et al., 2023). The microbial communities are also dynamic in response to the environmental conditions, plant species and soil management practices and their nutrient transformation activities are also adjusted to maximize the nutrient balance in the ecosystem (Dou et al., 2025). All in all, the microbial mediation process brings about constant provision of important nutrients that facilitate the growth of plants, soil fitness and the sustainability of production in ecosystems (Imran and Ortas, 2025).

5.2 Influence of Plant Diversity on Soil Microbial Communities

Plant biodiversity is a decisive factor of the structure and functioning of the soil microbial communities (Zhang et al., 2025). With the existence of diverse plant communities, there is increased diversity in root exudates, litter materials and chemical cues, resulting in the formation of disparate niches that favor diverse taxa. Plant species are capable of hosting their own specific collection of microbial symbionts, which has an effect on the process of nutrient cycling in a non-uniform manner (Li et al., 2025). Legumes, such as the example of enhancing soils with nitrogen by symbiosis with rhizobia, and grasses, which might increase decomposers fungi, can cause faster carbon turnover (Schmid, et al., 2021). This diversity in the input of plants increases the diversity of microbes and functional redundancy, which subsequently stabilize nutrient cycling and ecosystem performance in changing environment conditions (Zhang et al., 2025).

In addition, species composition and richness have a strong effect on plant-microbe feedbacks. Large plant diversity has been generally associated with large microbial biomass, enzymatic activity, and mineralization of nutrients (Li et al., 2025). The positive feedback between specific microbial taxa and particular plant species can be used to increase soil fertility and plant productivity. An example of this is that often in diverse plant communities, mycorrhizal colonization rates are more effective in enhancing phosphorus acquisition and stress tolerance in multiple species (Zhang et al., 2025). Monocultures or low-diversity systems, on the contrary, are more likely to decrease microbial diversity, lower the efficiency of nutrient cycling, and predispose organisms to pathogenicity and nutrient deficiencies. Therefore, it is important that plant diversity is not only preserved to ensure the above ground biodiversity, but also the health and functionality of soil microbial communities that govern the availability of nutrients (Stefan et al., 2021).

To sum up, nutrient cycling is coupled to interact to the interactions of plant-microbes processes and depict the complexity in the connectivity of the terrestrial ecosystems (Sharma et al., 2025). Microbial community mediate nutrient availability and transformation of nutrients which are needed, and plant diversity determines microbial composition and functionality. Knowledge of such interactions is important in managing ecosystems, agricultural practices, and restoration activities because it demonstrates that maintenance of productivity and soil health requires activity of both microbes and variety of plant communities (Jia et al., 2025). By combining the knowledge of the mediation of microbials and the plant-based effect it is possible to have more resilient and nutrient-efficient ecosystems, and the question of the need for holistic perspectives in ecology and land management becomes clear (Xing et al., 2025).

6. ENVIRONMENTAL AND ANTHROPOGENIC DRIVERS

The interplay between environmental conditions and humans is a great impact on the ecosystem processes, such as nutrient cycling and microbial dynamics (Hedberg and Smith, 2025). The interaction of natural pressures, such as climate, temperature, rainfall, and seasonal fluctuations, and anthropogenic pressures, such as the change in land use, pollution, and resource exploitation determine the soil health and ecosystems (Deng et al., 2025). Microbial communities, which play a major role in the transformation and availability of nutrients, are sensitive to these drivers and typically cause rates of decomposition, mineralization and nutrient retention to change (Guo et al., 2025). On the same note, the environmental variation and human activities impact on plant communities, which are a part of nutrient input in the form of litter and root exudates (Wang et al., 2021). The interaction of these natural and human drivers may have a promoting or disturbing effect on nutrient cycling, which has implications for the productivity, resilience and biodiversity of the ecosystem. These influences are critical factors to understand in order to make predictions of the ecological responses to global change and to search for sustainable management approaches (Hossain et al., 2025).

6.1 Effects of Climate Change on Nutrient Cycling and Microbial Communities

Climate change is one of the most important environmental drivers in the world that has an impact on the processes in the soil (Sehgal et al., 2025). Due to the higher temperatures, alteration of precipitation and the frequency of extreme events such as droughts and flood impact directly and indirectly the nutrient cycling. Higher temperatures tend to speed up the decomposition and mineralization of organic matter because the metabolic rates of microbes tend to be higher (Wu et al., 2025). This has the potential to increase the nutrient supply in the short-term but over the long-term, this can lead to a depletion in the stocks of soil organic matter, and decrease the soil fertility. The changes in soil moisture also respond to changes in temperature to affect microbial behavior (Ding and Yu, 2025). Droughts usually decrease the microbial biomass and will retard the nutrient cycling and excess rainfall may also result in nutrient leaching especially nitrogen and phosphorus hence resulting in reduced nutrient retention in the soils (Iqbal et al., 2025).

The composition and diversity of microbial communities is also influenced by climate change. Some microbial taxa, especially those which are attuned to constant conditions of moisture and temperature, could decrease during changing conditions, whereas opportunistic species which can withstand stress can increase (Yasir et al., 2025). These changes may modify the process of mineralization and immobilization of nutrients, which affects their acquisition and growth of plants. As an example, drought can cause a negative effect on nitrogen-fixing bacteria, which restricts biological sources of nitrogen, and can cause fungi that decompose food to increase in warm and moist environments. These alterations may have the ability to disproportion the nutrient stoichiometry and plant-microbe feedbacks that could eventually affect the productivity and resilience of the ecosystems (Li et al., 2025).

Moreover, climate change that is linked to high carbon dioxide (CO_2) concentrations may have an indirect influence on nutrient cycling (Lukac et al. 2010). High levels of CO_2 tend to encourage the growth of plants, leading to more litter production, and exudation from the roots providing aided food for the microbial populations. Nevertheless, this may be subject to imbalances of nutrients if the mineralization rate through microbes is not able to maintain itself with the increased demand of plants (Xu et al., 2025). Consequently, there can be changes in the ratio of carbon to nitrogen in ecosystems, variation in the activities of microbial enzymes, and variation in the patterns of nutrient limitation (Shen et al., 2022). In general, all the discussed effects of climatic change present intricate, interacting impacts on microbial-mediated nutrient cycling, reflecting the susceptibility of soil processes to any global environmental disturbance (Murtaza et al., 2025).

6.2 Land Use Change, Grazing, and Soil Disturbance

The anthropogenic alterations in the land use and agricultural activities such as agriculture, urbanization, deforestation, and heavy grazing have far reaching impacts on the nutrient cycling and microbial communities (Abile et al., 2025). The transformation of natural ecosystems to croplands is usually done by means of tillage, fertilizer application and monoculture planting that alters soil structures, microbial diversity and nutrient availability. Tillage breaks the soil aggregates, which exposes the organic matter and results in rapid decomposition and loss of nutrients (Das et al., 2024). Conversely, conservation practices and no-till prevent the destruction of microbial habitats, loss of organic matter and long-term cycling of nutrients. Another way in which crop rotation and intercropping encourage different microbial communities is by facilitating a variety of nutrient transformation routes (Lai et al., 2022).

Another important anthropogenic driver is grazing by livestock. Moderate grazing would help in stimulating plant regeneration; enhance root exudation and microbial activity which may lead to improved nutrient cycling (Fartyal et al., 2025). But, due to overgrazing, the soil is frequently compacted, vegetation cover is often less and more soil is often eroded which disturbs the habitation of microbes and reduces nutrient retention. Trampling increases the soil solidity that decreases soil porosity and consequently reduces the availability of oxygen and other microbial aerobic processes, including nitrification and decomposition (Shang et al., 2024). Grazing induced changes in the plant community composition that can change both litter and root exudates into the soil and consequently alter the structure and dynamics of microbes (Guo et al., 2023).Soil disturbances caused by machinery, construction, and land clearing further exacerbate nutrient loss. Disturbances disrupt microbial communities, reduce soil organic matter, and increase susceptibility to erosion (Bidast and Golchin, 2025). They can also trigger pulses of nutrient mineralization followed by depletion, leading to temporal nutrient imbalances. Additionally, the interaction between disturbance and climatic variability can intensify nutrient leaching or volatilization, particularly for nitrogen and phosphorus, undermining ecosystem productivity and resilience. Hence, land use change, grazing, and soil disturbance collectively shape microbial-mediated nutrient cycling, and their effects depend on the intensity, frequency, and type of human activity (Bitew et al., 2025).

6.3 Deposition of Pollutants and Nutrient Imbalances

Industrialization and urbanization have enhanced deposition of pollutants such as the heavy metals, the acidifying compounds and the excess nutrients that have significant effects on the microbial communities of the soil and their nutrient cycling (Penuelas & Sardans, 2023). The fertilizers and burning fossil fuels can cause nutrient imbalance whereby the microbial taxa that need high nitrogen conditions can dominate and cause the extinction of the microbial taxa that have adapted to low-nitrogen soils. Too much nitrogen may also lead to acidification of soil, change the activity of enzymes of microbes and make other nutrient essential like phosphorus and potassium less soluble and available (Abd et al., 2025).

Microbial population has toxic impacts on heavy metals like cadmium, lead and mercury which cause loss in diversity and interfere with the important processes of nutrient transformation. Some microbes can endure or even paralyze heavy metals, yet the whole community performance is frequently affected and as a result of these effects there are slower rates of decomposition, slower nutrient recycling, and lowered soil fertility (Penuelas & Sardans, 2023). Organic contaminants may also interfere with microbial activity by inhibiting the sensitive taxa (e.g. pesticides and hydrocarbons) and boosting pollutant-degrading species, which can change the nutrient balance (Singh et al., 2021). Another method through which pollutant deposition may have indirect impacts on nutrient cycling is through altering plant-microbes interactions. Stressed plants typically emit altered root exudates and secondary metabolites, and this contact might change the structure and function of microbial populations (Dietrich et al., 2025). This can contribute to a decrease in symbiont efficiency, e.g. mycorrhizal colonization or fixation of nitrogen, which has a further-increasing effect on the nutrient imbalances. The interaction of pollution, nutrient enrichment and anthropogenic activation point to the vulnerability of soil ecologies to the contemporary environmental requirements and the need to monitor and minimize anthropogenic impacts on the environmental system and maintain its productivity and resilience (Kamanmalek et al., 2024). To conclude, the effects of environmental and anthropogenic driver are immense on nutrient cycling and microbial community dynamics which include effects from climate change, change in land use, grazing, soil disturbance, and pollutants deposition (Medhi et al., 2021). Climate variability affects microbial activity, decomposition rates, and nutrient availability and land management and perturbations affect the structure, diversity and functional capacity of the microbial habitat. This is in addition to the presence of pollutants and nutrient imbalances which causes greater disturbances in the soil processes and the interactions between plants and microbes, which highlight the sensitivity of ecosystems to interacting stressors (Mansoor et al., 2025). These drivers are important in understanding the ecological responses to global change, sustainable land management practices and ensuring maintenance of the balance between nutrient supply and demand in the ecosystem. The integration of information of the environmental factors and human-induced stressors provides holistic platform of protection of soil health, nutrient cycling involved with microbes and sustenance of healthy terrestrial ecosystem (Macci et al., 2025).

7. ECOLOGICAL AND FUNCTIONAL SIGNIFICANCE

7.1 Maintenance of Soil Fertility and Ecosystem Productivity

Soil fertility in alpine ecosystems relies on a close and dynamic interaction between biological activity, environmental limitations and limited nutrient resources (Donhauser et al., 2018). Because of the limited depth, low nutrient status and severe climatic conditions, biological processes that underpin functionality in alpine soils operate under strong constraints. Within this difficult environment, nutrient cycling becomes a right-minded opinion to maintain the functioning of these ecosystems

(Sen et al., 2025). Microbial communities within the rhizosphere and litter layer and within soil aggregates function as the primary decomposers of organic matter which release nutrients that may be utilized by the plant. Cold-adapted bacteria, fungi and Cryptosporodontia (archaebacteria) use specific enzymes to break down hard substances such as lignin, cellulose and dead mass of other microbes (mainly bacteria) (Wan et al. 2025). Their capacity to decompose these compounds into plant available nutrients during the brief seasonal period favorable for biological activity is key to maintaining the growth of the alpine plant (Xu et al., 2025).

Other nutrient-taking advantages in these nutrient-limited environments are offered by symbiotic relationships with microbes. Mycorrhizal fungi (arbuscular and ectomycorrhizal) produce mycorrhizal networks to extend the reach of plant roots and aid them to reach phosphorus, nitrogen and important micronutrients beyond the immediate plant roots. Mycorrhizal fungi both arbuscular and ectomycorrhizal fungi create mycorrhizal networks to extend out past plant roots and help them access phosphorus, nitrogen and important micronutrients beyond the plant roots. Their hyphal networks are also associated with strengthening the soil structure through an effect on aggregation and consequently enhancing stability, porosity, and water retention characteristics that are important in soils in high elevations that are subject to drying. Biological nitrogen fixation plays another role in promoting soil fertility: rhizobia associated with legumes, actinorhizal bacteria in some shrubs and free-living nitrogen-fixing cyanobacteria make important nitrogen contributions to soils in areas where there is low atmospheric deposition. These contributions are particularly significant in early successional or disturbed alpine sites, where nitrogen can frequently limit plant establishment (Imran and Ortas, 2025).

Alpine plants are also important in conservation and nutrient cycling. Many species are growing slowly, have long-lived leaves and are efficient in the resorption of nutrients before leaf drop, thus minimizing losses from the system (Sandra et al., 2025). Their extensive root networks provide them with the ability to probe nutrients in patchy soils, in addition to root exudates stimulating the beneficial microbes and producing 'biogeochemical hotspots' where nutrient turnover is faster, locally. These hotspots help the plant to obtain nutrients fast during the short growing season (Napoletano et al., 2025). Because nutrient cycling in alpine soils is slow by natural standards, the productivity of these ecosystems relies to a heavy degree on stable interactions of plants and micro-organisms. Disturbances like overgrazing, soil compaction, frost heaving, or warm climate driven change these processes causing a reduction in nutrient availability and changes in plant communities. The close interaction between the plants and the soil microbes is, therefore, at the core of sustaining the soil fertility, ecosystem productivity, and biodiversity in alpian areas (Singh et al., 2025).

7.2 Role in Carbon Sequestration and Greenhouse Gas Emissions

The current application discusses the theme "alpine ecosystems and their role in global carbon sequestration" which specifically addresses "Alpine ecosystems play a globally important role in carbon sequestration." (Bright et al., 2025) Their cold temperatures, mountainous topography and more inherently slow course of decomposing processes enable soil organic carbon (SOC) accumulation over extended periods of time. Plant-microbe interactions influence to a large extent the amount of carbon that gets stored and how much is released (Hao et al., 2025). Alpine vegetation usually invests a high proportion of its biomass below ground, where it is stored as carbon in soils in the form of roots, exudates and leaf litter. Mycorrhizal fungi, particularly ectomycorrhizal fungi, help to add stable organic matter like glomalin-like protein and fungal necro mass which help to bind soil particles and enhance long-term carbon storage (Pang et al., 2025).

The activity of microorganisms is largely responsible for the fate of this carbon. Although cold-adapted microbes can operate at low temperatures, the overall metabolic rates are limited so organic carbon can accumulate in the soil. Microbial necro mass has become a major, long-lived component of SOC recognition has been attributed to the low decomposition rate and negative relationship to soil minerals (Hou et al., 2024). However, greenhouse gases are also generated by microbes as discussed in our metabolic activities. CO_2 released during microbial and root respiration, CH_4 as anaerobic conditions in waterlogged meadows or thawing permafrost and N_2O during microbial nitrification and denitrification in particular freezing and thawing (Qasim et al, 2025). Climate warming puts alpines in carbon jeopardy Higher temperatures increase microbial respiration, which enables decomposition to occur more quickly and the carbon that has been preserved for decades or even centuries to be released. Thawing permafrost exposes long-stored (deep) carbon to microbial decomposition, which may shift these ecosystems from functioning as carbon sinks to becoming carbon sources. Changes in snowpack depth and timing of melt further affect patterns of soil temperature and are likely to produce more frequent cycles of freeze-thawing and destroy soil structure and stimulate microbes.

Changes in the makeup of vegetation from slower-growing species to faster-growing ones may also change the amount and type of carbon going into soils (Broadbent et al., 2021). It is the plant-microbe interactions that have the potential to moderate these changes and possibly intensify it (Kaur et al., 2025). For example, mycorrhizal fungi can help spiritualize carbon by competing against saprotrophs and slow down the process of decomposition (Nema et al., 2025). In contrast, warming-induced changes to fast-growing groups of microbes may accelerate the rate of carbon turnover and increase losses of carbon. Root exudate patterns could change as well: some of the changes may increase carbon storage via mutualism among the microbial communities whereas others may activate decomposers and promote greater carbon dioxide release. The understanding of how the interactions between these factors respond to climate change is of high importance for estimating under which climate change the alpine ecosystems will contribute to global greenhouse gas balances in future (Cai et al., 2025).

7.3 Contribution to Alpine Plant Adaptation and Resilience

Alpine plants experience some of the harshest growing conditions on the planet such as extreme cold, short growing seasons, extreme UV radiation, strong winds and nutrient poor soils. Microbial associates have significant roles to help plants survive and thrive under these limitations (Chandra et al., 2025). Mycorrhizal fungi promote nutrient, particularly phosphorus and nitrogen, entry to plants while nitrogen-fixing bacteria supply plants with biologically-fallible nitrogen in nutrient-deficient soils. Endophytic microbes also have great aiding plant growth because they produce plant hormones, enhance nutrient uptake, and maintain metabolic balance (Kang et al, 2025). Microbes play a major role in enhancing plant tolerance to abiotic stress. Cold-adapted fungi and bacteria secrete proteins of antifreeze, osmoprotectants, etc. that help plants maintain cell stability in freezing temperature. Some endophytes plays crucial roles in drought resistance by improve water use efficiency, alter the root and reduce transpiration (Rosbakh et al, 2025). Others help to activate antioxidant defenses that protect the plant cells from the effects of UV radiation and from excessive and low temperatures. These microbial contributions keep the alpine plants functional despite the predictability of these plants with potentially detrimental climates (Hou et al., 2024).

In a Swiss Alps case study, Pütz et al. (2011) investigated how declining snow reliability affects winter tourism and whether technical snowmaking is an effective adaptation. They compared three destinations (Davos, Scuol, Braunwald) and combined (i) tourist surveys on preferences for snow reliability and acceptance of artificial snow with (ii) stakeholder interviews and destination data to evaluate adaptation strategies and regional economic implications. The authors note that in Switzerland the share of ski pistes covered by artificial snow rose rapidly from <10% (2000) to ~33% (2010), reflecting growing reliance on snowmaking as a response to warming winters. Importantly, their findings emphasize that while snow reliability matters, tourists' overall destination choice is also strongly influenced by non-snow attractions (winter + summer), implying that diversifying toward snow-independent activities can be an economically rational pathway in some cases, especially where snowmaking costs, water/energy demand, and ecological trade-offs become limiting. (Pütz et al., 2011).

The rhizosphere is an important area where the microbial populations are actively influenced by the plant according to plant stress conditions (Binter et al. 2025). Through exudates from roots, plants are able to selectively recruit the beneficial microbes in the presence of nutrient shortage or malnutrition conditions resulting from environmental pressure. Beneficial microbes also protect the plants from pathogens because they either outcompete harmful species, produce antimicrobial compounds or induce plant immune responses and thus also reduce the energy plants need to invest in defense (Feng et al., 2025). Over evolutionary time scales, plant-microbe interactions have an impact on species persistence, colonization and community assembly. Microbial facilitation may give species the ability to move into higher/marginal habitats through enhancing soil conditions. These interactions are also important in ecosystem recovery after a disturbance such as erosion or avalanches as they will help stabilize the soils and assist plant succession (Losapio et al., 2025). Climate change, however, endangers many of these longstanding interactions. Changes in community composition of microbial organisms, mismatches in the timing between plant and microbial activity and lowered efficiency of nutrient acquisition may pose a challenge to plant resilience in a warming world (Deb et al., 2023). Still, beneficial microbial life is currently at the forefront of plant adaptability and may be natural buffers that will assist the adaptability of alpine vegetation to future climatic shifts. These microbe partnerships are the biological basis of the persistence in long-term stability, adaptation and existence of alpine ecosystems (Du et al., 2025).

7.4 Socio-economic implications of alpine ecosystem change for local communities and tourism

Alpine ecosystems underpin local livelihoods and regional economies through freshwater regulation, hazard mitigation, and cultural ecosystem services. As climate warming reduces snow cover and glacier mass, mountain "water-tower" functions can shift runoff timing and reduce dry-season water availability, creating risks for drinking water, irrigation, and hydropower, with impacts felt both in mountain settlements and far downstream (IPCC, 2022). Evidence syntheses also highlight that climate-driven ecosystem change affects multiple services simultaneously (food/forage provision, water regulation, recreation), so community vulnerability often reflects combined ecological and economic pressures rather than a single stressor (Palomo, 2017; Grêt-Regamey et al., 2012).

Tourism is a particularly sensitive sector because it depends on snow reliability, landscape quality, and safe infrastructure. Large-scale assessments project substantial declines in natural snow cover days across global ski regions under all emissions scenarios, implying shorter and less reliable seasons and increasing pressure to expand infrastructure and interventions such as snowmaking (Mitterwallner et al., 2024). A comprehensive review of ski tourism research similarly concludes that climate risk is widespread across destinations and that adaptation (including snowmaking) can raise costs and sustainability trade-offs (Steiger et al., 2019). Beyond skiing, changing land use and climate can intensify conflicts among users and limit access to cultural ecosystem services in mountain regions, affecting visitor experiences and local revenues (Schirpke et al., 2020).

Finally, alpine socio-economic impacts extend to land management and risk reduction. Mountain agriculture and grazing help maintain valued landscapes and provide regulating services (e.g., slope stability), and studies show tourists can place measurable value on these services—suggesting a direct link between ecological condition, visitor preferences, and rural income diversification (Mazzocchi & Sali, 2022). Together, these findings support integrating ecological indicators (snow, soil stability, vegetation change) with local livelihood planning and tourism adaptation strategies (e.g., four-season tourism, water-saving operations, and conservation-compatible infrastructure) to sustain both ecosystems and economies (IPCC, 2022; Palomo, 2017).

8. FUTURE DIRECTIONS

Although many studies on nutrient cycling and plant-microbe interactions in alpine ecosystems have significantly increased knowledge base for these ecosystems, critical gaps in our understanding of alpine ecosystem responses to rapid environmental change still exist. Alpine environments are likely to warm more rapidly than many lowlands and precipitation, snowpack, and human activity changes are poised to become a threat to sustainability and stability of long-established biogeochemical processes. Moving forward, the research needs to use a combination of advanced analysis tools, long-term data collection, and multi-scale approaches in order to better anticipate and manage these changes. A very important step is a enhanced functional insight on alpine microbial communities. High-throughput sequencing, metagenomics and metabolomics are powerful means to detect gene and metabolic pathways of control of nutrient transformations, and adaptation under low temperatures, and symbiotic relationships with the plant/microbe. Such work will help to clarify microbial shifts in response to warming temperatures, altered snow regimes, functions and more frequent freeze-thaw cycles.

Contributed equally to an enhanced understanding of these issues will be an improvement in understanding how above- and below-ground processes interact. While the dynamics of above-ground vegetation are well documented, less well studied are the dynamics of below ground phenomena such as the exudation of soil organic substances by plant roots, the networking function of mycorrhizal associations, and the chemistry of rhizospheres in the context of nutrient cycling. Understanding how plant communities and herbivory, or soil disturbance, or climate stress changes, microbial networks are altered is crucial to predict ecosystem resilience. Long-term climatic manipulation experiments are also important. Most of the alpine studies are short term studies, which limits our ability to project future changes. Experiments preparing for warming, drought, changes in snowpack, and accomplishment of freezes and lasts can expose boundaries and feedback in biogeochemical cycles that are contrarily hard to notice. Such studies will help identify processes that are most susceptible to environmental change. Another priority here is also scaling up local observations to broader landscapes. Predictive models that use attributes of microorganisms mixed with remote sensing technologies can help with predictive models of how the nutrient and carbon cycles may respond to various climate scenarios (Du et al., 2025; Hooshyarkhah et al., 2025). These models are needed to identify possible ecological tipping points and to provide input for conservation strategies at the scale of regions, even at a global scale. Finally, the transformation of ecological knowledge into practical restoration and management attempts is of increasing urgency. Plant-microbe interactions provide promising tools for rehabilitation such as, for example, cold-adapted mycorrhizal fungi or nitrogen-fixing bacteria for supporting establishment of plants at degraded alpine slopes. Different types of soil microbiome engineering and nature-based solutions will be beneficial to help restore nutrient cycling, stabilize vulnerable soils and enhance the resilience of our ecosystems (Beattie et al., 2025). Together, these are the future directions in the area requiring interdisciplinary research that will span from molecular biology to soil science to ecology and climate science. Such integration will be mandatory to ensure that alpine ecosystems continue to be functional, resilient and thus able to sustain biodiversity in a rapidly changing world.

9. CONCLUSION

Alpine ecosystems, in spite of their small geographic size, and due to their range of precarious conditions are disproportionately large contributors to the global biodiversity, function of water systems and storage of carbon in soil. Harsh climatic conditions, nutrient impoverished soils and short growing seasons necessitate the close interplay of plants, micro-organisms and the physical environment. This chapter emphasizes that nutrient cycling and plant-microbe interactions help to maintain soil fertility, plant productivity and ecological stability in these fragile landscapes. Microbial communities have the biochemical powerhouse for cycling nutrients for alpine soils. Through decomposition, mineralization, nitrogen-fixation and mycorrhizal nutrient uptake, microbes moderate the supply of essential nutrients were low temperature limits biological activity. Plants, in turn, select for the structure and function of microbes by exudates released from plant roots, through symbiotic relationships and through characteristics for improved resource use efficiency. This feedback is useful for maintaining alpine vegetation in nutrient limited and climatically extreme habitats. Alpine regions are also important stores of carbon that, because of the cold soil, store organic carbon slowly, due to slow decomposition. Plant-microbe interactions influence the balance between carbon sequestration and greenhouse gas emissions and therefore, plant-microbe interactions connect alpine soils to the global climate regulation. However, warming, changing snowmelt, permafrost thaw and changing hydrology can lead to unfolding of long-lived carbon pools, faster decomposition and a greater amount of CO_2, CH_4 and N_2O emissions. Understanding these mechanisms is critical to make for the carbon dynamics and information for mitigation.

Microbes are even the basis of plant adaptation and resilience. Mycorrhizal fungi as well as endophytes and nitrogen fixing bacteria which aid plants in resisting drought, cold, nutritional draw down and pathogens, thereby affecting community assembly and ecosystem stability. Yet massive gaps exist between the known and the unknown most alpine microbes have yet to be cultured, long-term climate experiments are seldom conducted, and some models and conservation strategies do not consider soil microbiomes and their recovery and resilience. Addressing these gaps require interdisciplinary efforts involving molecular tools, ecological theory, climate science, and applied management. Nutrient cycling process and plant-microbe interaction are the basis for the function, stability and resiliency of alpine ecosystems. Protecting these biological networks is important if alpine environments are to continuously support biodiversity, water regulation and climate buffering under a scenario of increasing global change. Continued research and conservation planning and ecosystem management that is based on a holistic plant-soil-microbe perspective will be crucial for protection of these high-altitude landscapes and the contributions of these ecosystems to global environmental stability.

REFERENCES

Abd, G. I., Salih, M. A. J., Ahmed, H. E., Alani, M. H., & Abed, B. M. (2025, February). The role of chemical pesticides in environmental pollution and ecological imbalance. [). IOP Publishing.]. *IOP Conference Series. Earth and Environmental Science, 1449*(1), 012145. DOI: 10.1088/1755-1315/1449/1/012145

Abile, H., Fituma, K., & Mammo, S. (2025). Impact of land use types on selected soil physicochemical parameters in the case of Liben Jawi district, Ethiopia. *Scientific Reports, 15*(1), 27944. DOI: 10.1038/s41598-025-12784-z PMID: 40744989

Ahmad, A., Ahmed, M. M., Akhtar, A., Shuihong, C., Zafar, Z., Ullah, R., Asim, M., He, Z., & Khan, M. B. (2026). New Insight into Endophytic Fungi–Plant Symbioses Under Climate Change: Molecular Crosstalk, Nutrient Exchange, and Ecosystem Resilience. *Applied Microbiology, 6*(3), 47. https://doi.org/10.3390/applmicrobiol6030047

Alzahrani, A. J., Alghamdi, A. G., & Ibrahim, H. M. (2024). Assessment of soil loss due to wind erosion and dust deposition: Implications for sustainable management in arid regions. *Applied Sciences (Basel, Switzerland), 14*(23), 10822. DOI: 10.3390/app142310822

Åqvist, J., Isaksen, G. V., & Brandsdal, B. O. (2017). Computation of enzyme cold adaptation. *Nature Reviews Chemistry, 1*(7), 0051.

Beattie, G. A., Edlund, A., Esiobu, N., Gilbert, J., Nicolaisen, M. H., Jansson, J. K., Jensen, P., Keiluweit, M., Lennon, J. T., Martiny, J., Minnis, V. R., Newman, D., Peixoto, R., Schadt, C., & Van der Meer, J. R. (2025). Soil microbiome interventions for carbon sequestration and climate mitigation. *mSystems, 10*(1), e01129–e24. DOI: 10.1128/msystems.01129-24 PMID: 39692482

Bidast, S., & Golchin, A. (2025). Soil carbon storage in rangeland ecosystems: The role of land management, climate change, and soil chemical transformations. *Geoderma Regional, 42*, 00984. DOI: 10.1016/j.geodrs.2025.e00984

Binter, J., Macek, M., & Dolezal, J. (2025). Integrating morphological, anatomical, and physiological traits to explain elevational distributions in Himalayan steppe and alpine plants. *Journal of Integrative Plant Biology, 67*(10), 2643–2657. DOI: 10.1111/jipb.13971 PMID: 40662507

Bitew, A., Molla, E., Meshesha, D. T., Wato, T., Getnet, K., & Miju, C. (2025). Land use dynamics and its effect on selected soil physicochemical properties in Necho watershed, northwest Ethiopia. *Agrosystems, Geosciences & Environment, 8*(1), e70076. DOI: 10.1002/agg2.70076

Bonfanti, N., Clément, J. C., Münkemüller, T., Barré, P., Baudin, F., & Poulenard, J. (2025). Prolonged warming leads to carbon depletion and increases nutrient availability in alpine soils. *Applied Soil Ecology, 213*, 106239. DOI: 10.1016/j.apsoil.2025.106239

Bright, K., Dienes, B., Keiluweit, M., Rixen, C., & Aeppli, M. (2025). Climate change impacts on organic carbon cycling in European alpine soils. *Soil Biology & Biochemistry, 109891*, 109891. Advance online publication. DOI: 10.1016/j.soilbio.2025.109891

Broadbent, A. A., Newbold, L. K., Pritchard, W. J., Michas, A., Goodall, T., Cordero, I., Giunta, A., Snell, H. S. K., Pepper, V. V. L. H., Grant, H. K., Soto, D. X., Kaufmann, R., Schloter, M., Griffiths, R. I., Bahn, M., & Bardgett, R. D. (2024). Climate change disrupts the seasonal coupling of plant and soil microbial nutrient cycling in an alpine ecosystem. *Global Change Biology, 30*(3), e17245. DOI: 10.1111/gcb.17245 PMID: 38511487

Broadbent, A. A., Snell, H. S., Michas, A., Pritchard, W. J., Newbold, L., Cordero, I., Goodall, T., Schallhart, N., Kaufmann, R., Griffiths, R. I., Schloter, M., Bahn, M., & Bardgett, R. D. (2021). Climate change alters temporal dynamics of alpine soil microbial functioning and biogeochemical cycling via earlier snowmelt. *The ISME Journal, 15*(8), 2264–2275. DOI: 10.1038/s41396-021-00922-0 PMID: 33619353

Cai, Y., Chen, H., Zhao, W., Sun, Y., Wang, M., Chen, Y., & Ma, J. (2025). Investigation of greenhouse gas emissions in soil remediation activities based on the integration of machine learning models and life cycle assessment. *Journal of Cleaner Production, 487*, 144630. DOI: 10.1016/j.jclepro.2024.144630

Chandra, S., Trivedi, V., Purohit, V. K., Nautiyal, M. C., & Chandra, S. (2025). Climatic change and physiological adaptations in alpine medicinal plants. In *High altitude medicinal plants: Botany, conservation and cultivation* (pp. 11–35). Springer Nature Switzerland.

Dalling, J. W., Heineman, K., González, G., & Ostertag, R. (2016). Geographic, environmental and biotic sources of variation in the nutrient relations of tropical montane forests. *Journal of Tropical Ecology, 32*(5), 368–383. DOI: 10.1017/S0266467415000619

Das, D., Ray, P., & Datta, S. P. (2024). Nutrient cycling and climate change. In *Climate change impacts on soil–plant–atmosphere continuum* (pp. 37–62). Springer Nature Singapore. DOI: 10.1007/978-981-99-7935-6_2

Deb, P., & Mandal, A. Harendra, Choudhury, S. R., Das, A., Hazra, S., ... & Moulick, D. (2023). Improving plant nutrient use efficiency for climate-resilient agriculture. In *Climate-resilient agriculture, Vol. 2: Agro-biotechnological advancement for crop production* (pp. 209–243). Springer International Publishing.

Deng, Y., Lu, Y., Du, X., Ye, X., & Feng, J. (2025). Identifying spatial patterns and driving factors of anthropogenic impacts on the groundwater environment based on groundwater chemical kinetics. *Journal of Cleaner Production*, *486*, 144436. DOI: 10.1016/j.jclepro.2024.144436

Dietrich, V., Niederberger, J., Frank, J., & Hauck, M. (2025). Does atmospheric nitrogen deposition cause critical nutrient imbalances in temperate tree regeneration? *Trees (Berlin)*, *39*(2), 38. DOI: 10.1007/s00468-025-02615-z

Ding, J., & Yu, S. (2025). Impacts of land use on soil nitrogen-cycling microbial communities: Insights from community structure, functional gene abundance, and network complexity. *Life (Chicago, Ill.)*, *15*(3), 466. DOI: 10.3390/life15030466 PMID: 40141811

Donhauser, J., & Frey, B. (2018). Alpine soil microbial ecology in a changing world. *FEMS Microbiology Ecology*, *94*(9), fiy099. DOI: 10.1093/femsec/fiy099 PMID: 30032189

Dou, T., Zhang, K., Shi, X., Liu, W., Yu, F., & Liu, D. (2025). Crop–mushroom rotation: A comprehensive review of its multifaceted impacts on soil quality, agricultural sustainability, and ecosystem health. *Agronomy (Basel)*, *15*(3), 563. DOI: 10.3390/agronomy15030563

Du, K., Shao, Y., Yao, N., Yu, H., Ma, S., Mao, X., Wang, L., & Wang, J. (2025). Alpine Meadow Fractional Vegetation Cover Estimation Using UAV-Aided Sentinel-2 Imagery. *Sensors (Basel)*, *25*(14), 4506. DOI: 10.3390/s25144506 PMID: 40732634

Du, Y., Yang, Y., Wu, S., Gao, X., He, X., & Dong, S. (2025). Core microbes regulate plant–soil resilience by maintaining network resilience during long-term restoration of alpine grasslands. *Nature Communications*, *16*(1), 3116. DOI: 10.1038/s41467-025-58080-2 PMID: 40169576

Fartyal, A., Bargali, S. S., Bargali, K., & Negi, B. (2025). Changes in soil properties, organic carbon, and nutrient stocks after land-use change from forests to grasslands in Kumaun Himalaya, India. *Land Degradation & Development*, *36*(7), 2438–2457. DOI: 10.1002/ldr.5507

Feng, J., Yin, X., Wang, C., Hu, L., & Zhou, T. (2025). Effects of snowpack changes on functional traits and reproductive allocation of four perennial plants in an alpine meadow. *BMC Plant Biology*, *25*(1), 834. DOI: 10.1186/s12870-025-06855-6 PMID: 40604470

Garcia, J., & Kao-Kniffin, J. (2018). Microbial group dynamics in plant rhizospheres and their implications on nutrient cycling. *Frontiers in Microbiology*, *9*, 1516. DOI: 10.3389/fmicb.2018.01516 PMID: 30050510

Gavini, S. S., Ezcurra, C., & Aizen, M. A. (2019). Plant–plant interactions promote alpine diversification. *Evolutionary Ecology*, *33*(2), 195–209. DOI: 10.1007/s10682-019-09972-5

Gojon, A., Nussaume, L., Luu, D. T., Murchie, E. H., Baekelandt, A., Rodrigues Saltenis, V. L., Cohan, J.-P., Desnos, T., Inzé, D., Ferguson, J. N., Guiderdonni, E., Krapp, A., Klein Lankhorst, R., Maurel, C., Rouached, H., Parry, M. A. J., Pribil, M., Scharff, L. B., & Nacry, P. (2023). Approaches and determinants to sustainably improve crop production. *Food and Energy Security*, *12*(1), e369. DOI: 10.1002/fes3.369

Grêt-Regamey, A., Brunner, S. H., & Kienast, F. (2012). Mountain ecosystem services: Who cares? *Mountain Research and Development*, *32*(S1), S23–S34. DOI: 10.1659/MRD-JOURNAL-D-10-00115.S1

Guo, Y., Liao, H. L., Boughton, E. H., Martens-Habbena, W., & Qiu, J. (2023). Effects of land-use intensity, grazing and fire disturbances on soil bacterial and fungal communities in subtropical wetlands. *Agriculture, Ecosystems & Environment*, *345*, 108314. DOI: 10.1016/j.agee.2022.108314

Guo, Y., Peng, H., Wang, Q., Wang, J., Wu, Z., Shao, B., Xing, G., Huang, Z., Zhao, F., Cui, H., Cui, X., & Tong, Y. (2025). Unveiling the global dynamics of dissolved organic carbon in aquatic ecosystems: Climatic and anthropogenic impact, and future predictions. *The Science of the Total Environment*, *958*, 178109. DOI: 10.1016/j.scitotenv.2024.178109 PMID: 39693645

Hao, Y., Mao, J., Bachmann, C. M., Hoffman, F. M., Koren, G., Chen, H., ... & Dai, Y. (2025). Soil moisture controls over carbon sequestration and greenhouse gas emissions: A review. *npj Climate and Atmospheric Science*, *8*(1), 16.

Hassan, N., Rafiq, M., Hayat, M., Shah, A. A., & Hasan, F. (2016). Psychrophilic and psychrotrophic fungi: A comprehensive review. *Reviews in Environmental Science and Biotechnology*, *15*(2), 147–172. DOI: 10.1007/s11157-016-9395-9

Hedberg, C. P., & Smith, F. A. (2025). Unraveling the role of environmental and anthropogenic drivers in shaping global patterns in mammal diversity. *Ecography*, *2025*(10), 07674. DOI: 10.1002/ecog.07674

Helmi, N. R. (2025). Exploring the diversity and antimicrobial potential of actinomycetes isolated from different environments in Saudi Arabia: A systematic review. *Frontiers in Microbiology*, *16*, 1568899. DOI: 10.3389/fmicb.2025.1568899 PMID: 40207161

Hooshyarkhah, B., Johnson, D. L., Spencer, L., Ryait, H. S., & Chegoonian, A. (2025). Mapping Four Decades of Treeline Ecotone Migration: Remote Sensing of Alpine Ecotone Shifts on the Eastern Slopes of the Canadian Rocky Mountains. *Remote Sensing (Basel)*, *17*(24), 4004. DOI: 10.3390/rs17244004

Hossain, M. M., Sultana, F., Mostafa, M., Ferdus, H., Kundu, M., Adhikary, S., Akter, N., Saha, A., & Sabbir, M. A. A. (2025). Stress responses and mechanisms of phytopathogens infecting humans: Threats, drivers, and recommendations. *Stresses*, *5*(2), 28. DOI: 10.3390/stresses5020028

Hou, M., Leng, C., Zhu, J., Yang, M., Yin, Y., Xing, Y., & Chen, J. (2024). Alpine and subalpine plant microbiome mediated plants adapt to the cold environment: A systematic review. *Environmental Microbiome*, *19*(1), 82. DOI: 10.1186/s40793-024-00614-0 PMID: 39487507

Hou, Z., Wang, R., Chang, S., Zheng, Y., Ma, T., Xu, S., Zhang, X., Shi, X., Lu, J., Luo, D., Wang, B., Du, Z., & Wei, Y. (2024). The contribution of microbial necromass to soil organic carbon and influencing factors along a variation of habitats in alpine ecosystems. *The Science of the Total Environment*, *921*, 171126. DOI: 10.1016/j.scitotenv.2024.171126 PMID: 38387574

Hui, M. L. Y., Tan, L. T. H., Letchumanan, V., He, Y. W., Fang, C. M., Chan, K. G., Law, J. W.-F., & Lee, L. H. (2021). The extremophilic actinobacteria: From microbes to medicine. *Antibiotics (Basel, Switzerland)*, *10*(6), 682. DOI: 10.3390/antibiotics10060682 PMID: 34201133

Imran, & Ortas, I. (2025). Mechanism in soil health improvement with soil microbial actions and its role in potential agriculture. *Communications in Soil Science and Plant Analysis*, 1–22.

IPCC. (2022). *Climate Change 2022: Impacts, Adaptation and Vulnerability. Contribution of Working Group II to the Sixth Assessment Report of the Intergovernmental Panel on Climate Change* (Pörtner, H.-O., Eds.). Cambridge University Press., DOI: 10.1017/9781009325844

Iqbal, S., Begum, F., Nguchu, B. A., Claver, U. P., & Shaw, P. (2025). The invisible architects: Microbial communities and their transformative role in soil health and global climate changes. *Environmental Microbiome*, *20*(1), 36. DOI: 10.1186/s40793-025-00694-6 PMID: 40133952

Ishfaq, M., Tam, N. F. Y., Lang, T., Hussain, M., & Zhou, H. (2025). Nitrogen–phosphorus conservation and trade-offs in mangroves. *Plant and Soil*, •••, 1–20.

Jia, F., Li, W., Zhou, B., Xiao, Y., Si, B., Saikeshan, A., Lv, T., & Wang, Z. (2025). Consecutive high-efficient water-saving irrigation increase crop yield and decrease soil salinity through reconstructing rhizosphere soil bacterial communities. *The Science of the Total Environment*, *959*, 178238. DOI: 10.1016/j.scitotenv.2024.178238 PMID: 39742583

Jiang, C., Yu, G., Li, Y., Cao, G., Yang, Z., Sheng, W., & Yu, W. (2012). Nutrient resorption of coexistence species in alpine meadow of the Qinghai–Tibetan Plateau explains plant adaptation to nutrient-poor environment. *Ecological Engineering*, *44*, 1–9. DOI: 10.1016/j.ecoleng.2012.04.006

Johnstone, J. (2021). *Alpine plant life: Functional plant ecology of high mountain ecosystems*.

Juan, Y., Tian, L., Sun, W., Qiu, W., Curtin, D., Gong, L., & Liu, Y. (2020). Simulation of soil freezing–thawing cycles under typical winter conditions: Implications for nitrogen mineralization. *Journal of Soils and Sediments*, *20*(1), 143–152. DOI: 10.1007/s11368-019-02374-8

Kamanmalek, S., & Alamdari, N. (2024). Advancing equitable stormwater management: A decision support tool integrating best practices for nutrient removal and environmental justice. *Ecological Informatics*, *80*, 102496. DOI: 10.1016/j.ecoinf.2024.102496

Kang, D., Lungu, S. E., Danso, F., Dzou, C. F., Chen, Y., Zheng, X., Nie, F., Lin, H., Chen, J., & Zhou, G. (2025). Animal health and nutrition: Metabolic disorders in cattle and improvement strategies. *Frontiers in Veterinary Science*, *12*, 1470391. DOI: 10.3389/fvets.2025.1470391 PMID: 40201075

Kaur, S., Das, A., Dwivedi, S., Tripathi, S., & Shrivastava, N. (2025). Effect of climate change on beneficial plant–microbe interaction in the rhizosphere. In *Plant–microbiome interactions for climate-resilient agriculture* (pp. 99–120). DOI: 10.1007/978-981-96-3534-4_5

Khan, M. T., Supronienė, S., Žvirdauskienė, R., & Aleinikovienė, J. (2025). Climate, soil, and microbes: Interactions shaping organic matter decomposition in croplands. *Agronomy (Basel)*, *15*(8), 1928. DOI: 10.3390/agronomy15081928

Lai, H., Gao, F., Su, H., Zheng, P., Li, Y., & Yao, H. (2022). Nitrogen distribution and soil microbial community characteristics in a legume–cereal intercropping system: A review. *Agronomy (Basel)*, *12*(8), 1900. DOI: 10.3390/agronomy12081900

Lallemand, F., Martin-Magniette, M. L., Gilard, F., Gakière, B., Launay-Avon, A., Delannoy, É., & Selosse, M. A. (2019). In situ transcriptomic and metabolomic study of the loss of photosynthesis in the leaves of mixotrophic plants exploiting fungi. *The Plant Journal*, *98*(5), 826–841. DOI: 10.1111/tpj.14276 PMID: 30735596

Lazzaro, A., Hilfiker, D., & Zeyer, J. (2015). Structures of microbial communities in alpine soils: Seasonal and elevational effects. *Frontiers in Microbiology*, *6*, 1330. DOI: 10.3389/fmicb.2015.01330 PMID: 26635785

Li, J., Wang, X., Yuan, M., Duan, W., Xia, J., Zhang, X., Zhao, Y., & Wang, J. (2025). Effect of soil microbial community on ecosystem multifunctionality in an alpine grassland. *Catena*, *249*, 108714. DOI: 10.1016/j.catena.2025.108714

Li, R., He, Z., He, P., Yang, R., Ma, D., Li, Y., Xiang, Y., Sun, W., Zhu, X., & Zhang, Z. (2025). Crop diversification improves productivity and soil health by regulating fungal diversity. *Soil & Tillage Research*, *252*, 106628. DOI: 10.1016/j.still.2025.106628

Li, S., Zhou, X., Liu, L., Su, Z., Zhao, J., Zhang, J., Cai, Z., Peñuelas, J., & Huang, X. (2025). Plant diversity reduces the risk of antibiotic resistance genes in agroecosystems. *Advancement of Science*, *12*(11), 2410990. DOI: 10.1002/advs.202410990 PMID: 39874208

Li, W., Lu, Q., Alharbi, S. A., Soromotin, A. V., Kuzyakov, Y., & Lei, Y. (2023). Plant–soil–microbial interactions mediate vegetation succession in retreating glacial forefields. *The Science of the Total Environment*, *873*, 162393. DOI: 10.1016/j.scitotenv.2023.162393 PMID: 36841408

Li, W., Zheng, T. D., Cheng, X. P., & He, S. Q. (2023). Changes in functional traits and diversity of typical alpine grasslands after a short-term trampling disturbance. *Frontiers in Ecology and Evolution*, *11*, 1154911. DOI: 10.3389/fevo.2023.1154911

Li, X., Zeng, R., & Liao, H. (2016). Improving crop nutrient efficiency through root architecture modifications. *Journal of Integrative Plant Biology*, *58*(3), 193–202. DOI: 10.1111/jipb.12434 PMID: 26460087

Liebelt, D. J., Jordan, J. T., & Doherty, C. J. (2019). Only a matter of time: The impact of daily and seasonal rhythms on phytochemicals. *Phytochemistry Reviews*, *18*(6), 1409–1433. DOI: 10.1007/s11101-019-09617-z

Liu, F., Zhang, W., & Li, S. (2025). Effects of freeze–thaw cycles on uptake preferences of plants for nutrient: A review. *Plants*, *14*(7), 1122. DOI: 10.3390/plants14071122 PMID: 40219190

Liu, M., Wen, J. H., Chen, Y. M., Xu, W. J., Wang, Q., & Ma, Z. L. (2022). Warming increases soil carbon input in a *Sibiraea angustata*-dominated alpine shrub ecosystem. *Journal of Plant Ecology*, *15*(2), 335–346. DOI: 10.1093/jpe/rtab101

Liu, X., Cong, J., Lu, H., Xue, Y., Wang, X., Li, D., & Zhang, Y. (2017). Community structure and elevational distribution pattern of soil Actinobacteria in alpine grasslands. *Acta Ecologica Sinica*, *37*(4), 213–218. DOI: 10.1016/j.chnaes.2017.02.010

Liu, Y., Evans, S. E., Friesen, M. L., & Tiemann, L. K. (2022). Root exudates shift how N mineralization and N fixation contribute to the plant-available N supply in low fertility soils. *Soil Biology & Biochemistry*, *165*, 108541. DOI: 10.1016/j.soilbio.2021.108541

Losapio, G., Lee, J. R., Fraser, C. I., Gillespie, M. A., Kerr, N. R., Zawierucha, K., ... & Brown, L. E. (2025). Impacts of deglaciation on biodiversity and ecosystem function. *Nature Reviews Biodiversity*, 1–15.

Lukac, M., Calfapietra, C., Lagomarsino, A., & Loreto, F. (2010). Global climate change and tree nutrition: Effects of elevated CO_2 and temperature. *Tree Physiology*, *30*(9), 1209–1220. DOI: 10.1093/treephys/tpq040 PMID: 20571150

Macci, C., Vannucchi, F., Scartazza, A., Masciandaro, G., Doni, S., & Peruzzi, E. (2025). Soil–plant indicators for assessing nutrient cycling and ecosystem functionality in urban forestry. *Urban Science (Basel, Switzerland)*, *9*(3), 82. DOI: 10.3390/urbansci9030082

Mage, S. M., & Porder, S. (2013). Parent material and topography determine soil phosphorus status in the Luquillo Mountains of Puerto Rico. *Ecosystems (New York, N.Y.)*, *16*(2), 284–294. DOI: 10.1007/s10021-012-9612-5

Mansoor, S., Manzoor, S., Gull, R., Gani, G., Wani, O. A., Farooq, S., Popescu, S. M., Arya, V. M., Park, W.-P., & Chung, Y. S. (2025). Biochar mediated remediation of emerging inorganic pollutants and their toxicological effects on plant and soil health. *Journal of Soil Science and Plant Nutrition*, *25*(1), 1612–1642. DOI: 10.1007/s42729-025-02227-z

Mazzocchi, C., & Sali, G. (2022). Tourists' perception of ecosystem services provided by mountain agriculture. *Sustainability (Basel), 14*(19), 12171. DOI: 10.3390/su141912171

Medhi, K., Bhardwaj, R., & Laxmi, R. (2021). Climate change with its impacts on soil and soil microbiome regulating biogeochemical nutrient transformations. In *Climate change and the microbiome: Sustenance of the ecosphere* (pp. 95–138). Springer International Publishing. DOI: 10.1007/978-3-030-76863-8_6

Mitterwallner, V., Steinbauer, M., Mathes, G., & Walentowitz, A. (2024). Global reduction of snow cover in ski areas under climate change. *PLoS One, 19*(3), e0299735. DOI: 10.1371/journal.pone.0299735 PMID: 38478484

Monson, R. K., Rosenstiel, T. N., Forbis, T. A., Lipson, D. A., & Jaeger, C. H.III. (2006). Nitrogen and carbon storage in alpine plants. *Integrative and Comparative Biology, 46*(1), 35–48. DOI: 10.1093/icb/icj006 PMID: 21672721

Mou, Z., Kuang, L., Zhang, J., Li, Y., Wu, W., Liang, C., Hui, D., Lambers, H., Sardans, J., Peñuelas, J., Liu, J., Ren, H., & Liu, Z. (2023). Nutrient availability and stoichiometry mediate microbial effects on soil carbon sequestration in tropical forests. *Soil Biology & Biochemistry, 186*, 109186. DOI: 10.1016/j.soilbio.2023.109186

Mukhia, S., Kumar, A., Kumari, P., & Kumar, R. (2022). Psychrotrophic plant beneficial bacteria from the glacial ecosystem of Sikkim Himalaya: Genomic evidence for the cold adaptation and plant growth promotion. *Microbiological Research, 260*, 127049. DOI: 10.1016/j.micres.2022.127049 PMID: 35504236

Murtaza, G., Ahmed, Z., Iqbal, R., & Deng, G. (2025). Biochar from agricultural waste as a strategic resource for promotion of crop growth and nutrient cycling of soil under drought and salinity stress conditions: A comprehensive review with context of climate change. *Journal of Plant Nutrition, 48*(11), 1832–1883. DOI: 10.1080/01904167.2025.2460769

Napoletano, M., Bellino, A., & Baldantoni, D. (2025). Depletion and recovery of soil organic matter: Ecological, economic and social implications. *Ecological Civilization, 2*(2), 10002. DOI: 10.70322/ecolciviliz.2025.10002

Nema, A., Kumar, A., & Warudkar, V. (2025). An in-depth critical review of different carbon capture techniques: Assessing their effectiveness and role in reducing climate change emissions. *Energy Conversion and Management, 323*, 119244. DOI: 10.1016/j.enconman.2024.119244

Njeru, C. M., Ekesi, S., Mohamed, S. A., Kinyamario, J. I., Kiboi, S., & Maeda, E. E. (2017). Assessing stock and thresholds detection of soil organic carbon and nitrogen along an altitude gradient in an East Africa mountain ecosystem. *Geoderma Regional*, *10*, 29–38. DOI: 10.1016/j.geodrs.2017.04.002

Oliver, T. H., Heard, M. S., Isaac, N. J., Roy, D. B., Procter, D., Eigenbrod, F., Freckleton, R., Hector, A., Orme, C. D. L., Petchey, O. L., Proença, V., Raffaelli, D., Suttle, K. B., Mace, G. M., Martín-López, B., Woodcock, B. A., & Bullock, J. M. (2015). Biodiversity and resilience of ecosystem functions. *Trends in Ecology & Evolution*, *30*(11), 673–684. DOI: 10.1016/j.tree.2015.08.009 PMID: 26437633

Palomo, I. (2017). Climate change impacts on ecosystem services in high mountain areas: A literature review. *Mountain Research and Development*, *37*(2), 179–187. DOI: 10.1659/MRD-JOURNAL-D-16-00110.1

Pandey, S. N. (2020). Role of micronutrients in biochemical responses of crops under abiotic stresses. In *Sustainable agriculture in the era of climate change* (pp. 93–112). Springer International Publishing. DOI: 10.1007/978-3-030-45669-6_4

Pang, J., Tian, Z., Zhang, M., Wang, Y., Qi, T., Zhang, Q., & Zhang, P. (2025). Enhancing carbon sequestration and greenhouse gas mitigation in semiarid farmland: The promising role of biochar application with biodegradable film mulching. *Journal of Integrative Agriculture*, *24*(2), 517–535.

Penuelas, J., & Sardans, J. (2023). Human-driven global nutrient imbalances increase risks to health. *Eco-Environment & Health*, *2*(4), 246–251. DOI: 10.1016/j.eehl.2023.08.003 PMID: 38435356

Pütz, M., Gallati, D., Kytzia, S., Elsasser, H., Lardelli, C., Teich, M., Waltert, F., & Rixen, C. (2011). Winter tourism, climate change, and snowmaking in the Swiss Alps: Tourists' attitudes and regional economic impacts. *Mountain Research and Development*, *31*(4), 357–362. DOI: 10.1659/MRD-JOURNAL-D-11-00039.1

Qasim, M., Babar, S., Younas, M. U., Rajput, N., Xia, X., Wang, J., Wang, X., & Jiang, C. (2025). Exploration of biochars for enhancing soil health, carbon sequestration, and greenhouse gas emission reductions under citrus cultivation. *European Journal of Agronomy*, *168*, 127661. DOI: 10.1016/j.eja.2025.127661

Rosbakh, S., Porro, F., Abeli, T., Di Cecco, V., Erschbamer, B., Fernández-Calzado, R., Jimenez-Alfaro, B., Lodetti, S., Lorite, J., Moser, D., Orsenigo, S., Pauli, H., Petraglia, A., Rossi, G., Saccone, P., Stanisci, A., White, F. J., Winkler, M., & Mondoni, A. (2025). Understanding long-term abundance shifts in European alpine plants through the lenses of functional seed trait ecology. *Diversity & Distributions*, *31*(7), e70047. DOI: 10.1111/ddi.70047

Saeed, L., Zuhra, N., & Saeed, S. (2025). Nitrogen cycling and its dynamics in soil–plant systems, future perspective and research directions: A comprehensive review. *The Critical Review of Social Sciences Studies, 3*(4), 933–954. DOI: 10.59075/8wkz9d78

Sandra, P. R., Shankar, A., Garkoti, S. C., & Adarsh, C. K. (2025). Understanding the effects of forest types, vegetation structural diversity, and soil properties on above- and below-ground carbon stock of moist deciduous forest ecosystems in Western Ghats, India. *Catena, 257*, 109198. DOI: 10.1016/j.catena.2025.109198

Schirpke, U., Scolozzi, R., Dean, G., Haller, A., Jäger, H., Kister, J., Kovács, B., Sarmiento, F. O., Sattler, B., & Schleyer, C. (2020). Cultural ecosystem services in mountain regions: Conceptualising conflicts among users and limitations of use. *Ecosystem Services, 46*, 101210. DOI: 10.1016/j.ecoser.2020.101210

Schmid, M. W., van Moorsel, S. J., Hahl, T., De Luca, E., De Deyn, G. B., Wagg, C., Niklaus, P. A., & Schmid, B. (2021). Effects of plant community history, soil legacy and plant diversity on soil microbial communities. *Journal of Ecology, 109*(8), 3007–3023. DOI: 10.1111/1365-2745.13714

Seastedt, T. R., & Oldfather, M. F. (2021). Climate change, ecosystem processes and biological diversity responses in high elevation communities. *Climate (Basel), 9*(5), 87. DOI: 10.3390/cli9050087

Sehgal, H., Sharma, E., Pant, C., Malhotra, S., & Joshi, M. (2025). Impact of climate change on soil health and nutrient cycling. In *Soil health and nutrition management* (pp. 109–128). CABI. DOI: 10.1079/9781800624597.0006

Sen, S. K., Jain, P., Singh, R. K., & Singh, A. K. (2025). Role of soil algae in the maintenance of soil fertility and soil health. In *Soil algae: Morphology, ecology and biotechnological applications* (pp. 293–304). Springer Nature Singapore. DOI: 10.1007/978-981-97-9873-5_9

Shafi, Z., Shahid, M., Ilyas, T., & Singh, K. (2025). Next-generation perspectives on microbially synthesized siderophores: Molecular engineering, multi-omics insights, and applications for smart climate-resilient crops. *World Journal of Microbiology & Biotechnology, 41*(10), 393. DOI: 10.1007/s11274-025-04575-5 PMID: 41099996

Shahid, M., Singh, U. B., Khan, M. S., Singh, P., Kumar, R., Singh, R. N., Kumar, A., & Singh, H. V. (2023). Bacterial ACC deaminase: Insights into enzymology, biochemistry, genetics, and potential role in amelioration of environmental stress in crop plants. *Frontiers in Microbiology, 14*, 1132770. DOI: 10.3389/fmicb.2023.1132770 PMID: 37180266

Shang, Q., Li, C., & Liu, Y. (2024). The negative effects of tourist trampling on the soil physical properties and microbial community composition in a natural oak forest. *Forests*, *15*(8), 1419. DOI: 10.3390/f15081419

Sharma, D., Astapati, A. D., Dey, S., Bhattacharjee, D., Singha, D. M., & Nath, S. (2025). Microbial interactions in soil ecosystems: Facilitating plant growth, nutrient cycling, and environmental dynamics. In *Microbial allies: Understanding plant–microbe interactions in sustainable agriculture* (pp. 21–48). Springer Nature Switzerland. DOI: 10.1007/978-3-031-90530-8_2

Shen, M., Song, B., Zhou, C., Almatrafi, E., Hu, T., Zeng, G., & Zhang, Y. (2022). Recent advances in impacts of microplastics on nitrogen cycling in the environment: A review. *The Science of the Total Environment*, *815*, 152740. DOI: 10.1016/j.scitotenv.2021.152740 PMID: 34974017

Sindhu, S. S., Parmar, P., & Phour, M. (2013). Nutrient cycling: Potassium solubilization by microorganisms and improvement of crop growth. In *Geomicrobiology and biogeochemistry* (pp. 175–198). Springer.

Singh, G., Singh, A., Singh, P., Shukla, R., Tripathi, S., & Mishra, V. K. (2021). The fate of organic pollutants and their microbial degradation in water bodies. In *Pollutants and water management: Resources, strategies and scarcity* (pp. 210–240). DOI: 10.1002/9781119693635.ch9

Singh, P., Choudhary, K. K., Chaudhary, N., Gupta, S., Sahu, M., Tejaswini, B., & Sarkar, S. (2022). Salt stress resilience in plants mediated through osmolyte accumulation and its crosstalk mechanism with phytohormones. *Frontiers in Plant Science*, *13*, 1006617. DOI: 10.3389/fpls.2022.1006617 PMID: 36237504

Singh, S. S., Lamare, E., Chopade, B. A., Chanda, R., Devi, N. R., Upadhyay, K. K., & Tripathi, S. K. (2025). The hidden players: Soil microbes in ecosystems sustainability. In *Climate change and soil microorganisms for environmental sustainability* (pp. 295–320). Springer Nature Singapore. DOI: 10.1007/978-981-96-3425-5_13

Song, X., Fang, C., Yuan, Z. Q., & Li, F. M. (2021). Long-term growth of alfalfa increased soil organic matter accumulation and nutrient mineralization in a semi-arid environment. *Frontiers in Environmental Science*, *9*, 649346. DOI: 10.3389/fenvs.2021.649346

Stefan, L., Hartmann, M., Engbersen, N., Six, J., & Schöb, C. (2021). Positive effects of crop diversity on productivity driven by changes in soil microbial composition. *Frontiers in Microbiology*, *12*, 660749. DOI: 10.3389/fmicb.2021.660749 PMID: 33936016

Steiger, R., Scott, D., Abegg, B., Pons, M., & Aall, C. (2019). A critical review of climate change risk for ski tourism. *Current Issues in Tourism, 22*(11), 1343–1379. DOI: 10.1080/13683500.2017.1410110

Sun, H., Wang, L., Kumar, A., Auwal, M., Van Zwieten, L., Ge, T., & Kuzyakov, Y. (2025). Nutrient availability mediates organic carbon turnover in paddy soils through regulating microbial metabolism. *Geoderma, 458*, 117313. DOI: 10.1016/j.geoderma.2025.117313

Talal, M., Chen, X., Iqbal, I., & Ali, I. (2025). A critical review: Unearthing the hidden players—The role of extremophilic fungi in forest ecosystems. *Forests, 16*(5), 855. DOI: 10.3390/f16050855

Tian, L., Zhao, L., Wu, X., Fang, H., Zhao, Y., Yue, G., Liu, G., & Chen, H. (2017). Vertical patterns and controls of soil nutrients in alpine grassland: Implications for nutrient uptake. *The Science of the Total Environment, 607*, 855–864. DOI: 10.1016/j.scitotenv.2017.07.080 PMID: 28711847

Tiwari, P., Bose, S. K., Park, K. I., Dufossé, L., & Fouillaud, M. (2024). Plant–microbe interactions under the extreme habitats and their potential applications. *Microorganisms, 12*(3), 448. DOI: 10.3390/microorganisms12030448 PMID: 38543499

Tsai, H. H., Tang, Y., Jiang, L., Xu, X., Dénervaud Tendon, V., Pang, J., Jia, Y., Wippel, K., Vacheron, J., Keel, C., Andersen, T. G., Geldner, N., & Zhou, F. (2025). Localized glutamine leakage drives the spatial structure of root microbial colonization. *Science, 390*(6768), eadu4235. DOI: 10.1126/science.adu4235 PMID: 41037624

Wahab, A., Muhammad, M., Munir, A., Abdi, G., Zaman, W., Ayaz, A., Khizar, C., & Reddy, S. P. P. (2023). Role of arbuscular mycorrhizal fungi in regulating growth, enhancing productivity, and potentially influencing ecosystems under abiotic and biotic stresses. *Plants, 12*(17), 3102. DOI: 10.3390/plants12173102 PMID: 37687353

Wan, L., Liu, G., & Su, X. (2025). Organic fertilization balances biodiversity maintenance, grass production, soil storage, nutrient cycling and greenhouse gas emissions for sustainable grassland development in China: A meta-analysis. *Agriculture, Ecosystems & Environment, 381*, 109473. DOI: 10.1016/j.agee.2025.109473

Wang, J., He, Q., Wu, Y., Zhu, H., Sun, H., Zhou, J., Wang, D., Li, J., & Bing, H. (2021). Effects of pioneer N_2-fixing plants on the resource status and establishment of neighboring non-N_2-fixing plants in a newly formed glacier floodplain, eastern Tibetan Plateau. *Plant and Soil, 458*(1), 261–276. DOI: 10.1007/s11104-020-04462-y

Wang, J., Luo, P., Yang, H., Mou, C., & Mo, L. (2016). Different responses of alpine plants to nitrogen addition: Effects on plant–plant interactions. *Scientific Reports*, *6*(1), 38320. DOI: 10.1038/srep38320 PMID: 27922131

Wang, W., Shu, X., Zhang, Q., & Guénon, R. (2015). Effects of freeze–thaw cycles on the soil nutrient balances, infiltration, and stability of cyanobacterial soil crusts in northern China. *Plant and Soil*, *386*(1), 263–272. DOI: 10.1007/s11104-014-2263-x

Wang, X. M., Zhao, G. Q., Chen, J., Jing, F. Z., Li, L., Wang, Y. Y., Ma, P., Wu, Y.-H., Xu, S.-J., & He, W. L. (2025). Adaptation of plants to UV-B radiation with altitude in Tuha Basin: Synergistic regulation of epidermal structure, secondary metabolites, and organic element allocation. *Life (Chicago, Ill.)*, *15*(9), 1375. DOI: 10.3390/life15091375 PMID: 41010317

Wang, Z., Meng, Q., Allam, M., Hu, D., Zhang, L., & Menenti, M. (2021). Environmental and anthropogenic drivers of surface urban heat island intensity: A case-study in the Yangtze River Delta, China. *Ecological Indicators*, *128*, 107845. DOI: 10.1016/j.ecolind.2021.107845

Wipf, S., Gottfried, M., & Nagy, L. (2013). Climate change and extreme events – Their impacts on alpine and arctic ecosystem structure and function. *Plant Ecology & Diversity*, *6*(3–4), 303–306. DOI: 10.1080/17550874.2013.819533

Wipf, S., Sommerkorn, M., Stutter, M. I., Wubs, E. J., & Van der Wal, R. (2015). Snow cover, freeze–thaw, and the retention of nutrients in an oceanic mountain ecosystem. *Ecosphere*, *6*(10), 1–16. DOI: 10.1890/ES15-00099.1

Wu, H., Wu, N., Ling, Q., Tang, T., Hu, P., Duan, P., Zhang, Q., Xiao, J., Zhao, J., Zhang, W., Chen, H., & Wang, K. (2025). Potential roles of soil viruses in Karst forest soil carbon and nitrogen cycles. *Forests*, *16*(5), 735. DOI: 10.3390/f16050735

Xing, Y., Wang, X., & Mustafa, A. (2025). Exploring the link between soil health and crop productivity. *Ecotoxicology and Environmental Safety*, *289*, 117703. DOI: 10.1016/j.ecoenv.2025.117703 PMID: 39808880

Xu, H., Liu, N., Yang, G., Zhang, H., Badgery, W. B., & Zhang, Y. (2025). Snow interacts with defoliation height to drive grassland sustainability via grass biomass maintenance. *Rangeland Ecology and Management*, *98*, 332–343. DOI: 10.1016/j.rama.2024.09.003

Xu, Y., Cui, K., Zhang, X., Diwu, G., Zhu, Y., Deng, L., Zhong, Y., & Yan, W. (2025). Shifts in fungal communities drive soil profile nutrient cycling during grassland restoration. *mBio*, *16*(3), e02834–e24. DOI: 10.1128/mbio.02834-24 PMID: 39853100

Yang, C., Dong, Y., Friman, V. P., Jousset, A., Wei, Z., Xu, Y., & Shen, Q. (2019). Carbon resource richness shapes bacterial competitive interactions by alleviating growth–antibiosis trade-off. *Functional Ecology, 33*(5), 868–875. DOI: 10.1111/1365-2435.13292

Yasir, M., Hossain, A., & Pratap-Singh, A. (2025). Pesticide degradation: Impacts on soil fertility and nutrient cycling. *Environments (Basel, Switzerland), 12*(8), 272. DOI: 10.3390/environments12080272

Zhang, E., Wang, Y., Crowther, T. W., Sun, W., Chen, S., Zhou, D., Shangguan, Z., Huang, J., He, J.-S., Wang, Y., Sheng, J., Tang, L., Li, X., Dong, M., Wu, Y., Hu, S., Bai, Y., & Yu, G. (2025). Mycorrhiza increases plant diversity and soil carbon storage in grasslands. *Proceedings of the National Academy of Sciences of the United States of America, 122*(7), e2412556122. DOI: 10.1073/pnas.2412556122 PMID: 39937867

Zhang, H., Yang, X., Zhu, L., Ahmed, N., & Gao, Y. (2025). Laboratory investigation on the effects of root-secreted organic acids on carbon and nitrogen mineralization in alpine soils caused by invasive shrubs. *Journal of Environmental Chemical Engineering, 13*(4), 117271. DOI: 10.1016/j.jece.2025.117271

Zhang, M. X., Hu, J. P., Li, J. L., Che, Z., Li, L., Lü, Z. L., Dong, W.-Q., Zhang, J.-Q., Yao, T., Duan, T.-Y., & Zhang, J. L. (2025). Responses of legume-associated rhizobacterial communities to plant diversity and soil traits in alpine grassland. *Land Degradation & Development, 36*(2), 441–453. DOI: 10.1002/ldr.5369

Zhang, X., Zhang, C., Liu, Y., Huo, L., Yang, Z., Tong, Y., Zhang, X., Yu, Z., Yang, X., Cao, Q., & Dong, Q. (2025). Effects of planting patterns on physicochemical properties, metabolites and microbial community structure of rhizosphere soil in perennial cultivated grassland. *Scientific Reports, 15*(1), 12047. DOI: 10.1038/s41598-025-94366-7 PMID: 40199986

Zhao, Y. F., Wang, X., Jiang, S. L., Zhou, X. H., Liu, H. Y., Xiao, J. J., Hao, Z. G., & Wang, K. C. (2021). Climate and geochemistry interactions at different altitudes influence soil organic carbon turnover times in alpine grasslands. *Agriculture, Ecosystems & Environment, 320*, 107591. DOI: 10.1016/j.agee.2021.107591

Zhou, X., Chen, X., Qi, X., Zeng, Y., Guo, X., Zhuang, G., & Ma, A. (2023). Soil bacterial communities associated with multi-nutrient cycling under long-term warming in the alpine meadow. *Frontiers in Microbiology, 14*, 1136187. DOI: 10.3389/fmicb.2023.1136187 PMID: 36910214

Zou, L., Tian, F., Liang, T., Fensholt, R., He, T., & Schaepman-Strub, G. (2025). Topographic effects on vegetation greening and area expansion in global alpine zones under climate change. *International Journal of Applied Earth Observation and Geoinformation, 142*, 104727. DOI: 10.1016/j.jag.2025.104727

Compilation of References

Abatzoglou, J. T., Redmond, K. T., & Edwards, L. M. (2009). Classification of Regional Climate Variability in the State of California. *Journal of Applied Meteorology and Climatology*, *48*(8), 1527–1541. DOI: 10.1175/2009JAMC2062.1

Abbass, K., Qasim, M. Z., Song, H., Murshed, M., Mahmood, H., & Younis, I. (2022). A review of the global climate change impacts, adaptation, and sustainable mitigation measures. *Environmental Science and Pollution Research International*, *29*(28), 42539–42559. DOI: 10.1007/s11356-022-19718-6 PMID: 35378646

Abbet, C., Mayor, R., Roguet, D., Spichiger, R., Hamburger, M., & Potterat, O. (2014). Ethnobotanical survey on wild alpine food plants in Lower and Central Valais (Switzerland). *Journal of Ethnopharmacology*, *151*(1), 624–634. DOI: 10.1016/j.jep.2013.11.022 PMID: 24269246

Abdelsalam, M. I. (2021). Effects of Overgrazing on Rangeland Resources in Semi-arid Areas and Rangel and Management: A review Article. *Agrica*, *10*(2), 144–151. DOI: 10.5958/2394-448X.2021.00022.5

Abd, G. I., Salih, M. A. J., Ahmed, H. E., Alani, M. H., & Abed, B. M. (2025, February). The role of chemical pesticides in environmental pollution and ecological imbalance. []. IOP Publishing.]. *IOP Conference Series. Earth and Environmental Science*, *1449*(1), 012145. DOI: 10.1088/1755-1315/1449/1/012145

Abeli, T., Vamosi, J. C., & Orsenigo, S. (2018). The importance of marginal population hotspots of cold-adapted species for research on climate change and conservation. *Journal of Biogeography*, *45*(5), 977–985. DOI: 10.1111/jbi.13196

Abile, H., Fituma, K., & Mammo, S. (2025). Impact of land use types on selected soil physicochemical parameters in the case of Liben Jawi district, Ethiopia. *Scientific Reports*, *15*(1), 27944. DOI: 10.1038/s41598-025-12784-z PMID: 40744989

Adhikari, P., Joshi, K., Singh, M., & Pandey, A. (2022). Influence of altitude on secondary metabolites, antioxidants, and antimicrobial activities of Himalayan yew (Taxus wallichiana). *Plant Biosystems, 156*(1), 187–195. DOI: 10.1080/11263504.2020.1845845

Afsari, R., Nazari-Sharabian, M., Hosseini, A., & Karakouzian, M. (2024). A CMIP6 multi-model analysis of the impact of climate change on severe meteorological droughts through multiple drought indices—Case study of Iran's metropolises. *Water (Basel), 16*(5), 711. DOI: 10.3390/w16050711

Agard, P., Omrani, J., Jolivet, L., Whitechurch, H., Vrielynck, B., Spakman, W., Monié, P., Meyer, B., & Wortel, R. (2011). Zagros orogeny: A subduction-dominated process. *Geological Magazine, 148*(5-6), 692–725. DOI: 10.1017/S001675681100046X

Agee, J. K. (1993). *Fire ecology of Pacific Northwest forests* (Vol. 499). Island press Washington, DC.

Ahmad, A., Ahmed, M. M., Akhtar, A., Shuihong, C., Zafar, Z., Ullah, R., Asim, M., He, Z., & Khan, M. B. (2026). New Insight into Endophytic Fungi–Plant Symbioses Under Climate Change: Molecular Crosstalk, Nutrient Exchange, and Ecosystem Resilience. *Applied Microbiology, 6*(3), 47. https://doi.org/10.3390/applmicrobiol6030047

Ahmad, L., Biswas, A., Warland, J., & Anjum, I. (2023). Atmospheric Pressure and Solar Radiation. In *Climate Change and Agrometeorology* (pp. 17–51). Springer. DOI: 10.1007/978-981-99-4863-5_3

Ahmed, M. J., Murtaza, G., Shaheen, H., & Habib, T. (2020). Distribution pattern and associated flora of Jurinea dolomiaea in the western Himalayan highlands of Kashmir: An indicator endemic plant of alpine phytodiversity. *Ecological Indicators, 116*, 106461. DOI: 10.1016/j.ecolind.2020.106461

Ahmed, N., Li, J., Li, Y., Deng, L., Deng, L., Chachar, M., Chachar, Z., Chachar, S., Hayat, F., Raza, A., Umrani, J. H., Gong, L., & Tu, P. (2025). Symbiotic synergy: How Arbuscular Mycorrhizal Fungi enhance nutrient uptake, stress tolerance, and soil health through molecular mechanisms and hormonal regulation. *IMA Fungus, 16*, e144989. DOI: 10.3897/imafungus.16.144989 PMID: 40162002

Ahmed, Z., & Nihei, T. (2024). Assessing the environmental impacts of adventure tourism in the world's highest mountains: A comprehensive review for promoting sustainable tourism in high-altitude areas. *Journal of Advanced Research in Social Sciences and Humanities, 9*(2), 1–15. DOI: 10.26500/JARSSH-09-2024-0201

Akhani, H. (2006). Biodiversity of halophytic and sabkha ecosystems in Iran. In *Sabkha Ecosystems: Volume II: West and Central Asia* (pp. 71-88). Springer. DOI: 10.1007/978-1-4020-5072-5_6

Akhani, H., Djamali, M., Ghorbanalizadeh, A., & Ramezani, E. (2010). Plant biodiversity of Hyrcanian relict forests, N Iran: An overview of the flora, vegetation, palaeoecology and conservation. *Pakistan Journal of Botany*, *42*(1), 231–258.

Alatalo, J. M., Erfanian, M. B., Molau, U., Chen, S., Bai, Y., & Jägerbrand, A. K. (2022). Changes in plant composition and diversity in an alpine heath and meadow after 18 years of experimental warming. *Alpine Botany*, *132*(2), 181–193. DOI: 10.1007/s00035-021-00272-9

Alatalo, J. M., Jägerbrand, A. K., & Molau, U. (2016). Impacts of different climate change regimes and extreme climatic events on an alpine meadow community. *Scientific Reports*, *6*(1), 21720. DOI: 10.1038/srep21720 PMID: 26888225

Alexander, J. M., Diez, J. M., & Levine, J. M. (2015). Novel competitors shape species' responses to climate change. *Nature*, *525*(7570), 515–518. DOI: 10.1038/nature14952 PMID: 26374998

Ali Shah, I., Muhammad, Z., Khan, H. J. J. o. W., & Change, C. (2022). Impact of climate change on spatiotemporal variations in the vegetation cover and hydrology of district Nowshera. *13*(11), 3867-3882.

Ali, A., Khan, M. A., Ashraf, M. I., & Goheer, M. A. J. P. J. o. B. (2021). Economic valuation of ecosystem services in alpine pastures of western Himalayas: a case study of Kaghan valley, Pakistan. *53*(5), 1801-1806.

Ali, M. F., Ahmad, M., Mukhtar, L., Ali, N., Raza, M. Y., Qadir, G., . . . Ahmad, S. (2025). Climate Change and Biodiversity. In *Climate Resilient and Sustainable Agriculture: Volume 1: Adaptation and Mitigation Strategies* (pp. 23-47). Springer. DOI: 10.1007/978-3-032-00190-0_2

Ali, I., Shah, A., Alotaibi, B., & Ali, A. (2024). Assessing the impacts of climate change on high mountain land-based livelihoods: An empirical investigation in District Nagar, Gilgit-Baltistan, Pakistan. *Heliyon*, *10*(21), e39877. DOI: 10.1016/j.heliyon.2024.e39877 PMID: 39553690

Allbed, A., & Kumar, L. (2013). Soil salinity mapping and monitoring in arid and semi-arid regions using remote sensing technology: A review. *Advances in Remote Sensing*, *2*(4), 373–385. DOI: 10.4236/ars.2013.24040

Allen, M. B., & Talebian, M. (2011). Structural variation along the Zagros and the nature of the Dezful Embayment. *Geological Magazine*, *148*(5-6), 911–924. DOI: 10.1017/S0016756811000318

Almalki, R., Khaki, M., Saco, P. M., & Rodriguez, J. F. J. R. S. (2022). Monitoring and mapping vegetation cover changes in arid and semi-arid areas using remote sensing technology. *RE:view*, *14*(20), 5143.

Almusaed, A., Almssad, A., Alasadi, A., Yitmen, I., & Al-Samaraee, S. (2023). Assessing the role and efficiency of thermal insulation by the "BIO-GREEN PANEL" in enhancing sustainability in a built environment. *Sustainability (Basel)*, *15*(13), 10418. DOI: 10.3390/su151310418

Alnes, V. (2024). *Net primary production of alpine lichen in a changing environment*. NTNU.

Altaf, M. J. E. (2022)... *Ecology of Mountain Ran.*, *6*, 141–149.

Altesor, A., Oesterheld, M., Leoni, E., Lezama, F., & Rodríguez, C. (2005). Effect of grazing on community structure and productivity of a Uruguayan grassland. *Plant Ecology*, *179*(1), 83–91. DOI: 10.1007/s11258-004-5800-5

Althiab-Almasaud, R., Teyssier, E., Chervin, C., Johnson, M. A., & Mollet, J.-C. (2024). Pollen viability, longevity, and function in angiosperms: Key drivers and prospects for improvement. *Plant Reproduction*, *37*(3), 273–293. DOI: 10.1007/s00497-023-00484-5 PMID: 37926761

Alzahrani, A. J., Alghamdi, A. G., & Ibrahim, H. M. (2024). Assessment of soil loss due to wind erosion and dust deposition: Implications for sustainable management in arid regions. *Applied Sciences (Basel, Switzerland)*, *14*(23), 10822. DOI: 10.3390/app142310822

Anderegg, W. R., Klein, T., Bartlett, M., Sack, L., Pellegrini, A. F., Choat, B., & Jansen, S. (2016). Meta-analysis reveals that hydraulic traits explain cross-species patterns of drought-induced tree mortality across the globe. *Proceedings of the National Academy of Sciences of the United States of America*, *113*(18), 5024–5029. DOI: 10.1073/pnas.1525678113 PMID: 27091965

Anthelme, F., & Dangles, O. (2012). Plant–plant interactions in tropical alpine environments. *Perspectives in Plant Ecology, Evolution and Systematics*, *14*(5), 363–372. DOI: 10.1016/j.ppees.2012.05.002

Antonsson, H. (2012). *Plant species composition and diversity in cliff and mountain ecosystems*.

Åqvist, J., Isaksen, G. V., & Brandsdal, B. O. (2017). Computation of enzyme cold adaptation. *Nature Reviews Chemistry, 1*(7), 0051.

Arias, P. A., Garreaud, R., Poveda, G., Espinoza, J. C., Molina-Carpio, J., Masiokas, M., Viale, M., Scaff, L., & Van Oevelen, P. J. (2021). Hydroclimate of the Andes part II: Hydroclimate variability and sub-continental patterns. *Frontiers in Earth Science (Lausanne), 8*, 505467. DOI: 10.3389/feart.2020.505467

Arosio, C., Giupponi, L., Giorgi, A., Cislaghi, A., & D'Amico, M. E. (2025). Recovery of Soil-Based Ecosystem Services in Abandoned Ski Resorts: The Valcanale Case Study (Bergamo, Italian Alps). In *Sustainability* (Vol. 17, Issue 12). DOI: 10.3390/su17125418

Asghar, A., Ameer, A., Ahmad, F., Hameed, M., Ahmad, M. S. A., Qin, M., Asghar, A., Khan, S. A., Li, H. Y., & Zhang, Z. F. (2025). Adaptive response of rare mountain sedge (Carex Ecostata CB Clarke) to hyper-cold environments. *Plant and Soil, 512*(1), 751–775. DOI: 10.1007/s11104-024-07108-5

Ashby, E., & Stevens, W. (1935). The quantitative analysis of vegetation. *Annals of Botany, 49*(196), 779–802. DOI: 10.1093/oxfordjournals.aob.a090535

Asimakopoulos, D. (2013). Climate and climate change. In *Energy and climate in the urban built environment* (pp. 19–32). Routledge.

Austin, M. W., Smith, A. B., Olsen, K. M., Hoch, P. C., Krakos, K. N., Schmocker, S. P., & Miller-Struttmann, N. E. (2024). Climate change increases flowering duration, driving phenological reassembly and elevated co-flowering richness. *The New Phytologist, 243*(6), 2486–2500. DOI: 10.1111/nph.19994 PMID: 39049577

BAHUGUNA, R., PRAKASH, V., & BISHT, H. (2013). QUANTITATIVE ENHANCEMENT OF ACTIVE CONTENT AND BIOMASS OF TWO ACONITUM SPECIES THROUGH SUITABLE CULTIVATION TECHNOLOGY. *International Journal of Conservation Science, 4*(1).

Bai, T., Liu, J., Liu, H., Ni, F., Han, X., Qiao, X., & Sun, X. (2024). Elevation-dependent patterns of temporally asymmetrical vegetation response to climate in an alpine basin on the Qinghai-Tibet Plateau. *Ecological Indicators, 159*, 111736. DOI: 10.1016/j.ecolind.2024.111736

Bai, Y., Peng, Y., Zhang, D., Yang, G., Chen, L., Kang, L., Zhou, W., Wei, B., Xie, Y., & Yang, Y. (2024). Heating up the roof of the world: Tracing the impacts of in-situ warming on carbon cycle in alpine grasslands on the Tibetan Plateau. *National Science Review, 12*(2), nwae371. Advance online publication. DOI: 10.1093/nsr/nwae371 PMID: 39872222

Bandyopadhyay, M. (2022). *Response Of Small Carnivore Community To Landscape And Climatic Variability Along The Elevation Gradient In The Great Himalayan National Park Conservation Area, The Western Himalaya*. Wildlife Institute of India.

Barber, A. L. (2013). *Physiology and early life-history associated with extreme longevity: an investigation of Pinus longaeva (Great Basin bristlecone pine)*. University of California.

Bardgett, R. D., Bullock, J. M., Lavorel, S., Manning, P., Schaffner, U., Ostle, N., Chomel, M., Durigan, G., & Fry, L., E., Johnson, D., Lavallee, J. M., Le Provost, G., Luo, S., Png, K., Sankaran, M., Hou, X., Zhou, H., Ma, L., Ren, W., … Shi, H. (. (2021). Combatting global grassland degradation. *Nature Reviews. Earth & Environment, 2*(10), 720–735. DOI: 10.1038/s43017-021-00207-2

Barrett, S. C. (2015). Influences of clonality on plant sexual reproduction. *Proceedings of the National Academy of Sciences of the United States of America, 112*(29), 8859–8866. DOI: 10.1073/pnas.1501712112 PMID: 26195747

Barrows, T. T., Mills, S. C., Fitzsimmons, K., Wasson, R., & Galloway, R. (2022). Low-altitude periglacial activity in southeastern Australia during the late Pleistocene. *Quaternary Research, 107*, 125–146. DOI: 10.1017/qua.2021.72

Basak, N., Rai, A. K., Barman, A., Mandal, S., Sundha, P., Bedwal, S., & Sharma, P. C. (2022). Salt affected soils: Global perspectives. In *Soil health and environmental sustainability: Application of Geospatial Technology* (pp. 107–129). Springer. DOI: 10.1007/978-3-031-09270-1_6

Bascuñán-Godoy, L., García-Plazaola, J. I., Bravo, L. A., & Corcuera, L. J. (2010). Leaf functional and micro-morphological photoprotective attributes in two ecotypes of Colobanthus quitensis from the Andes and Maritime Antarctic. *Polar Biology, 33*(7), 885–896. DOI: 10.1007/s00300-010-0765-4

Battaglini, L., Bovolenta, S., Gusmeroli, F., Salvador, S., & Sturaro, E. (2014). Environmental Sustainability of Alpine Livestock Farms. *Italian Journal of Animal Science, 13*(2), 3155. DOI: 10.4081/ijas.2014.3155

Bavcon, J., Malovrh, K., Tomšič, M., & Ravnjak, B. (2024). In Situ Conservation of Dry Meadows. In *Land* (Vol. 13, Issue 3). DOI: 10.3390/land13030315

Bay, S. M., Zeng, E. Y., Lorenson, T. D., Tran, K., & Alexander, C. (2003). Temporal and spatial distributions of contaminants in sediments of Santa Monica Bay, California. *Marine Environmental Research, 56*(1), 255–276. DOI: 10.1016/S0141-1136(02)00334-3 PMID: 12648959

Beattie, G. A., Edlund, A., Esiobu, N., Gilbert, J., Nicolaisen, M. H., Jansson, J. K., Jensen, P., Keiluweit, M., Lennon, J. T., Martiny, J., Minnis, V. R., Newman, D., Peixoto, R., Schadt, C., & Van der Meer, J. R. (2025). Soil microbiome interventions for carbon sequestration and climate mitigation. *mSystems*, *10*(1), e01129–e24. DOI: 10.1128/msystems.01129-24 PMID: 39692482

Beaumont, P., Blake, G. H., & Wagstaff, J. M. (1976). *The Middle East: a Geographical Study*. Wiley.

Beckage, B., Osborne, B., Gavin, D. G., Pucko, C., Siccama, T., & Perkins, T. (2008). A rapid upward shift of a forest ecotone during 40 years of warming in the Green Mountains of Vermont. *Proceedings of the National Academy of Sciences of the United States of America*, *105*(11), 4197–4202. DOI: 10.1073/pnas.0708921105 PMID: 18334647

Behroozeh, S., Khoorani, A., Damaneh, H. E., & Jokar Arsanjani, J. (2025). Monitoring vegetation dynamics across land use types in Iran: Spatiotemporal relationships with soil temperature and water volume. *Earth Science Informatics*, *18*(2), 261. DOI: 10.1007/s12145-024-01645-4

Behroozian, M., Amini, T., Zare, H., & Ejtehadi, H. (2025). Assessing climate change impacts on the geographical distribution of Cupressus sempervirens in the Mediterranean region. *Scientific Reports*, *15*(1), 40127. DOI: 10.1038/s41598-025-23930-y PMID: 41249316

Bektaş, B. (2024). *Transient dynamics of Alpine grasslands linking biodiversity and ecosystem functions in a warming world*. Universite Grenoble Alpes.

Belasri, A., & Lakhouili, A. (2016). Estimation of soil erosion risk using the universal soil loss equation (USLE) and geo-information technology in Oued El Makhazine Watershed, Morocco. *Journal of Geographic Information System*, *8*(01), 98–107. DOI: 10.4236/jgis.2016.81010

Bell-Butler, P. (2024). *Moss Matters: Unravelling the intricate web of interactions between mosses, microclimate, and seed germination in New Zealand's alpine ecosystems*. Open Access Te Herenga Waka-Victoria University of Wellington.

Bell, N., Griffin, P. C., Hoffmann, A. A., & Miller, A. D. (2018). Spatial patterns of genetic diversity among Australian alpine flora communities revealed by comparative phylogenomics. *Journal of Biogeography*, *45*(1), 177–189. DOI: 10.1111/jbi.13120

Benateau, S., Gaudard, A., Stamm, C., & Altermatt, F. (2019). Climate change and freshwater ecosystems: Impacts on water quality and ecological status.

Benito Alonso, J. L., & Villar Pérez, L. (2012). Continent-wide response of mountain vegetation to climate change.

Berberian, M., & King, G. (1981). Towards a paleogeography and tectonic evolution of Iran. *Canadian Journal of Earth Sciences*, *18*(2), 210–265. DOI: 10.1139/e81-019

Berkes, F., Colding, J., & Folke, C. (2000). Rediscovery of traditional ecological knowledge as adaptive management. *Ecological Applications*, *10*(5), 1251–1262. DOI: 10.1890/1051-0761(2000)010[1251:ROTEKA]2.0.CO;2

Berlepsch, H. (2022). *The Alps*. BoD–Books on Demand.

Bermúdez Villanueva, R. (2013). *Spatio-temporal variability in ecophysiological determinants of performance: a study in perennial herbs of coastal foredunes.*

Bernal, M., Llorens, L., Julkunen-Tiitto, R., Badosa, J., & Verdaguer, D. (2013). Altitudinal and seasonal changes of phenolic compounds in Buxus sempervirens leaves and cuticles. *Plant Physiology and Biochemistry*, *70*, 471–482. DOI: 10.1016/j.plaphy.2013.06.012 PMID: 23845826

Bettoni, M., Maerker, M., Bosino, A., Conedera, M., Simoncelli, L., & Vogel, S. (2023). Land use effects on surface runoff and soil erosion in a southern Alpine valley. *Geoderma*, *435*, 116505. DOI: 10.1016/j.geoderma.2023.116505

Beyschlag, W., & Ryel, R. J. (2007). Plant physiological ecology: An essential link for integrating across disciplines and scales in plant ecology. *Flora-Morphology, Distribution. Flora (Jena)*, *202*(8), 608–623. DOI: 10.1016/j.flora.2007.05.001

Bhattacharya, A. (2022). Effect of low temperature stress on photosynthesis and allied traits: A review. *Physiological processes in plants under low temperature stress*, 199-297.

Bhattacharya, A. (2022). *Physiological processes in plants under low temperature stress*. Springer. DOI: 10.1007/978-981-16-9037-2

Bhatt, P., Giri, P. R., & Napari, B. (2024). Soil Compaction And Its Impact On Soil Properties, Microbiome, Greenhouse Gas Emission And Plant Root Growth. [BDA]. *Big Data In Agriculture*, *6*(1), 66–69. DOI: 10.26480/bda.01.2024.66.69

Bidast, S., & Golchin, A. (2025). Soil carbon storage in rangeland ecosystems: The role of land management, climate change, and soil chemical transformations. *Geoderma Regional*, *42*, 00984. DOI: 10.1016/j.geodrs.2025.e00984

Bigard, C., Thiriet, P., Pioch, S., & Thompson, J. D. (2020). Strategic landscape-scale planning to improve mitigation hierarchy implementation: An empirical case study in Mediterranean France. *Land Use Policy*, *90*, 104286. DOI: 10.1016/j.landusepol.2019.104286

Billings, W. D., & Mooney, H. A. (1968). The ecology of arctic and alpine plants. *Biological Reviews of the Cambridge Philosophical Society*, *43*(4), 481–529. DOI: 10.1111/j.1469-185X.1968.tb00968.x

Binter, J., Macek, M., & Dolezal, J. (2025). Integrating morphological, anatomical, and physiological traits to explain elevational distributions in Himalayan steppe and alpine plants. *Journal of Integrative Plant Biology*, *67*(10), 2643–2657. DOI: 10.1111/jipb.13971 PMID: 40662507

Bird, O. (2024). *Alpine Plant Communities and Ecosystem Services Under Drivers of Global Ecosystem Change*. Open Access Te Herenga Waka-Victoria University of Wellington.

Birkeland, S., Gustafsson, A. L. S., Brysting, A. K., Brochmann, C., & Nowak, M. D. (2020). Multiple Genetic Trajectories to Extreme Abiotic Stress Adaptation in Arctic Brassicaceae. *Molecular Biology and Evolution*, *37*(7), 2052–2068. DOI: 10.1093/molbev/msaa068 PMID: 32167553

Bita-Nicolae, C., Yildiz, F., & Kaya, O. J. S. (2023). Exploring the biodiversity and conservation value of alpine grasslands in the Bucegi Massif. *Romanian Carpathians.*, *15*(16), 12643.

Bitew, A., Molla, E., Meshesha, D. T., Wato, T., Getnet, K., & Miju, C. (2025). Land use dynamics and its effect on selected soil physicochemical properties in Necho watershed, northwest Ethiopia. *Agrosystems, Geosciences & Environment*, *8*(1), e70076. DOI: 10.1002/agg2.70076

Blaschke, T. (2010). Object based image analysis for remote sensing. *ISPRS Journal of Photogrammetry and Remote Sensing*, *65*(1), 2–16. DOI: 10.1016/j.isprsjprs.2009.06.004

Blondel, J. (2010). *The Mediterranean region: biological diversity in space and time*. Oxford University Press.

Bockheim, J. G. (2015). Global Distribution of Cryosols with Mountain Permafrost: An Overview. *Permafrost and Periglacial Processes*, *26*(1), 1–12. DOI: 10.1002/ppp.1830

Bonanomi, G., Idbella, M., Allegrezza, M., & Tesei, G. (2023). Dieback of the cushion plant Silene acaulis at its southern limit of distribution in the Apennines. *Alpine Botany*, *133*(1), 57–62. DOI: 10.1007/s00035-023-00292-7

Bonfanti, N., Clément, J. C., Münkemüller, T., Barré, P., Baudin, F., & Poulenard, J. (2025). Prolonged warming leads to carbon depletion and increases nutrient availability in alpine soils. *Applied Soil Ecology*, *213*, 106239. DOI: 10.1016/j.apsoil.2025.106239

BOSTANABAD, A. S. Biodiversity Report: Alpine Plants of the Southern Alps in New Zealand–Preserving High-Altitude Ecosystems.

Boulangeat, I., Lavergne, S., Van Es, J., Garraud, L., & Thuiller, W. (2012). Niche breadth, rarity and ecological characteristics within a regional flora spanning large environmental gradients. *Journal of Biogeography*, *39*(1), 204–214. DOI: 10.1111/j.1365-2699.2011.02581.x

Bowden, D. A., Rowden, A. A., Chin, C. C., Hempel, S., Wood, B., Hart, A., & Clark, M. R. (2020). *Best practice in seabed image analysis for determining taxa, habitat, or substrata distributions*. Ministry for Primary Industries.

Braziunas, K. H., Geres, L., Richter, T., Glasmann, F., Senf, C., Thom, D., Seibold, S., & Seidl, R. (2024). Projected climate and canopy change lead to thermophilization and homogenization of forest floor vegetation in a hotspot of plant species richness. *Global Change Biology*, *30*(1), e17121. DOI: 10.1111/gcb.17121 PMID: 38273493

Briceño, V. F., Harris-Pascal, D., Nicotra, A. B., Williams, E., & Ball, M. C. (2014). Variation in snow cover drives differences in frost resistance in seedlings of the alpine herb Aciphylla glacialis. *Environmental and Experimental Botany*, *106*, 174–181. DOI: 10.1016/j.envexpbot.2014.02.011

Brighenti, S., Tolotti, M., Bruno, M. C., Wharton, G., Pusch, M. T., & Bertoldi, W. (2019). Ecosystem shifts in Alpine streams under glacier retreat and rock glacier thaw: A review. *The Science of the Total Environment*, *675*, 542–559. DOI: 10.1016/j.scitotenv.2019.04.221 PMID: 31030160

Bright, K., Dienes, B., Keiluweit, M., Rixen, C., & Aeppli, M. (2025). Climate change impacts on organic carbon cycling in European alpine soils. *Soil Biology & Biochemistry*, *109891*, 109891. Advance online publication. DOI: 10.1016/j.soilbio.2025.109891

Broadbent, A. A., Newbold, L. K., Pritchard, W. J., Michas, A., Goodall, T., Cordero, I., Giunta, A., Snell, H. S. K., Pepper, V. V. L. H., Grant, H. K., Soto, D. X., Kaufmann, R., Schloter, M., Griffiths, R. I., Bahn, M., & Bardgett, R. D. (2024). Climate change disrupts the seasonal coupling of plant and soil microbial nutrient cycling in an alpine ecosystem. *Global Change Biology*, *30*(3), e17245. DOI: 10.1111/gcb.17245 PMID: 38511487

Broadbent, A. A., Snell, H. S., Michas, A., Pritchard, W. J., Newbold, L., Cordero, I., Goodall, T., Schallhart, N., Kaufmann, R., Griffiths, R. I., Schloter, M., Bahn, M., & Bardgett, R. D. (2021). Climate change alters temporal dynamics of alpine soil microbial functioning and biogeochemical cycling via earlier snowmelt. *The ISME Journal*, *15*(8), 2264–2275. DOI: 10.1038/s41396-021-00922-0 PMID: 33619353

Brodribb, T. J., Carriquí, M., Delzon, S., McAdam, S. A., & Holbrook, N. (2020). Advanced vascular function discovered in a widespread moss. *Nature Plants*, *6*(3), 273–279. DOI: 10.1038/s41477-020-0602-x PMID: 32170283

Brown, L., & Elliman, T. (2020). *Grasses, sedges, rushes: an identification guide.* Yale University Press.

Bulluck, L., Lin, B., & Schold, E. J. R. S. (2022). Fine resolution imagery and LIDAR-derived canopy heights accurately classify land cover with a focus on shrub/sapling cover in a mountainous landscape. *14*(6), 1364.

Burgess, T. I., Oliva, J., Sapsford, S. J., Sakalidis, M. L., Balocchi, F., & Paap, T. (2022). Anthropogenic Disturbances and the Emergence of Native Diseases: A Threat to Forest Health. *Current Forestry Reports*, *8*(2), 111–123. DOI: 10.1007/s40725-022-00163-0

Burton, P. J., Jentsch, A., & Walker, L. R. (2020). The Ecology of Disturbance Interactions. *Bioscience*, *70*(10), 854–870. DOI: 10.1093/biosci/biaa088

Buyens, I. P. R., Raath-Krüger, M. J., Haddad, W. A., & Le Roux, P. C. (2023). Fine-scale variation in the effect of the cushion plant Azorella selago on vascular plants, mosses, hepatics and lichens in the sub-Antarctic. *Journal of Vegetation Science*, *34*(4), e13200. DOI: 10.1111/jvs.13200

Buytaert, W., Cuesta-Camacho, F., & Tobón, C. (2011). Potential impacts of climate change on the environmental services of humid tropical alpine regions. *Global Ecology and Biogeography*, *20*(1), 19–33. DOI: 10.1111/j.1466-8238.2010.00585.x

Buytaert, W., Deckers, J., & Wyseure, G. (2006). Description and classification of nonallophanic Andosols in south Ecuadorian alpine grasslands (páramo). *Geomorphology*, *73*(3), 207–221. DOI: 10.1016/j.geomorph.2005.06.012

Cáceres-Mago, K., Cáceres, A., & Llambí, L. D. J. A. B. (2022). Effects of nurse shrubs on symbioses between soil fungi and associated plants along a tropical alpine elevation gradient. *132*(2), 285-300.

Cache, T., Ramirez, J. A., Molnar, P., Ruiz-Villanueva, V., & Peleg, N. (2023). Increased erosion in a pre-Alpine region contrasts with a future decrease in precipitation and snowmelt. *Geomorphology, 436*, 108782. DOI: 10.1016/j.geomorph.2023.108782

Cai, S., & Li, Q. (2024). Snow cover dynamics: Impacts on soil moisture and plant growth in temperate ecosystems. *Molecular Soil Biology, 15.*

Cai, Y., Chen, H., Zhao, W., Sun, Y., Wang, M., Chen, Y., & Ma, J. (2025). Investigation of greenhouse gas emissions in soil remediation activities based on the integration of machine learning models and life cycle assessment. *Journal of Cleaner Production, 487*, 144630. DOI: 10.1016/j.jclepro.2024.144630

Callaghan, T. V., Björn, L. O., Chernov, Y., Chapin, T., Christensen, T. R., Huntley, B., Ims, R. A., Johansson, M., Jolly, D., Jonasson, S., Matveyeva, N., Panikov, N., Oechel, W., Shaver, G., Elster, J., Henttonen, H., Laine, K., Taulavuori, K., Taulavuori, E., & Zöckler, C. (2004). Biodiversity, distributions and adaptations of Arctic species in the context of environmental change. *Ambio, 33*(7), 404–417. DOI: 10.1579/0044-7447-33.7.404 PMID: 15573569

Callaghan, T. V., & Johansson, M. (2021). Snow, ice, and the biosphere. In *Snow and ice-related hazards, risks, and disasters* (pp. 137–164). Elsevier. DOI: 10.1016/B978-0-12-817129-5.00012-3

Callaway, R. M. (2007). *Positive interactions and interdependence in plant communities* (Vol. 415). Springer.

Canavan, K., Canavan, S., Clark, V. R., Gwate, O., Richardson, D. M., Sutton, G. F., & Martin, G. D. (2021). The alien plants that threaten South Africa's mountain ecosystems. *Land (Basel), 10*(12), 1393. DOI: 10.3390/land10121393

Cannone, N., Guglielmin, M., Casiraghi, C., & Malfasi, F. (2022). Salix shrub encroachment along a 1000 m elevation gradient triggers a major ecosystem change in the European Alps. *Ecography, 2022*(2), ecog.06007. DOI: 10.1111/ecog.06007

Carboni, L. J., Yahdjian, L., & Oñatibia, G. R. (2023). Effects of livestock grazing intensification on plant communities of Patagonian drylands increase with increasing aridity. *Applied Vegetation Science, 26*(4), e12754. DOI: 10.1111/avsc.12754

Cavender-Bares, J., Schneider, F. D., Santos, M. J., Armstrong, A., Carnaval, A., Dahlin, K. M., Fatoyinbo, L., Hurtt, G. C., Schimel, D., Townsend, P. A., Ustin, S. L., Wang, Z., & Wilson, A. M. (2022). Integrating remote sensing with ecology and evolution to advance biodiversity conservation. *Nature Ecology & Evolution*, *6*(5), 506–519. DOI: 10.1038/s41559-022-01702-5 PMID: 35332280

Cavieres, L. A., Sanhueza, C. E., & Hernández-Fuentes, C. (2024). Warming had contrasting effects on the importance of facilitative interactions with a cushion nurse species on native and non-native species in the high-Andes of central Chile. *Oikos*, *2024*(8), e10296. DOI: 10.1111/oik.10296

Cayan, D. R., Redmond, K. T., & Riddle, L. G. (1999). ENSO and Hydrologic Extremes in the Western United States. *Journal of Climate*, *12*(9), 2881–2893. DOI: 10.1175/1520-0442(1999)012<2881:EAHEIT>2.0.CO;2

Centeri, C. (2022). Effects of Grazing on Water Erosion, Compaction and Infiltration on Grasslands. In *Hydrology* (Vol. 9, Issue 2). DOI: 10.3390/hydrology9020034

Chai, Y., Xu, L., Xu, Y., Yang, K., Zhu, R., Zhang, R., & Li, X. (2025). Impact of Plateau Grassland Degradation on Ecological Suitability: Revealing Degradation Mechanisms and Dividing Potential Suitable Areas with Multi Criteria Models. *Remote Sensing (Basel)*, *17*(15), 2539. https://www.mdpi.com/2072-4292/17/15/2539. DOI: 10.3390/rs17152539

Chala, D., Brochmann, C., Psomas, A., Ehrich, D., Gizaw, A., Masao, C. A., Bakkestuen, V., & Zimmermann, N. E. (2016). Good-bye to tropical alpine plant giants under warmer climates? Loss of range and genetic diversity in *Lobelia rhynchopetalum. Ecology and Evolution*, *6*(24), 8931–8941. DOI: 10.1002/ece3.2603 PMID: 28035281

Chandra, S., Trivedi, V., Purohit, V. K., Nautiyal, M. C., & Chandra, S. (2025). Climatic change and physiological adaptations in alpine medicinal plants. In *High altitude medicinal plants: Botany, conservation and cultivation* (pp. 11–35). Springer Nature Switzerland.

Chang, X., Wang, W., & Zhou, H. (2025). Nitrogen acquisition by invasive plants: Species preferential N uptake matching with soil N dynamics contribute to its fitness and domination. *Plants*, *14*(5), 748. DOI: 10.3390/plants14050748 PMID: 40094724

Chang, Y., Ding, Y., Zhao, Q., Qin, J., & Zhang, S. (2022). Optimization of canopy resistance models for alpine meadow in the northeastern Tibetan Plateau. *Journal of Hydrology (Amsterdam)*, *610*, 128007. DOI: 10.1016/j.jhydrol.2022.128007

Chauhan, H. K., Gallacher, D., Bhatt, A., & Bisht, A. K. (2023). The Himalayas: A climate change laboratory. *Environmental Development, 45*, 100814. DOI: 10.1016/j.envdev.2023.100814

Chemini, C., & Rizzoli, A. (2003). Land use change and biodiversity conservation in the Alps. *Journal of Mountain Ecology, 7*(1), 1–7.

Chen, J., Chen, S., Fu, R., Li, D., Jiang, H., Wang, C., Peng, Y., Jia, K., & Hicks, B. J. (2022). Remote sensing big data for water environment monitoring: Current status, challenges, and future prospects. *Earth's Future, 10*(2), e2021EF002289.

Chen, J. G., Yang, Y., Wang, S. W., Stöcklin, J., Peng, D. L., & Sun, H. (2020). Recruitment of the high elevation cushion plant Arenaria polytrichoides is limited by competition, thus threatened by currently established vegetation. *Journal of Systematics and Evolution, 58*(1), 59–68. DOI: 10.1111/jse.12481

Chen, J., Chen, X., Qian, L., Zhang, Y., Li, B., Shi, H., Sun, L., Schöb, C., & Sun, H. (2024). Degeneration of foundation cushion species induced by ecological constraints can cause massive changes in alpine plant communities. *Science China. Life Sciences, 67*(4), 789–802. DOI: 10.1007/s11427-022-2383-6 PMID: 38057621

Chen, J., Huang, Y., Brachi, B., Yun, Q., Zhang, W., Lu, W., Li, H., Li, W., Sun, X., Wang, G.-y., He, J., Zhou, Z., Chen, K.-y., Ji, Y., Shi, M., Sun, W., Yang, Y., Zhang, R., Abbott, R. J., & Sun, H. (2019). Genome-wide analysis of Cushion willow provides insights into alpine plant divergence in a biodiversity hotspot. *Nature Communications, 10*(1), 5230. DOI: 10.1038/s41467-019-13128-y PMID: 31745089

Chen, J., Li, Y., Yang, Y., & Sun, H. (2017). How cushion communities are maintained in alpine ecosystems: A review and case study on alpine cushion plant reproduction. *Plant Diversity, 39*(4), 221–228. DOI: 10.1016/j.pld.2017.07.002 PMID: 30159515

Chen, L., Zeng, X., & Gao, Y. (2025). The impact of grazing exclusion on phosphorus dynamics in alpine grassland ecosystems. *Plant and Soil, 513*(1), 1215–1230. DOI: 10.1007/s11104-025-07238-4

Chen, Q., Song, Y., An, Y., Lu, Y., & Zhong, G. (2024). Soil microorganisms: Their role in enhancing crop nutrition and health. *Diversity, 16*(12), 734. DOI: 10.3390/d16120734

Chen, Y., Qin, W., Zhang, Q., Wang, X., Feng, J., Han, M., Hou, Y., Zhao, H., Zhang, Z., He, J.-S., Torn, M. S., & Zhu, B. (2024). Whole-soil warming leads to substantial soil carbon emission in an alpine grassland. *Nature Communications, 15*(1), 4489. DOI: 10.1038/s41467-024-48736-w PMID: 38802385

Chettri, N., Shrestha, A. B., & Sharma, E. (2019). Climate change trends and ecosystem resilience in the Hindu Kush Himalayas. In *Himalayan Weather and Climate and their Impact on the Environment* (pp. 525–552). Springer.

Chu, L., Hou, M., & Jiang, Z. (2022a). How does the fragmentation of pasture affect grassland ecology?——Evidence from typical pastoral areas in China. *Ecological Indicators, 136*, 108701. DOI: 10.1016/j.ecolind.2022.108701

Corlett, R. T. (2016). Plant diversity in a changing world: Status, trends, and conservation needs. *Plant Diversity, 38*(1), 10–16. DOI: 10.1016/j.pld.2016.01.001 PMID: 30159445

Cortés, A. J., & Wheeler, J. (2018). The environmental heterogeneity of mountains at a fine scale in a changing world. *Mountains, Climate, and Biodiversity, 584*.

Cox, G. W. (1996). Laboratory manual of general ecology. Wm. C. *Brown, Dubuque, 1*, 232.

Crawford, R. M. (2008). *Plants at the margin: ecological limits and climate change.* Cambridge University Press. DOI: 10.1017/CBO9780511754906

Cubas, J., Martín-Esquivel, J. L., Marrero-Gómez, M., Díaz, J. R. D., Rodríguez, F., & González-Mancebo, J. M. (2022). Climate change causes rapid collapse of a keystone shrub from insular Alpine ecosystems. *Journal for Nature Conservation, 69*, 126263. DOI: 10.1016/j.jnc.2022.126263

Cui, X., Gibbes, C., Southworth, J., & Waylen, P. (2013). Using remote sensing to quantify vegetation change and ecological resilience in a semi-arid system. *Land (Basel), 2*(2), 108–130. DOI: 10.3390/land2020108

Cui, Z., Yang, W.-S., Cheng, Z., Zhang, Z., Li, S., Zhao, J., López-Vicente, M., & Wu, G.-L. (2022). Top-down degradation of alpine meadow in the Qinghai-Tibetan Plateau: Gravelization initiate hillside surface aridity and meadow community disappearance. *Catena, 210*, 105933. DOI: 10.1016/j.catena.2021.105933

D'Odorico, P., Porporato, A., & Runyan, C. W. (2006). *Dryland ecohydrology* (Vol. 9). Springer. DOI: 10.1007/1-4020-4260-4

Dai, L., Fu, R., Guo, X., Du, Y., Zhang, F., & Cao, G. (2023). Variations in and factors controlling soil hydrological properties across different slope aspects in alpine meadows. *Journal of Hydrology (Amsterdam), 616*, 128756. DOI: 10.1016/j.jhydrol.2022.128756

Dainese, M., Lepš, J., & de Bello, F. (2015). Different effects of elevation, habitat fragmentation and grazing management on the functional, phylogenetic and taxonomic structure of mountain grasslands. *Perspectives in Plant Ecology, Evolution and Systematics, 17*(1), 44–53. DOI: 10.1016/j.ppees.2014.09.002

Dalling, J. W., Heineman, K., González, G., & Ostertag, R. (2016). Geographic, environmental and biotic sources of variation in the nutrient relations of tropical montane forests. *Journal of Tropical Ecology, 32*(5), 368–383. DOI: 10.1017/S0266467415000619

Damgaard, C. (2022). Adaptive management plans rooted in quantitative ecological predictions of ecosystem processes: Putting monitoring data to practical use. *Environmental Conservation, 49*(1), 27–32. DOI: 10.1017/S0376892921000357

Dansereau, P. (1951). Description and recording of vegetation upon a structural basis. *Ecology, 32*(2), 172–229. DOI: 10.2307/1930415

Das, D., Ray, P., & Datta, S. P. (2024). Nutrient cycling and climate change. In *Climate change impacts on soil–plant–atmosphere continuum* (pp. 37–62). Springer Nature Singapore. DOI: 10.1007/978-981-99-7935-6_2

Davidson, J., Hassanzadeh, J., Berzins, R., Stockli, D. F., Bashukooh, B., Turrin, B., & Pandamouz, A. (2004). The geology of Damavand volcano, Alborz Mountains, northern Iran. *Geological Society of America Bulletin, 116*(1-2), 16–29. DOI: 10.1130/B25344.1

Davis, M. B., & Shaw, R. G. (2001). Range shifts and adaptive responses to Quaternary climate change. *Science, 292*(5517), 673–679. DOI: 10.1126/science.292.5517.673 PMID: 11326089

De Cesare, G. (2025). Sedimentation of a high Alpine Hydropower Reservoir under Climate Change: What will disappear first, the Glacier or the Reservoir? *Authorea Preprints*. DOI: 10.1002/9781394216727.ch4

De Toma, A., Malavasi, M., Marzialetti, F., & Cutini, M. (2025). Unveiling spatial patterns and trajectories of shrub dynamics in Mediterranean alpine ecosystems. *Plant Ecology, 226*(2), 149–160. DOI: 10.1007/s11258-024-01479-6

Deb, P., & Mandal, A. Harendra, Choudhury, S. R., Das, A., Hazra, S., ... & Moulick, D. (2023). Improving plant nutrient use efficiency for climate-resilient agriculture. In *Climate-resilient agriculture, Vol. 2: Agro-biotechnological advancement for crop production* (pp. 209–243). Springer International Publishing.

Decocq, G. (2016). Moving from Patterns to Processes: A Challenge for the Phytosociology of the Twenty-First Century? In *Vegetation Structure and Function at Multiple Spatial, Temporal and Conceptual Scales* (pp. 407-424). Springer. DOI: 10.1007/978-3-319-21452-8_18

Deilmann, T., Bernhardt-Römermann, M., Hentschel, J., Gros, P., & Römermann, C. (2025). Temperate Forest Floor Bryophytes Functionally Respond to Small-Scale Variability in Water, Light and Nutrient Availability. *Ecology and Evolution*, *15*(8), e71839. DOI: 10.1002/ece3.71839 PMID: 40746384

Deng, Y., Lu, Y., Du, X., Ye, X., & Feng, J. (2025). Identifying spatial patterns and driving factors of anthropogenic impacts on the groundwater environment based on groundwater chemical kinetics. *Journal of Cleaner Production*, *486*, 144436. DOI: 10.1016/j.jclepro.2024.144436

Dierking, R. M., & Kallenbach, R. L. (2012). Mediterranean and Continental Tall Fescue: II. Effects of Cold, Nonfreezing Temperatures on Leaf Extension, Proline, Fructan, and Abscisic Acid. *Crop Science*, *52*(1), 460–469. DOI: 10.2135/cropsci2011.03.0160

Dietrich, V., Niederberger, J., Frank, J., & Hauck, M. (2025). Does atmospheric nitrogen deposition cause critical nutrient imbalances in temperate tree regeneration? *Trees (Berlin)*, *39*(2), 38. DOI: 10.1007/s00468-025-02615-z

Ding, J., & Yu, S. (2025). Impacts of land use on soil nitrogen-cycling microbial communities: Insights from community structure, functional gene abundance, and network complexity. *Life (Chicago, Ill.)*, *15*(3), 466. DOI: 10.3390/life15030466 PMID: 40141811

Dirnböck, T., Dullinger, S., Gottfried, M., Ginzler, C., & Grabherr, G. (2003a). Mapping alpine vegetation. 209-219. DOI: 10.1007/978-3-642-18967-8_10

Dirnböck, T., Dullinger, S., Gottfried, M., Ginzler, C., & Grabherr, G. (2003b). Mapping alpine vegetation based on image analysis, topographic variables and Canonical Correspondence Analysis. *Applied Vegetation Science*, *6*(1), 85–96. DOI: 10.1111/j.1654-109X.2003.tb00567.x

Dirnböck, T., Essl, F., & Rabitsch, W. (2011). Disproportional risk for habitat loss of high-altitude endemic species under climate change. *Global Change Biology*, *17*(2), 990–996. DOI: 10.1111/j.1365-2486.2010.02266.x

Diyaolu, C. O., & Folarin, I. O. (2024). The role of biodiversity in agricultural resilience: Protecting ecosystem services for sustainable food production. *Int. J. Res. Publ. Rev*, *5*(10), 1560–1573. DOI: 10.55248/gengpi.5.1024.2741

Djamali, M., Akhani, H., Khoshravesh, R., Andrieu-Ponel, V., Ponel, P., & Brewer, S. (2011). Application of the global bioclimatic classification to Iran: Implications for understanding the modern vegetation and biogeography. *Ecologia Mediterranea, 37*(1), 91–114. DOI: 10.3406/ecmed.2011.1350

Djamali, M., Baumel, A., Brewer, S., Jackson, S. T., Kadereit, J. W., López-Vinyallonga, S., Mehregan, I., Shabanian, E., & Simakova, A. (2012). Ecological implications of Cousinia Cass.(Asteraceae) persistence through the last two glacial–interglacial cycles in the continental Middle East for the Irano-Turanian flora. *Review of Palaeobotany and Palynology, 172*, 10–20. DOI: 10.1016/j.revpalbo.2012.01.005

Djamali, M., Brewer, S., Breckle, S. W., & Jackson, S. T. (2012). Climatic determinism in phytogeographic regionalization: A test from the Irano-Turanian region, SW and Central Asia. *Flora-Morphology, Distribution. Flora (Jena), 207*(4), 237–249. DOI: 10.1016/j.flora.2012.01.009

Dobbertin, M., Hilker, N., Rebetez, M., Zimmermann, N. E., Wohlgemuth, T., & Rigling, A. (2005). The upward shift in altitude of pine mistletoe (Viscum album ssp. austriacum) in Switzerland—The result of climate warming? *International Journal of Biometeorology, 50*(1), 40–47. DOI: 10.1007/s00484-005-0263-5 PMID: 15875222

Dobos, A., Hegyi, T. P., Bujtor, D. L., Tolnai, Z., & Hegyi, B. (2024). Description of the Characteristic Soil Profiles and Indication of the Degree of Sheet Erosion in Verpelét. *Journal of Environmental Geography, 17*(1–4), 73–90. DOI: 10.14232/jengeo-2024-45762

Dodonov, A. E., & Baiguzina, L. (1995). Loess stratigraphy of Central Asia: Palaeoclimatic and palaeoenvironmental aspects. *Quaternary Science Reviews, 14*(7-8), 707–720. DOI: 10.1016/0277-3791(95)00054-2

Dolanc, C. R., Thorne, J. H., & Safford, H. D. (2013). Widespread shifts in the demographic structure of subalpine forests in the Sierra Nevada, California, 1934 to 2007. *Global Ecology and Biogeography, 22*(3), 264–276. DOI: 10.1111/j.1466-8238.2011.00748.x

Doležal, J., Korznikov, K., Bazant, V., Ruka, A. T., & Gago, J. (2025). Elevational Shifts in Tropical Tree Leaf Traits: Interactions Between Soil, Climate, Light, and Phylogeny. BioRxiv, 2025.2007. 2030.667609.

Dong, S., Wen, L., Li, Y., Wang, X., Zhu, L., & Li, X. (2012). Soil-quality effects of grassland degradation and restoration on the Qinghai-Tibetan Plateau. *Soil Science Society of America Journal, 76*(6), 2256–2264. DOI: 10.2136/sssaj2012.0092

Donhauser, J., & Frey, B. (2018). Alpine soil microbial ecology in a changing world. *FEMS Microbiology Ecology*, *94*(9), fiy099. DOI: 10.1093/femsec/fiy099 PMID: 30032189

Donovan, V. M., Roberts, C. P., Wonkka, C. L., Beck, J. L., Popp, J. N., Allen, C. R., & Twidwell, D. (2020). Range-wide monitoring of population trends for Rocky Mountain bighorn sheep. *Biological Conservation*, *248*, 108639. DOI: 10.1016/j.biocon.2020.108639

Dou, T., Zhang, K., Shi, X., Liu, W., Yu, F., & Liu, D. (2025). Crop–mushroom rotation: A comprehensive review of its multifaceted impacts on soil quality, agricultural sustainability, and ecosystem health. *Agronomy (Basel)*, *15*(3), 563. DOI: 10.3390/agronomy15030563

Duchicela, S. A., Cuesta, F., Tovar, C., Muriel, P., Jaramillo, R., Salazar, E., & Pinto, E. (2021). Microclimatic warming leads to a decrease in species and growth form diversity: Insights from a tropical alpine grassland. *Frontiers in Ecology and Evolution*, *9*, 673655. DOI: 10.3389/fevo.2021.673655

Du, K., Shao, Y., Yao, N., Yu, H., Ma, S., Mao, X., Wang, L., & Wang, J. (2025). Alpine Meadow Fractional Vegetation Cover Estimation Using UAV-Aided Sentinel-2 Imagery. *Sensors (Basel)*, *25*(14), 4506. DOI: 10.3390/s25144506 PMID: 40732634

Dumas, K., Marais, G., Paitreault, S., Avrillier, J.-N., Rosa, Z., Hedde, M., Rasplus, J.-Y., Chérasse, S., Ouvrard, D., Reynaud, P., Trap, J., Filali Alaoui, I., Noël, F., Perrier, C., Yannic, G., Gallet, C., Bernier, N., Lavergne, S., & Ibanez, S. (2025). Cushion plants in the Alps are swarming with invertebrate life. *Alpine Botany*, *135*(2), 1–12. DOI: 10.1007/s00035-025-00335-1

Du, Y., Yang, Y., Wu, S., Gao, X., He, X., & Dong, S. (2025). Core microbes regulate plant–soil resilience by maintaining network resilience during long-term restoration of alpine grasslands. *Nature Communications*, *16*(1), 3116. DOI: 10.1038/s41467-025-58080-2 PMID: 40169576

Eckert, N., Corona, C., Giacona, F., Gaume, J., Mayer, S., van Herwijnen, A., Hagenmuller, P., & Stoffel, M. (2024). Climate change impacts on snow avalanche activity and related risks. *Nature Reviews. Earth & Environment*, *5*(5), 369–389. DOI: 10.1038/s43017-024-00540-2

Egerton, F. N. (2013). History of ecological sciences, part 48: Formalizing plant ecology, about 1870 to mid-1920s. *Bulletin of the Ecological Society of America*, *94*(4), 341–378. DOI: 10.1890/0012-9623-94.4.341

Ekka, P., Patra, S., Upreti, M., Kumar, G., Kumar, A., & Saikia, P. (2023). Land Degradation and Its Impacts on Biodiversity and Ecosystem Services. In Raj, A., Jhariya, M. K., Banerjee, A., Nema, S., & Bargali, K. (Eds.), *Land and Environmental Management through Forestry* (1st ed., pp. 77–101). Wiley., DOI: 10.1002/9781119910527.ch4

Eldridge, D. J., Guirado, E., Reich, P. B., Ochoa-Hueso, R., Berdugo, M., Sáez-Sandino, T., Blanco-Pastor, J. L., Tedersoo, L., Plaza, C., Ding, J., Sun, W., Mamet, S., Cui, H., He, J.-Z., Hu, H.-W., Sokoya, B., Abades, S., Alfaro, F., Bamigboye, A. R., & Delgado-Baquerizo, M. (2023). The global contribution of soil mosses to ecosystem services. *Nature Geoscience*, *16*(5), 430–438. DOI: 10.1038/s41561-023-01170-x

Elevation-dependent warming in mountain regions of the world. (2015). *Nature Climate Change, 5*(5), 424-430.

Elton, C. S., & Miller, R. S. (1954). The ecological survey of animal communities: With a practical system of classifying habitats by structural characters. *Journal of Ecology*, *42*(2), 460–496. DOI: 10.2307/2256872

Engler, R., Randin, C. F., Thuiller, W., Dullinger, S., Zimmermann, N. E., Araujo, M. B., & Albert, C. H. (2011). 21st century climate change threatens mountain flora unequally across Europe. *Global Change Biology*, *17*(7), 2330–2341. DOI: 10.1111/j.1365-2486.2010.02393.x

Enquist, B. J., Feng, X., Boyle, B., Maitner, B., Newman, E. A., Jørgensen, P. M., & Corlett, R. T. (2019). The commonness of rarity: Global and future distribution of rarity across land plants. *Science Advances*, *5*(11), eaaz0414. DOI: 10.1126/sciadv.aaz0414 PMID: 31807712

Entezami, H., Mojarrad, F., Shahabi, H., & Ghaderpour, E. (2024). Spatiotemporal Variability in Snow and Land Cover in Sefid-Rud Basin, Iran. *Sustainability (Basel)*, *16*(21), 9381. DOI: 10.3390/su16219381

Erschbamer, B., Niederfriniger Schlag, R., Carnicero, P., & Kaufmann, R. (2023). Long-term monitoring confirms limitations of recruitment and facilitation and reveals unexpected changes of the successional pathways in a glacier foreland of the Central Austrian Alps. *Plant Ecology*, *224*(4), 373–386. DOI: 10.1007/s11258-023-01308-2

Esposito, M., Palma, L., Belli, A., Sabbatini, L., & Pierleoni, P. (2022). Recent Advances in Internet of Things Solutions for Early Warning Systems: A Review. *Sensors (Basel)*, *22*(6), 2124. DOI: 10.3390/s22062124 PMID: 35336296

Evangelista, A., Frate, L., Carranza, M. L., Attorre, F., Pelino, G., & Stanisci, A. (2016). Changes in composition, ecology and structure of high-mountain vegetation: A re-visitation study over 42 years. *AoB Plants*, *8*, plw004. DOI: 10.1093/aobpla/plw004 PMID: 26819258

Fallah, B., Didovets, I., Rostami, M., & Hamidi, M. (2024). Climate change impacts on Central Asia: Trends, extremes and future projections. *International Journal of Climatology*, *44*(10), 3191–3213. DOI: 10.1002/joc.8519

Fan, B., Ding, N., Tian, T., Gao, P., Wan, Y., Ma, M., & Sun, K. (2025). Different water and photosynthetic resource use strategies explain the widespread distribution of Dasiphora fruticosa in Qinghai-Tibet Plateau alpine meadows. *Environmental and Experimental Botany*, *231*, 106106. DOI: 10.1016/j.envexpbot.2025.106106

Fan, S., Qin, J., Sun, H., Dan, Z., Li, Z., & Yang, J. (2025). Different Responses of Soil Nutrient Dynamics and Microbial Activities to Soil Moisture Changes in Alpine Wetlands and Meadows on the Tibetan Plateau. *Land Degradation & Development*, *36*(6), 1928–1938. DOI: 10.1002/ldr.5472

Farinotti, D., Immerzeel, W. W., de Kok, R. J., Quincey, D. J., & Dehecq, A. (2020). Manifestations and mechanisms of the Karakoram glacier Anomaly. *Nature Geoscience*, *13*(1), 8–16. DOI: 10.1038/s41561-019-0513-5 PMID: 31915463

Fartyal, A., Bargali, S. S., Bargali, K., & Negi, B. (2025). Changes in soil properties, organic carbon, and nutrient stocks after land-use change from forests to grasslands in Kumaun Himalaya, India. *Land Degradation & Development*, *36*(7), 2438–2457. DOI: 10.1002/ldr.5507

Fassnacht, F. E., Latifi, H., Stereńczak, K., Modzelewska, A., Lefsky, M., Waser, L. T.,…Ghosh, A. J. R. s. o. e. (2016b). Review of studies on tree species classification from remotely sensed data. *186*, 64-87.

Fassnacht, F. E., Latifi, H., Stereńczak, K., Modzelewska, A., Lefsky, M., Waser, L. T., Straub, C., & Ghosh, A. (2016a). Review of studies on tree species classification from remotely sensed data. *Remote Sensing of Environment*, *186*, 64–87. DOI: 10.1016/j.rse.2016.08.013

Fathizad, H., Ardakani, M. A. H., Sodaiezadeh, H., Kerry, R., & Taghizadeh-Mehrjardi, R. (2020). Investigation of the spatial and temporal variation of soil salinity using random forests in the central desert of Iran. *Geoderma*, *365*, 114233. DOI: 10.1016/j.geoderma.2020.114233

Fattahi, Z., Beheshtian, M., Mohseni, M., Poustchi, H., Sellars, E., Nezhadi, S. H., & Jamali, P. (2019). Iranome: A catalog of genomic variations in the Iranian population. *Human Mutation, 40*(11), 1968–1984. DOI: 10.1002/humu.23880 PMID: 31343797

Fazlioglu, F., & Wan, J. S. (2021). Warming matters: Alpine plant responses to experimental warming. *Climatic Change, 164*(3), 56. DOI: 10.1007/s10584-021-02996-3

Feizizadeh, B., Omarzadeh, D., Mohammadzadeh Alajujeh, K., Blaschke, T., & Makki, M. (2022). Impacts of the Urmia Lake drought on soil salinity and degradation risk: An integrated geoinformatics analysis and monitoring approach. *Remote Sensing (Basel), 14*(14), 3407. DOI: 10.3390/rs14143407

Feng, J., Yin, X., Wang, C., Hu, L., & Zhou, T. (2025). Effects of snowpack changes on functional traits and reproductive allocation of four perennial plants in an alpine meadow. *BMC Plant Biology, 25*(1), 834. DOI: 10.1186/s12870-025-06855-6 PMID: 40604470

Feng, L., Gan, M., & Tian, F.-P. (2019a). Effects of Grassland Tourism on Alpine Meadow Community and Soil Properties in the Qinghai-Tibetan Plateau [journal article]. *Polish Journal of Environmental Studies, 28*(6), 4147–4152. DOI: 10.15244/pjoes/99065

Feng, T., Pucker, B., Kuang, T., Song, B., Yang, Y., Lin, N., Zhang, H., Moore, M. J., Brockington, S. F., Wang, Q., Deng, T., Wang, H., & Sun, H. (2023). The genome of the glasshouse plant noble rhubarb (Rheum nobile) provides a window into alpine adaptation. *Communications Biology, 6*(1), 706. DOI: 10.1038/s42003-023-05044-1 PMID: 37429977

Ferdush, J. (2022). *Determining the effects of elevated carbon dioxide on soil acidification, cation depletion, and soil inorganic carbon and mapping soil carbons using artificial intelligence*. Mississippi State University.

Ficetola, G. F., Marta, S., Guerrieri, A., Cantera, I., Bonin, A., Cauvy-Fraunié, S., Ambrosini, R., Caccianiga, M., Anthelme, F., Azzoni, R. S., Almond, P., Alviz Gazitúa, P., Ceballos Lievano, J. L., Chand, P., Chand Sharma, M., Clague, J. J., Cochachín Rapre, J. A., Compostella, C., Encarnación, R. C., & Carteron, A. (2024). The development of terrestrial ecosystems emerging after glacier retreat. *Nature, 632*(8024), 336–342. DOI: 10.1038/s41586-024-07778-2 PMID: 39085613

Fickert, T. (2020). Common patterns and diverging trajectories in primary succession of plants in eastern alpine glacier forelands. *Diversity, 12*(5), 191. DOI: 10.3390/d12050191

Filimonenko, E., & Kuzyakov, Y. (2025). Activation energy of organic matter decomposition in soil and consequences of global warming. *Global Change Biology, 31*(9), e70472. DOI: 10.1111/gcb.70472 PMID: 40905536

Finne, E. A., Bjerke, J. W., Stordal, F., & Tallaksen, L. M. (2025). Lichens are more tolerant against winter warming stress than vascular and non-vascular plants: Insights from an alpine field experiment. *Journal of Ecology, 113*(3), 518–530. DOI: 10.1111/1365-2745.14482

Fites-Kaufman, J., Rundel, P., Stephenson, N., & Weixelman, D. (2007). *Montane and Subalpine Vegetation of the Sierra Nevada and Cascade Ranges.* Terrestrial Vegetation of California., DOI: 10.1525/california/9780520249554.003.0017

Foley, W. J., McIlwee, A., Lawler, I., Aragones, L., Woolnough, A. P., & Berding, N. (1998). Ecological applications of near infrared reflectance spectroscopy–a tool for rapid, cost-effective prediction of the composition of plant and animal tissues and aspects of animal performance. *Oecologia, 116*(3), 293–305. DOI: 10.1007/s004420050591 PMID: 28308060

Ford, K. R., Ettinger, A. K., Lundquist, J. D., Raleigh, M. S., & Hille Ris Lambers, J. (2013). Spatial heterogeneity in ecologically important climate variables at coarse and fine scales in a high-snow mountain landscape. *PLoS One, 8*(6), e65008. DOI: 10.1371/journal.pone.0065008 PMID: 23762277

Forster, P. M., Smith, C. J., Walsh, T., Lamb, W. F., Lamboll, R., Hauser, M., & Palmer, M. D. (2023). Indicators of Global Climate Change 2022: Annual update of large-scale indicators of the state of the climate system and human influence. *Earth System Science Data, 15*(6), 2295–2327. DOI: 10.5194/essd-15-2295-2023

Freeman, B. G., Scholer, M. N., Ruiz-Gutierrez, V., & Fitzpatrick, J. W. (2018). Climate change causes upslope shifts and mountaintop extirpations in a tropical bird community. *Proceedings of the National Academy of Sciences of the United States of America, 115*(47), 11982–11987. DOI: 10.1073/pnas.1804224115 PMID: 30373825

Freppaz, M., Godone, D., Filippa, G., Maggioni, M., Lunardi, S., Williams, M. W., & Zanini, E. (2010). Soil Erosion Caused by Snow Avalanches: A Case Study in the Aosta Valley (NW Italy). *Arctic, Antarctic, and Alpine Research, 42*(4), 412–421. DOI: 10.1657/1938-4246-42.4.412

Freshwater, D., Sherwood, G., & Drury, V. (2006). International research collaboration: Issues, benefits and challenges of the global network. *Journal of Research in Nursing : JRN, 11*(4), 295–303. DOI: 10.1177/1744987106066304

Friedlingstein, P., O'sullivan, M., Jones, M. W., Andrew, R. M., Gregor, L., Hauck, J., Le Quéré, C., Luijkx, I. T., Olsen, A., & Peters, G. P. J. E. s. s. d. (2022). Global carbon budget 2022. *14*(11), 4811-4900.

Frodin, D. G. (2001). *Guide to standard floras of the world: an annotated, geographically arranged systematic bibliography of the principal floras, enumerations, checklists and chorological atlases of different areas.* Cambridge University Press. DOI: 10.1017/CBO9780511541803

Gaffney, P. P. J., Tang, Q., Wang, J., Zhang, C., Xu, X., Xu, X., Li, Y., Pap, S., Ratcliffe, J. L., & Li, Q. (2025). The high-altitude peatland carbon cycle: A review of the impacts of climate change, human disturbance and management. *Geography and Sustainability, 6*(6).

Gallagher, R., Hughes, L., & Leishman, M. (2009). Phenological trends among Australian alpine species: Using herbarium records to identify climate-change indicators. *Australian Journal of Botany*, *57*(1), 1–9. DOI: 10.1071/BT08051

Gambo, A., Fullen, M. A., & Baldwin, T. C. (2025). Indigenous farmers' knowledge and perception of desertification and soil erosion in Jigawa State, Nigeria. *Discover Soil*, *2*(1), 22. DOI: 10.1007/s44378-025-00028-5

Gang, C., Zhou, W., Chen, Y., Wang, Z., Sun, Z., Li, J., Qi, J., & Odeh, I. J. E. E. S. (2014). Quantitative assessment of the contributions of climate change and human activities on global grassland degradation. *72*(11), 4273-4282.

Gao, E., Ma, H., Yang, T., Kaiser-Bunbury, C. N., & Zhao, Z. (2023). Meadow transformations alter above-and below-ground ecological networks and ecosystem multifunctionality. *Functional Ecology*, *37*(6), 1703–1716. DOI: 10.1111/1365-2435.14333

Gao, Y., Xing, F., Jin, Y., Nie, D., & Wang, Y. (2012). Foraging responses of clonal plants to multi-patch environmental heterogeneity: Spatial preference and temporal reversibility. *Plant and Soil*, *359*(1), 137–147. DOI: 10.1007/s11104-012-1148-0

García Lino, M. C., Cavieres, L. A., Zotz, G., & Bader, M. Y. (2017). Carbohydrate reserves in the facilitator cushion plant Laretia acaulis suggest carbon limitation at high elevation and no negative effects of beneficiary plants. *Oecologia*, *183*(4), 997–1006. DOI: 10.1007/s00442-017-3840-5 PMID: 28233055

Garcia, J., & Kao-Kniffin, J. (2018). Microbial group dynamics in plant rhizospheres and their implications on nutrient cycling. *Frontiers in Microbiology*, *9*, 1516. DOI: 10.3389/fmicb.2018.01516 PMID: 30050510

García-Plazaola, J. I., Rojas, R., Christie, D. A., & Coopman, R. E. (2015). Photosynthetic responses of trees in high-elevation forests: Comparing evergreen species along an elevation gradient in the Central Andes. *AoB Plants*, *7*, plv058. Advance online publication. DOI: 10.1093/aobpla/plv058 PMID: 26002745

García-Ruiz, J. M., Arnáez, J., Lasanta, T., Nadal-Romero, E., & López-Moreno, J. I. (2024). *The Changing World of Mountains in the Long Term BT - Mountain Environments: Changes and Impacts: Natural Landscapes and Human Adaptations to Diversity* (J. M. García-Ruiz, J. Arnáez, T. Lasanta, E. Nadal-Romero, & J. I. López-Moreno (eds.); pp. 27–50). Springer Nature Switzerland. DOI: 10.1007/978-3-031-51955-0_3

Gavini, S. S., Ezcurra, C., & Aizen, M. A. (2019). Plant–plant interactions promote alpine diversification. *Evolutionary Ecology*, *33*(2), 195–209. DOI: 10.1007/s10682-019-09972-5

Gentili, R., Armiraglio, S., Sgorbati, S., & Baroni, C. (2013). Geomorphological disturbance affects ecological driving forces and plant turnover along an altitudinal stress gradient on alpine slopes. *Plant Ecology*, *214*(4), 571–586. DOI: 10.1007/s11258-013-0190-1

Gentili, R., Bacchetta, G., Fenu, G., Cogoni, D., Abeli, T., Rossi, G., Salvatore, M. C., Baroni, C., & Citterio, S. (2015). From cold to warm-stage refugia for boreo-alpine plants in southern European and Mediterranean mountains: The last chance to survive or an opportunity for speciation? *Biodiversity (Nepean)*, *16*(4), 247–261. DOI: 10.1080/14888386.2015.1116407

Gentry, A. H. (1988). Changes in plant community diversity and floristic composition on environmental and geographical gradients. *Annals of the Missouri Botanical Garden*, *75*(1), 1–34. DOI: 10.2307/2399464

Ge, P., Xue, J., Ru, Y., Li, Y., Li, D., Han, P., Li, A., & Huang, J. (2025). Intensive rotational grazing has positive effects on productivity of rangeland. *Agriculture, Ecosystems & Environment*, *384*, 109558. DOI: 10.1016/j.agee.2025.109558

Germino, M. J., & Smith, W. K. (2000). High resistance to low-temperature photoinhibition in two alpine, snowbank species. *Physiologia Plantarum*, *110*(1), 89–95. DOI: 10.1034/j.1399-3054.2000.110112.x

Getnet, M., & Asseffa, S. (2025). *Agriculture in the Abbay Basin BT - Abbay River Basin: Biophysical Setting, Environmental Degradation, Hydropolitics and Development Potential* (A. Melesse, B. Gessesse, & W. Zewdie (eds.); pp. 67–87). Springer Nature Switzerland. DOI: 10.1007/978-3-031-65241-7_4

Ghorbanian, A., Mohammadzadeh, A., & Jamali, S. (2022). Linear and non-linear vegetation trend analysis throughout Iran using two decades of MODIS NDVI imagery. *Remote Sensing (Basel)*, *14*(15), 3683. DOI: 10.3390/rs14153683

Gibbon, A., Silman, M. R., Malhi, Y., Fisher, J. B., Meir, P., Zimmermann, M., Dargie, G. C., Farfan, W. R., & Garcia, K. C. (2010). Ecosystem carbon storage across the grassland–forest transition in the high Andes of Manu National Park, Peru. *Ecosystems (New York, N.Y.)*, *13*(7), 1097–1111. DOI: 10.1007/s10021-010-9376-8

Gibbons, B. (2025). *Europe's Alpine Flowers: A Field Guide*. Princeton University Press.

Giguet-Covex, C., Bajard, M., Chen, W., Walsh, K. J., Rey, P.-J., Messager, E., Etienne, D., Sabatier, P., Ficetola, F. G., Gielly, L., Blanchet, C., Guffond, C., Chiquet, P., Arnaud, F., & Poulenard, J. (2023). Long-term trajectories of mountain agro-ecosystems in the North-Western Alps. *Regional Environmental Change*, *23*(2), 58. DOI: 10.1007/s10113-023-02030-5

Gillani, S. W., Ahmad, M., Manzoor, M., Al-Andal, A., Khan, R. W. A., Ayoola, J. O., & Waheed, M. (2025). Integrating environmental variables and biotic interactions to predict the future distribution of the endangered Aconitum heterophyllum Wall. ex Royle. *BMC Plant Biology*, *26*(1), 45. DOI: 10.1186/s12870-025-07859-y PMID: 41339780

Gillani, S. W., Ahmad, M., Manzoor, M., Waheed, M., Khan, R. W. A., & Sohail, A. (2026). Geographic distribution patterns, population dynamics, and habitat vulnerability of endangered and endemic Trillium govanianum. *Environmental Monitoring and Assessment*, *198*(2), 162. DOI: 10.1007/s10661-026-15013-w PMID: 41569443

Gillespie, T., Brock, J., & Wright, C. (2004). Prospects for quantifying structure, floristic composition and species richness of tropical forests. *International Journal of Remote Sensing*, *25*(4), 707–715. DOI: 10.1080/01431160310001598917

Gillis, L., Maza, M., Garcia-Maribona, J., Lara, J., Suzuki, T., Cierco, M. A., Paul, M., Folkard, A., & Balke, T. (2022). Living on the edge: How traits of ecosystem engineers drive bio-physical interactions at coastal wetland edges. [Gong, L., & Ivanovna, V. I. Water holding capacity of some bryophyte species in connection to their anatomy and morphology.]. *Advances in Water Resources*, *166*, 104257. DOI: 10.1016/j.advwatres.2022.104257

Gillson, L., Biggs, H., Smit, I. P., Virah-Sawmy, M., & Rogers, K. (2019). Finding common ground between adaptive management and evidence-based approaches to biodiversity conservation. *Trends in Ecology & Evolution*, *34*(1), 31–44. DOI: 10.1016/j.tree.2018.10.003 PMID: 30447939

Giorgini, D., Chelli, S., Campetella, G., Chiarucci, A., & Canullo, R. (2015). Predicting the habitat type distribution to assess the representativeness of the Natura 2000 network. 58th annual symposium of the International Association for Vegetable Science: Understanding Broad-Scale Vegetation Patterns: 19-24 July 2015, Brno, Czech Republic: Abstracts., Glime, J. M. (2013). Field taxonomy and collection methods.

Giupponi, L., Panza, R., Pedrali, D., Sala, S., & Giorgi, A. (2025). Ecology, Floristic–Vegetational Features, and Future Perspectives of Spruce Forests Affected by Ips typographus: Insight from the Southern Alps. In *Plants* (Vol. 14, Issue 11). DOI: 10.3390/plants14111681

Giupponi, L., Leoni, V., Pedrali, D., Zuccolo, M., & Cislaghi, A. (2023). Plant cover is related to vegetation and soil features in limestone screes colonization: A case study in the Italian Alps. *Plant and Soil*, *483*(1), 495–513. DOI: 10.1007/s11104-022-05760-3

GOAJK. (2019). *2019. AJK at Glance; A report by Planning & Development Department Muzaffarabad.* Azad Jammu & Kashmir.

Gojon, A., Nussaume, L., Luu, D. T., Murchie, E. H., Baekelandt, A., Rodrigues Saltenis, V. L., Cohan, J.-P., Desnos, T., Inzé, D., Ferguson, J. N., Guiderdonni, E., Krapp, A., Klein Lankhorst, R., Maurel, C., Rouached, H., Parry, M. A. J., Pribil, M., Scharff, L. B., & Nacry, P. (2023). Approaches and determinants to sustainably improve crop production. *Food and Energy Security*, *12*(1), e369. DOI: 10.1002/fes3.369

Golvan, Y.-J., & Rioux, J.-A. (1961). Écologie des mérions du Kurdistan Iranien Relations avec l'épidémiologie de la Peste rurale-(Enquête de l'Institut Pasteur de l'Iran. Rapport préliminaire). *Annales de Parasitologie Humaine et Comparee*, *36*(4), 449–588. DOI: 10.1051/parasite/1961364449 PMID: 13900030

Gonzalez, S. L., & Ghermandi, L. (2021). Overgrazing causes a reduction in the vegetation cover and seed bank of Patagonian grasslands. *Plant and Soil*, *464*(1–2), 75–87. DOI: 10.1007/s11104-021-04931-y

Gorenflo, L. J., & Brandon, K. (2005). Agricultural capacity and conservation in high biodiversity forest ecosystems. *Ambio*, *34*(3), 199–204. DOI: 10.1579/0044-7447-34.3.199 PMID: 16042277

Grabherr, G. (2009). Biodiversity in the high ranges of the Alps: Ethnobotanical and climate change perspectives. *Global Environmental Change*, *19*(2), 167–172. DOI: 10.1016/j.gloenvcha.2009.01.007

Grabherr, G., Gottfried, M., & Pauli, H. (2000). GLORIA: A global observation research initiative in alpine environments. *Mountain Research and Development*, *20*(2), 190–191. DOI: 10.1659/0276-4741(2000)020[0190:GAGORI]2.0.CO;2

Grabherr, G., Gottfried, M., & Pauli, H. (2010). Climate change impacts in alpine environments. *Geography Compass*, *4*(8), 1133–1153. DOI: 10.1111/j.1749-8198.2010.00356.x

Grace, J., Berninger, F., & Nagy, L. (2002). Impacts of climate change on the tree line. *Annals of Botany*, *90*(4), 537–544. DOI: 10.1093/aob/mcf222 PMID: 12324278

Grange, M. C., Munoz, F., Moretti, M., Varona-Y-Varona, S., Renaud, J., Colace, M. P., Guéguen, M., Ssymank, A., & Gallien, L. (2025). The golden threat—Solidago invasion alters native plant–pollinator interactions by vegetative crowding. *Journal of Ecology*, *113*(12), 3762–3772. DOI: 10.1111/1365-2745.70194

Gray, A. (1889). *Darwiniana: essays and reviews pertaining to Darwinism*. D. Appleton.

Grêt-Regamey, A., Brunner, S. H., & Kienast, F. (2012). Mountain ecosystem services: Who cares? *Mountain Research and Development*, *32*(S1), S23–S34. DOI: 10.1659/MRD-JOURNAL-D-10-00115.S1

Grêt-Regamey, A., Walz, A., & Bebi, P. (2008). Valuing ecosystem services for sustainable landscape planning in Alpine regions. *Mountain Research and Development*, *28*(2), 156–165. DOI: 10.1659/mrd.0951

Grissinger, E. H. (2023). Rill and gullies erosion. *Soil Erosion, Conservation, and Rehabilitation*, 153–168.

Growcock, A. J. W. (2006). *Impacts of camping and trampling on Australian alpine and subalpine vegetation and soils*.

Güllü, A., Danquah, J. O., & Dilibal, S. (2021). Characterization of energy dissipative cushions made of Ni–Ti shape memory alloy. *Smart Materials and Structures*, *31*(1), 015018. DOI: 10.1088/1361-665X/ac383d

Guo, X., Hu, Q., Hao, G., Wang, X., Zhang, D., Ma, T., & Liu, J. (2018). The genomes of two Eutrema species provide insight into plant adaptation to high altitudes. *DNA Research : an International Journal for Rapid Publication of Reports on Genes and Genomes*, *25*(3), 307–315. DOI: 10.1093/dnares/dsy003 PMID: 29394339

Guo, Y., Liao, H. L., Boughton, E. H., Martens-Habbena, W., & Qiu, J. (2023). Effects of land-use intensity, grazing and fire disturbances on soil bacterial and fungal communities in subtropical wetlands. *Agriculture, Ecosystems & Environment, 345*, 108314. DOI: 10.1016/j.agee.2022.108314

Guo, Y., Peng, H., Wang, Q., Wang, J., Wu, Z., Shao, B., Xing, G., Huang, Z., Zhao, F., Cui, H., Cui, X., & Tong, Y. (2025). Unveiling the global dynamics of dissolved organic carbon in aquatic ecosystems: Climatic and anthropogenic impact, and future predictions. *The Science of the Total Environment, 958*, 178109. DOI: 10.1016/j.scitotenv.2024.178109 PMID: 39693645

Gurau, S., Imran, M., & Ray, R. L. (2025). Algae: A cutting-edge solution for enhancing soil health and accelerating carbon sequestration – A review. *Environmental Technology & Innovation, 37*, 103980. DOI: 10.1016/j.eti.2024.103980

Haider, S., Palm, S., Bruelheide, H., de Villemereuil, P., Menzel, A., & Lachmuth, S. (2022). Disturbance and indirect effects of climate warming support a plant invader in mountains. *Oikos, 2022*(4), e08783. DOI: 10.1111/oik.08783

Hamid, M., Khuroo, A. A., Malik, A. H., Ahmad, R., & Singh, C. P. (2020). Assessment of alpine summit flora in Kashmir Himalaya and its implications for long-term monitoring of climate change impacts. *Journal of Mountain Science, 17*(8), 1974–1988. DOI: 10.1007/s11629-019-5924-7

Han, H., Zeeshan, Z., Talpur, B. A., Sadiq, T., Bhatti, U. A., Awwad, E. M., Al-Razgan, M., & Ghadi, Y. Y. (2024). Studying long term relationship between carbon Emissions, Soil, and climate Change: Insights from a global Earth modeling Framework. *International Journal of Applied Earth Observation and Geoinformation, 130*, 103902. DOI: 10.1016/j.jag.2024.103902

Hanson, J. H. (2025). Access to Assets and Influence: Governance and Livelihoods in Protected Areas of the Annapurna and Everest Regions, Nepal. *Human Ecology, 53*(5), 1069–1085. Advance online publication. DOI: 10.1007/s10745-025-00643-4 PMID: 41623595

Hao, Y., Mao, J., Bachmann, C. M., Hoffman, F. M., Koren, G., Chen, H., ... & Dai, Y. (2025). Soil moisture controls over carbon sequestration and greenhouse gas emissions: A review. *npj Climate and Atmospheric Science, 8*(1), 16.

Haq, I. U., & Al Fouzan, K. J. J. o. R. M. C. (2017). Research productivity at King Saud bin Abdul Aziz University for health sciences, Kingdom of Saudi Arabia: A bibliometric appraisal. *21*(2).

Haq, S. M., Yaqoob, U., Calixto, E. S., Kumar, M., Rahman, I. U., & Hashem, A., Abd_Allah, E. F., Alakeel, M. A., Alqarawi, A. A., Abdalla, M., Lone, F. A., Khan, M. A., Khan, U., & Ijaz, F. (. (2021). Long-Term Impact of Transhumance Pastoralism and Associated Disturbances in High-Altitude Forests of Indian Western Himalaya. *Sustainability, 13*(22), 12497. DOI: 10.3390/su132212497

Hassan, N., Rafiq, M., Hayat, M., Shah, A. A., & Hasan, F. (2016). Psychrophilic and psychrotrophic fungi: A comprehensive review. *Reviews in Environmental Science and Biotechnology, 15*(2), 147–172. DOI: 10.1007/s11157-016-9395-9

Haussmann, N. S., Boelhouwers, J. C., & McGeoch, M. A. (2009). Fine scale variability in soil frost dynamics surrounding cushions of the dominant vascular plant species (azorella selago) on sub-antarctic marion island. *Geografiska Annaler. Series A. Physical Geography, 91*(4), 257–268. DOI: 10.1111/j.1468-0459.2009.00368.x

Hawke, D. (2022). The biogeochemistry and ecological impact of Westland petrels (Procellaria westlandica) on terrestrial ecosystems. *New Zealand Journal of Ecology, 46*(1), 1–16. DOI: 10.20417/nzjecol.46.3

Hayat, S., Hayat, Q., Alyemeni, M. N., Wani, A. S., Pichtel, J., & Ahmad, A. (2012). Role of proline under changing environments. *Plant Signaling & Behavior, 7*(11), 1456–1466. DOI: 10.4161/psb.21949 PMID: 22951402

He, J.-S., Dong, S., Shang, Z., Sundqvist, M. K., Wu, G., Yang, Y. J. P., & Soil. (2021). Above-belowground interactions in alpine ecosystems on the roof of the world. *458*(1), 1-6.

Hedberg, O. (1964). *Features of afroalpine plant ecology*. Sv. växtgeografiska sällsk.

Hedberg, C. P., & Smith, F. A. (2025). Unraveling the role of environmental and anthropogenic drivers in shaping global patterns in mammal diversity. *Ecography, 2025*(10), 07674. DOI: 10.1002/ecog.07674

He, F., Dong, S., Yang, J., Hao, X., Ma, C., Shen, H., Xiao, J., Li, Y., Zhang, R., Shi, H., Wang, Z., Zhang, Y., & Zuo, H. (2025). Rotational grazing can enhance the soil available nutrients rather than plant diversity to promote ANPP of alpine steppe. *Agriculture, Ecosystems & Environment, 387*, 109617. DOI: 10.1016/j.agee.2025.109617

Helmi, N. R. (2025). Exploring the diversity and antimicrobial potential of actinomycetes isolated from different environments in Saudi Arabia: A systematic review. *Frontiers in Microbiology, 16*, 1568899. DOI: 10.3389/fmicb.2025.1568899 PMID: 40207161

Helm, N., Chytrý, K., Hülber, K., Moser, D., Wessely, J., Gattringer, A., Hausharter, J., Pauli, H., Winkler, M., Saccone, P., Lamprecht, A., Rutzinger, M., Mayr, A., Kollert, A., & Dullinger, S. (2024). Fine-scale alpine plant community assembly: Relative roles of environmental sorting, dispersal processes and species interactions. *Journal of Ecology, 112*(12), 2745–2757. DOI: 10.1111/1365-2745.14401

Herman, F., De Doncker, F., Delaney, I., Prasicek, G., & Koppes, M. (2021). The impact of glaciers on mountain erosion. *Nature Reviews. Earth & Environment, 2*(6), 422–435. DOI: 10.1038/s43017-021-00165-9

Hess, C. (2022). *Insulation effect of moss versus vascular plant cover on active layer depth in the High Arctic Svalbard* Itä-Suomen yliopisto].

Hester, A., & Brooker, R. (2007). Threatened habitats: Marginal vegetation in upland areas. *Issues in Environmental Science and Technology, 25*, 107–134. DOI: 10.1039/9781847557650-00107

Heywood, V. (2001). Floristics and monography—An uncertain future? *Taxon, 50*(2), 361–380. DOI: 10.2307/1223886

Hock, R., Rasul, G., Adler, C., Cáceres, B., Gruber, S., Hirabayashi, Y., . . . Kutuzov, S. (2019). High mountain areas. In *IPCC special report on the ocean and cryosphere in a changing climate* (pp. 131-202). H.-O. Pörtner, DC Roberts, V. Masson-Delmotte, P. Zhai, M. Tignor, E….

Hoersch, B., Braun, G., & Schmidt, U. (2002). Relation between landform and vegetation in alpine regions of Wallis, Switzerland. A multiscale remote sensing and GIS approach. *Computers, Environment and Urban Systems, 26*(2-3), 113–139. DOI: 10.1016/S0198-9715(01)00039-4

Holzmann, K. L., Walls, R. L., & Wiens, J. J. (2023). Accelerating local extinction associated with very recent climate change. *Ecology Letters, 26*(11), 1877–1886. DOI: 10.1111/ele.14303 PMID: 37721806

Hooker, J. D. (1854). X. On the functions and structure of the rostellum of Listera ovata. *Philosophical Transactions of the Royal Society of London*, (144), 259–263. DOI: 10.1098/rstl.1854.0011

Hooker, J. D., & Thomson, T. (1855). *Flora indica: being a systematic account of the plants of British India, together with observations on the structure and affinities of their natural orders and genera* (Vol. 1). Pamplin. DOI: 10.5962/bhl.title.50109

Hooshyarkhah, B., Johnson, D. L., Spencer, L., Ryait, H. S., & Chegoonian, A. (2025). Mapping Four Decades of Treeline Ecotone Migration: Remote Sensing of Alpine Ecotone Shifts on the Eastern Slopes of the Canadian Rocky Mountains. *Remote Sensing (Basel), 17*(24), 4004. DOI: 10.3390/rs17244004

Hossain, M. M., Sultana, F., Mostafa, M., Ferdus, H., Kundu, M., Adhikary, S., Akter, N., Saha, A., & Sabbir, M. A. A. (2025). Stress responses and mechanisms of phytopathogens infecting humans: Threats, drivers, and recommendations. *Stresses, 5*(2), 28. DOI: 10.3390/stresses5020028

Hossein-Panahi, B., Bozorg-Haddad, O., Loáiciga, H., Meghwar, S. M., & Zeleňáková, M. (2022). Case Studies Around the World. In *Climate Change in Sustainable Water Resources Management* (pp. 361–414). Springer. DOI: 10.1007/978-981-19-1898-8_10

Hotaling, S., Finn, D. S., Joseph Giersch, J., Weisrock, D. W., & Jacobsen, D. (2017). Climate change and alpine stream biology: Progress, challenges, and opportunities for the future. *Biological Reviews of the Cambridge Philosophical Society, 92*(4), 2024–2045. DOI: 10.1111/brv.12319 PMID: 28105701

Hou DaBin, H. D., & Shu, GuangMing, S. G., & Ren ZhenLong, R. Z. (2006). Effect of nitrogen, phosphorus, and potassium on several nitrogen metabolism enzymes of Aconitum carmichaeli Dexb. (Ranunculaceae). *Xi Nan Nong Ye Xue Bao, 19*(5), 857–862.

Hou, M., Leng, C., Zhu, J., Yang, M., Yin, Y., Xing, Y., & Chen, J. (2024). Alpine and subalpine plant microbiome mediated plants adapt to the cold environment: A systematic review. *Environmental Microbiome, 19*(1), 82. DOI: 10.1186/s40793-024-00614-0 PMID: 39487507

Hou, Q., Ji, Z., Yang, H., & Yu, X. (2022). Impacts of climate change and human activities on different degraded grassland based on NDVI. *Scientific Reports, 12*(1), 15918. DOI: 10.1038/s41598-022-19943-6 PMID: 36151254

Hou, Y., Zhao, W., Zhan, T., Hua, T., & Pereira, P. (2024). Alpine meadow and alpine steppe plant-soil network in Qinghai-Tibet Plateau, China. *Catena, 244*, 108235. DOI: 10.1016/j.catena.2024.108235

Hou, Z., Wang, R., Chang, S., Zheng, Y., Ma, T., Xu, S., Zhang, X., Shi, X., Lu, J., Luo, D., Wang, B., Du, Z., & Wei, Y. (2024). The contribution of microbial necromass to soil organic carbon and influencing factors along a variation of habitats in alpine ecosystems. *The Science of the Total Environment, 921*, 171126. DOI: 10.1016/j.scitotenv.2024.171126 PMID: 38387574

Hughes, C. E., & Atchison, G. W. (2015). The ubiquity of alpine plant radiations: From the Andes to the Hengduan Mountains. *The New Phytologist*, *207*(2), 275–282. DOI: 10.1111/nph.13230 PMID: 25605002

Hui, M. L. Y., Tan, L. T. H., Letchumanan, V., He, Y. W., Fang, C. M., Chan, K. G., Law, J. W.-F., & Lee, L. H. (2021). The extremophilic actinobacteria: From microbes to medicine. *Antibiotics (Basel, Switzerland)*, *10*(6), 682. DOI: 10.3390/antibiotics10060682 PMID: 34201133

Hülber, K., Sonnleitner, M., Haider, J., Schwentenwein, M., Winkler, M., Schneeweiss, G. M., & Schönswetter, P. (2018). Reciprocal transplantations reveal strong niche differentiation among ploidy-differentiated species of the Senecio carniolicus aggregate (Asteraceae) in the easternmost Alps. *Alpine Botany*, *128*(2), 107–119. DOI: 10.1007/s00035-018-0209-2

Hu, S., Hsu, P.-C., Li, W., Wang, L., Chen, H., & Zhou, B. (2023). Mechanisms of Tibetan Plateau warming amplification in recent decades and future projections. *Journal of Climate*, *36*(17), 5775–5792. DOI: 10.1175/JCLI-D-22-0471.1

Huttunen, S., Bell, N., & Hedenäs, L. (2018). The evolutionary diversity of mosses–taxonomic heterogeneity and its ecological drivers. *Critical Reviews in Plant Sciences*, *37*(2-3), 128–174. DOI: 10.1080/07352689.2018.1482434

Hu, X., Sun, Q., Zhang, S., Li, W., Wang, X., Long, R., Jin, G., & Zhang, B. (2025). Effects of No-Till Seeding and Fertilization on Vegetation Restoration and Soil Physicochemical Properties in Alpine Degraded Grazing Grasslands. *Agronomy (Basel)*, *15*(3), 578. DOI: 10.3390/agronomy15030578

Ianni, E., Geneletti, D., & Ciolli, M. (2015). Revitalizing traditional ecological knowledge: A study in an alpine rural community. *Environmental Management*, *56*(1), 144–156. DOI: 10.1007/s00267-015-0479-z PMID: 25847107

Ibrahim, M., Khan, M., Ali, S., Razzaq, A., Zaman, A., Iqbal, M., & Jan, F. (2019). Floristic composition and species diversity of plant resources of rural area "Takht Bhai" district Mardan, Khyber Pakhtunkhwa, Pakistan. *Medicinal and aromatic Plants (Los Angeles)*, *8*(338), 2167. DOI: 10.35248/2167-0412.19.8.338

Imran, & Ortas, I. (2025). Mechanism in soil health improvement with soil microbial actions and its role in potential agriculture. *Communications in Soil Science and Plant Analysis*, 1–22.

Inouye, D. W. (2008). Effects of climate change on phenology, frost damage, and floral abundance of montane wildflowers. *Ecology*, *89*(2), 353–362. DOI: 10.1890/06-2128.1 PMID: 18409425

Inouye, D. W. (2020). Effects of climate change on alpine plants and their pollinators. *Annals of the New York Academy of Sciences, 1469*(1), 26–37. DOI: 10.1111/nyas.14104 PMID: 31025387

IPCC. (2022). *Climate Change 2022: Impacts, Adaptation and Vulnerability. Contribution of Working Group II to the Sixth Assessment Report of the Intergovernmental Panel on Climate Change* (Pörtner, H.-O., Eds.). Cambridge University Press., DOI: 10.1017/9781009325844

Iqbal, M. A., & Grichar, W. J. (2024). *Grasslands: Conservation and Development.* BoD–Books on Demand. DOI: 10.5772/intechopen.107797

Iqbal, M., Tian, Y., Zhang, Y., Liu, M., Song, M., Zhou, J., Jin, P., Ma, D., Yu, C., & Xu, X. (2025). Grazing directly drives shifts in species composition with minimal mediation by edaphic properties in a temperate grassland. *Plant and Soil*, 1–22. DOI: 10.1007/s11104-025-08114-x

Iqbal, S., Begum, F., Nguchu, B. A., Claver, U. P., & Shaw, P. (2025). The invisible architects: Microbial communities and their transformative role in soil health and global climate changes. *Environmental Microbiome, 20*(1), 36. DOI: 10.1186/s40793-025-00694-6 PMID: 40133952

Ishaq, S., Khan, M., Begum, F., Hussain, K., Amir, R., Hussain, A., & Ali, S. J. J. o. M. A. R. (2016). Climate change impact on mountain biodiversity: a special reference to Gilgit-Baltistan of Pakistan. *1*, 53-63.

Ishfaq, M., Tam, N. F. Y., Lang, T., Hussain, M., & Zhou, H. (2025). Nitrogen–phosphorus conservation and trade-offs in mangroves. *Plant and Soil*, •••, 1–20.

Izyumov, S., & Kuzmin, Y. O. (2014). Study of the recent geodynamic processes in the Kopet-Dag region. *Izvestiya. Izvestiya, Russian Academy of Sciences. Physics of the Solid Earth, 50*(6), 719–731. DOI: 10.1134/S1069351314060019

Jaccard, P. (1912). The distribution of the flora in the alpine zone. 1. *The New Phytologist, 11*(2), 37–50. DOI: 10.1111/j.1469-8137.1912.tb05611.x

Janiczek, A., Suchan, T., Paul, W., Consortium, I., & Ronikier, M. (2025). Spatial Discontinuity of Mountain Systems and Genetic Structure of Alpine Plants: The Alps–Carpathians Disjunction in a Comparative Phylogeographical Context. *Journal of Biogeography, 52*(6), e15122. DOI: 10.1111/jbi.15122

Jansson, J. K., & Hofmockel, K. S. (2020). Soil microbiomes and climate change. *Nature Reviews. Microbiology, 18*(1), 35–46. DOI: 10.1038/s41579-019-0265-7 PMID: 31586158

Jędrych, M., Zagajewski, B., & Marcinkowska-Ochtyra, A. (2017a). Application of Sentinel-2 and EnMAP new satellite data to the mapping of alpine vegetation of the Karkonosze Mountains. *Polish Cartographical Review, 49.*

Jędrych, M., Zagajewski, B., & Marcinkowska-Ochtyra, A. J. P. C. R. (2017b). Application of Sentinel-2 and EnMAP new satellite data to the mapping of alpine vegetation of the Karkonosze Mountains. *49.*

Jensen, J. R. (2009). *Remote sensing of the environment: An earth resource perspective 2/e*. Pearson Education India. 1-239.

Jia, F., Li, W., Zhou, B., Xiao, Y., Si, B., Saikeshan, A., Lv, T., & Wang, Z. (2025). Consecutive high-efficient water-saving irrigation increase crop yield and decrease soil salinity through reconstructing rhizosphere soil bacterial communities. *The Science of the Total Environment, 959*, 178238. DOI: 10.1016/j.scitotenv.2024.178238 PMID: 39742583

Jiang, C., Yu, G., Li, Y., Cao, G., Yang, Z., Sheng, W., & Yu, W. (2012). Nutrient resorption of coexistence species in alpine meadow of the Qinghai–Tibetan Plateau explains plant adaptation to nutrient-poor environment. *Ecological Engineering, 44*, 1–9. DOI: 10.1016/j.ecoleng.2012.04.006

Jiao, J., Zou, H., Jia, Y., & Wang, N. (2009). Research progress on the effects of soil erosion on vegetation. *Acta Ecologica Sinica, 29*(2), 85–91. DOI: 10.1016/j.chnaes.2009.05.001

Jin, X., Deng, A., Fan, Y., Ma, K., Zhao, Y., Wang, Y., Zheng, K., Zhou, X., & Lu, G. (2025). Diversity, functionality, and stability: Shaping ecosystem multifunctionality in the successional sequences of alpine meadows and alpine steppes on the Qinghai-Tibet Plateau. *Frontiers in Plant Science, 16*, 1436439. DOI: 10.3389/fpls.2025.1436439 PMID: 40182548

Johannesson, K. (2024). Climate Change and Baltic Sea Genetic Diversity. In *Oxford Research Encyclopedia of Climate Science*. DOI: 10.1093/acrefore/9780190228620.013.925

Johnstone, J. (2021). *Alpine plant life: Functional plant ecology of high mountain ecosystems.*

Johnston, S. R., Boddy, L., & Weightman, A. J. (2016). Bacteria in decomposing wood and their interactions with wood-decay fungi. *FEMS Microbiology Ecology, 92*(11), fiw179. DOI: 10.1093/femsec/fiw179 PMID: 27559028

Joshi, S., Shrestha, R. M., Ismail, M., Qamer, F. M., Jasra, A. W., Dan, Z., Ning, W. J. H.-a. r., & Himalayas, t. i. i. t. H. K. (2013). High-altitude rangelands and land use practices in the Karakoram-Pamir landscape. 37.

Juan, Y., Tian, L., Sun, W., Qiu, W., Curtin, D., Gong, L., & Liu, Y. (2020). Simulation of soil freezing–thawing cycles under typical winter conditions: Implications for nitrogen mineralization. *Journal of Soils and Sediments*, *20*(1), 143–152. DOI: 10.1007/s11368-019-02374-8

Jungblut, A. D., & Vincent, W. F. (2017). Cyanobacteria in polar and alpine ecosystems. In *Psychrophiles: from biodiversity to biotechnology* (pp. 181–206). Springer. DOI: 10.1007/978-3-319-57057-0_9

Kalvoda, J., & Emmer, A. (2021). Mass wasting and erosion in different morphoclimatic zones of the Makalu Barun region, Nepal Himalaya. *Geografiska Annaler. Series A. Physical Geography*, *103*(4), 368–396. DOI: 10.1080/04353676.2021.2000816

Kamanmalek, S., & Alamdari, N. (2024). Advancing equitable stormwater management: A decision support tool integrating best practices for nutrient removal and environmental justice. *Ecological Informatics*, *80*, 102496. DOI: 10.1016/j.ecoinf.2024.102496

Kameyama, Y., Kasagi, T., & Kudo, G. (2008). A hybrid zone dominated by fertile F1s of two alpine shrub species, Phyllodoce caerulea and Phyllodoce aleutica, along a snowmelt gradient. *Journal of Evolutionary Biology*, *21*(2), 588–597. DOI: 10.1111/j.1420-9101.2007.01476.x PMID: 18205785

Kameyama, Y., & Kudo, G. (2009). Flowering phenology influences seed production and outcrossing rate in populations of an alpine snowbed shrub, Phyllodoce aleutica: Effects of pollinators and self-incompatibility. *Annals of Botany*, *103*(9), 1385–1394. DOI: 10.1093/aob/mcp037 PMID: 19228703

Kameyama, Y., & Kudo, G. (2015). Intrinsic and extrinsic factors acting on the reproductive process in alpine-snowbed plants: Roles of phenology, biological interaction, and breeding system. *Plant Species Biology*, *30*(1), 3–15. DOI: 10.1111/1442-1984.12070

Kang, D., Lungu, S. E., Danso, F., Dzou, C. F., Chen, Y., Zheng, X., Nie, F., Lin, H., Chen, J., & Zhou, G. (2025). Animal health and nutrition: Metabolic disorders in cattle and improvement strategies. *Frontiers in Veterinary Science*, *12*, 1470391. DOI: 10.3389/fvets.2025.1470391 PMID: 40201075

Kang, M., Wu, H., Liu, H., Liu, W., Zhu, M., Han, Y., Liu, W., Chen, C., Song, Y., Tan, L., Yin, K., Zhao, Y., Yan, Z., Lou, S., Zan, Y., & Liu, J. (2023). The pan-genome and local adaptation of Arabidopsis thaliana. *Nature Communications, 14*(1), 6259. DOI: 10.1038/s41467-023-42029-4 PMID: 37802986

Karabulut, M. (2018). An examination of spectral reflectance properties of some wetland plants in Göksu Delta, Turkey. *Journal of International Environmental Application and Science, 13*(4), 194–203.

Karl, T. R., Derr, V. E., Easterling, D. R., Folland, C. K., Hofmann, D. J., Levitus, S., Nicholls, N., Parker, D. E., & Withee, G. W. (1995). Critical issues for long-term climate monitoring. *Climatic Change, 31*(2), 185–221. DOI: 10.1007/BF01095146

Kartini, N. L., Saifulloh, M., Trigunasih, N. M., Sukmawati, N. M. S., & Mega, I. (2024). Impact of long-term continuous cropping on soil nutrient depletion. *Ecological Engineering & Environmental Technology, 25.*

Kasagi, T., & Kudo, G. (2003). Variations in bumble bee preference and pollen limitation among neighboring populations: Comparisons between Phyllodoce caerulea and Phyllodoce aleutica (Ericaceae) along snowmelt gradients. *American Journal of Botany, 90*(9), 1321–1327. DOI: 10.3732/ajb.90.9.1321 PMID: 21659231

Kaštovská, E., Choma, M., Čapek, P., Kaňa, J., Tahovská, K., & Kopáček, J. (2022). Soil warming during winter period enhanced soil N and P availability and leaching in alpine grasslands: A transplant study. *PLoS One, 17*(8), e0272143. DOI: 10.1371/journal.pone.0272143 PMID: 35917373

Kauer, L., Imholt, C., Jacob, J., & Kuehn, R. (2024). Assessing the effects of land-use intensity on small mammal community composition and genetic variation in Myodes glareolus and Microtus arvalis across grassland and forest habitats. *Landscape Ecology, 40*(1), 1. DOI: 10.1007/s10980-024-02021-9

Kaur, S., Das, A., Dwivedi, S., Tripathi, S., & Shrivastava, N. (2025). Effect of climate change on beneficial plant–microbe interaction in the rhizosphere. In *Plant–microbiome interactions for climate-resilient agriculture* (pp. 99–120). DOI: 10.1007/978-981-96-3534-4_5

Kayani, S., Ahmad, M., Sultana, S., Shinwari, Z. K., Zafar, M., Yaseen, G., & Bibi, T. (2015). Ethnobotany of medicinal plants among the communities of Alpine and Sub-alpine regions of Pakistan. *Journal of Ethnopharmacology, 164*, 186–202. DOI: 10.1016/j.jep.2015.02.004 PMID: 25680839

Kearney, M., & Porter, W. (2009). Mechanistic niche modelling: Combining physiological and spatial data to predict species' ranges. *Ecology Letters, 12*(4), 334–350. DOI: 10.1111/j.1461-0248.2008.01277.x PMID: 19292794

Kehl, M. (2009). Quaternary climate change in Iran—The state of knowledge. *Erdkunde, 63*(1), 1–17. DOI: 10.3112/erdkunde.2009.01.01

Kennedy, R. E., Cohen, W. B., & Schroeder, T. A. (2007). Trajectory-based change detection for automated characterization of forest disturbance dynamics. *Remote Sensing of Environment, 110*(3), 370–386. DOI: 10.1016/j.rse.2007.03.010

Kent, M. (2011). *Vegetation description and data analysis: a practical approach.* John Wiley & Sons.

Kevan, P. G., & Coates, C. (2024). Heliocaminiform structures: Plant organs that function as microgreenhouses. *Facets, 9,* 1–20. DOI: 10.1139/facets-2023-0099

Khalili, K., Tahoudi, M. N., Mirabbasi, R., & Ahmadi, F. (2016). Investigation of spatial and temporal variability of precipitation in Iran over the last half century. *Stochastic Environmental Research and Risk Assessment, 30*(4), 1205–1221. DOI: 10.1007/s00477-015-1095-4

Khan, G., Khan, B., Qamer, F. M., Abbas, S., Khan, A., & Xi, C. J. J. o. K. S. U.-S. (2016). Himalayan ibex (Capra ibex sibirica) habitat suitability and range resource dynamics in the Central Karakorum National Park, Pakistan. *28*(3), 245-254.

Khan, M. N., Ali, S., Yaseen, T., Adnan, M., Ullah, S., Zaman, A., Iqbal, M., Shah, S. N., Ali, A., & Razzaq, A. (2020). Assessment of proximate and nutritional contents in selected weedy grasses for potential use as fodder in District Charsadda, KP: Assessment of proximate and nutritional contents in selected weedy grasses. *Proceedings of the Pakistan Academy of Sciences: B. Life and Environmental Sciences, 57*(2), 83-94.

Khan, M. T., Supronienė, S., Žvirdauskienė, R., & Aleinikovienė, J. (2025). Climate, soil, and microbes: Interactions shaping organic matter decomposition in croplands. *Agronomy (Basel), 15*(8), 1928. DOI: 10.3390/agronomy15081928

Khan, R. W. A., Shaheen, H., Qayyum, S., Awan, S. N., Shah, T. A., Alsahli, A. A., Younous, Y. A., & El-Sheikh, M. A. (2025). Predicting habitat refugia of the medicinal plant Dolomiaea costus under two climate change scenarios. *Discover Life, 55*(1), 2. DOI: 10.1007/s11084-024-09670-5

Khan, S. M., Page, S. E., Ahmad, H., & Harper, D. M. (2013). Sustainable utilization and conservation of plant biodiversity in montane ecosystems: The western Himalayas as a case study. *Annals of Botany*, *112*(3), 479–501. DOI: 10.1093/aob/mct125 PMID: 23825353

Khan, S., Khatoon, S., & Hussain, A. (2018). Inventorying and monitoring of alpine specific genera and species of haramosh and bagraote valleys., (karakoram range) gilgit-baltistan, Pakistan. *Pakistan Journal of Agricultural Sciences. Pakistan Journal of Agricultural Sciences*, *6*(3), 595–601. Advance online publication. DOI: 10.21162/PAKJAS/18.4998

Khan, S., Shaheen, H., Aziz, S., & Nasar, S. (2021). Diversity and distribution of Genus Primula in Kashmir region: An indicator genus of the western Himalayan mountain wetlands and glacial forelands. *Biodiversity and Conservation*, *30*(6), 1673–1688. DOI: 10.1007/s10531-021-02163-1

Khoso, R. B. (2024). Global drivers, local consequences: Assessing geodiversity and geosystem services for global change adaptation and sustainable development of mountain regions–Examples from the Monte Rosa W-Alps.

Kiabi, B. H., Abdoli, A., & Naderi, M. (1999). Status of the fish fauna in the South Caspian Basin of Iran. *Zoology in the Middle East*, *18*(1), 57–65. DOI: 10.1080/09397140.1999.10637782

Kiabi, B. H., Dareshouri, B. F., Ghaemi, R. A., & Jahanshahi, M. (2002). Population status of the Persian leopard (Panthera pardus saxicolor Pocock, 1927) in Iran. *Zoology in the Middle East*, *26*(1), 41–47. DOI: 10.1080/09397140.2002.10637920

Kikuzawa, K. (1995). Leaf phenology as an optimal strategy for carbon gain in plants. *Canadian Journal of Botany*, *73*(2), 158–163. DOI: 10.1139/b95-019

Kim, J.-S., Kim, J.-H., Hyun, J.-O., Jang, C.-G., Choi, H.-J., Lee, J.-H., Gwon, J.-H., Son, S.-W., So, S., Kim, J.-W., Park, H.-C., & Myeong, H.-H. (2025). A checklist of alpine and subalpine plants in the Korean Peninsula. *Journal of Asia-Pacific Biodiversity*. 10.1016/j.japb.2025.06.010

Kleier, C., & Lambrinos, J. G. (2005). The Importance of Nurse Associations for Three Tropical Alpine Life Forms. *Arctic, Antarctic, and Alpine Research*, *37*(3), 331–336. DOI: 10.1657/1523-0430(2005)037[0331:TIONAF]2.0.CO;2

Klein, J. A., Harte, J., & Zhao, X. (2004a). Experimental warming causes large and rapid species loss, dampened by simulated grazing, on the Tibetan Plateau. *Ecology Letters*, *7*(12), 1170–1179. DOI: 10.1111/j.1461-0248.2004.00677.x

Klimešová, J., & Martínková, J. (2022). Clonal growth, resprouting, and vegetative propagation of weeds. *Persistence strategies of weeds*, 200-218.

Knežević, M., Topalović, A., Fuštić, B., & Životić, L. (2025). In Knežević, M., Topalović, A., Fuštić, B., & Životić, L. (Eds.), *Soil Types—Properties and Distribution BT - The Soils of Montenegro* (pp. 117–196). Springer Nature Switzerland., DOI: 10.1007/978-3-031-82919-2_5

Kok, P. J., Broholm, T. L., van Doorn, L., Fiorillo, B. F., & Smith, C. (2024). Lost in space and time: Robust demography and enhanced resilience buffer adverse environmental effects in a highly isolated and sedentary pre-pleistocene landscape vertebrate. *BMC Ecology and Evolution*, *24*(1), 125. DOI: 10.1186/s12862-024-02314-2 PMID: 39394073

Komarkova, V., & Wielgolaski, F. (1999). Stress and disturbance in cold region ecosystems. *Ecosystems of the world*, 39-122.

Körner, C. (2021). The alpine life zone. In *Alpine Plant Life: Functional Plant Ecology of High Mountain Ecosystems* (pp. 23-51). Springer. DOI: 10.1007/978-3-030-59538-8_2

Körner, C., & Change, G. (2018). Alpine ecosystems and the high-elevation treeline. *Encyclopedia of ecology.*, 407-413.

Körner, C. (1995). Alpine plant diversity: a global survey and functional interpretations. In *Arctic and alpine biodiversity: patterns, causes and ecosystem consequences* (pp. 45–62). Springer. DOI: 10.1007/978-3-642-78966-3_4

Körner, C. (2007). The use of 'altitude'in ecological research. *Trends in Ecology & Evolution*, *22*(11), 569–574. DOI: 10.1016/j.tree.2007.09.006 PMID: 17988759

Körner, C. (2016). Plant adaptation to cold climates. *F1000 Research*, 5. PMID: 27990251

Körner, C. (2019). Mountain Biodiversity, its Causes and Function: An Overview. In Körner, Ch., & Spehn, E. M. (Eds.), *Mountain Biodiversity* (1st ed., pp. 3–20). Routledge., DOI: 10.4324/9780429342585-1

Körner, C. (2021). *Alpine Plant Life: Functional Plant Ecology of High Mountain Ecosystems*. Springer International Publishing., DOI: 10.1007/978-3-030-59538-8

Körner, C. (2021). Alpine treelines. In *Alpine plant life: functional plant ecology of high mountain ecosystems* (pp. 141–173). Springer.

Körner, C. (2021a). The alpine life zone. In *Alpine plant life: Functional plant ecology of high mountain ecosystems* (pp. 23–51). Springer.

Körner, C. (2023). Concepts in alpine plant ecology. *Plants, 12*(14), 2666. DOI: 10.3390/plants12142666 PMID: 37514280

Körner, C., Berninger, U.-G., Daim, A., Eberl, T., Mendoza, F. F., Füreder, L., Grube, M., Hainzer, E., Kaiser, R., Meyer, E., Newesely, C., Niedrist, G., Niedrist, G. H., Petermann, J. S., Seeber, J., Tappeiner, U., & Wickham, S. (2022). Long-term monitoring of high-elevation terrestrial and aquatic ecosystems in the Alps–a five-year synthesis. *Eco.Mont, 14*(2), 48–69. DOI: 10.1553/eco.mont-14-2s48

Körner, C., & Hiltbrunner, E. (2021). Why is the alpine flora comparatively robust against climatic warming? *Diversity, 13*(8), 383. DOI: 10.3390/d13080383

Körner, C., Jetz, W., Paulsen, J., Payne, D., Rudmann-Maurer, K., & Spehn, M., E. (. (2017). A global inventory of mountains for bio-geographical applications. *Alpine Botany, 127*(1), 1–15. DOI: 10.1007/s00035-016-0182-6

Körner, C., Nakhutsrishvili, G., & Spehn, E. (2006). High-elevation land use, biodiversity, and ecosystem functioning. In *Land use change and mountain biodiversity* (pp. 3–36). CRC Press. DOI: 10.1201/9781420002874-1

Körner, C., & Paulsen, J. (2004). A world-wide study of high altitude treeline temperatures. *Journal of Biogeography, 31*(5), 713–732. DOI: 10.1111/j.1365-2699.2003.01043.x

Korner, C., & Spehn, E. M. (2024). *Mountain biodiversity: a global assessment.* CRC Press. DOI: 10.1201/9781003579519

Kosick, H. (2023). *Drivers of change in the temperate-boreal refugium of Cape Breton Highlands National Park.* Memorial University of Newfoundland.

Kotlarski, S., Gobiet, A., Morin, S., Olefs, M., Rajczak, J., & Samacoïts, R. (2023). 21st Century alpine climate change. *Climate Dynamics, 60*(1), 65–86. DOI: 10.1007/s00382-022-06303-3

Koyanagi, K., Andreoli, A., Nordio, G., Pitscheider, F., Tomelleri, E., & Comiti, F. (2025). Disturbance-intervention interactions on post-windthrow hillslope runoff and erosion processes in the Eastern Italian Alps. *Journal of Hydrology (Amsterdam), 656*, 133001. DOI: 10.1016/j.jhydrol.2025.133001

Kudo, G., & Hirao, A. S. (2006). Habitat-specific responses in the flowering phenology and seed set of alpine plants to climate variation: Implications for global-change impacts. *Population Ecology, 48*(1), 49–58. DOI: 10.1007/s10144-005-0242-z

Kullman, L. (2010). Alpine flora dynamics–a critical review of responses to climate change in the Swedish Scandes since the early 1950s. *Nordic Journal of Botany, 28*(4), 398–408. DOI: 10.1111/j.1756-1051.2010.00812.x

Kumar, R., Mondal, B., & Bordoloi, N. (2024). Application of straw-derived biochar: A sustainable approach to improve soil quality and crop yield and reduce N2O emissions in paddy soil. *Environmental Science and Pollution Research International, 31*(51), 60804–60818. DOI: 10.1007/s11356-024-35269-4 PMID: 39392575

Kumar, S., & Khanduri, V. P. (2024). Impact of climate change on the Himalayan alpine treeline vegetation. *Heliyon, 10*(23), e40797. DOI: 10.1016/j.heliyon.2024.e40797 PMID: 39698095

Kuo, C. C., Liu, Y. C., Su, Y., Liu, H. Y., & Lin, C. T. (2022). Responses of alpine summit vegetation under climate change in the transition zone between subtropical and tropical humid environment. *Scientific Reports, 12*(1), 13352. DOI: 10.1038/s41598-022-17682-2 PMID: 35922458

Kurepa, J., & Smalle, J. A. (2023). Plant hormone modularity and the survival-reproduction trade-off. *Biology (Basel), 12*(8), 1143. DOI: 10.3390/biology12081143 PMID: 37627027

Lagerholm, V. K., Sandoval-Castellanos, E., Vaniscotte, A., Potapova, O. R., Tomek, T., Bochenski, Z. M., Shepherd, P., Barton, N., Van Dyck, M. C., Miller, R., Höglund, J., Yoccoz, N. G., Dalén, L., & Stewart, J. R. (2017). Range shifts or extinction? Ancient DNA and distribution modelling reveal past and future responses to climate warming in cold-adapted birds. *Global Change Biology, 23*(4), 1425–1435. DOI: 10.1111/gcb.13522 PMID: 27762483

Lai, H., Gao, F., Su, H., Zheng, P., Li, Y., & Yao, H. (2022). Nitrogen distribution and soil microbial community characteristics in a legume–cereal intercropping system: A review. *Agronomy (Basel), 12*(8), 1900. DOI: 10.3390/agronomy12081900

Lallemand, F., Martin-Magniette, M. L., Gilard, F., Gakière, B., Launay-Avon, A., Delannoy, É., & Selosse, M. A. (2019). In situ transcriptomic and metabolomic study of the loss of photosynthesis in the leaves of mixotrophic plants exploiting fungi. *The Plant Journal, 98*(5), 826–841. DOI: 10.1111/tpj.14276 PMID: 30735596

Lambert, A. M., Miller-Rushing, A. J., & Inouye, D. W. (2010). Changes in snowmelt date and summer precipitation affect the flowering phenology of Erythronium grandiflorum (glacier lily; Liliaceae). *American Journal of Botany, 97*(9), 1431–1437. DOI: 10.3732/ajb.1000095 PMID: 21616897

Lamouille-Hébert, M., Arthaud, F., & Datry, T. (2024). Climate change and the biodiversity of alpine ponds: Challenges and perspectives. *Ecology and Evolution, 14*(2), e10883. DOI: 10.1002/ece3.10883 PMID: 38327685

Lamprecht, A., Semenchuk, P. R., Steinbauer, K., Winkler, M., & Pauli, H. (2018). Climate change leads to accelerated transformation of high-elevation vegetation in the central Alps. *The New Phytologist, 220*(2), 447–459. DOI: 10.1111/nph.15290 PMID: 29938796

Lavergne, S., Mouquet, N., Thuiller, W., & Ronce, O. (2010). Biodiversity and climate change: Integrating evolutionary and ecological responses of species and communities. *Annual Review of Ecology, Evolution, and Systematics, 41*(1), 321–350. DOI: 10.1146/annurev-ecolsys-102209-144628

Lavergne, S., Pouchon, C., Roquet, C., Alberti, A., Boleda, M., Perrier, C., Renaud, J., Boyer, F., Bzeznik, B., Rome, M., Douzet, R., Lammers, Y., Orvain, C., Gavory, F., Labadie, K., Aubert, S., Consortium, P. N., Consortium, P. A., Wincker, P., & Coissac, E. (2025). Towards a comprehensive barcoding and phylogenomic reference for the European Arctic–alpine flora. *Botany Letters, 172*(4), 1–13. DOI: 10.1080/23818107.2025.2549565

Lavoie, C. (2013). Biological collections in an ever changing world: Herbaria as tools for biogeographical and environmental studies. *Perspectives in Plant Ecology, Evolution and Systematics, 15*(1), 68–76. DOI: 10.1016/j.ppees.2012.10.002

Lavorel, S., & Garnier, E. (2002). Predicting changes in community composition and ecosystem functioning from plant traits: Revisiting the Holy Grail. *Functional Ecology, 16*(5), 545–556. DOI: 10.1046/j.1365-2435.2002.00664.x

Lawson, E. (2024). *Moss and Lichen*. Reaktion Books.

Lazzaro, A., Hilfiker, D., & Zeyer, J. (2015). Structures of microbial communities in alpine soils: Seasonal and elevational effects. *Frontiers in Microbiology, 6*, 1330. DOI: 10.3389/fmicb.2015.01330 PMID: 26635785

Lee, H., Calvin, K., Dasgupta, D., Krinner, G., Mukherji, A., Thorne, P., ... & Park, Y. (2023). *IPCC, 2023: Climate change 2023: Synthesis report, summary for policymakers. Contribution of working groups i, II and III to the sixth assessment report of the intergovernmental panel on climate change [core writing team, h. Lee and j. Romero (eds.)]. IPCC, geneva, Switzerland* (pp. 1-34). Intergovernmental Panel on Climate Change (IPCC).

Legg, S. (2021). IPCC, 2021: Climate change 2021-the physical science basis. *Interaction, 49*(4), 44–45.

Legros, J. P. (1992). Chapter 7 - Soils of Alpine mountains. In I. P. Martini & W. B. T.-D. in E. S. P. Chesworth (Eds.), *Weathering, Soils & Paleosols* (Vol. 2, pp. 155–181). Elsevier. 10.1016/B978-0-444-89198-3.50012-X

Lelli, C., Nascimbene, J., Alberti, D., Agostini, N., Zoccola, A., Piovesan, G., & Chiarucci, A. (2021). Long-term changes in Italian mountain forests detected by resurvey of historical vegetation data. *Journal of Vegetation Science, 32*(1), e12939. DOI: 10.1111/jvs.12939

Leng, R., Harrison, S., & Anderson, K. (2023). Himalayan alpine ecohydrology: An urgent scientific concern in a changing climate. *Ambio, 52*(2), 390–410. DOI: 10.1007/s13280-022-01792-2 PMID: 36324019

Lepcha, P. (2024). Climate change and its impact on mountainous plant species. *RE:view.*

Leppänen, S. M., Rissanen, A. J., & Tiirola, M. (2015). Nitrogen fixation in Sphagnum mosses is affected by moss species and water table level. *Plant and Soil, 389*(1), 185–196. DOI: 10.1007/s11104-014-2356-6

Leuschner, C., & Ellenberg, H. (2017). Vegetation of the alpine and nival belts. In *Ecology of Central European Non-Forest Vegetation: Coastal to Alpine, Natural to Man-Made Habitats: Vegetation Ecology of Central Europe, Volume II* (pp. 271-431). Springer. DOI: 10.1007/978-3-319-43048-5_5

Li, B. V., Jenkins, C. N., & Xu, W. (2022). Strategic protection of landslide vulnerable mountains for biodiversity conservation under land-cover and climate change impacts. *Proceedings of the National Academy of Sciences of the United States of America, 119*(2), e2113416118. DOI: 10.1073/pnas.2113416118 PMID: 34983877

Liebelt, D. J., Jordan, J. T., & Doherty, C. J. (2019). Only a matter of time: The impact of daily and seasonal rhythms on phytochemicals. *Phytochemistry Reviews, 18*(6), 1409–1433. DOI: 10.1007/s11101-019-09617-z

Li, J., Li, X., Gao, J., Kazhaocairang, , Ma, G., & Qi, X. (2021). Micro-scale fragmentation of the alpine meadow landscape on the Qinghai-Tibet Plateau under external disturbances. *Catena, 201*, 105220. DOI: 10.1016/j.catena.2021.105220

Li, J., Wang, X., Yuan, M., Duan, W., Xia, J., Zhang, X., Zhao, Y., & Wang, J. (2025). Effect of soil microbial community on ecosystem multifunctionality in an alpine grassland. *Catena, 249*, 108714. DOI: 10.1016/j.catena.2025.108714

Lin, L., Cao, G., Xu, X., Li, C., Fan, B., Li, B., Lan, Y., Si, M., & Dai, L. (2022). Changes and Relationships between Components in the Plant-Soil System and the Dominant Plant Functional Groups in Alpine Kobresia Meadows Due to Overgrazing. *Diversity, 14*(3), 183. DOI: 10.3390/d14030183

Lin, L., Xu, X., Cao, G., Zhang, F., Li, Y., Fan, B., Li, Q., & Huang, J. (2023a). Synergies between microsites of plant communities and steady-stage alpine meadows on the Qinghai–Tibetan Plateau. *Grassland Research, 2*(4), 289–298. DOI: 10.1002/glr2.12057

Linqing, Z., Tian, F., Liang, T., Fensholt, R., He, T., & Schaepman-Strub, G. (2025). Topographic effects on vegetation greening and area expansion in global alpine zones under climate change. *International Journal of Applied Earth Observation and Geoinformation, 142*, 104727. DOI: 10.1016/j.jag.2025.104727

Liping, C., Yujun, S., & Saeed, S. (2018). Monitoring and predicting land use and land cover changes using remote sensing and GIS techniques—A case study of a hilly area, Jiangle, China. *PLoS One, 13*(7), e0200493. DOI: 10.1371/journal.pone.0200493 PMID: 30005084

Li, R., He, Z., He, P., Yang, R., Ma, D., Li, Y., Xiang, Y., Sun, W., Zhu, X., & Zhang, Z. (2025). Crop diversification improves productivity and soil health by regulating fungal diversity. *Soil & Tillage Research, 252*, 106628. DOI: 10.1016/j.still.2025.106628

Li, R., Luo, T., Tang, Y., Du, M., & Zhang, X. (2013). The altitudinal distribution center of a widespread cushion species is related to an optimum combination of temperature and precipitation in the central Tibetan Plateau. *Journal of Arid Environments, 88*, 70–77. DOI: 10.1016/j.jaridenv.2012.07.018

Li, R., Wang, P., Liu, Q., Luo, R., & Pang, X. (2025). Microbial functional groups govern soil multifunctionality during alpine grassland restoration via turf transplantation. *Catena, 258*, 109269. DOI: 10.1016/j.catena.2025.109269

Li, S., Zhou, X., Liu, L., Su, Z., Zhao, J., Zhang, J., Cai, Z., Peñuelas, J., & Huang, X. (2025). Plant diversity reduces the risk of antibiotic resistance genes in agroecosystems. *Advancement of Science, 12*(11), 2410990. DOI: 10.1002/advs.202410990 PMID: 39874208

Li, T., Kamran, M., Chang, S., Peng, Z., Wang, Z., Ran, L., Jiang, W. Q., Jin, Y., Zhang, X., You, Y., Li, L., & Hou, F. (2022). Climate-soil interactions improve the stability of grassland ecosystem by driving alpine plant diversity. *Ecological Indicators, 141*, 109002. DOI: 10.1016/j.ecolind.2022.109002

Liu, P., Jing, X., Degen, A., & Shang, Z. (2024a). Conservation and Sustainable Utilization of Rangelands in the Hindu Kush Himalayan Region. 190–205. DOI: 10.1079/9781800622579.0014

Liu, Y., Zhang, Z., Tong, L., Khalifa, M., Wang, Q., Gang, C., Wang, Z., Li, J., & Sun, Z. J. E. I. (2019). Assessing the effects of climate variation and human activities on grassland degradation and restoration across the globe. *106*, 105504.

Liu, F., Zhang, W., & Li, S. (2025). Effects of freeze–thaw cycles on uptake preferences of plants for nutrient: A review. *Plants*, *14*(7), 1122. DOI: 10.3390/plants14071122 PMID: 40219190

Liu, J.-Q., Li, J.-L., & Lai, Y.-J. (2021). Plant diversity and ecology on the Qinghai–Tibet Plateau. *Journal of Systematics and Evolution*, *59*(6), 1139–1141. DOI: 10.1111/jse.12813

Liu, J., Qi, X., Ma, C., Wang, Z., & Li, H. (2022). Response of Sheet Erosion to the Characteristics of Physical Soil Crusts for Loessial Soils. *Frontiers in Environmental Science*, *10*, 905045. https://www.frontiersin.org/journals/environmental-science/articles/10.3389/fenvs.2022.905045. DOI: 10.3389/fenvs.2022.905045

Liu, M., Wen, J. H., Chen, Y. M., Xu, W. J., Wang, Q., & Ma, Z. L. (2022). Warming increases soil carbon input in a *Sibiraea angustata*-dominated alpine shrub ecosystem. *Journal of Plant Ecology*, *15*(2), 335–346. DOI: 10.1093/jpe/rtab101

Liu, X., Cong, J., Lu, H., Xue, Y., Wang, X., Li, D., & Zhang, Y. (2017). Community structure and elevational distribution pattern of soil Actinobacteria in alpine grasslands. *Acta Ecologica Sinica*, *37*(4), 213–218. DOI: 10.1016/j.chnaes.2017.02.010

Liu, Y., Evans, S. E., Friesen, M. L., & Tiemann, L. K. (2022). Root exudates shift how N mineralization and N fixation contribute to the plant-available N supply in low fertility soils. *Soil Biology & Biochemistry*, *165*, 108541. DOI: 10.1016/j.soilbio.2021.108541

Liu, Y.-F., Fang, H., Huang, Z., Leite, P. A. M., Liu, Y., López-Vicente, M., Zhao, J., Shi, Z.-H., & Wu, G.-L. (2023). Shrub encroachment increases soil erosion risk in hillside alpine meadows of the Qinghai-Tibetan Plateau, NW China. *Catena*, *222*, 106842. DOI: 10.1016/j.catena.2022.106842

Liu, Y., Xie, Y., Guo, Z., & Xi, G. (2023). Effects of Climate Variability and Human Activities on Vegetation Dynamics across the Qinghai–Tibet Plateau from 1982 to 2020. *Remote Sensing (Basel)*, *15*(20), 4988. https://www.mdpi.com/2072-4292/15/20/4988. DOI: 10.3390/rs15204988

Li, W., Lu, Q., Alharbi, S. A., Soromotin, A. V., Kuzyakov, Y., & Lei, Y. (2023). Plant–soil–microbial interactions mediate vegetation succession in retreating glacial forefields. *The Science of the Total Environment*, *873*, 162393. DOI: 10.1016/j.scitotenv.2023.162393 PMID: 36841408

Li, W., Zheng, T. D., Cheng, X. P., & He, S. Q. (2023). Changes in functional traits and diversity of typical alpine grasslands after a short-term trampling disturbance. *Frontiers in Ecology and Evolution*, *11*, 1154911. DOI: 10.3389/fevo.2023.1154911

Li, X., Zeng, R., & Liao, H. (2016). Improving crop nutrient efficiency through root architecture modifications. *Journal of Integrative Plant Biology*, *58*(3), 193–202. DOI: 10.1111/jipb.12434 PMID: 26460087

Li, Y.-Y., Dong, S.-K., Wen, L., Wang, X.-X., & Wu, Y. (2014). Soil carbon and nitrogen pools and their relationship to plant and soil dynamics of degraded and artificially restored grasslands of the Qinghai–Tibetan Plateau. *Geoderma*, *213*, 178–184. DOI: 10.1016/j.geoderma.2013.08.022

Li, Y., Zhang, H., Xue, X., Jiang, Y., & Shen, Q. (2018). Deep learning for remote sensing image classification: A survey. *Wiley Interdisciplinary Reviews. Data Mining and Knowledge Discovery*, *8*(6), e1264. DOI: 10.1002/widm.1264

Llambí, L. D., Melfo, A., Gámez, L. E., Pelayo, R. C., Cárdenas, M., Rojas, C., Torres, J. E., Ramírez, N., Huber, B., & Hernández, J. (2021). Vegetation assembly, adaptive strategies and positive interactions during primary succession in the forefield of the last Venezuelan glacier. *Frontiers in Ecology and Evolution*, *9*, 657755. DOI: 10.3389/fevo.2021.657755

Löbmann, M. T. (2020). *The surface-mat effect and its measurement for estimating shallow erosion resistance of herbaceous vegetated alpine slopes*. Free University of Bozen-Bolzano.

Lobréaux, S., & Miquel, C. (2020). Identification of Arabis alpina genomic regions associated with climatic variables along an elevation gradient through whole genome scan. *Genomics*, *112*(1), 729–735. DOI: 10.1016/j.ygeno.2019.05.008 PMID: 31085222

Lockwood, M., Mitchell, M., Moore, S. A., & Clement, S. (2014). Biodiversity governance and social-ecological system dynamics: Transformation in the Australian Alps. *Ecology and Society*, *19*(2), art13. DOI: 10.5751/ES-06393-190213

Loik, M. E., Still, C. J., Huxman, T. E., & Harte, J. (2004). In situ photosynthetic freezing tolerance for plants exposed to a global warming manipulation in the Rocky Mountains, Colorado, USA. *The New Phytologist, 162*(2), 331–341. DOI: 10.1111/j.1469-8137.2004.01002.x

Long, Y., & Shen, Z. (2013). Disaggregating heterogeneous agent attributes and location. *Computers, Environment and Urban Systems, 42*, 14–25. DOI: 10.1016/j.compenvurbsys.2013.09.002

Lopez, G., Ahmadi, S. H., Amelung, W., Athmann, M., Ewert, F., Gaiser, T., Gocke, M. I., Kautz, T., Postma, J., Rachmilevitch, S., Schaaf, G., Schnepf, A., Stoschus, A., Watt, M., Yu, P., & Seidel, S. J. (2023). Nutrient deficiency effects on root architecture and root-to-shoot ratio in arable crops. *Frontiers in Plant Science, 13*, 1067498. DOI: 10.3389/fpls.2022.1067498 PMID: 36684760

Losapio, G., Lee, J. R., Fraser, C. I., Gillespie, M. A., Kerr, N. R., Zawierucha, K., ... & Brown, L. E. (2025). Impacts of deglaciation on biodiversity and ecosystem function. *Nature Reviews Biodiversity*, 1–15.

Łuczyński, P. (2022). Tsunamites versus tempestites: Various types of redeposited stromatoporoid beds in the Devonian of the Holy Cross Mountains (Poland), a case study from the Ołowianka Quarry. *PLoS One, 17*(5), e0268349. DOI: 10.1371/journal.pone.0268349 PMID: 35580129

Ludovicy, S., Noroozi, J., Semenchuk, P., Moser, D., Wessely, J., Talebi, A., & Dullinger, S. (2022). Protected area network insufficiently represents climatic niches of endemic plants in a Global Biodiversity Hotspot. *Biological Conservation, 275*, 109768. DOI: 10.1016/j.biocon.2022.109768

Lukac, M., Calfapietra, C., Lagomarsino, A., & Loreto, F. (2010). Global climate change and tree nutrition: Effects of elevated CO_2 and temperature. *Tree Physiology, 30*(9), 1209–1220. DOI: 10.1093/treephys/tpq040 PMID: 20571150

Lyu, Z.-Y., Zhou, X.-L., Wang, S.-Q., Yang, G.-M., Sun, W.-G., Zhang, J.-Y., Zhang, R., & Shen, S.-K. (2024). The first high-altitude autotetraploid haplotype-resolved genome assembled (Rhododendron nivale subsp. boreale) provides new insights into mountaintop adaptation. *GigaScience, 13*, giae052. DOI: 10.1093/gigascience/giae052 PMID: 39110622

Macci, C., Vannucchi, F., Scartazza, A., Masciandaro, G., Doni, S., & Peruzzi, E. (2025). Soil–plant indicators for assessing nutrient cycling and ecosystem functionality in urban forestry. *Urban Science (Basel, Switzerland), 9*(3), 82. DOI: 10.3390/urbansci9030082

Maccioni, S., d'Angella, F., De Carlo, M., & Sfogliarini, B. (2024). Stakeholder Engagement and Triggers for Sustainable Development in Complex Fragile Ecosystems: Evidence from Alpine Trentino Region. In *Sustainability* (Vol. 16, Issue 22). DOI: 10.3390/su16229879

MacGregor, A. (2018). *Naturalists in the field: collecting, recording and preserving the natural world from the fifteenth to the twenty-first century* (Vol. 2). Brill.

MacMahon, J. A. (1980). Ecosystems over time: succession and other types of change. *Forests: fresh perspectives from ecosystem analysis*, 27-58.

Macphie, K. H., & Phillimore, A. B. (2024). Phenology. *Current Biology*, *34*(5), R183–R188. DOI: 10.1016/j.cub.2024.01.007 PMID: 38471442

Magaña Ugarte, R., Escudero, A., & Gavilán, R. G. (2019). Metabolic and physiological responses of Mediterranean high-mountain and alpine plants to combined abiotic stresses. *Physiologia Plantarum*, *165*(2), 403–412. DOI: 10.1111/ppl.12898 PMID: 30536685

Mage, S. M., & Porder, S. (2013). Parent material and topography determine soil phosphorus status in the Luquillo Mountains of Puerto Rico. *Ecosystems (New York, N.Y.)*, *16*(2), 284–294. DOI: 10.1007/s10021-012-9612-5

Magurran, A. E., Baillie, S. R., Buckland, S. T., Dick, J. M., Elston, D. A., Scott, E. M., Smith, R. I., Somerfield, P. J., & Watt, A. D. (2010). Long-term datasets in biodiversity research and monitoring: Assessing change in ecological communities through time. *Trends in Ecology & Evolution*, *25*(10), 574–582. DOI: 10.1016/j.tree.2010.06.016 PMID: 20656371

Maharjan, S. K., Sterck, F. J., Raes, N., Zhao, Y., & Poorter, L. (2023). Climate change induced elevational range shifts of Himalayan tree species. *Biotropica*, *55*(1), 53–69. DOI: 10.1111/btp.13159

Mahdi, S. S., & Kazmi, S. A.. (2024). Climate Vulnerability and Adaptation Prospects for Mountain Ecosystems in Kashmir Himalaya. *Scientific Reports*.

Mahmoudabadi, E., Karimi, A., Haghnia, G. H., & Sepehr, A. (2017). Digital soil mapping using remote sensing indices, terrain attributes, and vegetation features in the rangelands of northeastern Iran. *Environmental Monitoring and Assessment*, *189*(10), 500. DOI: 10.1007/s10661-017-6197-7 PMID: 28894961

Mainetti, A., D'Amico, M., Probo, M., Quaglia, E., Ravetto Enri, S., Celi, L., & Lonati, M. (2021). Successional Herbaceous Species Affect Soil Processes in a High-Elevation Alpine Proglacial Chronosequence. *Frontiers in Environmental Science, Volume 8*-. https://www.frontiersin.org/journals/environmental-science/articles/10.3389/fenvs.2020.615499

Makki, T., Mostafavi, H., Matkan, A. A., Aghighi, H., Valavi, R., Chee, Y. E., & Teimori, A. (2023). Impacts of climate change on the distribution of riverine endemic fish species in Iran, a biodiversity hotspot region. *Freshwater Biology, 68*(6), 1007–1019. DOI: 10.1111/fwb.14081

Ma, L., Yang, L., Zhao, J., Wei, J., Kong, X., Wang, C., Zhang, X., Yang, Y., & Hu, X. (2015). Comparative proteomic analysis reveals the role of hydrogen sulfide in the adaptation of the alpine plant Lamiophlomis rotata to altitude gradient in the Northern Tibetan Plateau. *Planta, 241*(4), 887–906. DOI: 10.1007/s00425-014-2209-9 PMID: 25526962

Malik, M. (1988). Feed Availability, Requirements for Animals and Current Patterns of Utilization in Pakistan. *Non-conventional Feed Resources and Fibrous Agricultural Residues*, 158.

Manafzadeh, S., Staedler, Y. M., & Conti, E. (2017). Visions of the past and dreams of the future in the O rient: The I rano-T uranian region from classical botany to evolutionary studies. *Biological Reviews of the Cambridge Philosophical Society, 92*(3), 1365–1388. DOI: 10.1111/brv.12287 PMID: 27349491

Manes, S., Costello, M. J., Beckett, H., Debnath, A., Devenish-Nelson, E., Grey, K.-A., & Krause, C. (2021). Endemism increases species' climate change risk in areas of global biodiversity importance. *Biological Conservation, 257*, 109070. DOI: 10.1016/j.biocon.2021.109070

Mansoor, S., Manzoor, S., Gull, R., Gani, G., Wani, O. A., Farooq, S., Popescu, S. M., Arya, V. M., Park, W.-P., & Chung, Y. S. (2025). Biochar mediated remediation of emerging inorganic pollutants and their toxicological effects on plant and soil health. *Journal of Soil Science and Plant Nutrition, 25*(1), 1612–1642. DOI: 10.1007/s42729-025-02227-z

Mantua, N. J., Hare, S. R., Zhang, Y., Wallace, J. M., & Francis, R. C. (1997). A Pacific Interdecadal Climate Oscillation with Impacts on Salmon Production*. *Bulletin of the American Meteorological Society, 78*(6), 1069–1080. DOI: 10.1175/1520-0477(1997)078<1069:APICOW>2.0.CO;2

Manzoor, M., Ahmad, M., Gillani, S. W., Waheed, M., Grießinger, J., Shaheen, H., & Meier, W. J.-H. (2025). Biodiversity patterns and environmental drivers of fragile alpine glacial lake ecosystems in the Himalaya: Implications for conservation and management. *Ecological Indicators, 178*, 114028. DOI: 10.1016/j.ecolind.2025.114028

Manzoor, M., Ahmad, M., Gillani, S. W., Waheed, M., Shaheen, H., Haq, S. M., & Almutairi, K. F. (2026). Alpine Endemism under Threat: Ecological Patterns and Conservation Strategies in the Changing Himalayas. *Global Ecology and Conservation*, 04051. DOI: 10.1016/j.gecco.2025.e04051

Manzoor, M., Ahmad, M., Gillani, S. W., Waheed, M., Shaheen, H., Mehmood, A. B., Fonge, B. A., & Al-Andal, A. (2025). Population dynamics, threat assessment, and conservation strategies for critically endangered Meconopsis aculeata in alpine zone. *BMC Plant Biology, 25*(1), 358. DOI: 10.1186/s12870-025-06361-9 PMID: 40102752

Manzoor, M., Ahmad, M., Zafar, M., Gillani, S. W., Shaheen, H., Pieroni, A., & Makhkamov, T. (2023). The local medicinal plant knowledge in Kashmir Western Himalaya: A way to foster ecological transition via community-centred health seeking strategies. *Journal of Ethnobiology and Ethnomedicine, 19*(1), 56. DOI: 10.1186/s13002-023-00631-2 PMID: 38037066

Marian, M., Licciardello, G., Vicelli, B., Pertot, I., & Perazzolli, M. (2022). Ecology and potential functions of plant-associated microbial communities in cold environments. *FEMS Microbiology Ecology, 98*(1), fiab161. Advance online publication. DOI: 10.1093/femsec/fiab161 PMID: 34910139

Martín-Esquivel, J., Marrero-Gómez, M., Cubas, J., González-Mancebo, J., Olano, J., & del Arco, M. (2020). Climate warming and introduced herbivores disrupt alpine plant community of an oceanic island (Tenerife, Canary Islands). *Plant Ecology, 221*(11), 1117–1131. DOI: 10.1007/s11258-020-01066-5

Martin, J. B., Pain, A. J., & Martin, E. E. (2024). Geochemistry of glacial, proglacial, and deglaciated environments. *Treatise on Geochemistry, 3*, 251–299.

Mason, N., Ward, M., Watson, J. E., Venter, O., & Runting, R. K. (2020). Global opportunities and challenges for transboundary conservation. *Nature Ecology & Evolution, 4*(5), 694–701. DOI: 10.1038/s41559-020-1160-3 PMID: 32203481

Massah, B. A., & Saadi, T. (2021). Detection and attribution of changing in seasonal variability cause of climate change (Case study: hillsides of central Southern Alborz Mountains).

Masters, N. (2021). *For the love of soil: Strategies to regenerate our food production systems*. Integrity Soils Ltd.

Matinzadeh, Z., López-Angulo, J., Escudero, A., Palacio, S., Abedi, M., & Akhani, H. (2022). Functional structure of plant communities along salinity gradients in Iranian salt marshes. *Plant-Environment Interactions*, *3*(1), 16–27. DOI: 10.1002/pei3.10070 PMID: 37283692

Matyukira, C., & Mhangara, P. (2024). Advances in vegetation mapping through remote sensing and machine learning techniques: A scientometric review. *European Journal of Remote Sensing*, *57*(1), 2422330. DOI: 10.1080/22797254.2024.2422330

Ma, X., Ma, W., Liu, F., Chen, Q., Niu, B., & Zhou, Q. (2025). Delineation and Intensity Classification of Freeze–Thaw Erosion in the Western Section of Northern China's Agro-Pastoral Transition Zone Under Controlling Factors. *Land (Basel)*, *14*(11), 2278. DOI: 10.3390/land14112278

Maxwell, A. E., Warner, T. A., & Fang, F. (2018). Implementation of machine-learning classification in remote sensing: An applied review. *International Journal of Remote Sensing*, *39*(9), 2784–2817. DOI: 10.1080/01431161.2018.1433343

Ma, Y., Huang, X., Feng, Q., & Liang, T. (2022). Alpine grassland reviving response to seasonal snow cover on the Tibetan Plateau. *Remote Sensing (Basel)*, *14*(10), 2499. DOI: 10.3390/rs14102499

Ma, Y., Wang, J., Hu, Q., Li, J., Sun, Y., Zhang, L., Abbott, R. J., Liu, J., & Mao, K. (2019). Ancient introgression drives adaptation to cooler and drier mountain habitats in a cypress species complex. *Communications Biology*, *2*(1), 213. DOI: 10.1038/s42003-019-0445-z PMID: 31240251

Mazzocchi, C., & Sali, G. (2022). Tourists' perception of ecosystem services provided by mountain agriculture. *Sustainability (Basel)*, *14*(19), 12171. DOI: 10.3390/su141912171

McIntire, E. J., & Fajardo, A. (2014). Facilitation as a ubiquitous driver of biodiversity. *The New Phytologist*, *201*(2), 403–416. DOI: 10.1111/nph.12478 PMID: 24102266

McLaughlin, S. P. (1994). Floristic plant geography: The classification of floristic areas and floristic elements. *Progress in Physical Geography*, *18*(2), 185–208. DOI: 10.1177/030913339401800202

Médail, F., & Baumel, A. (2018). Using phylogeography to define conservation priorities: The case of narrow endemic plants in the Mediterranean Basin hotspot. *Biological Conservation*, *224*, 258–266. DOI: 10.1016/j.biocon.2018.05.028

Medhi, K., Bhardwaj, R., & Laxmi, R. (2021). Climate change with its impacts on soil and soil microbiome regulating biogeochemical nutrient transformations. In *Climate change and the microbiome: Sustenance of the ecosphere* (pp. 95–138). Springer International Publishing. DOI: 10.1007/978-3-030-76863-8_6

Mehta, P. (2024). The impact of climate change on the environment, water resources, and agriculture: a comprehensive review. *Climate, Environment and Agricultural Development: A Sustainable Approach Towards Society*, 189-201.

Melis, M., Dessì, F., Loddo, P., Maccioni, A., Gallo, M., Ul Hassan, R., & Aurang Zaib, M. (2020). ESA Sentinel 2 Imagery and GBGEOApp: Integrated Tools for the Deosai National Park Management Plan. *The International Archives of the Photogrammetry, Remote Sensing and Spatial Information Sciences, 43*, 145–151. DOI: 10.5194/isprs-archives-XLIII-B3-2020-145-2020

Melis, M., Dessì, F., Loddo, P., Maccioni, A., Gallo, M., Ul Hassan, R., & Sciences, S. I. (2020). ESA Sentinel 2 Imagery and GBGEOApp. *Integrated Tools for the Deosai National Park Management Plan., 43*, 145–151.

Memariani, F., Akhani, H., Joharchi, M. R., Khoshravesh, R., & Malekmohammadi, M. (2016). Plant diversity of the Khorassan-Kopet Dagh Floristic Province (Irano-Turanian Region). *Phytotaxa, 249*(1), 001-002.

Memariani, F., Zarrinpour, V., & Akhani, H. (2016). A review of plant diversity, vegetation, and phytogeography of the Khorassan-Kopet Dagh floristic province in the Irano-Turanian region (northeastern Iran–southern Turkmenistan). *Phytotaxa, 249*(1), 8–30. DOI: 10.11646/phytotaxa.249.1.4

Mendes, C. B., & Prevedello, J. A. (2020). Does habitat fragmentation affect landscape-level temperatures? A global analysis. *Landscape Ecology, 35*(8), 1743–1756. DOI: 10.1007/s10980-020-01041-5

Menke, S. B., & Holway, D. A. (2020). Historical resurvey indicates no decline in Argentine ant site occupancy in coastal southern California. *Biological Invasions, 22*(5), 1669–1679. DOI: 10.1007/s10530-020-02211-x

Meyer, S. E., Callaham, M. A.Jr, Stewart, J. E., & Warren, S. D. (2021). Invasive species response to natural and anthropogenic disturbance. In *Invasive species in forests and rangelands of the United States: A comprehensive science synthesis for the United States forest sector* (pp. 85–110). Springer International Publishing Cham. DOI: 10.1007/978-3-030-45367-1_5

Meza-Joya, F. L., Morgan-Richards, M., Koot, E. M., & Trewick, S. A. (2023). Global warming leads to habitat loss and genetic erosion of alpine biodiversity. *Journal of Biogeography*, *50*(5), 961–975. DOI: 10.1111/jbi.14590

Mia, M. A. R., Dey, D., Sakib, M. R., Biswas, M. Y., Prottay, A. A. S., Paul, N., Rimti, F. H., Abdullah, Y., Biswas, P., Iftehimul, M., Paul, P., Sarkar, C., El-Nashar, H. A. S., El-Shazly, M., & Islam, M. T. (2023). The efficacy of natural bioactive compounds against prostate cancer: Molecular targets and synergistic activities. *Phytotherapy Research*, *37*(12), 5724–5754. DOI: 10.1002/ptr.8017 PMID: 37786304

Miao, S., Zhang, Y., Li, B., Yuan, X., & Zuo, J. (2024). Antibiotic intermediates and antibiotics synergistically promote the development of multiple antibiotic resistance in antibiotic production wastewater. *Journal of Hazardous Materials*, *479*, 135601. DOI: 10.1016/j.jhazmat.2024.135601 PMID: 39243543

Midgley, G. F., Chown, S. L., & Kgope, B. S. (2007). Monitoring effects of anthropogenic climate change on ecosystems: A role for systematic ecological observation? [commentary]. *South African Journal of Science*, *103*(7), 282–286.

Miehe, G., Schleuss, P.-M., Seeber, E., Babel, W., Biermann, T., Braendle, M., Chen, F., Coners, H., Foken, T., Gerken, T., Graf, H.-F., Guggenberger, G., Hafner, S., Holzapfel, M., Ingrisch, J., Kuzyakov, Y., Lai, Z., Lehnert, L., Leuschner, C., & Wesche, K. (2019). The Kobresia pygmaea ecosystem of the Tibetan highlands–Origin, functioning and degradation of the world's largest pastoral alpine ecosystem: Kobresia pastures of Tibet. *The Science of the Total Environment*, *648*, 754–771. DOI: 10.1016/j.scitotenv.2018.08.164 PMID: 30134213

Miehe, S., Miehe, G., Miehe, S., Böhner, J., Bäumler, R., Ghimire, S., Bhattarai, K., Chaudhary, R., & Subedi, M. (2015). *16. Vegetation ecology. Nepal: an introduction to the natural history, ecology and human environment in the Himalayas*. Royal Botanic Garden Edinburgh.

Millar, C. I. (2012). Geologic, climatic, and vegetation history of California. *In: Baldwin, Bruce G.; Goldman, Douglas H.; Keil, David J.; Patterson, Robert; Rosatti, Thomas J.; Wilken, Dieter H., editors. The Jepson Manual: Vascular Plants of California. 2nd Edition. University of California Press. pp. 49-67.*, 49-67.

Miller, D. J. (1999). Herders of forty centuries: nomads of Tibetan rangelands in western China.

Mina, U., Kumar, A., Chaturvedi, A. K., & Kumar, P. (2021). Climate change and plant diversity: Threats and opportunities. In *Climate Change and the Microbiome: Sustenance of the Ecosphere* (pp. 369–396). Springer. DOI: 10.1007/978-3-030-76863-8_20

Mishra, A. P., Chandra, N., Kumar, A., Pande, C. B., & Singh, G. (2024). *Threatened Medicinal Plants in the Indian Himalayan Region*. Springer. DOI: 10.1007/978-3-031-73687-2

Mitra, A., & Paul, S. (2025). Balancing Mountain Tourism, Cultural Heritage, and Environmental Stability: Strategies and Challenges in Sustainable Development. In *Balancing Mountain Tourism, Cultural Heritage, and Environmental Stability* (pp. 31-54). IGI Global Scientific Publishing.

Mittermeier, R. A., Da Fonseca, G. A., Rylands, A. B., & Brandon, K. (2005). A brief history of biodiversity conservation in Brazil. *Conservation Biology*, *19*(3), 601–607. DOI: 10.1111/j.1523-1739.2005.00709.x

Mitterwallner, V., Steinbauer, M., Mathes, G., & Walentowitz, A. (2024). Global reduction of snow cover in ski areas under climate change. *PLoS One*, *19*(3), e0299735. DOI: 10.1371/journal.pone.0299735 PMID: 38478484

Mi, X., Feng, G., Hu, Y., Zhang, J., Chen, L., Corlett, R. T., Hughes, A. C., Pimm, S., Schmid, B., Shi, S., Svenning, J.-C., & Ma, K. (2021). The global significance of biodiversity science in China: An overview. *National Science Review*, *8*(7), nwab032. DOI: 10.1093/nsr/nwab032 PMID: 34694304

Mizuno, K. (1998). Succession Processes of Alpine Vegetation in Response to Glacial Fluctuations of Tyndall Glacier, Mt. Kenya, Kenya. *Arctic and Alpine Research*, *30*(4), 340–348. DOI: 10.1080/00040851.1998.12002909

Mohseni, P. (2024). *Mapping snow and vegetation coverage using multitemporal open satellite Imagery: the case study of the Maritime Alps*. Politecnico di Torino.

Monson, R. K., Rosenstiel, T. N., Forbis, T. A., Lipson, D. A., & Jaeger, C. H.III. (2006). Nitrogen and carbon storage in alpine plants. *Integrative and Comparative Biology*, *46*(1), 35–48. DOI: 10.1093/icb/icj006 PMID: 21672721

Montagna, M., Mereghetti, V., Lencioni, V., & Rossaro, B. (2016). Integrated taxonomy and DNA barcoding of alpine midges (Diptera: Chironomidae). *PLoS One*, *11*(3), e0149673. DOI: 10.1371/journal.pone.0149673 PMID: 26938660

Mou, X. M., Yu, Y. W., Zhao, C., Soromotin, A., Kuzyakov, Y., & Li, X. G. (2022). Sedge replacement by grasses accelerates litter decomposition and decreases organic matter formation in alpine meadow soils. *Land Degradation & Development*, *33*(16), 3260–3270. DOI: 10.1002/ldr.4386

Mou, Z., Kuang, L., Zhang, J., Li, Y., Wu, W., Liang, C., Hui, D., Lambers, H., Sardans, J., Peñuelas, J., Liu, J., Ren, H., & Liu, Z. (2023). Nutrient availability and stoichiometry mediate microbial effects on soil carbon sequestration in tropical forests. *Soil Biology & Biochemistry*, *186*, 109186. DOI: 10.1016/j.soilbio.2023.109186

Mucina, L., Bültmann, H., Dierßen, K., Theurillat, J. P., Raus, T., Čarni, A., Šumberová, K., Willner, W., Dengler, J., García, R. G., Chytrý, M., Hájek, M., Di Pietro, R., Iakushenko, D., Pallas, J., Daniëls, F. J. A., Bergmeier, E., Santos Guerra, A., Ermakov, N., & Tichý, L. (2016). Vegetation of Europe: Hierarchical floristic classification system of vascular plant, bryophyte, lichen, and algal communities. *Applied Vegetation Science*, *19*(S1), 3–264. DOI: 10.1111/avsc.12257

Mueller-Dombois, D., & Ellenberg, H. (1974). *Aims and methods of vegetation ecology*. Wiley.

Mueller, M., Egg, L., Ruff, T., Haas, A., Schubert, M., & Gum, B. (2025). Evidence of Natural Reproduction of North American Rainbow Trout (Oncorhynchus mykiss) in Three Alpine Rivers in Bavaria, Germany. *Fisheries Management and Ecology*, *32*(2), e12765. DOI: 10.1111/fme.12765

Muellner-Riehl, A. N. (2019). Mountains as evolutionary arenas: Patterns, emerging approaches, paradigm shifts, and their implications for plant phylogeographic research in the Tibeto-Himalayan region. *Frontiers in Plant Science*, *10*, 195. DOI: 10.3389/fpls.2019.00195 PMID: 30936883

Mufungizi, A. A., Musakwa, W., & Chanza, N. (2023). Shifting ecosystems, past, current, and emerging trends: A bibliometric analysis and systematic review of literature. *Ecological Indicators*, *156*, 111175. DOI: 10.1016/j.ecolind.2023.111175

Muhammed, A., & Elias, E. (2021). Class and landscape level habitat fragmentation analysis in the Bale mountains national park, southeastern Ethiopia. *Heliyon*, *7*(7), e07642. DOI: 10.1016/j.heliyon.2021.e07642 PMID: 34377861

Mukhia, S., Kumar, A., Kumari, P., & Kumar, R. (2022). Psychrotrophic plant beneficial bacteria from the glacial ecosystem of Sikkim Himalaya: Genomic evidence for the cold adaptation and plant growth promotion. *Microbiological Research*, *260*, 127049. DOI: 10.1016/j.micres.2022.127049 PMID: 35504236

Murtaza, G., Ahmed, Z., Iqbal, R., & Deng, G. (2025). Biochar from agricultural waste as a strategic resource for promotion of crop growth and nutrient cycling of soil under drought and salinity stress conditions: A comprehensive review with context of climate change. *Journal of Plant Nutrition*, *48*(11), 1832–1883. DOI: 10.1080/01904167.2025.2460769

Musa, I. O., Samuel, J. O., Adams, M., Abdulsalam, M., Nathaniel, V., Maude, A. M., Adedayo, O. A., & Tiamiyu, A. T. (2024). In Aransiola, S. A., Babaniyi, B. R., Aransiola, A. B., & Maddela, N. R. (Eds.), *Soil Erosion, Mineral Depletion and Regeneration BT - Prospects for Soil Regeneration and Its Impact on Environmental Protection* (pp. 159–172). Springer Nature Switzerland., DOI: 10.1007/978-3-031-53270-2_7

Muthukrishnan, R., Sullivan, L. L., Shaw, A. K., & Forester, J. D. (2020). Trait plasticity alters the range of possible coexistence conditions in a competition–colonisation trade-off. *Ecology Letters*, *23*(5), 791–799. DOI: 10.1111/ele.13477 PMID: 32086876

Nadeau, C. P., Urban, M. C., & Bridle, J. R. (2017). Climates past, present, and yet-to-come shape climate change vulnerabilities. *Trends in Ecology & Evolution*, *32*(10), 786–800. DOI: 10.1016/j.tree.2017.07.012 PMID: 28844791

Naeimi, A., Alavi, S. A., & Madanipour, S. (2022). Tectonic evolution of the foreland basin of the SE Alborz Mountains, northern Iran. *Journal of Asian Earth Sciences*, *223*, 104981. DOI: 10.1016/j.jseaes.2021.104981

Nagelmüller, S., Hiltbrunner, E., & Körner, C. (2017). Low temperature limits for root growth in alpine species are set by cell differentiation. *AoB Plants*, *9*(6), plx054. Advance online publication. DOI: 10.1093/aobpla/plx054 PMID: 29218137

Nagendra, H., Lucas, R., Honrado, J. P., Jongman, R. H., Tarantino, C., Adamo, M., & Mairota, P. J. E. I. (2013a). Remote sensing for conservation monitoring: Assessing protected areas, habitat extent, habitat condition, species diversity, and threats. *33*, 45-59.

Nagendra, H., Lucas, R., Honrado, J. P., Jongman, R. H. G., Tarantino, C., Adamo, M., & Mairota, P. (2013b). Remote sensing for conservation monitoring: Assessing protected areas, habitat extent, habitat condition, species diversity, and threats. *Ecological Indicators*, *33*, 45–59. 10.1016/j.ecolind.2012.09.014. DOI: 10.1016/j.ecolind.2012.09.014

Nagy, L., & Grabherr, G. (2009). *The biology of alpine habitats*. Oxford University Press. DOI: 10.1093/oso/9780198567035.001.0001

Najafi, S., Sharafati, A., & Moghaddam, H. K. (2023). Impact of climate change adaptation strategies on groundwater resources: A case study of Sari-Neka coastal aquifer, Northern Iran. *Environmental Earth Sciences*, *82*(23), 571. DOI: 10.1007/s12665-023-11205-6

Napoletano, M., Bellino, A., & Baldantoni, D. (2025). Depletion and recovery of soil organic matter: Ecological, economic and social implications. *Ecological Civilization, 2*(2), 10002. DOI: 10.70322/ecolciviliz.2025.10002

Naz, A., Amjad, M. S., Umair, M., Altaf, M., & Ni, J. (2022). Medicinal plants used as therapeutic medicine in the Himalayan region of Azad Jammu and Kashmir. *South African Journal of Botany, 150*, 194–216. 10.1016/j.sajb.2022.07.007. DOI: 10.1016/j.sajb.2022.07.007

Nazari, Z., Ejtehadi, H., Mirshamsi, O., & Memariani, F. (2022). Modelling potential distribution of an endemic plant (Ferula Microcolea) based on bioclimatic variables in Iran. *Acta Botanica Hungarica, 64*(1-2), 157–172. DOI: 10.1556/034.64.2022.1-2.8

Nema, A., Kumar, A., & Warudkar, V. (2025). An in-depth critical review of different carbon capture techniques: Assessing their effectiveness and role in reducing climate change emissions. *Energy Conversion and Management, 323*, 119244. DOI: 10.1016/j.enconman.2024.119244

Nepal, S., Krause, P., Flügel, W. A., Fink, M., & Fischer, C. (2014). Understanding the hydrological system dynamics of a glaciated alpine catchment in the Himalayan region using the J2000 hydrological model. *Hydrological Processes, 28*(3), 1329–1344. DOI: 10.1002/hyp.9627

Nethmini, R. T., Zhao, H., Pan, L., Qin, X., Huang, J., He, Q., Shi, X., Jiang, G., Hou, Q., Chen, Q., Li, X., Dong, K., Xie, L., & Li, N. (2025). Thermal sensitivity and niche plasticity of generalist and specialist leaf-endophytic bacteria in Mangrove Kandelia obovata. *Communications Biology, 8*(1), 5. DOI: 10.1038/s42003-024-07446-1 PMID: 39753754

Nila, M. U. S., Bobrowski, M., & Schickhoff, U. (2025). Growing Season Normalized Difference Vegetation Index in the Nepal Himalaya and Adjacent Areas, 2000–2019: Sensitivity to Climate Change and Terrain Factors. *Land (Basel), 14*(4), 749. DOI: 10.3390/land14040749

Ning, W., Rawat, G. S., Sharma, E. J. H.-a. r., & Himalayas, t. i. i. t. H. K. (2014). High-altitude ecosystem interfaces in the Hindu Kush Himalayan region. 3.

Ning, W., Jian, L., & Zhaoli, Y. (2004). Grazing intensity on the plant diversity of alpine meadow in the eastern Tibetan plateau. *Rangifer*, •••, 9–15. DOI: 10.7557/2.24.4.1664

Njeru, C. M., Ekesi, S., Mohamed, S. A., Kinyamario, J. I., Kiboi, S., & Maeda, E. E. (2017). Assessing stock and thresholds detection of soil organic carbon and nitrogen along an altitude gradient in an East Africa mountain ecosystem. *Geoderma Regional, 10*, 29–38. DOI: 10.1016/j.geodrs.2017.04.002

Nobis, M., Ebel, A. L., Nowak, A., Paszko, B., Bobrov, A. A., Kotukhov, Y. A., & Olonova, M. V. (2015). Contribution to the flora of Asian and European countries: New national and regional vascular plant records, 4. *Acta Botanica Gallica, 162*(4), 301–316. DOI: 10.1080/12538078.2015.1090329

Nobis, M., Wróbel, S., Klichowska, E., Nowak, A., Wróbel, A., Nobis, A., Paszko, B., Świerszcz, S., Chen, W. L., Kauzal, P., Krzempek, M., Liu, B., Nowak, S., Piwowarczyk, R., Pedraja, Ó. S., & Zięba, A. (2023). New national and regional plant records: Contribution to the flora of the Old World countries. *Acta Societatis Botanicorum Poloniae, 92*(1), 1–21. DOI: 10.5586/asbp/162050

Nogales, M., Mendes, S. B., Rumeu, B., Marrero, P., Romero, J., Guerrero-Campos, M., Mallorquín, Á., Trujillo, O., Sánchez, C., & Vargas, P. (2025). The potential ecological collapse of the seed dispersal network in an insular high mountain: Teide National Park. *Global Ecology and Conservation, 62*, e03793. DOI: 10.1016/j.gecco.2025.e03793

Noori, R., Maghrebi, M., Jessen, S., Bateni, S. M., Heggy, E., Javadi, S., Noury, M., Pistre, S., Abolfathi, S., & AghaKouchak, A. (2023). Decline in Iran's groundwater recharge. *Nature Communications, 14*(1), 6674. DOI: 10.1038/s41467-023-42411-2 PMID: 37865681

Noori, S., Zahiri, R., Yusefi, G. H., Rajabizadeh, M., Hawlitschek, O., Rakhshani, E., Husemann, M., & Rajaei, H. (2024). Patterns of zoological diversity in Iran—A review. *Diversity, 16*(10), 621. DOI: 10.3390/d16100621

Noroozi, J., Akhani, H., & Breckle, S.-W. (2008). Biodiversity and phytogeography of the alpine flora of Iran. *Biodiversity and Conservation, 17*(3), 493–521. DOI: 10.1007/s10531-007-9246-7

Noroozi, J., & Körner, C. (2018). A bioclimatic characterization of high elevation habitats in the Alborz mountains of Iran. *Alpine Botany, 128*(1), 1–11. DOI: 10.1007/s00035-018-0202-9 PMID: 29576762

Noroozi, J., Talebi, A., Doostmohammadi, M., Manafzadeh, S., Asgarpour, Z., & Schneeweiss, G. M. (2019). Endemic diversity and distribution of the Iranian vascular flora across phytogeographical regions, biodiversity hotspots and areas of endemism. *Scientific Reports, 9*(1), 12991. DOI: 10.1038/s41598-019-49417-1 PMID: 31506504

Noroozi, J., Talebi, A., Doostmohammadi, M., Rumpf, S. B., Linder, H. P., & Schneeweiss, G. M. (2018). Hotspots within a global biodiversity hotspot-areas of endemism are associated with high mountain ranges. *Scientific Reports*, *8*(1), 10345. DOI: 10.1038/s41598-018-28504-9 PMID: 29985437

Norouzi Nazar, M. S., Asadolahi, Z., Rabbani, F., Abbaspour, K. C., & Sakieh, Y. (2023). Modeling the integrated effects of landuse and climate change on the hydrologic response of Gorganroud watershed in Iran. *Theoretical and Applied Climatology*, *151*(3-4), 151. DOI: 10.1007/s00704-022-04345-5

Notarnicola, R. F., Nicotra, A. B., Kruuk, L. E. B., & Arnold, P. A. (2023). Effects of warming temperatures on germination responses and trade-offs between seed traits in an alpine plant. *Journal of Ecology*, *111*(1), 62–76. DOI: 10.1111/1365-2745.14014

O'Brien, A. M., Ginnan, N. A., Rebolleda-Gómez, M., & Wagner, M. R. (2021). Microbial effects on plant phenology and fitness. *American Journal of Botany*, *108*(10), 1824–1837. DOI: 10.1002/ajb2.1743 PMID: 34655479

Oldfather, M. F., Ennis, A., Miller, B. W., Clark-Wolf, K., Rangwala, I., Robe, H., & Littlefield, C. (2025). Climate change impacts and adaptation in U.S. Rocky Mountain high-elevation ecosystems. *Arctic, Antarctic, and Alpine Research*, *57*(1), 2450089. DOI: 10.1080/15230430.2025.2450089

Oliver, T. H., Heard, M. S., Isaac, N. J., Roy, D. B., Procter, D., Eigenbrod, F., Freckleton, R., Hector, A., Orme, C. D. L., Petchey, O. L., Proença, V., Raffaelli, D., Suttle, K. B., Mace, G. M., Martín-López, B., Woodcock, B. A., & Bullock, J. M. (2015). Biodiversity and resilience of ecosystem functions. *Trends in Ecology & Evolution*, *30*(11), 673–684. DOI: 10.1016/j.tree.2015.08.009 PMID: 26437633

Opedal, Ø. H., Armbruster, W. S., & Graae, B. J. (2015). Linking small-scale topography with microclimate, plant species diversity and intra-specific trait variation in an alpine landscape. *Plant Ecology & Diversity*, *8*(3), 305–315. DOI: 10.1080/17550874.2014.987330

Orekhova, A., Barták, M., Hájek, J., & Morkusová, J. (2022). Species-specific responses of spectral reflectance and the photosynthetic characteristics in two selected Antarctic mosses to thallus desiccation. *Acta Physiologiae Plantarum*, *44*(1), 6. DOI: 10.1007/s11738-021-03339-6

Ortega-Solis, G., Díaz, I. A., Mellado-Mansilla, D., Tejo, C., Tello, F., Craven, D., Kreft, H., & Armesto, J. J. (2021). Trash-basket epiphytes as secondary foundation species: a review of their distribution and effects on biodiversity and ecosystem functions. bioRxiv, 2021.2006. 2022.449473.

Pahlevani, A., & Falatoury, A. N. (2024). Rediscover of the Iranian endemic alpine Arenaria bulica after 139 years and note of the related species (Caryophyllaceae). *Phytotaxa*, *646*(1), 47–57. DOI: 10.11646/phytotaxa.646.1.3

Pak-Met. (2018). The normals of climatic data of Azad Jammu & Kashmir. In: Pakistan Meteorological Department, Regional Centre Muzaffarabad, Pakistan.

Palomo, I. (2017). Climate change impacts on ecosystem services in high mountain areas: A literature review. *Mountain Research and Development*, *37*(2), 179–187. DOI: 10.1659/MRD-JOURNAL-D-16-00110.1

Pandey, S. N. (2020). Role of micronutrients in biochemical responses of crops under abiotic stresses. In *Sustainable agriculture in the era of climate change* (pp. 93–112). Springer International Publishing. DOI: 10.1007/978-3-030-45669-6_4

Pandian, K., Vijayakumar, S., Mustaffa, M. R. A. F., Subramanian, P., & Chitraputhirapillai, S. (2024). Biochar – a sustainable soil conditioner for improving soil health, crop production and environment under changing climate: A review. *Frontiers in Soil Science*, *4*, 1376159. DOI: 10.3389/fsoil.2024.1376159

Pang, J., Tian, Z., Zhang, M., Wang, Y., Qi, T., Zhang, Q., & Zhang, P. (2025). Enhancing carbon sequestration and greenhouse gas mitigation in semiarid farmland: The promising role of biochar application with biodegradable film mulching. *Journal of Integrative Agriculture*, *24*(2), 517–535.

Panthi, J., Vetaas, O. R., & Shrestha, B. B. (2020). Tree growth responses to climatic drivers across an elevation gradient in Pinus wallichiana forests in the central Himalayas. *Forest Ecology and Management*, *477*, 118526.

Parisod, C. (2025). The niche of Alpine Botany. *Alpine Botany*, *135*(1), 1–5. DOI: 10.1007/s00035-025-00329-z

Park, M., Guzmán, Q. J. A., Dyalsingh, H., Scott, A., & Cavender-Bares, J. (2025). *Forest composition and diversity buffer microclimates and enhance productivity*. Life Sciences. DOI: 10.32942/X2ZP87

Park, S., Kim, J. H., & Lee, E. J. (2022). Resistance of plant communities to invasion by tall fescue: An experimental study combining species diversity, functional traits and nutrient levels. *Basic and Applied Ecology*, *58*, 39–49. DOI: 10.1016/j.baae.2021.12.002

Parmesan, C., & Yohe, G. (2003). A globally coherent fingerprint of climate change impacts across natural systems. *Nature*, *421*(6918), 37–42. DOI: 10.1038/nature01286 PMID: 12511946

Patty, L., Halloy, S. R. P., Hiltbrunner, E., & Körner, C. (2010). Biomass allocation in herbaceous plants under grazing impact in the high semi-arid Andes. *Flora - Morphology, Distribution, Functional Ecology of Plants, 205*(10), 695-703. DOI: 10.1016/j.flora.2009.12.039

Pauli, H., & Halloy, S. R. (2019). High mountain ecosystems under climate change. In *Oxford research encyclopedia of climate science*. DOI: 10.1093/acrefore/9780190228620.013.764

Pauli, H., Gottfried, M., & Grabherr, G. (2003). Effects of climate change on the alpine and nival vegetation of the Alps. *Journal of mountain ecology, 7*(Suppl), 9-12.

Pauli, H., Gottfried, M., Dullinger, S., Abdaladze, O., Akhalkatsi, M., Alonso, J. L. B., & Calzado, R. F. (2012). Recent plant diversity changes on Europe's mountain summits. *Science*, *336*(6079), 353–355. DOI: 10.1126/science.1219033 PMID: 22517860

Payne, D., Snethlage, M., Geschke, J., Spehn, E. M., & Fischer, M. (2020). Nature and people in the Andes, East African Mountains, European Alps, and Hindu Kush Himalaya: Current research and future directions. *Mountain Research and Development*, *40*(2), A1. DOI: 10.1659/MRD-JOURNAL-D-19-00075.1

Pedrotti, F. (2018). Three-dimensional maps of vegetation series and geoseries with the use of stereograms. *52*, 55-67. DOI: 10.24193/Contrib.Bot.52.4

Pelliza, Y. I., Fernandez, A., Saiz, H., & Tadey, M. (2021). Together we stand, divided we fall: Effects of livestock grazing on vegetation patches in a desert community. *Journal of Vegetation Science*, *32*(2), e13015. DOI: 10.1111/jvs.13015

Penuelas, J., & Sardans, J. (2023). Human-driven global nutrient imbalances increase risks to health. *Eco-Environment & Health*, *2*(4), 246–251. DOI: 10.1016/j.eehl.2023.08.003 PMID: 38435356

Pepin, N., Bradley, R., Diaz, H., Baraer, M., Caceres, E., Forsythe, N., & Liu, X. (2015). Elevation-dependent warming in mountain regions of the world. [In.]. *Nature Climate Change*, *5*(5), 424–430. DOI: 10.1038/nclimate2563

Pérez, F. L. (2020). Andean rolling mosses gather on stone pavements: Geoecology of Grimmia longirostris Hook. in a high periglacial páramo. *Catena*, *187*, 104389. DOI: 10.1016/j.catena.2019.104389

Pérez-Hernández, J., & Gavilán, R. G. (2021). Impacts of land-use changes on vegetation and ecosystem functioning: Old-field secondary succession. *Plants*, *10*(5), 990. DOI: 10.3390/plants10050990 PMID: 34065656

Peringer, A., Frank, V., & Snell, R. S. (2022). Climate change simulations in Alpine summer pastures suggest a disruption of current vegetation zonation. *Global Ecology and Conservation, 37*, e02140. DOI: 10.1016/j.gecco.2022.e02140

Pescador, D. S., Sierra-Almeida, Á., Torres, P. J., & Escudero, A. (2016). Summer Freezing Resistance: A Critical Filter for Plant Community Assemblies in Mediterranean High Mountains [Original Research]. *Frontiers in Plant Science, Volume 7 - 2016*. DOI: 10.3389/fpls.2016.00194

Pescador, D. S., de Bello, F., Valladares, F., & Escudero, A. (2015). Plant Trait Variation along an Altitudinal Gradient in Mediterranean High Mountain Grasslands: Controlling the Species Turnover Effect. *PLoS One*, *10*(3), e0118876. DOI: 10.1371/journal.pone.0118876 PMID: 25774532

Petrychenko, V. F., Korniichuk, O. V., Buhaiov, V. D., Horenskyi, V. M., & Veklenko, Y. A. (2022). Influence of hydrothermal factors on feed and seed productivity of alfalfa in the conditions of the Right-Bank Forest-Steppe. *Редакційна колегія, 13*(3), 49-59.

Pettorelli, N., Laurance, W. F., O'Brien, T. G., Wegmann, M., Nagendra, H., & Turner, W. (2014). Satellite remote sensing for applied ecologists: Opportunities and challenges. *Journal of Applied Ecology*, *51*(4), 839–848. DOI: 10.1111/1365-2664.12261

Pieroni, A., & Giusti, M. E. (2009). Alpine ethnobotany in Italy: Traditional knowledge of gastronomic and medicinal plants among the Occitans of the upper Varaita valley, Piedmont. *Journal of Ethnobiology and Ethnomedicine*, *5*(1), 32. DOI: 10.1186/1746-4269-5-32 PMID: 19895681

Piredda, R., Simeone, M. C., Attimonelli, M., Bellarosa, R., & Schirone, B. (2011). Prospects of barcoding the Italian wild dendroflora: Oaks reveal severe limitations to tracking species identity. *Molecular Ecology Resources*, *11*(1), 72–83. DOI: 10.1111/j.1755-0998.2010.02900.x PMID: 21429102

Piria, M., Radočaj, T., Vilizzi, L., & Britvec, M. (2022). Climate change may exacerbate the risk of invasiveness of non-native aquatic plants: The case of the Pannonian and Mediterranean regions of Croatia. *NeoBiota*, *76*, 25–52. DOI: 10.3897/neobiota.76.83320

Plunger, J., Guariento, E., Steinwandter, M., Colla, F., Rief, A., & Seeber, J. (2022). Shifts in ground-dwelling predator communities in response to changes in management intensity in Alpine meadows. *Soil Organisms*, *94*(3), 149–161.

Poff, N. L. (1997). Landscape filters and species traits: Towards mechanistic understanding and prediction in stream ecology. *Journal of the North American Benthological Society, 16*(2), 391–409. DOI: 10.2307/1468026

Polechová, J., & Barton, N. H. (2015). Limits to adaptation along environmental gradients. *Proceedings of the National Academy of Sciences of the United States of America, 112*(20), 6401–6406. DOI: 10.1073/pnas.1421515112 PMID: 25941385

Poncet, A., Vogl, C. R., & Weckerle, C. S. (2015). Folkbotanical classification: Morphological, ecological and utilitarian characterization of plants in the Napf region, Switzerland. *Journal of Ethnobiology and Ethnomedicine, 11*(1), 13. DOI: 10.1186/1746-4269-11-13 PMID: 25971246

Poni, S., Sabbatini, P., & Palliotti, A. (2022). Facing spring frost damage in grapevine: Recent developments and the role of delayed winter pruning–a review. *American Journal of Enology and Viticulture, 73*(4), 211–226. DOI: 10.5344/ajev.2022.22011

Pornaro, C., Schneider, M., Leinauer, B., & Macolino, S. (2017). Above-and below-ground patterns in a subalpine grassland-shrub mosaic. *151*(3), 493-503.

Pornaro, C., Schneider, M., Leinauer, B., & Macolino, S. (2017). Above-and below-ground patterns in a subalpine grassland-shrub mosaic. *Plant Biosystems, 151*(3), 493–503. DOI: 10.1080/11263504.2016.1187679

Porro, F. (2020). Comparative plant diversity and seed germination patterns of alpine species in the context of climate change.

Porro, F., Tomaselli, M., Abeli, T., Gandini, M., Gualmini, M., Orsenigo, S., Petraglia, A., Rossi, G., & Carbognani, M. (2019). Could plant diversity metrics explain climate-driven vegetation changes on mountain summits of the GLORIA network? *Biodiversity and Conservation, 28*(13), 3575–3596. DOI: 10.1007/s10531-019-01837-1

Prescott, C. E., Frouz, J., Grayston, S. J., Quideau, S. A., & Straker, J. (2019). Rehabilitating forest soils after disturbance. In *Developments in Soil Science* (Vol. 36, pp. 309–343). Elsevier., DOI: 10.1016/B978-0-444-63998-1.00013-6

Probo, M., Lonati, M., Ravetto Enri, S., Mariotte, P., & Pauler, C. (2025). Influence de la gestion pastorale sur la conservation de la biodiversité des pâturages de montagne. *INRAE Productions Animales, 8436*, 8436. Advance online publication. DOI: 10.20870/productions-animales.2025.38.2.8436

Profile, E. (2017). Mountains of Central Asia Biodiversity Hotspot. *Draft for Submission to the CEPF Donor Council, 13.*

Pugnaire, F. I., Losapio, G., & Schöb, C. (2021). Species interactions involving cushion plants in high-elevation environments under a changing climate. *Ecosistemas (Madrid), 30*(1), 2186. DOI: 10.7818/ECOS.2186

Pütz, M., Gallati, D., Kytzia, S., Elsasser, H., Lardelli, C., Teich, M., Waltert, F., & Rixen, C. (2011). Winter tourism, climate change, and snowmaking in the Swiss Alps: Tourists' attitudes and regional economic impacts. *Mountain Research and Development, 31*(4), 357–362. DOI: 10.1659/MRD-JOURNAL-D-11-00039.1

Pyke, G. H., & Ehrlich, P. R. (2010). Biological collections and ecological/environmental research: A review, some observations and a look to the future. *Biological Reviews of the Cambridge Philosophical Society, 85*(2), 247–266. DOI: 10.1111/j.1469-185X.2009.00098.x PMID: 19961469

Qasim, M., Babar, S., Younas, M. U., Rajput, N., Xia, X., Wang, J., Wang, X., & Jiang, C. (2025). Exploration of biochars for enhancing soil health, carbon sequestration, and greenhouse gas emission reductions under citrus cultivation. *European Journal of Agronomy, 168*, 127661. DOI: 10.1016/j.eja.2025.127661

Qiu, Y., Yan, Z., Feng, J., Hua, L., Fan, L., Li, Z., Wang, J., & Qian, C. (2023). Robust historical and future drying trends in Central Asia evidenced by the latest observation and modeling datasets. *Atmospheric Research, 295*, 107033. DOI: 10.1016/j.atmosres.2023.107033

Qi, Y., Wei, W., Chen, C., & Chen, L. (2019). Plant root-shoot biomass allocation over diverse biomes: A global synthesis. *Global Ecology and Conservation, 18*, e00606. DOI: 10.1016/j.gecco.2019.e00606

Quadrana, L., Bortolini Silveira, A., Mayhew, G. F., LeBlanc, C., Martienssen, R. A., Jeddeloh, J. A., & Colot, V. (2016). The Arabidopsis thaliana mobilome and its impact at the species level. *elife, 5*, e15716.

Quan, Q., He, N., Zhang, R., Wang, J., Luo, Y., Ma, F., Pan, J., Wang, R., Liu, C., Zhang, J., Wang, Y., Song, B., Li, Z., Zhou, Q., Yu, G., & Niu, S. (2024). Plant height as an indicator for alpine carbon sequestration and ecosystem response to warming. *Nature Plants, 10*(6), 890–900. DOI: 10.1038/s41477-024-01705-z PMID: 38755277

Quinn, J. A. (2008). Arctic and alpine biomes.

Qureshi, R. A., Ghufran, M. A., Gilani, S. A., Sultana, K., & Ashraf, M. (2007). Ethnobotanical studies of selected medicinal plants of Sudhan Gali and Ganga Chotti hills, district Bagh, Azad Kashmir. *Pakistan Journal of Botany, 39*(7), 2275–2283.

Raes, N. (2009). *A quantitative analysis of botanical richness, endemicity and floristic regions based on herbarium records*. Leiden University.

Rahim, N. H., Razak, S. A., Chang, X., Saun, F. C., Khan, M. N., Hamzah, S. N., Rohman, F., Ali, B., Kaplan, A., & Iqbal, M. (2024). Ecosystem services by urban forest (UF) towards climate change adaptation: A review. *Polish Journal of Environmental Studies*, *33*(4), 3503–3513. DOI: 10.15244/pjoes/177147

Rahmonov, O., Rahmonov, M., Opała-Owczarek, M., Owczarek, P., Niedźwiedź, T., & Myga-Piątek, U. (2017). Ecological and cultural importance of juniper ecosystem in the area of Zeravshan valley (Tajikistan) on the background of environmental condition and anthropogenic hazards. *Geographia Polonica*, *90*(4), 441–461. DOI: 10.7163/GPol.0110

Rai, P., Bangelesa, F., Abel, D., Ziegler, K., Huang, J., Schaffhauser, T., Pollinger, F., Disse, M., & Paeth, H. (2024). Extreme precipitation and temperature indices under future climate change in central Asia based on CORDEX-CORE. *Theoretical and Applied Climatology*, *155*(7), 6015–6039. DOI: 10.1007/s00704-024-04976-w

Rai, S., Breme, N., Jandova, V., Lanta, V., Altman, J., Ruka, A. T., Rixen, C., & Dolezal, J. (2025). Growth dynamics and climate sensitivities in alpine cushion plants: Insights from Silene acaulis in the Swiss Alps. *Alpine Botany*, *135*(1), 79–89. DOI: 10.1007/s00035-024-00318-8

Ramzi, S., Noroozi, J., Gholizadeh, H., Hamzeh'ee, B., Asri, Y., Talebi, A., & Zarezadeh, S. (2024). IranVeg–the Vegetation Database of Iran: Current status and the way forward. *Vegetation Classification and Survey*, *5*, 237–256. DOI: 10.3897/VCS.114081

Rapuc, W., Guinoiseau, D., Arnaud, F., Dellinger, M., Sabatier, P., Gaillardet, J., Poulenard, J., & Bouchez, J. (2025). Human and climate impacts on the alpine Critical Zone over the past 10,000 y. *Proceedings of the National Academy of Sciences of the United States of America*, *122*(29), e2506030122. DOI: 10.1073/pnas.2506030122 PMID: 40658840

Rastegar-Pouyani, N., Gholamifard, A., Karamiani, R., Bahmani, Z., Mobaraki, A., Abtin, E., & Sayyadi, F. (2015). Sustainable management of the Herpetofauna of the Iranian plateau and coastal Iran. *Amphibian & Reptile Conservation*, *9*(1), 1–15.

Rathore, N., Thakur, D., & Chawla, A. (2018). Seasonal variations coupled with elevation gradient drives significant changes in eco-physiological and biogeochemical traits of a high altitude evergreen broadleaf shrub, Rhododendron anthopogon. *Plant Physiology and Biochemistry*, *132*, 708–719. DOI: 10.1016/j.plaphy.2018.08.009 PMID: 30150110

Rawat, G. (2007). *Alpine vegetation of the western Himalaya: species diversity, community structure, dynamics and aspects of conservation* D. Sc Thesis, Kumaun University, Nainital, 239pp].

Rawat, G. (2007). Pastoral practices, wild mammals and conservation status of alpine meadows in western Himalaya. *Journal of the Bombay Natural History Society*, *104*(1), 5.

Rawat, J. M., Rawat, B., Mishra, S., Negi, R. K., Mishra, S. N., Chandra, A., & Nautiyal, S. (2014). Altitudinal and seasonal variation in bioactive compound aconitine in Aconitum violaceum, a threatened medicinal plant of Indian Himalayan region. *International Journal of Advanced Research*, *2*(10), 981–988.

Razbani, E., Ghahremaninejad, F., & Joharchi, M. R. (2023). A Floristic Study of the Northern Slope of Bahar Mountain Region, North Khorasan Province, Iran. *Taxonomy and Biosystematics*, *15*(54), 21–40.

Rea, L. M., & Parker, R. A. (2014). *Designing and conducting survey research: A comprehensive guide*. John Wiley & Sons.

Reddy, C. S. (2017). Applications of GIS in plant taxonomy, species distribution and ecology. *Journal of Economic and Taxonomic Botany*, *41*(3-4).

Reed, B. C., Brown, J. F., VanderZee, D., Loveland, T. R., Merchant, J. W., & Ohlen, D. O. (1994). Measuring phenological variability from satellite imagery. *Journal of Vegetation Science*, *5*(5), 703–714. DOI: 10.2307/3235884

Reis, S. (2008). Analyzing land use/land cover changes using remote sensing and GIS in Rize, North-East Turkey. *Sensors (Basel)*, *8*(10), 6188–6202. DOI: 10.3390/s8106188 PMID: 27873865

Rellstab, C., Zoller, S., Sailer, C., Tedder, A., Gugerli, F., Shimizu, K. K., Holderegger, R., Widmer, A., & Fischer, M. C. (2020). Genomic signatures of convergent adaptation to Alpine environments in three Brassicaceae species. *Molecular Ecology*, *29*(22), 4350–4365. DOI: 10.1111/mec.15648 PMID: 32969558

Repe, B., Simončič, P., & Vrščaj, B. (2017). Factors of soil formation. In *The Soils of Slovenia* (pp. 19–60). Springer. DOI: 10.1007/978-94-017-8585-3_3

Retzer, J. L. (1956). Alpine soils of the Rocky Mountains 1. *Journal of Soil Science*, *7*(1), 22–32. DOI: 10.1111/j.1365-2389.1956.tb00858.x

Reveal, J. L. (1991). Botanical explorations in the American West-1889-1989: An essay on the last century of a floristic frontier. *Annals of the Missouri Botanical Garden*, *78*(1), 65–80. DOI: 10.2307/2399591

Richard, B., Dupouey, J. L., Corcket, E., Alard, D., Archaux, F., Aubert, M., Boulanger, V., Gillet, F., Langlois, E., Macé, S., Montpied, P., Beaufils, T., Begeot, C., Behr, P., Boissier, J.-M., Camaret, S., Chevalier, R., Decocq, G., Dumas, Y., & Lenoir, J. (2021). The climatic debt is growing in the understorey of temperate forests: Stand characteristics matter. *Global Ecology and Biogeography*, *30*(7), 1474–1487. DOI: 10.1111/geb.13312

Richter, R. (1997). Correction of atmospheric and topographic effects for high spatial resolution satellite imagery. *International Journal of Remote Sensing*, *18*(5), 1099–1111. DOI: 10.1080/014311697218593

Ritchie, J. (1995). Current trends in studies of long-term plant community dynamics. *The New Phytologist*, *130*(4), 469–494. DOI: 10.1111/j.1469-8137.1995.tb04325.x PMID: 33874480

Riva, F., Koper, N., & Fahrig, L. (2024). Overcoming confusion and stigma in habitat fragmentation research. *Biological Reviews of the Cambridge Philosophical Society*, *99*(4), 1411–1424. DOI: 10.1111/brv.13073 PMID: 38477434

Rixen, C., Høye, T. T., Macek, P., Aerts, R., Alatalo, J. M., Anderson, J. T., Arnold, P. A., Barrio, I. C., Bjerke, J. W., Björkman, M. P., Blok, D., Blume-Werry, G., Boike, J., Bokhorst, S., Carbognani, M., Christiansen, C. T., Convey, P., Cooper, E. J., Cornelissen, J. H. C., & Zong, S. (2022). Winters are changing: Snow effects on Arctic and alpine tundra ecosystems. *Arctic Science*, *8*(3), 572–608. DOI: 10.1139/as-2020-0058

Rixen, C., & Rolando, A. (Eds.). (2013). *The Impacts of Skiing and Related Winter Recreational Activities on Mountain Environments*. BENTHAM SCIENCE PUBLISHERS., DOI: 10.2174/97816080548861130101

Rodrigues, C. I. D., Brito, L. M., & Nunes, L. J. (2023). Soil carbon sequestration in the context of climate change mitigation: A review. *Soil Systems*, *7*(3), 64. DOI: 10.3390/soilsystems7030064

Rodriguez-Echeverria, S., Delgado-Baquerizo, M., Morillo, J. A., Gaxiola, A., Manzano, M., Marquet, P. A., González, L., Cavieres, L. A., Pugnaire, F. I., & Armas, C. (2021). Azorella cushion plants and aridity are important drivers of soil microbial communities in Andean ecosystems. *Ecosystems (New York, N.Y.)*, *24*(7), 1576–1590. DOI: 10.1007/s10021-021-00603-1

Romm, J. J. (2022). *Climate change: What everyone needs to know*. Oxford University Press. DOI: 10.1093/wentk/9780197647127.001.0001

Roozitalab, M. H., Siadat, H., & Farshad, A. (2018). *The soils of Iran*. Springer. DOI: 10.1007/978-3-319-69048-3

Rosbakh, S., Porro, F., Abeli, T., Di Cecco, V., Erschbamer, B., Fernández-Calzado, R., Jimenez-Alfaro, B., Lodetti, S., Lorite, J., Moser, D., Orsenigo, S., Pauli, H., Petraglia, A., Rossi, G., Saccone, P., Stanisci, A., White, F. J., Winkler, M., & Mondoni, A. (2025). Understanding long-term abundance shifts in European alpine plants through the lenses of functional seed trait ecology. *Diversity & Distributions*, *31*(7), e70047. DOI: 10.1111/ddi.70047

Rose-Person, A., Spasojevic, M. J., Forrester, C., Bowman, W. D., Suding, K. N., Oldfather, M. F., & Rafferty, N. E. (2024). Early snowmelt advances flowering phenology and disrupts the drivers of pollinator visitation in an alpine ecosystem. *Alpine Botany*, *134*(2), 141–150. DOI: 10.1007/s00035-024-00315-x

Roth-Nebelsick, A., & Krause, M. (2023). The plant leaf: A biomimetic resource for multifunctional and economic design. *Biomimetics*, *8*(2), 145. DOI: 10.3390/biomimetics8020145 PMID: 37092397

Rowe, R. J. (2005). Elevational gradient analyses and the use of historical museum specimens: A cautionary tale. *Journal of Biogeography*, *32*(11), 1883–1897. DOI: 10.1111/j.1365-2699.2005.01346.x

Rubinoff, D., Reil, J. B., Osborne, K. H., Gregory, C. J., Geib, S. M., & Dupuis, J. R. (2020). Phylogenomics reveals conservation challenges and opportunities for cryptic endangered species in a rapidly disappearing desert ecosystem. *Biodiversity and Conservation*, *29*(7), 2185–2200. DOI: 10.1007/s10531-020-01968-w

Ruchika, R. (2021). *Small-spatial scale ecophysiological and stress-induced (salicylic acid) biochemical investigations on desiccation-tolerant bryophyte SYNTRICHIA RURALIS*. Magyar Agrár-és Élettudományi Egyetem.

Rudnick, D., Ryan, S. J., Beier, P., Cushman, S. A., Dieffenbach, F., Epps, C., Gerber, L. R., Hartter, J. N., Jenness, J. S., & Kintsch, J. (2012). *The role of landscape connectivity in planning and implementing conservation and restoration priorities*. Issues in Ecology.

Rundel, P., & Millar, C. (2016). Alpine ecosystems. *In: Zavaleta, E.; Mooney, H., eds. Ecosystems of California. Berkeley, California: University of California Press: 613-634. Chapter 29.*, 613-634.

Rundquist, D. C. (2001). Field techniques in remote sensing: Learning by doing. *Geocarto International*, *16*(1), 85–90. DOI: 10.1080/10106040108542186

Rupp, A. I. (2020). *Leaf-inspired Design for Heat and Vapor Exchange*. The University of Akron.

Saeed, L., Zuhra, N., & Saeed, S. (2025). Nitrogen cycling and its dynamics in soil–plant systems, future perspective and research directions: A comprehensive review. *The Critical Review of Social Sciences Studies*, *3*(4), 933–954. DOI: 10.59075/8wkz9d78

Safdari, Z., Nahavandchi, H., & Joodaki, G. (2022). Estimation of groundwater depletion in Iran's catchments using well data. *Water (Basel)*, *14*(1), 131. DOI: 10.3390/w14010131

Salem, B. (2003). Application of GIS to biodiversity monitoring. *Journal of Arid Environments*, *54*(1), 91–114. DOI: 10.1006/jare.2001.0887

Salick, J., Fang, Z., & Hart, R. (2019). Rapid changes in eastern Himalayan alpine flora with climate change. *American Journal of Botany*, *106*(4), 520–530. DOI: 10.1002/ajb2.1263 PMID: 30934119

Salick, J., Ghimire, S. K., Fang, Z., Dema, S., & Konchar, K. M. (2014). Himalayan alpine vegetation, climate change and mitigation. *Journal of Ethnobiology*, *34*(3), 276–293. DOI: 10.2993/0278-0771-34.3.276

Samarin, M., Zweifel, L., Roth, V., & Alewell, C. (2020). Identifying Soil Erosion Processes in Alpine Grasslands on Aerial Imagery with a U-Net Convolutional Neural Network. In *Remote Sensing* (Vol. 12, Issue 24). DOI: 10.3390/rs12244149

Samson, S., Patnaik, S., Kannan, K. R., Prathima, T., & Sreedhar, B. (2023). Effect of Global Warming on Glaciers, Rivers and Water Towers. International Conference on Pervasive Knowledge and Collective Intelligence on Web and Social Media, Sekar, K. C., Thapliyal, N., Bhojak, P., Bisht, K., Pandey, A., Mehta, P., Negi, V. S., & Rawat, R. S. (2025). Early signals of climate change impacts on alpine plant diversity in Indian Himalaya. *Biodiversity and Conservation*, *34*(1), 207–233.

Sandra, P. R., Shankar, A., Garkoti, S. C., & Adarsh, C. K. (2025). Understanding the effects of forest types, vegetation structural diversity, and soil properties on above- and below-ground carbon stock of moist deciduous forest ecosystems in Western Ghats, India. *Catena*, *257*, 109198. DOI: 10.1016/j.catena.2025.109198

Sandwith, T. (2003). Overcoming barriers: Conservation and development in the Maloti-Drakensberg Mountains of Southern Africa. *Journal of Sustainable Forestry*, *17*(1-2), 149–169. DOI: 10.1300/J091v17n01_09

Santoianni, L. A., Innangi, M., Varricchione, M., Carboni, M., La Bella, G., Haider, S., & Stanisci, A. (2024). Ecological features facilitating spread of alien plants along Mediterranean mountain roads. *Biological Invasions*, *26*(11), 3879–3899. DOI: 10.1007/s10530-024-03418-y PMID: 39324107

Sarma, B., Sarmah, N., & Malakar, A. (2024). Degradation of Water and Land Resources in the Himalayan Mountain Ecosystems. In Borthakur, A., & Singh, P. (Eds.), *Addressing the Climate Crisis in the Indian Himalayas* (pp. 31–52). Springer Nature Switzerland., DOI: 10.1007/978-3-031-50097-8_2

Sati, P., Dimri, S., Trivedi, V. L., & Nautiyal, M. (2024). Environmental factors shaping alpine plant adaptations.

Scharn, R., Little, C. J., Bacon, C. D., Alatalo, J. M., Antonelli, A., Björkman, M. P., Molau, U., Nilsson, R. H., & Björk, R. G. (2021). Decreased soil moisture due to warming drives phylogenetic diversity and community transitions in the tundra. *Environmental Research Letters*, *16*(6), 064031. DOI: 10.1088/1748-9326/abfe8a

Scherrer, D., & Körner, C. (2011). Topographically controlled thermal-habitat differentiation buffers alpine plant diversity against climate warming. *Journal of Biogeography*, *38*(2), 406–416. DOI: 10.1111/j.1365-2699.2010.02407.x

Schild, A., & Sharma, E. (2011). Sustainable mountain development revisited. *Mountain Research and Development*, *31*(3), 237–241. DOI: 10.1659/MRD-JOURNAL-D-11-00069.1

Schirpke, U., Scolozzi, R., Dean, G., Haller, A., Jäger, H., Kister, J., Kovács, B., Sarmiento, F. O., Sattler, B., & Schleyer, C. (2020). Cultural ecosystem services in mountain regions: Conceptualising conflicts among users and limitations of use. *Ecosystem Services*, *46*, 101210. DOI: 10.1016/j.ecoser.2020.101210

Schloots, C.-L. (2022). *Water level fluctuation and vegetation patterns within an alpine wetland complex*. University of Otago.

Schmid, M. W., van Moorsel, S. J., Hahl, T., De Luca, E., De Deyn, G. B., Wagg, C., Niklaus, P. A., & Schmid, B. (2021). Effects of plant community history, soil legacy and plant diversity on soil microbial communities. *Journal of Ecology*, *109*(8), 3007–3023. DOI: 10.1111/1365-2745.13714

Seastedt, T. R., & Oldfather, M. F. (2021). Climate change, ecosystem processes and biological diversity responses in high elevation communities. *Climate (Basel)*, *9*(5), 87. DOI: 10.3390/cli9050087

Sedej, T. T., Erznožnik, T., & Rovtar, J. (2020). Effect of UV radiation and altitude characteristics on the functional traits and leaf optical properties in Saxifraga hostii at the alpine and montane sites in the Slovenian Alps. *Photochemical & Photobiological Sciences, 19*(2), 180–192. DOI: 10.1039/c9pp00032a PMID: 31956889

Sehgal, H., Sharma, E., Pant, C., Malhotra, S., & Joshi, M. (2025). Impact of climate change on soil health and nutrient cycling. In *Soil health and nutrition management* (pp. 109–128). CABI. DOI: 10.1079/9781800624597.0006

Semwal, R., Panda, T., & Tripathi, N. (2025). Empowering Voices: The Role of Local Communities in Mountain Tourism Development. In *Balancing Mountain Tourism, Cultural Heritage, and Environmental Stability* (pp. 185–200). IGI Global Scientific Publishing.

Senese, A., Maragno, D., Fugazza, D., Soncini, A., D'Agata, C., Azzoni, R. S., & Asif Khan, M. (2018). Inventory of glaciers and glacial lakes of the Central Karakoram National Park (CKNP–Pakistan). *Journal of Maps, 14*(2), 189–198. DOI: 10.1080/17445647.2018.1445561

Sen, S. K., Jain, P., Singh, R. K., & Singh, A. K. (2025). Role of soil algae in the maintenance of soil fertility and soil health. In *Soil algae: Morphology, ecology and biotechnological applications* (pp. 293–304). Springer Nature Singapore. DOI: 10.1007/978-981-97-9873-5_9

Senthilnathan, A., & D'Andrea, R. (2024). Coexistence of Competing Plants Under Plant–Soil Feedback. *Ecology Letters, 27*(9), e14503. DOI: 10.1111/ele.14503 PMID: 39354908

Šerá, B. (2021). Tree Regeneration by Seeds in Natural Forests. *Forests, 12*(10), 1346. DOI: 10.3390/f12101346

Serrano, A. R., Peco, B., Morillo, J. A., & Ochoa-Hueso, R. (2024). Abandonment of traditional livestock grazing reduces soil fertility and enzyme activity, alters soil microbial communities, and decouples microbial networks, with consequences for forage quality in Mediterranean grasslands. *Agriculture, Ecosystems & Environment, 366*, 108932. DOI: 10.1016/j.agee.2024.108932

Séverine Fauquette, J.-P. S., Médail, F., Muller[1], G. J.-M., & Bertini, E. (2018). The Alps: a geological, climatic and human perspective on vegetation history and modern plant diversity. *Mountains, climate and biodiversity*, 413.

Sexton, J. P., Clemens, M., Bell, N., Hall, J., Fyfe, V., & Hoffmann, A. A. (2024). Patterns and effects of gene flow on adaptation across spatial scales: Implications for management. *Journal of Evolutionary Biology, 37*(6), 732–745. DOI: 10.1093/jeb/voae064 PMID: 38888218

Shafi, Z., Shahid, M., Ilyas, T., & Singh, K. (2025). Next-generation perspectives on microbially synthesized siderophores: Molecular engineering, multi-omics insights, and applications for smart climate-resilient crops. *World Journal of Microbiology & Biotechnology, 41*(10), 393. DOI: 10.1007/s11274-025-04575-5 PMID: 41099996

Shaheen, H., Aziz, S., Nasar, S., Waheed, M., Manzoor, M., Siddiqui, M. H.,… Conservation. (2023). Distribution patterns of alpine flora for long-term monitoring of global change along a wide elevational gradient in the Western Himalayas. *48*, e02702.

Shaheen, H., Mashwani, Z. U. R., & Dar, M. E. U. I.(2015). Spatial patterns and diversity of alpine vegetation across Langer–Shandur Valley, Hindukush Himalayas. *Current Science,* 1534-1539.

Shaheen, H., Attique, A., Riaz, M. T., Manzoor, M., Khan, R. W. A., & Riaz, M. T. (2024). From biodiversity hotspot to conservation hotspot: Assessing distribution, population structure, associated flora and habitat geography of threatened Himalayan Yew in temperate forest ecosystems of Kashmir. *Biodiversity and Conservation, 33*(2), 1–25. DOI: 10.1007/s10531-023-02758-w

Shaheen, H., Azad, B., Mushtaq, A., & Khan, R. W. A. (2016). Fuelwood consumption pattern and its impact on forest structure in Kashmir Himalayas. *Bosque (Valdivia), 37*(2), 419–424. DOI: 10.4067/S0717-92002016000200020

Shaheen, H., Aziz, S., Nasar, S., Waheed, M., Manzoor, M., Siddiqui, M. H., Alamri, S., Haq, S. M., & Bussmann, R. W. (2023). Distribution patterns of alpine flora for long-term monitoring of global change along a wide elevational gradient in the Western Himalayas. *Global Ecology and Conservation, 48*, e02702. DOI: 10.1016/j.gecco.2023.e02702

Shaheen, H., Ibrahim, M., & Ullah, Z. (2019). Spatial patterns and diversity of the alpine flora of Deosai plateau, western Himalayas. *Pakistan Journal of Botany, 51*(1), 205–212. DOI: 10.30848/PJB2019-1(39)

Shaheen, H., Khan, S. M., Harper, D. M., Ullah, Z., & Qureshi, R. A. (2011). Species diversity, community structure, and distribution patterns in western Himalayan alpine pastures of Kashmir, Pakistan. *Mountain Research and Development, 31*(2), 153–159. DOI: 10.1659/MRD-JOURNAL-D-10-00091.1

Shahid, M., Singh, U. B., Khan, M. S., Singh, P., Kumar, R., Singh, R. N., Kumar, A., & Singh, H. V. (2023). Bacterial ACC deaminase: Insights into enzymology, biochemistry, genetics, and potential role in amelioration of environmental stress in crop plants. *Frontiers in Microbiology*, *14*, 1132770. DOI: 10.3389/fmicb.2023.1132770 PMID: 37180266

Shang, Q., Li, C., & Liu, Y. (2024). The negative effects of tourist trampling on the soil physical properties and microbial community composition in a natural oak forest. *Forests*, *15*(8), 1419. DOI: 10.3390/f15081419

Sharma, D., Astapati, A. D., Dey, S., Bhattacharjee, D., Singha, D. M., & Nath, S. (2025). Microbial interactions in soil ecosystems: Facilitating plant growth, nutrient cycling, and environmental dynamics. In *Microbial allies: Understanding plant–microbe interactions in sustainable agriculture* (pp. 21–48). Springer Nature Switzerland. DOI: 10.1007/978-3-031-90530-8_2

Sharma, E., Chettri, N., & Oli, K. P. (2010). Mountain biodiversity conservation and management: A paradigm shift in policies and practices in the Hindu Kush-Himalayas. *Ecological Research*, *25*(5), 909–923. DOI: 10.1007/s11284-010-0747-6 PMID: 32214652

Sharma, I., & Birman, S. (2024). Biodiversity loss, ecosystem services, and their role in promoting sustainable health. In *The climate-health-sustainability nexus: understanding the interconnected impact on populations and the environment* (pp. 163–188). Springer. DOI: 10.1007/978-3-031-56564-9_7

Sharma, K. P., Tiwari, A., & Shrestha, B. B. (2020). Changes in regeneration and leaf traits of Rhododendron campanulatum along a treeline ecotone in central Nepal. *Journal of Mountain Science*, *17*(3), 602–613. DOI: 10.1007/s11629-019-5386-y

Sharma, U. C., Datta, M., & Sharma, V. (2022). In Sharma, U. C., Datta, M., & Sharma, V. (Eds.), *Land Use and Management BT - Soils in the Hindu Kush Himalayas: Management for Agricultural Land Use* (pp. 295–462). Springer International Publishing., DOI: 10.1007/978-3-031-11458-8_7

Shen, M., Song, B., Zhou, C., Almatrafi, E., Hu, T., Zeng, G., & Zhang, Y. (2022). Recent advances in impacts of microplastics on nitrogen cycling in the environment: A review. *The Science of the Total Environment*, *815*, 152740. DOI: 10.1016/j.scitotenv.2021.152740 PMID: 34974017

Shen, N., Xu, W., Guo, F., Cui, Y., Chen, Y., & Su, X. (2023). Effect of fertilizer and harvesting stage on growth and bioactive compounds content of Rheum tanguticum. *Journal of Plant Nutrition*, *46*(18), 4462–4476. DOI: 10.1080/01904167.2023.2232407

Shepherd, D. A., Souitaris, V., & Gruber, M. (2021). Creating new ventures: A review and research agenda. *Journal of Management*, *47*(1), 11–42. DOI: 10.1177/0149206319900537

Shinoda, Y., & Akasaka, M. (2020). Interaction exposure effects of multiple disturbances: Plant population resilience to ungulate grazing is reduced by creation of canopy gaps. *Scientific Reports*, *10*(1), 1802. DOI: 10.1038/s41598-020-58672-6 PMID: 32020019

Shinwari, A. K., Ahmad, M., Alsahli, A. A., Israr, A., Shinwari, Z. K., Shinwari, S., & Yilmaz, M. E. (2024). Himalayan medicinal plants wealth, threats and conservation for securing the future of biodiversity. *Pakistan Journal of Botany*, *56*(6), 2327–2350. DOI: 10.30848/PJB2024-6(42)

Shinwari, Z. K., & Gilani, S. S. (2003). Sustainable harvest of medicinal plants at Bulashbar Nullah, Astore (northern Pakistan). *Journal of Ethnopharmacology*, *84*(2-3), 289–298. DOI: 10.1016/S0378-8741(02)00333-1 PMID: 12648828

Shipley, B., De Bello, F., Cornelissen, J. H. C., Laliberté, E., Laughlin, D. C., & Reich, P. B. (2016). Reinforcing loose foundation stones in trait-based plant ecology. *Oecologia*, *180*(4), 923–931. DOI: 10.1007/s00442-016-3549-x PMID: 26796410

Shi, S.-B., Zhu, W.-Y., Li, H.-M., Zhou, D.-W., Han, F., Zhao, X.-Q., & Tang, Y.-H. (2004). Photosynthesis of Saussurea superba and Gentiana straminea is not reduced after long-term enhancement of UV-B radiation. *Environmental and Experimental Botany*, *51*(1), 75–83. DOI: 10.1016/S0098-8472(03)00062-5

Shi, S., Shi, R., Li, T., & Zhou, D. (2022). UV-B Radiation Effects on the Alpine Plant Kobresia humilis in a Qinghai-Tibet Alpine Meadow. *Plants*, *11*(22), 3102. DOI: 10.3390/plants11223102 PMID: 36432831

Shrestha, B., Zhang, L., Shrestha, S., Khadka, N., & Maharjan, L. (2024). Spatiotemporal patterns, sustainability, and primary drivers of NDVI-derived vegetation dynamics (2003–2022) in Nepal. *Environmental Monitoring and Assessment*, *196*(7), 607. DOI: 10.1007/s10661-024-12754-4 PMID: 38858316

Shrestha, U. B., Gautam, S., & Bawa, K. S. (2012). Widespread climate change in the Himalayas and associated changes in local ecosystems. *PLoS One*, *7*(5), e36741. DOI: 10.1371/journal.pone.0036741 PMID: 22615804

Shteir, A. B. (1976). Albrecht von Haller's Botany and" Die Alpen. *Eighteenth-Century Studies*, *10*(2), 169–184. DOI: 10.2307/2737553 PMID: 11616915

Shukla, P. R., Skeg, J., Buendia, E. C., Masson-Delmotte, V., Pörtner, H.-O., Roberts, D.,… Van Diemen, S. (2019). Climate Change and Land: an IPCC special report on climate change, desertification, land degradation, sustainable land management, food security, and greenhouse gas fluxes in terrestrial ecosystems.

Sierra-Almeida, A., Cavieres, L. A., & Bravo, L. A. (2009). Freezing resistance varies within the growing season and with elevation in high-Andean species of central Chile. *The New Phytologist*, *182*(2), 461–469. DOI: 10.1111/j.1469-8137.2008.02756.x PMID: 19210722

Sierra-Almeida, A., Cavieres, L. A., & Bravo, L. A. (2010). Freezing resistance of high-elevation plant species is not related to their height or growth-form in the Central Chilean Andes. *Environmental and Experimental Botany*, *69*(3), 273–278. DOI: 10.1016/j.envexpbot.2010.04.015

Sierra-Almeida, A., Reyes-Bahamonde, C., & Cavieres, L. A. (2016). Drought increases the freezing resistance of high-elevation plants of the Central Chilean Andes. *Oecologia*, *181*(4), 1011–1023. DOI: 10.1007/s00442-016-3622-5 PMID: 27053321

Silva, H. C. H., Caraciolo, R. L. F., Marangon, L. C., Ramos, M. A., Santos, L. L., & Albuquerque, U. P. (2014). Evaluating different methods used in ethnobotanical and ecological studies to record plant biodiversity. *Journal of Ethnobiology and Ethnomedicine*, *10*(1), 48. DOI: 10.1186/1746-4269-10-48 PMID: 24916833

Sindhu, S. S., Parmar, P., & Phour, M. (2013). Nutrient cycling: Potassium solubilization by microorganisms and improvement of crop growth. In *Geomicrobiology and biogeochemistry* (pp. 175–198). Springer.

Singh, G., Singh, A., Singh, P., Shukla, R., Tripathi, S., & Mishra, V. K. (2021). The fate of organic pollutants and their microbial degradation in water bodies. In *Pollutants and water management: Resources, strategies and scarcity* (pp. 210–240). DOI: 10.1002/9781119693635.ch9

Singh, J. (2025). The Accelerating Melting of Glaciers: Impacts on Earth and Emerging Challenges. *International Journal of Advanced Research and Interdisciplinary Scientific Endeavours*, *2*(1), 401–414. DOI: 10.61359/11.2206-2501

Singh, P., Choudhary, K. K., Chaudhary, N., Gupta, S., Sahu, M., Tejaswini, B., & Sarkar, S. (2022). Salt stress resilience in plants mediated through osmolyte accumulation and its crosstalk mechanism with phytohormones. *Frontiers in Plant Science*, *13*, 1006617. DOI: 10.3389/fpls.2022.1006617 PMID: 36237504

Singh, S. S., Lamare, E., Chopade, B. A., Chanda, R., Devi, N. R., Upadhyay, K. K., & Tripathi, S. K. (2025). The hidden players: Soil microbes in ecosystems sustainability. In *Climate change and soil microorganisms for environmental sustainability* (pp. 295–320). Springer Nature Singapore. DOI: 10.1007/978-981-96-3425-5_13

Sklenář, P. (2017). Seasonal variation of freezing resistance mechanisms in north-temperate alpine plants. *Alpine Botany*, *127*(1), 31–39. DOI: 10.1007/s00035-016-0174-6

Sklenář, P., Dušková, E., & Balslev, H. (2011). Tropical and Temperate: Evolutionary History of Páramo Flora. *Botanical Review*, *77*(2), 71–108. DOI: 10.1007/s12229-010-9061-9

Smiti, K. (2023). *Chapter 11 - Tourism and the properties of mountainous soil: a dynamic relationship* (R. Bhadouria, S. Singh, S. Tripathi, & P. B. T.-U. S. of M. L. Singh (eds.); pp. 213–230). Elsevier. 10.1016/B978-0-323-95925-4.00011-X

Smiti, K. (2023). Tourism and the properties of mountainous soil: A dynamic relationship. In *Understanding Soils of Mountainous Landscapes* (pp. 213–230). Elsevier. DOI: 10.1016/B978-0-323-95925-4.00011-X

Soliveres, S., DeSoto, L., Maestre, F. T., & Olano, J. M. (2010). Spatio-temporal heterogeneity in abiotic factors modulate multiple ontogenetic shifts between competition and facilitation. *Perspectives in Plant Ecology, Evolution and Systematics*, *12*(3), 227–234. DOI: 10.1016/j.ppees.2010.02.003

Son, D. (2023). Plant co-occurrence patterns and soil environments associated with three dominant plants in the Arctic. *Journal of Ecology and Environment*, *47*(1), 1–13. DOI: 10.5141/jee.23.002

Song, X., Fang, C., Yuan, Z. Q., & Li, F. M. (2021). Long-term growth of alfalfa increased soil organic matter accumulation and nutrient mineralization in a semi-arid environment. *Frontiers in Environmental Science*, *9*, 649346. DOI: 10.3389/fenvs.2021.649346

Sosef, M. S., Dauby, G., Blach-Overgaard, A., van Der Burgt, X., Catarino, L., Damen, T., Deblauwe, V., Dessein, S., Dransfield, J., Droissart, V., Duarte, M. C., Engledow, H., Fadeur, G., Figueira, R., Gereau, R. E., Hardy, O. J., Harris, D. J., de Heij, J., Janssens, S., & Couvreur, T. L. P. (2017). Exploring the floristic diversity of tropical Africa. *BMC Biology*, *15*(1), 15. DOI: 10.1186/s12915-017-0356-8 PMID: 28264718

Speed, J. D. M., Austrheim, G., Hester, A. J., & Mysterud, A. (2012). Elevational advance of alpine plant communities is buffered by herbivory. *Journal of Vegetation Science*, *23*(4), 617–625. DOI: 10.1111/j.1654-1103.2012.01391.x

Spehn, E. M., Rudmann-Maurer, K., Korner, C., & MaseLLi, D. (2012). Mountain biodiversity and global change.

Spence, A. R., & Tingley, M. W. (2020). The challenge of novel abiotic conditions for species undergoing climate-induced range shifts. *Ecography*, *43*(11), 1571–1590. DOI: 10.1111/ecog.05170

Spencer, S. A., Anderson, A. E., Silins, U., & Collins, A. L. (2021). Hillslope and groundwater contributions to streamflow in a Rocky Mountain watershed underlain by glacial till and fractured sedimentary bedrock. *Hydrology and Earth System Sciences*, *25*(1), 237–255. DOI: 10.5194/hess-25-237-2021

Spitaler, R., Winkler, A., Lins, I., Yanar, S., Stuppner, H., & Zidorn, C. (2008). Altitudinal variation of phenolic contents in flowering heads of Arnica montana cv. ARBO: A 3-year comparison. *Journal of Chemical Ecology*, *34*(3), 369–375. DOI: 10.1007/s10886-007-9407-x PMID: 18259818

Stanchi, S., Freppaz, M., Godone, D., & Zanini, E. (2013). Assessing the susceptibility of alpine soils to erosion using soil physical and site indicators. *Soil Use and Management*, *29*(4), 586–596. DOI: 10.1111/sum.12063

Stefan, L., Hartmann, M., Engbersen, N., Six, J., & Schöb, C. (2021). Positive effects of crop diversity on productivity driven by changes in soil microbial composition. *Frontiers in Microbiology*, *12*, 660749. DOI: 10.3389/fmicb.2021.660749 PMID: 33936016

Steiger, R., Scott, D., Abegg, B., Pons, M., & Aall, C. (2019). A critical review of climate change risk for ski tourism. *Current Issues in Tourism*, *22*(11), 1343–1379. DOI: 10.1080/13683500.2017.1410110

Steinbauer, M. J., Grytnes, J.-A., Jurasinski, G., Kulonen, A., Lenoir, J., Pauli, H., & Barni, E. (2018). Accelerated increase in plant species richness on mountain summits is linked to warming. *Nature*, *556*(7700), 231–234. DOI: 10.1038/s41586-018-0005-6 PMID: 29618821

Stirling, G., & Wilsey, B. (2001). Empirical relationships between species richness, evenness, and proportional diversity. *American Naturalist*, *158*(3), 286–299. DOI: 10.1086/321317 PMID: 18707325

Stöckli, V., Wipf, S., Nilsson, C., & Rixen, C. (2011). Using historical plant surveys to track biodiversity on mountain summits. *Plant Ecology & Diversity*, *4*(4), 415–425. DOI: 10.1080/17550874.2011.651504

Streb, P., Aubert, S., Gout, E., & Bligny, R. (2003). Reversibility of cold- and light-stress tolerance and accompanying changes of metabolite and antioxidant levels in the two high mountain plant species Soldanella alpina and Ranunculus glacialis. *Journal of Experimental Botany*, *54*(381), 405–418. DOI: 10.1093/jxb/erg048 PMID: 12493869

Striker, G. G. (2024). Oxygen Transport and Plant Ventilation. In *Responses of Plants to Soil Flooding* (pp. 139–156). Springer. DOI: 10.1007/978-981-99-9112-9_9

Suhery, D., Idami, Z., & Wati, S. (2024). The Phonological Interference of Acehnese in Pronouncing Indonesian Language. *Journal of Languages and Language Teaching*, *12*(1), 120–135. DOI: 10.33394/jollt.v12i1.9554

Sullivan, J. R., & Nazaire, M. (2022). Specimen collection and preparation for a changing flora. *Rhodora*, *123*(993), 50–66. DOI: 10.3119/20-32

Sundriyal, M., & Sharma, B. (2016). Status of biodiversity in Central Himalaya. *Applied Ecology and Environmental Sciences*, *4*(2), 37–43.

Sun, H., Niu, Y., Chen, Y.-S., Song, B., Liu, C.-Q., Peng, D.-L., Chen, J.-G., & Yang, Y. (2014a). Survival and reproduction of plant species in the Qinghai–Tibet Plateau. *Journal of Systematics and Evolution*, *52*(3), 378–396. DOI: 10.1111/jse.12092

Sun, H., Wang, L., Kumar, A., Auwal, M., Van Zwieten, L., Ge, T., & Kuzyakov, Y. (2025). Nutrient availability mediates organic carbon turnover in paddy soils through regulating microbial metabolism. *Geoderma*, *458*, 117313. DOI: 10.1016/j.geoderma.2025.117313

Sun, J., Zhou, T. C., Liu, M., Chen, Y. C., Liu, G. H., Xu, M., Shi, P. L., Peng, F., Tsunekawa, A., Liu, Y., Wang, X.-D., Dong, S.-K., Zhang, Y.-J., & Li, Y.-N. (2020). Water and heat availability are drivers of the aboveground plant carbon accumulation rate in alpine grasslands on the Tibetan Plateau. *Global Ecology and Biogeography*, *29*(1), 50–64. DOI: 10.1111/geb.13006

Sun, L., Liu, Y.-F., Wang, X., Liu, Y., & Wu, G.-L. (2022). Soil nutrient loss by gully erosion on sloping alpine steppe in the northern Qinghai-Tibetan Plateau. *Catena*, *208*, 105763. DOI: 10.1016/j.catena.2021.105763

Sun, W., Zhou, S., Yu, B., Zhang, Y., Keenan, T., & Fu, B. (2025). Soil moisture-atmosphere interactions drive terrestrial carbon-water trade-offs. *Communications Earth & Environment*, *6*(1), 169. DOI: 10.1038/s43247-025-02145-z

Sun, Y., Zhang, A., Landis, J. B., Shi, W., Zhang, X., Sun, H., & Wang, H. (2023). Genome assembly of the snow lotus species Saussurea involucrata provides insights into acacetin and rutin biosynthesis and tolerance to an alpine environment. *Horticulture Research, 10*(10), uhad180. DOI: 10.1093/hr/uhad180 PMID: 37885817

Sveen, T. R., Hannula, S., & Bahram, M. (2024). Microbial regulation of feedbacks to ecosystem change. *Trends in Microbiology, 32*(1), 68–78. DOI: 10.1016/j.tim.2023.06.006 PMID: 37500365

Tabari, H., & Talaee, P. H. (2014). Sensitivity of evapotranspiration to climatic change in different climates. *Global and Planetary Change, 115*, 16–23. DOI: 10.1016/j.gloplacha.2014.01.006

Tai, S., Shangguan, D., Wu, J., Wang, R., & Li, D. (2025). Spatial–Temporal Heterogeneity of Wetlands in the Alpine Mountains of the Shule River Basin on the Northeastern Edge of the Qinghai–Tibet Plateau. *Remote Sensing (Basel), 17*(6), 976. DOI: 10.3390/rs17060976

Takhtajan, A. (1986). Florestic regions of the world.

Talal, M., Chen, X., Iqbal, I., & Ali, I. (2025). A critical review: Unearthing the hidden players—The role of extremophilic fungi in forest ecosystems. *Forests, 16*(5), 855. DOI: 10.3390/f16050855

Talebi, K. S., Sajedi, T., & Pourhashemi, M. (2014). Forests of Iran. *A treasure from the past, a hope for the future, 10*.

Tao, J., Xie, Y., Wang, W., Zhu, J., Zhang, Y., & Zhang, X. (2022). Elevational Gradient of Climate-Driving Effects on Cropland Ecosystem Net Primary Productivity in Alpine Region of the Southwest China. *Remote Sensing (Basel), 14*(13), 3069. DOI: 10.3390/rs14133069

Tchalenko, J. (1975). Seismicity and structure of the Kopet Dagh (Iran, USSR). *Philosophical Transactions of the Royal Society of London. Series A, Mathematical and Physical Sciences, 278*(1275), 1–28. DOI: 10.1098/rsta.1975.0019

Teague, R., & Kreuter, U. (2020). Managing Grazing to Restore Soil Health, Ecosystem Function, and Ecosystem Services. *Frontiers in Sustainable Food Systems, 4*, 534187. DOI: 10.3389/fsufs.2020.534187

Teng, Y., Zhan, J., Agyemang, F. B., & Sun, Y. (2020). The effects of degradation on alpine grassland resilience: A study based on meta-analysis data. *Global Ecology and Conservation, 24*, e01336. DOI: 10.1016/j.gecco.2020.e01336

Terzi, S., Torresan, S., Schneiderbauer, S., Critto, A., Zebisch, M., & Marcomini, A. (2019). Multi-risk assessment in mountain regions: A review of modelling approaches for climate change adaptation. *Journal of Environmental Management*, *232*, 759–771. DOI: 10.1016/j.jenvman.2018.11.100 PMID: 30529418

Testolin, R., Attorre, F., Borchardt, P., Brand, R. F., Bruelheide, H., Chytrý, M., & Haider, S. (2021). Global patterns and drivers of alpine plant species richness. *Global Ecology and Biogeography*, *30*(6), 1218–1231. DOI: 10.1111/geb.13297

Testolin, R., Attorre, F., & Jiménez-Alfaro, B. (2020). Global distribution and bioclimatic characterization of alpine biomes. *Ecography*, *43*(6), 779–788. DOI: 10.1111/ecog.05012

Thakur, D., & Chawla, A. (2019). Functional diversity along elevational gradients in the high altitude vegetation of the western Himalaya. *Biodiversity and Conservation*, *28*(8), 1977–1996. DOI: 10.1007/s10531-019-01728-5

Thapliyal, N., Bhojak, P., Arya, D., & Sekar, K. C. (2024). Ethnobotanical assessment of medicinal plants used in alpine regions of Uttarakhand, west Himalaya. *Ethnobotany Research and Applications*, *27*, 1–24. DOI: 10.32859/era.27.44.1-24

Theobald, D. M., Jacob, A. L., Elsen, P. R., Beever, E. A., Ehlers, L., & Hilty, J. (2024). Evaluating ecosystem protection and fragmentation of the world's major mountain regions. *Conservation Biology*, *38*(3), e14240. DOI: 10.1111/cobi.14240 PMID: 38407527

Thielen, S. M., Gall, C., Ebner, M., Nebel, M., Scholten, T., & Seitz, S. (2021). Water's path from moss to soil: A multi-methodological study on water absorption and evaporation of soil-moss combinations. *Journal of Hydrology and Hydromechanics*, *69*(4), 421–435. DOI: 10.2478/johh-2021-0021

Thomas, Frank, & Isard, S. (1986). Alpine vegetation classification using high resolution aerial imagery and topoclimatic index values. *Photogrammetric Engineering and Remote Sensing*, *52*, 381–388.

Tian, L., Zhao, L., Wu, X., Fang, H., Zhao, Y., Yue, G., Liu, G., & Chen, H. (2017). Vertical patterns and controls of soil nutrients in alpine grassland: Implications for nutrient uptake. *The Science of the Total Environment*, *607*, 855–864. DOI: 10.1016/j.scitotenv.2017.07.080 PMID: 28711847

Timofeeva, A. M., Galyamova, M. R., & Sedykh, S. E. (2023). Plant growth-promoting soil bacteria: Nitrogen fixation, phosphate solubilization, siderophore production, and other biological activities. *Plants*, *12*(24), 4074. DOI: 10.3390/plants12244074 PMID: 38140401

Tingley, M. W., & Beissinger, S. R. (2009). Detecting range shifts from historical species occurrences: New perspectives on old data. *Trends in Ecology & Evolution, 24*(11), 625–633. DOI: 10.1016/j.tree.2009.05.009 PMID: 19683829

Tito, R., Vasconcelos, H. L., & Feeley, K. J. (2020). Mountain ecosystems as natural laboratories for climate change experiments. *Frontiers in Forests and Global Change, 3*, 38. DOI: 10.3389/ffgc.2020.00038

Tiwari, P., Bose, S. K., Park, K. I., Dufossé, L., & Fouillaud, M. (2024). Plant–microbe interactions under the extreme habitats and their potential applications. *Microorganisms, 12*(3), 448. DOI: 10.3390/microorganisms12030448 PMID: 38543499

Tompolidi, A.-M., Mantovani, L., Frigeri, A., & Nazzareni, S. J. G. (2025). Remote Sensing Applications for Geological Mapping in the Mediterranean Region. A Review. *Geosciences, 15*(11), 425.

Tóth, E. G., Tremblay, F., Housset, J. M., Bergeron, Y., & Carcaillet, C. (2019). Geographic isolation and climatic variability contribute to genetic differentiation in fragmented populations of the long-lived subalpine conifer Pinus cembra L. in the western Alps. *BMC Evolutionary Biology, 19*(1), 190. DOI: 10.1186/s12862-019-1510-4 PMID: 31623551

Tourani, M., Isik, V., Saber, R., Caglayan, A., & Chitea, F. (2024). Evaluation of seismicity and seismotectonics in the Alborz Mountains: Insights from seismic parameters, Northern Iran. *Journal of Seismology, 28*(3), 675–706. DOI: 10.1007/s10950-024-10218-3

Tsai, H. H., Tang, Y., Jiang, L., Xu, X., Dénervaud Tendon, V., Pang, J., Jia, Y., Wippel, K., Vacheron, J., Keel, C., Andersen, T. G., Geldner, N., & Zhou, F. (2025). Localized glutamine leakage drives the spatial structure of root microbial colonization. *Science, 390*(6768), eadu4235. DOI: 10.1126/science.adu4235 PMID: 41037624

Tulloch, A. I. T., Healy, A., Silcock, J., Wardle, G. M., Dickman, C. R., Frank, A. S. K., Aubault, H., Barton, K., & Greenville, A. C. (2023). Long-term livestock exclusion increases plant richness and reproductive capacity in arid woodlands. *Ecological Applications, 33*(8), e2909. DOI: 10.1002/eap.2909 PMID: 37602895

Turrill, B. (1959). Plant geography. *Vistas in botany, 2*, 172.

Turrill, W. (1964). *Plant taxonomy, phytogeography, and plant ecology* (pp. 187-224).

Turrill, W. B. (2013). *Pioneer plant geography: the phytogeographical researches of Sir Joseph Dalton Hooker* (Vol. 4). Springer.

Ul Hassan Chaudhary, W., & Sivertun, Å. (2015). Object-Based Analysis of Multispectral RS Data and GIS for Detection of Climate Change Impact on the Karakoram Range Northern Pakistan. *Journal of Environmental Science and Engineering A, 4*(4), 303–310. DOI: 10.17265/2162-5298/2015.06.005

Ullah, Z., Ahmad, M., Sher, H., Shaheen, H., & Khan, S. M. J. P. J. B. (2015). Phytogeographic analysis and diversity of grasses and sedges (Poales) of Northern Pakistan. *47*, 93-104.

Urbach, D., Körner, C., & Hilpold, A. (2024). Mountain biodiversity under change. In *Safeguarding Mountain Social-Ecological Systems* (pp. 11–15). Elsevier. DOI: 10.1016/B978-0-12-822095-5.00002-4

Urban, M. C., Bocedi, G., Hendry, A. P., Mihoub, J.-B., Pe'er, G., Singer, A., Bridle, J. R., Crozier, L. G., De Meester, L., Godsoe, W., Gonzalez, A., Hellmann, J. J., Holt, R. D., Huth, A., Johst, K., Krug, C. B., Leadley, P. W., Palmer, S. C. F., Pantel, J. H., & Travis, J. M. J. (2016). Improving the forecast for biodiversity under climate change. *Science, 353*(6304), aad8466. DOI: 10.1126/science.aad8466 PMID: 27609898

Ustin, S., Gitelson, A., Jacquemoud, S., Schaepman, M. E., Asner, G., Gamon, J., & Zarco-Tejada, P. (2009). Retrieval of foliar information about plant pigment systems from high resolution spectroscopy. *Remote Sensing of Environment 113 (2009) Suppl.1, 113*. DOI: 10.5167/uzh-23317

Van Der Merwe, S. (2024). Investigating ecological drivers and impacts of vegetation change in sub Antarctic tundra.

Van Wilgen, B. W., Esler, K. J., Gaertner, M., Midgley, G. F., Richardson, D. M., Van Wilgen, N. J., Wilson, J. R., Carruthers, J., Cowling, R. M., & Forsyth, A. T. (2016). Ecological research and conservation management in the Cape Floristic Region between 1945 and 2015: History, current understanding and future challenges. *Transactions of the Royal Society of South Africa, 71*(3), 207–303. DOI: 10.1080/0035919X.2016.1225607

Vanneste, T., Michelsen, O., Graae, B. J., Kyrkjeeide, M. O., Holien, H., Hassel, K., Lindmo, S., Kapás, R. E., & De Frenne, P. (2017). Impact of climate change on alpine vegetation of mountain summits in Norway. *Ecological Research, 32*(4), 579–593. DOI: 10.1007/s11284-017-1472-1

Varga, K., Csízi, I., Monori, I., & Valkó, O. (2021). Threats and challenges related to grazing paddocks: Recovery of extremely overgrazed grassland after grazing exclusion. *Arid Land Research and Management, 35*(3), 346–357. DOI: 10.1080/15324982.2020.1869120

Vasconcelos, R. N., Cantillo-Pérez, T., Franca Rocha, W. J., Aguiar, W. M., Mendes, D. T., de Jesus, T. B., de Santana, C. O., de Santana, M. M., & Oliveira, R. P. (2024). Advances and Challenges in Species Ecological Niche Modeling: A Mixed Review. *Earth (Basel, Switzerland), 5*(4), 963–989. DOI: 10.3390/earth5040050

Vazirani, A., & Bhattacharjee, T. (2021). Entrepreneurial finance in the twenty-first century, a review of factors influencing venture capitalist's decision. *The Journal of Entrepreneurship, 30*(2), 306–335. DOI: 10.1177/09713557211025654

Venterink, H. O., Wassen, M. J., Belgers, J., & Verhoeven, J. J. J. o. E. (2001). Control of environmental variables on species density in fens and meadows: importance of direct effects and effects through community biomass. 1033-1040.

Verheijen, L., Brovkin, V., Aerts, R., Boenisch, G., Cornelissen, J., Kattge, J., Reich, P. B., Wright, I. J., & Van Bodegom, P. (2013). Impacts of trait variation through observed trait–climate relationships on performance of an Earth system model: A conceptual analysis. *Biogeosciences, 10*(8), 5497–5515. DOI: 10.5194/bg-10-5497-2013

Verma, S. K., Sahu, P. K., Kumar, K., Pal, G., Gond, S. K., Kharwar, R. N., & White, J. F. (2021). Endophyte roles in nutrient acquisition, root system architecture development and oxidative stress tolerance. *Journal of Applied Microbiology, 131*(5), 2161–2177. DOI: 10.1111/jam.15111 PMID: 33893707

Verrall, B., & Pickering, C. M. (2020). Alpine vegetation in the context of climate change: A global review of past research and future directions. *The Science of the Total Environment, 748*, 141344. DOI: 10.1016/j.scitotenv.2020.141344 PMID: 32814293

Victor, J. E., Smith, G. F., & Van Wyk, A. E. (2016). History and drivers of plant taxonomy in South Africa. *Phytotaxa, 269*(3), 193–208-193–208.

Vincent, A., Daigre, C., Fischer, O., Aðalgeirsdóttir, G., Violette, S., Hart, J., Guðmundsson, S., & Pálsson, F. (2024). A hydrogeological conceptual model of aquifers in catchments headed by temperate glaciers. *Hydrology and Earth System Sciences, 28*(14), 3475–3494. DOI: 10.5194/hess-28-3475-2024

Visakorpi, K., Block, S., Pellissier, L., Levine, J. M., & Alexander, J. (2023). Ecophysiological and morphological traits explain alpine plant species' response to warming. *Functional Ecology, 37*(2), 287–301. DOI: 10.1111/1365-2435.14228

Volberg, R. A. (1983). *Constraints and commitments in the development of American botany, 1880-1920*. UCSF.

von Christ, W. (1886). *Kritische Beiträge zur Metaphysik des Aristoteles*. Franz in Komm.

Von Haller, A. (1768). *Historia stirpium indigenarum Helvetiae inchoata* (Vol. 1). sumptibus Societatis typographicæ.

von Humboldt, A. (1819). *Voyage de Humboldt et Bonpland. Voyage aux régions équinoxiales du Nouveau continent, fait, en 1799, 1800, 1801, 1802, 1803, et 1804* (Vol. 2).

Von Ness, K. (2021). *Quantifying the effects of moisture change on the spectral signatures (350-2500 nm) of Arctic mosses, liverworts, and lichens*

Wagner, A. (2023). *Sleeping beauties: the mystery of dormant innovations in nature and culture*. Simon and Schuster.

Wahab, A., Muhammad, M., Munir, A., Abdi, G., Zaman, W., Ayaz, A., Khizar, C., & Reddy, S. P. P. (2023). Role of arbuscular mycorrhizal fungi in regulating growth, enhancing productivity, and potentially influencing ecosystems under abiotic and biotic stresses. *Plants, 12*(17), 3102. DOI: 10.3390/plants12173102 PMID: 37687353

Waheed, M., Haq, S. M., Arshad, F., Bussmann, R. W., Iqbal, M., Bukhari, N. A., & Hatamleh, A. A. (2022). Grasses in semi-arid lowlands—Community composition and spatial dynamics with special regard to the influence of edaphic factors. *Sustainability (Basel), 14*(22), 14964. DOI: 10.3390/su142214964

Walker, M. D., Walker, D., Theodose, T., Weber, P., Bowman, W., & Seastedt, T. (2001). *The vegetation hierarchical species-environment. Structure and function of an alpine ecosystem: Niwot Ridge, Colorado*. Oxford University Press.

Wallace, L., Lucieer, A., Malenovský, Z., Turner, D., & Vopěnka, P. J. F. (2016b). Assessment of forest structure using two UAV techniques: A comparison of airborne laser scanning and structure from motion (SfM) point clouds. *7*(3), 62.

Wallace, L., Lucieer, A., Malenovský, Z., Turner, D., & Vopěnka, P. (2016a). Assessment of forest structure using two UAV techniques: A comparison of airborne laser scanning and structure from motion (SfM) point clouds. *Forests, 7*(3), 62. DOI: 10.3390/f7030062

Wallner, M. (2022). Future effects of glacier retreat on downstream runoff and hydropower generation in Alpine regions of different catchment sizes/submitted by Mario Wallner, BSc.

Walther, G. R., Beißner, S., & Burga, C. A. (2005). Trends in the upward shift of alpine plants. *Journal of Vegetation Science, 16*(5), 541–548. DOI: 10.1111/j.1654-1103.2005.tb02394.x

Walther, G.-R., Post, E., Convey, P., Menzel, A., Parmesan, C., Beebee, T. J. C., Fromentin, J.-M., Hoegh-Guldberg, O., & Bairlein, F. (2002). Ecological responses to recent climate change. *Nature, 416*(6879), 389–395. DOI: 10.1038/416389a PMID: 11919621

Wang, S., Guo, L., He, B., & Li, T. (2020). The stability of Qinghai-Tibet Plateau ecosystem to climate change. *Physics and Chemistry of the Earth, Parts a/b/c, 115*, 102827.

Wang, X., Liu, S., Zuo, H., Zheng, W., Zhang, S., Huang, Y., Pingcuo, G., Ying, H., Zhao, F., & Li, Y. (2021). Genomic basis of high-altitude adaptation in Tibetan Prunus fruit trees. *Current Biology, 31*(17), 3848-3860. e3848.

Wang, F., Han, L., Liu, L., Bai, C., Ao, J., Hu, H., Li, R., Li, X., Guo, X., & Wei, Y. (2024). Advancements and perspective in the quantitative assessment of soil salinity utilizing remote sensing and machine learning algorithms: A review. *Remote Sensing (Basel), 16*(24), 4812. DOI: 10.3390/rs16244812

Wang, H., Li, J., Zhang, Q., Liu, J., Yi, B., Li, Y., Wang, J., & Di, H. (2019). Grazing and enclosure alter the vertical distribution of organic nitrogen pools and bacterial communities in semiarid grassland soils. *Plant and Soil, 439*(1), 525–539. DOI: 10.1007/s11104-019-04045-6

Wang, J., He, Q., Wu, Y., Zhu, H., Sun, H., Zhou, J., Wang, D., Li, J., & Bing, H. (2021). Effects of pioneer N_2-fixing plants on the resource status and establishment of neighboring non-N_2-fixing plants in a newly formed glacier floodplain, eastern Tibetan Plateau. *Plant and Soil, 458*(1), 261–276. DOI: 10.1007/s11104-020-04462-y

Wang, J., Li, M., Yu, C., & Fu, G. (2022). The change in environmental variables linked to climate change has a stronger effect on aboveground net primary productivity than does phenological change in alpine grasslands. *Frontiers in Plant Science, 12*, 798633. DOI: 10.3389/fpls.2021.798633 PMID: 35058958

Wang, J., Luo, P., Yang, H., Mou, C., & Mo, L. (2016). Different responses of alpine plants to nitrogen addition: Effects on plant–plant interactions. *Scientific Reports, 6*(1), 38320. DOI: 10.1038/srep38320 PMID: 27922131

Wang, J., Peng, J., Li, H., Yin, C., Liu, W., Wang, T., & Zhang, H. (2021). Soil salinity mapping using machine learning algorithms with the Sentinel-2 MSI in arid areas, China. *Remote Sensing (Basel), 13*(2), 305. DOI: 10.3390/rs13020305

Wang, J., Soininen, J., & Heino, J. (2021). Ecological indicators for aquatic biodiversity, ecosystem functions, human activities and climate change. *Ecological Indicators, 132*, 108250. DOI: 10.1016/j.ecolind.2021.108250

Wang, Q., Liu, H., Gao, A., Yang, X., Liu, W., Li, X., & Li, L. (2012). Intergenomic rearrangements after polyploidization of Kengyilia thoroldiana (Poaceae: Triticeae) affected by environmental factors. *PLoS One*, *7*(2), e31033. DOI: 10.1371/journal.pone.0031033 PMID: 22363542

Wang, W., Shu, X., Zhang, Q., & Guénon, R. (2015). Effects of freeze–thaw cycles on the soil nutrient balances, infiltration, and stability of cyanobacterial soil crusts in northern China. *Plant and Soil, 386*(1), 263–272. DOI: 10.1007/s11104-014-2263-x

Wang, X. M., Zhao, G. Q., Chen, J., Jing, F. Z., Li, L., Wang, Y. Y., Ma, P., Wu, Y.-H., Xu, S.-J., & He, W. L. (2025). Adaptation of plants to UV-B radiation with altitude in Tuha Basin: Synergistic regulation of epidermal structure, secondary metabolites, and organic element allocation. *Life (Chicago, Ill.)*, *15*(9), 1375. DOI: 10.3390/life15091375 PMID: 41010317

Wang, X., Dong, S., Yang, B., Li, Y., & Su, X. (2014). The effects of grassland degradation on plant diversity, primary productivity, and soil fertility in the alpine region of Asia's headwaters. *Environmental Monitoring and Assessment*, *186*(10), 6903–6917. DOI: 10.1007/s10661-014-3898-z PMID: 25023744

Wang, X., Li, Y., Gong, Y., Liu, Y., Zhao, J., & Li, K. (2025). Spatiotemporal dynamics and driving mechanism of aboveground biomass across three Alpine grasslands in Central Asia over the past 20 years using three algorithms. *Remote Sensing (Basel)*, *17*(3), 538. DOI: 10.3390/rs17030538

Wang, Y., Dang, N., Feng, K., Wang, J., Jin, X., Yao, S., Wang, L., Gu, S., Zheng, H., Lu, G., & Deng, Y. (2023). Grass-microbial inter-domain ecological networks associated with alpine grassland productivity. *Frontiers in Microbiology*, *14*, 1109128. DOI: 10.3389/fmicb.2023.1109128 PMID: 36760496

Wang, Y., Liu, Y., Ding, J., Chen, P., Zhan, T., Yao, Y., Song, J., & Fu, B. (2024). Impact of functional groups on aboveground biomass in alpine grassland communities. *Progress in Physical Geography*, *48*(5–6), 698–715. DOI: 10.1177/03091333241280095

Wang, Y., Xiao, J., Ma, Y., Ding, J., Chen, X., Ding, Z., & Luo, Y. (2023). Persistent and enhanced carbon sequestration capacity of alpine grasslands on Earth's Third Pole [In.]. *Science Advances*, *9*(20), eade6875. DOI: 10.1126/sciadv.ade6875 PMID: 37196073

Wang, Z., Kang, Y., Wang, Y., Tan, Y., Yao, B., An, K., & Su, J. (2023). Himalayan marmot (Marmota himalayana) redistribution to high latitudes under climate change. *Animals (Basel)*, *13*(17), 2736. DOI: 10.3390/ani13172736 PMID: 37684999

Wang, Z., Meng, Q., Allam, M., Hu, D., Zhang, L., & Menenti, M. (2021). Environmental and anthropogenic drivers of surface urban heat island intensity: A case-study in the Yangtze River Delta, China. *Ecological Indicators*, *128*, 107845. DOI: 10.1016/j.ecolind.2021.107845

Wani, I. A., Verma, S., Gupta, R., Ganaie, M. M., Nigam, G., Shafik, H. M., & Al-Misned, F. A. J. S. (2022). Nutrient analysis and species diversity of alpine grasslands: a comparative analysis of less studied biodiversity hotspots. *14*(2), 887.

Wan, L., Liu, G., & Su, X. (2025). Organic fertilization balances biodiversity maintenance, grass production, soil storage, nutrient cycling and greenhouse gas emissions for sustainable grassland development in China: A meta-analysis. *Agriculture, Ecosystems & Environment*, *381*, 109473. DOI: 10.1016/j.agee.2025.109473

Warren, R., Price, J., Graham, E., Forstenhaeusler, N., & VanDerWal, J. (2018). The projected effect on insects, vertebrates, and plants of limiting global warming to 1.5 C rather than 2 C. *Science*, *360*(6390), 791–795. DOI: 10.1126/science.aar3646 PMID: 29773751

Waterman, T. H. (2001). Evolutionary challenges of extreme environments (part 2). *The Journal of Experimental Zoology*, *291*(2), 130–168. DOI: 10.1002/jez.1065 PMID: 11479914

Watts, S. H., Mardon, D. K., Mercer, C., Watson, D., Cole, H., Shaw, R. F., & Jump, A. S. (2022). Riding the elevator to extinction: Disjunct arctic-alpine plants of open habitats decline as their more competitive neighbours expand. *Biological Conservation*, *272*, 109620. DOI: 10.1016/j.biocon.2022.109620

Webb, J. W., Khoury, T. A., & Hitt, M. A. (2020). The influence of formal and informal institutional voids on entrepreneurship. *Entrepreneurship Theory and Practice*, *44*(3), 504–526. DOI: 10.1177/1042258719830310

Weber, W. A. (2003). The Middle Asian element in the southern Rocky Mountain flora of the western United States: A critical biogeographical review. *Journal of Biogeography*, *30*(5), 649–685. DOI: 10.1046/j.1365-2699.2003.00864.x

Wei, J.-D., & Wang, W.-T. (2025). Vulnerability Assessment of Six Endemic Tibetan-Himalayan Plants Under Climate Change and Human Activities. *Plants*, *14*(15), 2424. DOI: 10.3390/plants14152424 PMID: 40805772

Wei, M., Sun, Q., & Liu, D. (2025). Moss Cover Modulates Soil Fungal Functional Communities and Nutrient Cycling in Alpine Forests. *Forests*, *16*(1), 138. DOI: 10.3390/f16010138

Werner, M., Weidinger, J., Böhner, J., Schickhoff, U., & Bobrowski, M. (2024). Instagram data for validating Nothofagus pumilio distribution mapping in the Southern Andes: A novel ground truthing approach. *Frontiers of Biogeography*, *17*, e140606. DOI: 10.21425/fob.17.140606

White, G. A. (2023). The Optimization of Tissue Culture Conditions for the Production of Cultivated Meat.

Whitesides, C. J., & Butler, D. R. (2010). Adequacies and deficiencies of alpine and subalpine treeline studies in the national parks of the western USA. *Progress in Physical Geography*, *35*(1), 19–42. DOI: 10.1177/0309133310384642

Wikström, S. A., Gubri, B., Asplund, M. E., Dahl, M., Gullström, M., Hansen, J. P., Kumblad, L., Rydin, E., Garbaras, A., & Björk, M. (2025). Influence of landscape characteristics and submerged aquatic vegetation on sediment carbon and nitrogen storage in shallow brackish water habitats. *Scientific Reports*, *15*(1), 7808. DOI: 10.1038/s41598-025-92217-z PMID: 40050425

Willis, C. G., Ellwood, E. R., Primack, R. B., Davis, C. C., Pearson, K. D., Gallinat, A. S., Yost, J. M., Nelson, G., Mazer, S. J., Rossington, N. L., Sparks, T. H., & Soltis, P. S. (2017). Old plants, new tricks: Phenological research using herbarium specimens. *Trends in Ecology & Evolution*, *32*(7), 531–546. DOI: 10.1016/j.tree.2017.03.015 PMID: 28465044

Wilson, B. R., Tulau, M., Kuginis, L., McInnes-Clarke, S., Grover, S., Milford, H., & Jenkins, B. R. (2022). Distribution, nature and threats to soils of the Australian Alps: A review. *Austral Ecology*, *47*(2), 166–188. DOI: 10.1111/aec.13115

Winkler, M., Lamprecht, A., Steinbauer, K., Hülber, K., Theurillat, J.-P., Breiner, F., Choler, P., Ertl, S., Gutiérrez Girón, A., Rossi, G., Vittoz, P., Akhalkatsi, M., Bay, C., Benito Alonso, J.-L., Bergström, T., Carranza, M. L., Corcket, E., Dick, J., Erschbamer, B., & Pauli, H. (2016). The rich sides of mountain summits – a pan-European view on aspect preferences of alpine plants. *Journal of Biogeography*, *43*(11), 2261–2273. DOI: 10.1111/jbi.12835

Wipf, S., Gottfried, M., & Nagy, L. (2013). Climate change and extreme events – Their impacts on alpine and arctic ecosystem structure and function. *Plant Ecology & Diversity*, *6*(3–4), 303–306. DOI: 10.1080/17550874.2013.819533

Wipf, S., Sommerkorn, M., Stutter, M. I., Wubs, E. J., & Van der Wal, R. (2015). Snow cover, freeze–thaw, and the retention of nutrients in an oceanic mountain ecosystem. *Ecosphere*, *6*(10), 1–16. DOI: 10.1890/ES15-00099.1

Wisz, M. S., Pottier, J., Kissling, W. D., Pellissier, L., Lenoir, J., Damgaard, C. F., Dormann, C. F., Forchhammer, M. C., Grytnes, J. A., Guisan, A., Heikkinen, R. K., Høye, T. T., Kühn, I., Luoto, M., Maiorano, L., Nilsson, M.-C., Normand, S., Öckinger, E., Schmidt, N. M., & Svenning, J.-C. (2013). The role of biotic interactions in shaping distributions and realised assemblages of species: Implications for species distribution modelling. *Biological Reviews of the Cambridge Philosophical Society*, *88*(1), 15–30. DOI: 10.1111/j.1469-185X.2012.00235.x PMID: 22686347

Wootton, L. M., Boucher, F. C., Pouchon, C., Roquet, C., Coissac, E., Renaud, J., Alsos, I. G., Valla, P. G., Husson, L., Bernet, M., Perrier, C., Douzet, R., Rome, M., Valay, J.-G., Alberti, A., Denoeud, F., Zimmermann, N. E., Wincker, P., Thuiller, W., & Lavergne, S. (2025). The late rise of sky-island vegetation in the European Alps. *Nature Plants*, *11*(6), 1–12. DOI: 10.1038/s41477-025-02001-0 PMID: 40457047

Wu, G., Xie, Y., Wei, J., & Yue, X. (2023). Freeze-thaw erosion mechanism and preventive actions of highway subgrade soil in an alpine meadow on the Qinghai-Tibet Plateau. *Engineering Failure Analysis*, *143*, 106933. DOI: 10.1016/j.engfailanal.2022.106933

Wu, H., Wu, N., Ling, Q., Tang, T., Hu, P., Duan, P., Zhang, Q., Xiao, J., Zhao, J., Zhang, W., Chen, H., & Wang, K. (2025). Potential roles of soil viruses in Karst forest soil carbon and nitrogen cycles. *Forests*, *16*(5), 735. DOI: 10.3390/f16050735

Wu, J., Li, M., Zhang, X., Fiedler, S., Gao, Q., Zhou, Y., Cao, W., Hassan, W., Mărgărint, M. C., Tarolli, P., & Tietjen, B. (2021). Disentangling climatic and anthropogenic contributions to nonlinear dynamics of alpine grassland productivity on the Qinghai-Tibetan Plateau. *Journal of Environmental Management*, *281*, 111875. DOI: 10.1016/j.jenvman.2020.111875 PMID: 33378737

Wu, L., Song, Z., Yang, X., Zhang, X., Hao, Q., Xia, S., & Zhao, X. (2022). Variation of soil silicon fractions and bioavailability during centennial-scale evolution of ecosystem after glacier retreat. *Soil Science Society of America Journal*, *86*(3), 552–565. DOI: 10.1002/saj2.20386

Wu, Q., Li, H., Wang, R., Paulussen, J., He, Y., Wang, M., Wang, B., & Wang, Z. (2006). Monitoring and predicting land use change in Beijing using remote sensing and GIS. *Landscape and Urban Planning*, *78*(4), 322–333. DOI: 10.1016/j.landurbplan.2005.10.002

Wu, S.-P., & Chen, Z.-S. (2005). Characteristics and genesis of Inceptisols with placic horizons in the subalpine forest soils of Taiwan. *Geoderma, 125*(3), 331–341. DOI: 10.1016/j.geoderma.2004.09.007

Xiao, L., Zhang, W., Hu, P., Vesterdal, L., Zhao, J., Tang, L., Xiao, D., & Wang, K. (2023). Mosses stimulate soil carbon and nitrogen accumulation during vegetation restoration in a humid subtropical area. *Soil Biology & Biochemistry, 184*, 109127. DOI: 10.1016/j.soilbio.2023.109127

Xie, H., Wu, Q., Hu, J., Yu, L., Bie, P., Wang, H., & Deng, D. (2018). Changes in soil physical and chemical properties during the process of alpine meadow degradation along the Eastern Qinghai-Tibet Plateau. *Eurasian Soil Science, 51*(12), 1440–1446. DOI: 10.1134/S1064229318130045

Xing, Y., Wang, X., & Mustafa, A. (2025). Exploring the link between soil health and crop productivity. *Ecotoxicology and Environmental Safety, 289*, 117703. DOI: 10.1016/j.ecoenv.2025.117703 PMID: 39808880

Xu, X., Wang, W., Ke, L., & Zhou, P. (2025). Weakening of regional contrasts in glacier changes around the Tarim Basin in the early 21st century. *npj Climate and Atmospheric Science, 8*(1), 289.

Xue, J., & Su, B. (2017). Significant remote sensing vegetation indices: A review of developments and applications. *Journal of Sensors, 2017*(1), 1353691. DOI: 10.1155/2017/1353691

Xu, H. P., Zhang, J., Pang, X. P., Wang, Q., Zhang, W. N., Wang, J., & Guo, Z. G. (2019). Responses of plant productivity and soil nutrient concentrations to different alpine grassland degradation levels. *Environmental Monitoring and Assessment, 191*(11), 678. DOI: 10.1007/s10661-019-7877-2 PMID: 31654145

Xu, H., Liu, N., Yang, G., Zhang, H., Badgery, W. B., & Zhang, Y. (2025). Snow interacts with defoliation height to drive grassland sustainability via grass biomass maintenance. *Rangeland Ecology and Management, 98*, 332–343. DOI: 10.1016/j.rama.2024.09.003

Xu, X., Ma, K., Fu, B.-J., Liu, X., Huang, Y., & Qi, J. (2006). Research review of the relationship between vegetation and soil loss. *Acta Ecologica Sinica, 26*, 3137–3143.

Xu, Y., Cui, K., Zhang, X., Diwu, G., Zhu, Y., Deng, L., Zhong, Y., & Yan, W. (2025). Shifts in fungal communities drive soil profile nutrient cycling during grassland restoration. *mBio, 16*(3), e02834–e24. DOI: 10.1128/mbio.02834-24 PMID: 39853100

Yan, B., Sun, Y., & Wei, Y. (2020). Potassium–calcium antagonistic interaction under tomato magnesium deficiency and magnesium fertiliser regulation in solar greenhouse. *Quality Assurance and Safety of Crops & Foods, 12*(3), 76–86. DOI: 10.15586/qas.v12i3.723

Yandi, S., Tao, M., Huakun, Z., Honglin, L., Zhonghua, Z., Li, M., Ruimin, Q., Hongye, S., Tao, C., Jingjing, W., & Xue, H. (2023). Characterization of the Plant–Soil feedback index in alpine meadow degradation and recovery: A field experiment. *Frontiers in Environmental Science, 10*, 1097030. https://www.frontiersin.org/journals/environmental-science/articles/10.3389/fenvs.2022.1097030. DOI: 10.3389/fenvs.2022.1097030

Yang, C., Dong, Y., Friman, V. P., Jousset, A., Wei, Z., Xu, Y., & Shen, Q. (2019). Carbon resource richness shapes bacterial competitive interactions by alleviating growth–antibiosis trade-off. *Functional Ecology, 33*(5), 868–875. DOI: 10.1111/1365-2435.13292

Yang, C., Qi, Y., Guo, J., Peng, L., Xiong, N., Zhang, W., & Zhao, W. (2025). Habitat fragmentation increases the risk of local extinction of small reptiles: A case study on Phrynocephalus przewalskii. *Ecotoxicology and Environmental Safety, 290*, 117717. DOI: 10.1016/j.ecoenv.2025.117717 PMID: 39823673

Yang, C., & Sun, J. (2021). Impact of soil degradation on plant communities in an overgrazed Tibetan alpine meadow. *Journal of Arid Environments, 193*, 104586. DOI: 10.1016/j.jaridenv.2021.104586

Yang, C., Xu, H., Li, Q., Wang, X., Tang, B., Chen, J., & Chen, M. (2025). Global loss of mountain vegetated landscapes and its impact on biodiversity conservation. *Nature Communications, 16*(1), 8971. DOI: 10.1038/s41467-025-64449-0 PMID: 41068100

Yang, F.-S., Liu, M., Guo, X., Xu, C., Jiang, J., Mu, W., Fang, D., Xu, Y.-C., Zhang, F.-M., Wang, Y.-H., Yang, T., Chen, H., Sahu, S. K., Li, R., Wang, G., Wang, Q., Xu, X., Ge, S., Liu, H., & Guo, Y.-L. (2024). Signatures of Adaptation and Purifying Selection in Highland Populations of Dasiphora fruticosa. *Molecular Biology and Evolution, 41*(6), msae099. Advance online publication. DOI: 10.1093/molbev/msae099 PMID: 38768215

Yang, H., Gou, X., Xue, B., Ma, W., Kuang, W., Tu, Z., Gao, L., Yin, D., & Zhang, J. (2023). Research on the change of alpine ecosystem service value and its sustainable development path. *Ecological Indicators, 146*, 109893. DOI: 10.1016/j.ecolind.2023.109893

Yang, Q., Bi, H., Yang, W., Li, T., Jiang, J., Zhang, L., Liu, J., & Hu, Q. (2020). The genome sequence of alpine Megacarpaea delavayi identifies species-specific whole-genome duplication. *Frontiers in Genetics*, *11*, 812. DOI: 10.3389/fgene.2020.00812 PMID: 32849811

Yang, Y., You, Q., Zuo, Z., Zhang, Y., Liu, Z., Kang, S., & Zhai, P. (2023). Elevation dependency of temperature trend over the Qinghai-Tibetan Plateau during 1901–2015. *Atmospheric Research*, *290*, 106791. DOI: 10.1016/j.atmosres.2023.106791

Yang, Z., Shen, Y., Jiang, H., Feng, F., & Dong, Q. (2021). Assessment of the environmental changes in arid and semiarid mining areas using long time-series Landsat images. *Environmental Science and Pollution Research International*, *28*(37), 52147–52156. DOI: 10.1007/s11356-021-14417-0 PMID: 33997931

Yao, H., Qin, R., & Chen, X. (2019). Unmanned aerial vehicle for remote sensing applications—A review. *Remote Sensing (Basel)*, *11*(12), 1443. DOI: 10.3390/rs11121443

Yaselyani, M., Bagheri, A., Saeidi, H., & Noroozi, J. (2024). Vascular plant diversity of the high mountains of Fereydunshahr, Central Zagros, Iran. *Vegetation Classification and Survey*, *5*, 329–346. DOI: 10.3897/VCS.110546

Yasir, M., Hossain, A., & Pratap-Singh, A. (2025). Pesticide degradation: Impacts on soil fertility and nutrient cycling. *Environments (Basel, Switzerland)*, *12*(8), 272. DOI: 10.3390/environments12080272

Ye, K., Li, J., Liang, Z., Zhao, X., Luo, H., Chen, Z., Du, P., Wu, K., Qiao, R., & Hu, J. (2025a). Overgrazing impacts plant species diversity in alpine wetlands indirectly by altering its environmental dependency. *Global Ecology and Conservation*, *59*, e03532. DOI: 10.1016/j.gecco.2025.e03532

Young, T. P., & Peacock, M. M. (1992). Giant Senecios and Alpine Vegetation of Mount Kenya. *Journal of Ecology*, *80*(1), 141–148. DOI: 10.2307/2261071

Yu, L., Chen, Y., Sun, W., & Huang, Y. (2019). Effects of grazing exclusion on soil carbon dynamics in alpine grasslands of the Tibetan Plateau. *Geoderma*, *353*, 133–143. DOI: 10.1016/j.geoderma.2019.06.036

Yusefi, G. H., Faizolahi, K., Darvish, J., Safi, K., & Brito, J. C. (2019). The species diversity, distribution, and conservation status of the terrestrial mammals of Iran. *Journal of Mammalogy*, *100*(1), 55–71. DOI: 10.1093/jmammal/gyz002

Zaman, A., Asadullah, L. B., Muhammad, Z., Razzaq, A., Jelani, G., Ali, U., Raees, N., & Khan, M. N. (2019). Floristics of weeds in Triticum aestivum L. fields of Tehsil Shabqadar, District Charsadda, KP, Pakistan. *International Journal of Botany Studies*, *4*(5), 37–44.

Zambrano, J., Garzon-Lopez, C. X., Yeager, L., Fortunel, C., Cordeiro, N. J., & Beckman, N. G. (2019). The effects of habitat loss and fragmentation on plant functional traits and functional diversity: What do we know so far? *Oecologia*, *191*(3), 505–518. DOI: 10.1007/s00442-019-04505-x PMID: 31515618

Zaqout, T., & Andradóttir, H. Ó. (2021). Hydrologic performance of grass swales in cold maritime climates: Impacts of frost, rain-on-snow and snow cover on flow and volume reduction. *Journal of Hydrology (Amsterdam)*, *597*, 126159. DOI: 10.1016/j.jhydrol.2021.126159

Zargar, S. A., Malik, R. A., Khuroo, A. A., Ganie, A. H., & Reshi, Z. A. (2025). Alien flora of the Himalayan highlands: Naturalised and invasive plants in the Trans-Himalayan region of Ladakh, India. *Biological Invasions*, *27*(4), 1–19. DOI: 10.1007/s10530-025-03562-z

Zavala, L. M., Jordán, A., Bellinfante, N., & Gil, J. (2010). Relationships between rock fragment cover and soil hydrological response in a Mediterranean environment. *Soil Science and Plant Nutrition*, *56*(1), 95–104. DOI: 10.1111/j.1747-0765.2009.00429.x

Zehzad, B., Kiabi, B. H., & Madjnoonian, H. (2002). The natural areas and landscape of Iran: An overview. *Zoology in the Middle East*, *26*(1), 7–10. DOI: 10.1080/09397140.2002.10637915

Zeng, M. (2024). The mutual effect of nutrients on plant–herbivore interactions. *Plant Ecology*, *225*(10), 1035–1045. DOI: 10.1007/s11258-024-01452-3

Zerbe, S. (2023). Subalpine and alpine grassland. In *Restoration of ecosystems–bridging nature and humans: A transdisciplinary approach* (pp. 185–208). Springer. DOI: 10.1007/978-3-662-65658-7_9

Zhan, C. (2024). Microbial decomposition and soil health: mechanisms and ecological implications. *Molecular Soil Biology, 15*.

Zhang, Q., Yao, J., Degen, A. A., & Zhou, H. (2025). Alpine meadow degradation: status, reasons, and trends. In *Grassland Degradation, Restoration and Sustainable Management of Global Alpine Area* (pp. 69-86). Elsevier. DOI: 10.1016/B978-0-443-21882-8.00004-4

Zhang, E., Wang, Y., Crowther, T. W., Sun, W., Chen, S., Zhou, D., Shangguan, Z., Huang, J., He, J.-S., Wang, Y., Sheng, J., Tang, L., Li, X., Dong, M., Wu, Y., Hu, S., Bai, Y., & Yu, G. (2025). Mycorrhiza increases plant diversity and soil carbon storage in grasslands. *Proceedings of the National Academy of Sciences of the United States of America, 122*(7), e2412556122. DOI: 10.1073/pnas.2412556122 PMID: 39937867

Zhang, H., Yang, X., Zhu, L., Ahmed, N., & Gao, Y. (2025). Laboratory investigation on the effects of root-secreted organic acids on carbon and nitrogen mineralization in alpine soils caused by invasive shrubs. *Journal of Environmental Chemical Engineering, 13*(4), 117271. DOI: 10.1016/j.jece.2025.117271

Zhang, H., Zhang, P., Niu, Y., Tao, T., Liu, G., Dong, C., Zheng, Z., Zhang, Z., Li, Y., Niu, Z., Liu, W., Guo, Z., Hu, S., Yang, Y., Li, M., Sun, H., Renner, S. S., & Liu, J. (2025). Genetic basis of camouflage in an alpine plant and its long-term co-evolution with an insect herbivore. *Nature Ecology & Evolution, 9*(4), 1–11. DOI: 10.1038/s41559-025-02653-3 PMID: 40065027

Zhang, J., Tian, Y., Yan, L., Zhang, G., Wang, X., Zeng, Y., Zhang, J., Ma, X., Tan, Y., Long, N., Wang, Y., Ma, Y., He, Y., Xue, Y., Hao, S., Yang, S., Wang, W., Zhang, L., Dong, Y., & Sheng, J. (2016). Genome of plant maca (Lepidium meyenii) illuminates genomic basis for high-altitude adaptation in the central Andes. *Molecular Plant, 9*(7), 1066–1077. DOI: 10.1016/j.molp.2016.04.016 PMID: 27174404

Zhang, J., Zhu, C., Xie, L., Wu, S., Cao, C., Wang, M., & Cui, S. (2025). Comparative study on landslide susceptibility assessment of different models: A case study of alpine mountainous region in Xinjiang. *Bulletin of Engineering Geology and the Environment, 84*(2), 95. DOI: 10.1007/s10064-025-04105-5

Zhang, M. X., Hu, J. P., Li, J. L., Che, Z., Li, L., Lü, Z. L., Dong, W.-Q., Zhang, J.-Q., Yao, T., Duan, T.-Y., & Zhang, J. L. (2025). Responses of legume-associated rhizobacterial communities to plant diversity and soil traits in alpine grassland. *Land Degradation & Development, 36*(2), 441–453. DOI: 10.1002/ldr.5369

Zhang, Q., Zhou, Q., Liu, F., Ma, W., Chen, Q., Wei, B., Li, L., & Zhi, Z. (2025). Progressive Shrinkage of the Alpine Periglacial Weathering Zone and Its Escalating Disaster Risks in the Gongga Mountains over the Past Four Decades. *Remote Sensing (Basel), 17*(14), 2462. DOI: 10.3390/rs17142462

Zhang, W. (2023). Influence of alpine meadow deterioration on soil microbial communities in the Yangtze River source region. *Frontiers in Environmental Science, 11*, 1210349. DOI: 10.3389/fenvs.2023.1210349

Zhang, W., Ren, C., Deng, J., Zhao, F., Yang, G., Han, X., Tong, X., & Feng, Y. (2018). Plant functional composition and species diversity affect soil C, N, and P during secondary succession of abandoned farmland on the Loess Plateau. *Ecological Engineering*, *122*, 91–99. DOI: 10.1016/j.ecoleng.2018.07.031

Zhang, X., Deng, T., Wang, H., & Sun, H. (2025). Genomic insights into alpine plant adaptation. *Plant Diversity*. Advance online publication. DOI: 10.1016/j.pld.2025.12.012

Zhang, X., Kuang, T., Dong, W., Qian, Z., Zhang, H., Landis, J. B., Feng, T., Li, L., Sun, Y., Huang, J., Deng, T., Wang, H., & Sun, H. (2023). Genomic convergence underlying high-altitude adaptation in alpine plants. *Journal of Integrative Plant Biology*, *65*(7), 1620–1635. DOI: 10.1111/jipb.13485 PMID: 36960823

Zhang, X., Sun, Y., Landis, J. B., Shen, J., Zhang, H., Kuang, T., Sun, W., Sun, J., Tiamiyu, B. B., Deng, T., Sun, H., & Wang, H. (2021). Transcriptomes of Saussurea (Asteraceae) provide insights into high-altitude adaptation. *Plants*, *10*(8), 1715. DOI: 10.3390/plants10081715 PMID: 34451759

Zhang, X., Zhang, C., Liu, Y., Huo, L., Yang, Z., Tong, Y., Zhang, X., Yu, Z., Yang, X., Cao, Q., & Dong, Q. (2025). Effects of planting patterns on physicochemical properties, metabolites and microbial community structure of rhizosphere soil in perennial cultivated grassland. *Scientific Reports*, *15*(1), 12047. DOI: 10.1038/s41598-025-94366-7 PMID: 40199986

Zhang, Y., Chen, J., & Sun, H. (2021). Alpine speciation and morphological innovations: Revelations from a species-rich genus in the northern hemisphere. *AoB Plants*, *13*(3), plab018. DOI: 10.1093/aobpla/plab018 PMID: 34025962

Zhang, Y., Lu, Y., & Song, X. (2023). Identifying the Main Factors Influencing Significant Global Vegetation Changes. *Forests*, *14*(8), 1607. DOI: 10.3390/f14081607

Zhang, Z., Zhao, Y., Lin, H., Li, Y., Fu, J., Wang, Y., Sun, J., & Zhao, Y. (2023). Comprehensive analysis of grazing intensity impacts alpine grasslands across the Qinghai-Tibetan Plateau: A meta-analysis. *Frontiers in Plant Science*, *13*, 1083709. DOI: 10.3389/fpls.2022.1083709 PMID: 36733589

Zhao, C., & Zhang, X. (2024). The Role of Polyploidy in the Genetic Structure and Expansion of Lepisorus clathratus in the Qinghai–Tibetan Plateau and Hengduan Mountains. *Plants*, *13*(22), 3181. https://www.mdpi.com/2223-7747/13/22/3181. DOI: 10.3390/plants13223181 PMID: 39599389

Zhao, P., Li, L., Lin, L., Zhai, G., Cruse, R. M., & Wang, E. (2022). Response of surface soil nutrients and organic carbon fractions to tillage erosion vs. water erosion in an agricultural landscape. *Soil Science Society of America Journal, 86*(6), 1470–1482. DOI: 10.1002/saj2.20461

Zhao, R. (2023). The influence of cushion plants on soil fertility islands under their canopies with ontogenesis in an extreme alpine ecosystem. *Plant and Soil, 484*(1), 379–392. DOI: 10.1007/s11104-022-05798-3

Zhao, Y. F., Wang, X., Jiang, S. L., Zhou, X. H., Liu, H. Y., Xiao, J. J., Hao, Z. G., & Wang, K. C. (2021). Climate and geochemistry interactions at different altitudes influence soil organic carbon turnover times in alpine grasslands. *Agriculture, Ecosystems & Environment, 320*, 107591. DOI: 10.1016/j.agee.2021.107591

Zhou, H., Yang, X., Zhou, C., Shao, X., Shi, Z., Li, H., Su, H., Qin, R., Chang, T., Hu, X., Yuan, F., Li, S., Zhang, Z., & Ma, L. (2023). Alpine grassland degradation and its restoration in the Qinghai–Tibet plateau. *Grasses, 2*(1), 31–46. DOI: 10.3390/grasses2010004

Zhou, T., Sun, J., & Shi, P. (2024). Mechanism of plant–soil feedback in a degraded alpine grassland on the Tibetan Plateau. *Journal of Plant Ecology, 17*(2), rtae025. Advance online publication. DOI: 10.1093/jpe/rtae025

Zhou, T., Yang, H., & Du, Y. (2025). Effects of grazing on alpine grassland soil available nutrients across the Tibetan Plateau. *Frontiers in Soil Science, 4*, 1525837. DOI: 10.3389/fsoil.2024.1525837

Zhou, W., Li, C., Wang, S., Ren, Z., & Stringer, L. C. (2023). Effects of vegetation restoration on soil properties and vegetation attributes in the arid and semi-arid regions of China. *Journal of Environmental Management, 343*, 118186. DOI: 10.1016/j.jenvman.2023.118186 PMID: 37224686

Zhou, X., Chen, X., Qi, X., Zeng, Y., Guo, X., Zhuang, G., & Ma, A. (2023). Soil bacterial communities associated with multi-nutrient cycling under long-term warming in the alpine meadow. *Frontiers in Microbiology, 14*, 1136187. DOI: 10.3389/fmicb.2023.1136187 PMID: 36910214

Zhou, Z., Su, P., Yang, J., Shi, R., & Ding, X. (2024). Warming affects leaf light use efficiency and functional traits in alpine plants: Evidence from a 4-year in-situ field experiment. *Frontiers in Plant Science, 15*, 1353762. DOI: 10.3389/fpls.2024.1353762 PMID: 38567127

Zhu, J., Zhang, Y., Wu, J., Zhang, X., Yu, G., Shen, Z., Yang, X., He, Y., Jiang, L., & Hautier, Y. (2024). Herbivore exclusion stabilizes alpine grassland biomass production across spatial scales. *Global Change Biology, 30*(1), e17155. DOI: 10.1111/gcb.17155 PMID: 38273528

Zhyvko, M. (2024). The cost of fragmentation: ecology and the impact on the global economy.

Ziegler, A. M. (1990). Phytogeographic patterns and continental configurations during the Permian Period. *Geological Society, London, Memoirs, 12*(1), 363-379.

Zimmermann, N., & Kienast, F. (1999). Predictive mapping of alpine grasslands in Switzerland: Species versus community approach. *Journal of Vegetation Science, 10*(4), 469–482. DOI: 10.2307/3237182

Zohary, M. (1973). *Geobotanical foundations of the Middle East.*

Zohary, D. (1973). The origin of cultivated cereals and pulses in the Near East. *Chromosomes Today, 4*, 307–320.

About the Contributors

Muhammad Nauman Khan holds a Ph.D. in Botany from Islamia College Peshawar, Pakistan, and is currently serving as an Assistant Professor at the University of Chakwal, Punjab, Pakistan. He is an accomplished researcher with more than 150 peer-reviewed international research articles and 40 book chapters published in reputed journals, achieving a cumulative impact factor of nearly 250. His academic activities include participation in international conferences through abstracts, oral presentations, and poster sessions. Dr. Khan is actively involved in scholarly publishing as a Subject Editor of Ethnobotany Research and Applications and an editorial board member of multiple journals. He has also served on organizing committees of major international conferences. Dr. Khan is a lifetime member of the Pakistan Botanical Society, the Weed Science Society of Pakistan, and the Khwarizmi Science Society.

***

Sara is a BS student of Radiology and Imaging Technology at Government College University, Faisalabad (GCUF), Pakistan. She completed her intermediate education from Superior College, Mianwali. She has a strong academic interest in medical imaging, diagnostic radiology, and healthcare sciences. Sara is motivated to enhance her knowledge through research, scientific writing, and academic activities, and aims to contribute positively to the field of radiology and medical sciences.

Nayab Ahmad has completed her MPhil from the Institute of Biotechnology and Genetic Engineering (IBGE) at The University of Agriculture, Peshawar, Pakistan. Her research interests focus on plant nanotechnology and its applications in improving plant growth, stress tolerance, and productivity. She has contributed to several research papers in the field of plant nanotechnology and related areas, reflecting her commitment to advancing innovative, sustainable solutions in agricultural science. She aims to continue her academic and research career by exploring novel nanotechnology-based approaches for crop improvement and environmental resilience.

Nouman Ahmad is an undergraduate Biotechnology student at the University of Mianwali, Punjab, Pakistan, and completed his intermediate education at PAEC Boys College Chashma. His main research interests are drug design and vaccine design using computational and molecular biotechnology approaches. He plans to expand his research into epigenetics and other areas of medical biotechnology in the future. Nouman is engaged in academic research writing and aims to pursue advanced studies and a research career in biomedical and molecular sciences.

Sidra Ahmad is a Visiting Lecturer at Khyber Medical University (KMU), Peshawar, Pakistan, a position she has held since 2020. She has previously worked as a Research Assistant at the University of Agriculture, Peshawar, where she gained valuable experience in academic research and project work. Through her teaching role, she is actively involved in guiding students and contributing to an engaging learning environment. Her academic interests include the application of scientific and evidence-based approaches to support development and improve community well-being. She has contributed to scholarly activities, including research and academic writing, and is committed to ongoing professional growth and meaningful participation in the academic community.

Amjad Ali did his Ph.D. in Agro-food system from University of Cattolica Sacred Heart, Piacenza, Italy. He is an accomplished researcher with more than 30 peer-reviewed international research articles and 5 book chapters published in reputed journals. His academic journey includes international experience in research as well as in teaching.

Hansa Gul is a researcher in the Department of Zoology at the University of Mianwali, Mianwali, Punjab, Pakistan. Affiliated with the University of Mianwali, she works at the interface of zoology and nanotechnology, exploring how modern nanoscale approaches can be applied to animal biology and related fields. Her academic and research interests broadly include animal science, innovative nanotechnological methods, and their potential applications in improving understanding of biological systems. Through her work, she aims to contribute to the development of interdisciplinary solutions addressing challenges in zoological research and environmental and biological sciences more generally.

Gul Nawaz holds an MS in Botany from the Institute of Botany, University of the Punjab and a BS in Botany from the University of Education, Lahore. He is currently working as a Biodiversity Specialist at SAMA^Verte, a knowledge-sharing partner organization for the international Better Cotton`s project. His research expertise lies in ecology, biodiversity conservation, endemism, plant taxonomy, flora and vegetation studies, ethnobotany, phytogeography and ecophysiology. He has authored a book and book chapter, published research papers, with several more in progress. A passionate and dedicated early-career researcher, Nawaz has completed over 80 e-learning courses through platforms such as Coursera, IUCN, GBCI and the United Nations, further strengthening his knowledge in his field. He is an active member of the Society for Ecological Restoration (America) and serves as a student ambassador for Oxford University's Nature Positive Universities initiative and the International Association of Vegetation Sciences. Beyond research, he enjoys exploring nature, engaging in critical thinking and developing innovative approaches to scientific inquiry.

Hafeeza Amna Saleem is a dedicated plant taxonomist with a Ph.D. in Botany from the University of the Punjab. She has extensive experience in plant identification, classification, and conservation, with a strong focus on field studies and botanical data analysis. Hafeeza has taught various courses in plant taxonomy, systematics, and cell biology at multiple universities, including the University of the Punjab and the University of Education. She is also an active peer reviewer for scientific journals and has presented research at several national and international conferences. Her research includes morphological and phylogenetic studies on Morus species, contributing significantly to botanical knowledge and biodiversity conservation.

Syed Nasar Shah is currently working as a Senior Science Teacher (SST SCI 1) at Government Higher Secondary School, Olandar, District Shangla, Khyber Pakhtunkhwa, Pakistan. He completed his undergraduate studies in Botany with honours from the University of Peshawar in 2015 and earned an M.Phil. in Plant Systematics and Biodiversity from Quaid-i-Azam University, Islamabad, in 2017. His research interests include plant systematics, taxonomy, and ecology, with a particular focus on pteridophytes, desiccation-tolerant ferns, angiosperm taxonomy, medicinal plants, and ethnobotany. He has published more than 40 research articles, four book chapters, and one book titled Ferns of Swat Revisited, and is best known for his work on the systematics and ecology of ferns in Malakand Division and Northern Pakistan.

Sana Wahab completed her M.Phil. studies from Quaid-i-Azam University, Islamabad, Pakistan. She has an impressive academic record, with more than 30 research articles published in her field, along with 10 book chapters. She is actively involved in the scientific community and has participated in various national and international conferences.

Index

A

B

C

E

F

G

H

I

M

N

P

R

S

V

W

www.ingramcontent.com/pod-product-compliance
Lightning Source LLC
LaVergne TN
LVHW062333300926
849161LV00007B/1675
* 9 7 9 8 3 3 7 3 6 7 5 6 9 *